THE POLISH AIR FORCE AT WAR

THE POLISH AIR FORCE AT WAR

The Official History
Vol.1 1939-1943

JERZY B. CYNK

Schiffer Military History
Atglen, PA

The dust jacket artwork by Piotr Górka of Kraków, Poland, was specially commissioned for these volumes by the Polish Air Force Association. Shown are two pairs of Spitfire VBs, one from No.303 Squadron and the other from No.317 Squadron, led by S/Ldr Jan Zumbach in BM144 RF•D "Donald Duck," peel off to attack German Fw 190 fighters over the Dieppe area in early June 1942.

Book Design by Robert Biondi.

Library of Congress Catalog Number: 97-81029.

Printed in the United States of America.
ISBN: 0-7643-0559-X

We are interested in hearing from authors with book ideas on related topics.

Published by Schiffer Publishing Ltd.
4880 Lower Valley Road
Atglen, PA 19310 USA
Phone: (610) 593-1777
FAX: (610) 593-2002
E-mail: Schifferbk@aol.com.
Please write for a free catalog.
This book may be purchased from the publisher.
Please include $3.95 postage.
Try your bookstore first.

CONTENTS

FOREWORD

by Air Vice-Marshal Aleksander Maisner CB, CBE, AFC,
President of the Polish Air Force Association in Great Britain

This comprehensive and detailed book is meant to act as a reminder of the contribution of the Polish Air Force to the Allied victory and to epitomize its entire war record. The highlights, such as the crucial part played by Polish fighter pilots in the Battle of Britain, have been covered in many works, but several other aspects have been sadly neglected up to now.

The campaign of September 1939 is important, as the Polish Air Force's highly creditable performance strengthened its faith in itself and gave it something to build on. Their extraordinary achievement of evacuating the overwhelming majority of the force's effectives to France and then to Britain provided the continuity and the experience that were to turn the Polish Air Force into a formidable force of 15 operational squadrons. This unprecedented renaissance of a complete Polish Air Force at the side of the Royal Air Force, also required the leadership and foresight of Poland's Prime Minister and Commander-in-Chief General Władysław Sikorski. It demanded and received a level of discipline and determination rarely met with, and it assured Poland an important place in the ranks of the Allies during the early stages of the war. The Poles carried this spirit into many theaters of war and fought with unflagging determination and dedication to the very end of the war.

Although the work puts its main emphasis on the operational aspect of the Polish Air Force's activities and on its links with the Polish underground forces which provided the Allies with invaluable intelligence – notably on the V 1 flying bomb and V 2 ballistic rocket – the vast input of Polish air and ground crews into ferry and transport duties, maintenance, repair and airfield construction and into training, is given its due. The Polish effort on the Eastern front is also covered.

The war ended, and Poland was free of German occupation. The Western Allies, however, under pressure from Stalin, withdrew their recognition of the Polish Government in London, and with their consent, Poland was left under the yoke of Soviet Communism for the following 45 years. On Stalin's insistence the Polish Armed Forces were not allowed to take part in the Victory parade in 1946. The Polish Government in Exile remained in London until 1990, but without diplomatic status. The majority of Poles decided not to return to their Fatherland. The 1947 Polish Resettlement Corps Bill was passed, enabling them and their immediate dependents to remain in freedom with basic rights in the United Kingdom, or to emigrate to the USA, Canada and other countries of the free world.

Finally, it is fitting that the work of the Polish Air Force Association is mentioned. This organization did sterling work in helping the members of the disbanded Polish Air Force to start a new life.

At the same time, it strove to keep alive the memory of their heroic deeds, through the erection of memorials and the publication of books, of which this is the ultimate. It also carried on the fight for a free Poland by denouncing the post-war settlement and reminding people of the injustice done to Poland in 1945.

More than fifty years have passed since the events here recorded. Judging by the generous public response to our appeals, and by the many moving letters of appreciation we continue to receive, the wartime achievements and sacrifices of the Polish Air Force still enjoy worldwide recognition. Indeed, support for Poland in present-day political activities, especially in Europe and in the USA, often stems from the recognition of Poland's wartime record in the air over half a century ago. With our wartime brothers-in-arms, the Royal Air Force, we continue to maintain a very close and friendly relationship.

Yet too many volumes of history, which have appeared in the West since the war, pay scant regard to the distinguished contribution of the Polish Air Force to the Allied victory in the air. To do it full justice, this task must be left to historians who are better able to place our story in its true settings, and to the Editorial Committee comprising wartime members of the Polish Air Force, whose painstaking, careful and critical scrutiny of the entire text, based on their own experience, ensured the validity and authenticity of the events recorded in these volumes.

London, January 1997

EDITORIAL NOTE

The Polish Air Force Association has, since its inception, considered itself to be in duty bound to produce a history of the Polish Air Force in the Second World War. In response to this commitment a history was published a mere four years after the end of the war, under the title "Destiny Can Wait." The Committee responsible for the production of that history were fully aware of its shortcomings due to scarcity of documentation, inaccuracies, and the lack of a proper perspective. However, as they made clear in their Editorial Note, the Committee did not intend it to be "the final history of the Polish Air Force in the Second World War. It has been rather designed as a preliminary account embodying first hand source material for those who will some day write a full history of the Polish Air Force during this period." Despite its shortcomings, "Destiny Can Wait" was a very successful publication; its many personalized accounts of events, and the proximity of those events to its publication, made a very readable and popular book. Its second edition, published in the USA in the late 1980s, was sold out within a year.

A great deal has happened in the 50 years or so since the publication of "Destiny Can Wait," which has had a considerable impact on the writing of this much more detailed history. Extensive official records relating to air operations during the Second World War have become accessible. Numerous books, memoirs, and articles on this and related subjects have been published; contentious issues and events have been subjected to detailed scrutiny and discussion, and judgments pronounced by various historians.

Accordingly, we have been most fortunate to be able to entrust the writing of this history to an experienced aviation historian, Mr. Jerzy B. Cynk, whose knowledge of the subject, painstaking research, and sound judgement enabled him to deal with this mass of material and to condense it in a very readable, and most informative manner.

The Editorial Committee appointed by the Council of the Polish Air Force Association, responsible for the production of this work, wishes to acknowledge the usual proviso concerning author's rights and its appreciation of his hard work and outstanding cooperation during the past seven years.

The Committee also wishes to record its thanks to Mr. T.C.G. James, CMG, MA who, having made a major contribution to the production of "Destiny Can Wait", has also very kindly and most effectively brought his experience to bear on this work.

ALEKSANDER MAISNER – P.A.F.A. President
TADEUSZ ANDERSZ – Chairman of the Committee
ANDRZEJ F.K. JEZIORSKI– Vice Chairman
STANISŁAW WANDZILAK – Secretary
THOMAS C.G. JAMES – Literary Consultant
ALICJA W. KALINIECKA
TADEUSZ Z. KWIATKOWSKI
LUDWIK A. MARTEL
JERZY A. POSZAJSKI
MIECZYSŁAW J. STACHIEWICZ

BRITISH TRIBUTE

by Air Chief Marshal Sir Frederick Rosier GCB, CBE, DSO

It is a privilege to have been invited by the Editorial Committee to give a British view of members of the Polish Air Force who came to this country to continue their fight against Germany in World War II and who joined forces with the Royal Air Force.

They were the descendants of people who at times enjoyed a leading position in Europe and at others endured great hardships, oppression and tyranny. Yet, throughout it all, they retained their language, their culture and their identity as Poles.

Many of these men were born whilst their country was still under foreign domination; a domination which lasted over 120 years. Not until 1918 did Poland again become independent – an independence which was to last only 21 years.

In 1939 the Polish Armed Forces had neither the strength, the equipment, nor the training, to counter major aggression. Nevertheless, when the Germans launched their attacks on the 1st of September the Poles resisted valiantly at high cost, but faced with overwhelming force and then stabbed in the back by Russia, further defence was impossible. The Polish Air Force, with their limited numbers of obsolete aircraft, had fought bravely, but the odds against them were so great that the fight did not last long.

As Sir Arthur Bryant put it "Poland chose to resist, and by resisting to be destroyed. Only a very great people would have done so under such circumstances, but the Poles, for all their political faults and failures, have proven again and again that they are a very great people." He went on "In the 17th century they saved European civilization: they did so again by their heroic sacrifice in the 20th Century."

In that autumn of 1939 great numbers of Poles, including experienced air and ground crews determined to continue the fight, made their way to France and to this country. As a wartime publication put it "They skied across Carpathians, they had been through the prisons of Hungary, they had stolen boats and rowed down the Drave river to Yugoslavia, they had come by steamer to Marseilles."

In France very few of the Polish pilots were given the opportunity to fight, and so it was that when their Commander in Chief, General Sikorski, called on them to get to this country by all possible means they obeyed with alacrity. They were warmly welcomed.

In the meantime about 2,000 who had arrived earlier in this country were undergoing training, the pilots amongst them being destined for bombers. However, as it became clear that more and more fighter pilots would be needed for the air defence of this country, it was decided that the Poles originally trained as fighter pilots would continue in their specialization. What a wise decision this was, for in the air war that followed, later to be known as The Battle of Britain, the Polish contribution was invaluable. Indeed, the margin between victory and defeat was so narrow that their contribution might be regarded as crucial.

It was put to me recently that a similarity exists between a military action which took place in Poland in 1920 and the Battle of Britain 20 years later. In the Polish action an outnumbered force under Marshal Pilsudski launched a daring flank attack on a Russian Bolshevik Army which had over-run much of Poland and was at that time laying siege to Warsaw. The Russians were soundly defeated and withdrew. Had it not been for this victory all Poland and probably much of Central and Eastern Europe would have come under Russian domination, with untold consequences for Western civilization. The outcome of that battle on 15 August 1920 is known as "The Miracle on the Vistula."

Similarly, in the Battle of Britain, a comparatively small force stopped a much larger German force from achieving the air superiority essential for a successful invasion. Had we lost – had the miracle over this Island not occurred – Europe would have entered a long dark age and the consequences for Western civilization would have been grave indeed. Yes, there are similarities.

It was not only in Fighter Command that the Poles were active. As early as September 1940 two of their bomber squadrons were operating against the invasion ports and by early 1941 the force had expanded to four squadrons flying Wellingtons.

It was said that there was a general rush to be posted to these squadrons. As one writer put it, "At the Polish depot in Blackpool a special notice was issued stating that all posts in these squadrons had been filled: there were no vacancies: Sorry, perhaps if someone dies. This did not help. Everybody wished to fly. Technical officers wanted to be pilots, skilled navigators were willing to go as air gunners, the blind recovered their sight, the deaf their hearing. Such was the determination of these men to get to grips with the enemy."

It was no flash in the pan. Time and casualties did not weaken their resolve. There were times when news of Poland or of what was likely to happen to Poland made them depressed and angry, and it would have been understandable had there been a slackening of their effort. But there was not. I know this was so in the fighter force for I was involved in it, and from all accounts it was the same in Polish units in other Commands.

As the war went on and more Poles arrived, having been freed from concentration camps in Russia, the force grew to a strength of 17,000 Polish units or individuals who were involved in most of the activities of the RAF: in Coastal Command, in the Tactical Air Force; in Special Duties supporting underground units in Europe (incidentally, during the Warsaw rising in 1944 the bravery of the aircrews was outstanding whilst their losses in this unit were appalling); and Poles were also in Training Command, in the Transport force, in Overseas Commands as well as in Fighter and Bomber Commands.

Out of that 17,000 of whom a large proportion were air crew, 1,900 lost their lives, over 1,200 of them in the Bomber Force alone.

Unfortunately, when the end of the war came and the free world rejoiced, the sacrifices of these men and their contribution to Victory did not bring to them what for so long they had fought for and longed for – to return to a free and independent Poland. So, rather than submit to Russian domination a majority decided to remain here in freedom. They and their families have been an asset to this country.

The events in Poland over the past few years must have given these men great satisfaction, for they showed that the desire for freedom was not dead and that the need to fight for it was not forgotten. The Poles led the way. Yes, they are a very great people.

This comprehensive historical work provides a full record of Poland's part in the final and complete Allied Air Victory in Europe in 1945. Its publication is in itself a tribute to the members of the Polish Air Force Association, for it will represent a most valuable addition to the heritage of the Polish nation.

General Władysław Sikorski, Leader of Fighting Poland, Prime Minister and War Minister in the Polish Government in Exile and Commander-in-Chief Polish Armed Forces, 30.09/07.11.1939 - 04.07.1943. (Painting by Cathleen Mann, 1943, courtesy The Polish Institute and Sikorski Museum, reproduction by Krzysztof Barbarski)

PREFACE

The task of presenting a complete history of the Polish Air Force under one cover – as required by the synopsis set out by the Editorial Committee – was a very challenging one. Almost every chapter of this work could be easily expanded into a full volume. A balance had to be struck to produce a book of reference, combined with a narrative which would be a fitting memento of the exploits of Polish airmen in the last war. It also had to appeal to a wider readership, including young aviation enthusiasts seeking information on specific combat engagements, and on camouflage, markings and serial numbers of aircraft flown by their Polish heroes. Incorporation of the political background was deemed essential to enable every reader to understand Polish attitudes and concerns. Space limitations made it necessary to concentrate the narrative on operations of the PAF and to treat other aspects of its widespread activities more briefly.

The text took me seven years to write and cross-check. My prime source material was the original documentation held in the Aviation Archives of the Instytut Polski i Muzeum Sikorskiego (IPMS, or Polish Institute and Sikorski Museum) in London, with the vast Air Files of the Public Records Office at Kew in Richmond serving as the invaluable complementary source.

The IPMS Aviation Archives included materials which began to be gathered by the Aviation Historical Commission, formed in Paris on 20 March 1940 under Col Franciszek Wiedeń in association with the Polish Military Historical Bureau. Steps to maintain the historical and archival record of the PAF in Great Britain were taken by S-Col Kazimierz G. Winicki in Blackpool towards the end of 1940, leading to the formation of a small Aviation Historical Office on 1 July 1941. This became the custodian of the aviation documentation evacuated from France. The orderly organization and coordination of historical and archival activities was undertaken by the HQ, Polish Armed Forces, in Great Britain in December 1941. This resulted in the expansion of the Aviation Historical Office, headed by S-Col Franciszek Haberek into a self-contained unit with an establishment of four staff in the spring of 1942, later named the Historical Office of the PAF. S-Col Stanisław Nazarkiewicz succeeded Haberek in October 1944.

The Office laid great stress on recreating the detailed records of operational activities of LW units in September 1939, which involved seeking personal accounts from the participants in order to fill large gaps in the surviving original documentation. Painstaking collation and scrutiny of all these materials by S-Col Haberek, S-Col Wajda, Mjr Pawłowski and Mjr Żaczkiewicz led to the compilation and publication of the first comprehensive list and statistics of Polish air victories and losses and of war diaries of LW combat squadrons and formations in the September Campaign. Similar treatment was extended to the PAF activities in France.

The official PAF documentation gathered by the Office provided rich source material relating to Polish air operations from Great Britain. Among the most important of these were the Operational Records Books (ORB) of individual squadrons, each consisting of two kinds of forms: RAF Form 540 – a daily record of operational activities (missions flown, results achieved, claims and losses) of the squadron, and RAF Form 541 – a complete record of all daily aircraft movements, including damage, loss, etc. The ORBs were supplemented by individual combat reports of aircrews given to the squadron intelligence officer immediately after a sortie, who wrote them down and made a preliminary assessment of the claims. Daily Orders, Authorization Books (official authority for every flight), squadrons' diaries, monthly reports, other periodical summaries of operations, and usually very well kept commemorative chronicles and histories of the units, furnished complementary documentation.

However, none of these documents, taken in isolation, could be regarded as absolutely reliable. Not all ORBs were completed and kept exactly as requested by RAF directives (this applying equally to the RAF and Allied units), and entries on Form 540 could not be taken as infallible. Fighter pilots' victory reports were forwarded – with supporting evidence, such as camera-gun films – to HQ Fighter Command, for final scrutiny and approval (which sometimes took several months to arrive), and on occasion the classification of the claim was changed from "destroyed" to "probable" or vice versa, or, more frequently, from "probable" to "damaged" or vice versa. Detailed victory scores of the fighter pilots, as ultimately approved by Fighter Command and the PAF, were recorded individually on the Form A/1 "Register of Flights, Victories and Decorations."

Similar re-assessments applied to various other data. In the early stages of the war considerable confusion existed between the terms "operational" and "combat" with reference to sorties flown. The same flights were in some instances described as "combat" or "operational" in the same documents, making the two terms effectively synonymous (and largely used in my text as such). Later all flights made over the sea or over land in connection with operations which did not involve any encounters with the enemy were defined as "operational;" all flights over enemy-held territories or along the enemy coastline within 10 miles from the coast, and all flights made over the sea or over land involving a contact with the enemy, were defined as "combat." Retrospective application of this ruling was in several cases virtually impossible, and some sorties, such as air/sea rescue searches, still remained difficult to classify. A subsequent reclassification of some sorties, after analysis of the courses flown, resulted in small discrepancies between the numbers of combat flights and hours flown as recorded in squadrons' monthly reports, those in the ORBs (on Forms 540), and later, final summaries of the units' overall efforts.

As the war approached its end, the PAF took steps aimed at reconciling inconsistencies and producing the ultimate authoritative records and sets of statistical data, to cover all its operational activities. Special commissions were instituted, headed by the Polish Liaison Officers at the HQ's of various RAF Commands. The most important of these was the Fighter Claims Commission, under Col Jerzy Bajan, the Polish Liaison Officer at Fighter Command, which issued the final PAF fighter score tables, "Polish Fighter Pilots' Achievements during the Second World War (1.9-1939 – 6.5.1945)" in March 1946.

The main task of preparing material for an official history of the PAF, which was to be the PAF's "Report to the Nation", was entrusted before the disbandment of the PAF to the Historical Commission under Col Olgierd Tuskiewicz. The Commission itself was able to complete only a scant preliminary script, which was issued in the form of a duplicated typescript in 1947 with only some 100 copies printed[1], but included the PAF final operational statistics and data of fundamental importance to all future historians and facilitated the appearance of an initial history of the PAF, entitled "Destiny Can Wait", two years later. During the liquidation of the PAF, all the source materials gathered by its Aviation Historical Office and special commissions were deposited in the IPMS, forming the nucleus of the present Aviation Archives.

Historians in Poland had little opportunity to access the IPMS Archives in London before the collapse of the communist regime. At first the rigorous state censorship put great constraints on all publications relating to the PAF. As restrictions were gradually eased, numerous books began to appear on the subject, but these unavoidably relied on personal reminiscences and secondary rather than prime historical documentation. This led to the growth of some preconceived ideas and myths, which might now be difficult to eradicate. It is only in recent years that some outstanding works of great historical interest have emerged, which I have leaned on, as indicated by the references in this volume.

Some specific problems, arising from the peculiarity of the PAF situation in Great Britain, were encountered during the writing of this book. Adoption of the most appropriate terminology in translation of operational and organizational terms from one language to the other posed a particular difficulty. Another was the question of ranks. Polish officers and NCOs were given new RAF ranks upon their recruitment to the service, which in most cases bore no relation to their Polish ranks. Although they were allowed to keep their Polish ranks as a matter of courtesy, these were not officially recognized by the British. Furthermore, RAF and PAF promotions were made separately, not in parallel. It was therefore agreed by the Editorial Committee that both ranks (RAF and PAF) should be given (if known) when names are first mentioned, and also when important references, such as casualties, occur in the text. As a rule the PAF ranks are shown second and in brackets (these are not, however, the equivalent ranks). In certain circumstances, generally with reference to the organization of the Inspectorate General of the PAF and other PAF Command structures, only Polish ranks are given, partly to indicate the hierarchy, and partly because the current RAF ranks of the officers involved were difficult to establish.

Dates in this work are given in the day-month-year order. This arrangement was adopted for the sake of consistency, as all original war documents quoted in the text were edited in such a manner.

Dates of formation of units, command and airfield changes and of rearmament with new equipment, frequently vary in official documents. I made an effort to apply constant criteria to these dates, as quoted in the book. The accepted date for the formation of a squadron is that on which the relevant order was issued by the HQ of an appropriate Group to the RAF Station concerned. The dates of command changes are those on which the actual takeover of a squadron by the new Commanding Officer took place. The dates of airfield changes are generally those on which the main air party of a squadron landed (or arrived by surface) at the new aerodrome. In the case of night arrivals, usually the following day's date is given. As far as rearmament is concerned, the data of delivery of the first

[1] This was published in slightly altered form by Col Franciszek Kalinowski (a former member of Tuskiewicz's commission), under his name, in Paris in 1969.

batch of new aircraft (as distinct from single aircraft) is usually quoted (on occasion a lengthy period elapsed before a unit became operational on a new type). There are some exceptions to these general rules when dates could not be precisely established.

The readers might find the numerous abbreviations used throughout the text tiresome and perhaps confusing; unfortunately these are unavoidable. A full list of abbreviations appears after the Preface and Acknowledgments.

A conscious effort was made on my part to present an objective and truthful historical work on the subject of the PAF's contribution to the Allied air victory in Europe, predominantly based on my own researches and free from any pressures, sentiments and established notions. I leave it to the readers to judge if I have achieved this goal.

Jerzy B. Cynk
London, January 1997

ACKNOWLEDGMENTS

Many individuals contributed to this book in a variety of ways and I wish to acknowledge my debt to them. I owe sincere thanks to my colleagues under the chairmanship of Tadeusz Andersz from the Editorial Committee for support, enlightening discourse and constructive advice, and to Cecil James in particular, for his careful reading and correcting of the manuscript.

I am deeply obliged to the staff of the IPMS: Ryszard Dembiński, Chairman, Krzysztof Barbarski, Honourary Curator, and Wacław Milewski and Andrzej Suchcitz, Keepers of Archives, and to the Polish Underground Movement Archives in London, for their invaluable help; also to the distinguished air historians and my longstanding friends in Poland: Krzysztof Chołoniewski, who provided me with his complete file of aircraft serial numbers, Andrzej Glass, Tomasz Kopański and Wojtek Matusiak, a leading member of the "Skrzydlata Polska" Editorial Team, and to M.J. Ingham in Lincoln, who made available to me his unpublished manuscript "History of No.300 squadron." I should like to express my gratitude to The Trustees of the Imperial War Museum and the Royal Air Force Museum, Hendon, for the permission to publish their photographs, and the same applies to Roy Grinnel, great American aviation artist, and Wolf Print Press, for allowing me to use the "First Kill" painting.

I am indebted to authors and correspondents who have assisted me by providing illusive pieces of information and rare photographs, including British historians: Chaz Bowyer, Charles W. Cain, Malcolm Passingham, Alfred Price and Christopher Shores; Jan Koniarek and James F. Lansdale in the USA; Józef Garliński, Tomasz Drecki, Wacław Klepacki, Wiktoria Puszczańska, and T. E. Willis in the UK; Volker Koos in Germany; Wacław Hołyś, Marian Krzyżan. Michał Mucha, Lechosław Musiałkowski, Witold Szewczyk and Tadeusz Żychiewicz in Poland; and ex-PAF personnel: Wojciech Kołaczkowski, late Marek J. Mażyński and late Witold Urbanowicz in the USA, Władysław Gnyś in Canada, and late Stanisław Cwynar, late Bolesław Drobiński, the late Aleksander Gabszewicz, Wacław Giermer, Jerzy Kaliniecki, Bolesław Klecha, Franciszek Kornicki, Witold Korzeniewicz-Kay, Tadeusz Kumiega, late Mieczyslaw Mümler, Borys Runc, Józef Rutkiewicz, Tadeusz Rybak, Mieczysław Sawicki, Mirosław Szelestowski-Shelton and Wacław Wąsik in the UK. All photographs are acknowledged individually in the captions, except for those from my own Polish Aviation Archives which appear without credits. I do apologize for any omissions, which, if occurred are unintentional.

My special thanks are due to Piotr Górka, the outstanding Polish aviation artist from Kraków, who prepared the painting for the dust jacket; to Robert Gretzyngier from Warsaw, who drew and re-drew all the maps and color profiles of the aircraft; and to Mike Gains from London, friend photographer, who produced splendid prints from faded photographs and private tiny snapshots for the use in this volume.

Lastly, and most importantly, I wish to express my heartfelt gratitude to my wife, Wiesia, for putting up with my total engrossment in this project for so long.

ABBREVIATIONS

AA Anti-Aircraft
ACSq Airfield Construction Squadron
ACTC Aircrew Training Centre
ADGB Air Defensce of Great Britain (reformed from Fighter Command 15 November 1943, returned to the former name 15 October 1944)
ADU Aircraft Delivery Unit
AEAF Allied Expeditionary Air Force
AFS Advanced Flying School
AI Airborne Interception
AK Armia Krajowa (Home Army), (reformed from ZWZ 14 February 1942)
AOC Air Officer Commanding
ASP Air Store Park
ATA Air Transport Auxiliary
ATC Air Training Corps
BOAC British Overseas Airways Corporation,
C.-in-C. Commander-in-Chief
c/n construction number
C.O. Commanding Officer
CWL Centrum Wyszkolenia Lotnictwa (Aviation Training Centre). Also (earlier) Centralne Warsztaty Lotnicze (Central Aviation Workshops), (reformed into PZL 1 January 1928)
CWZ Centrum Wyszkolenia Ziemnego (Groundcrew Training Centers)
DFC Distinguished Flying Cross (officers only)
DFM Distinguished Flying Medal (other ranks only)
DIAP Depot d'Instruction de l'Aviation Polonaise (Training Establishment of the Polish Aviation – the French equivalent to Aviation Training Centre)
ECD L'Escadrille de Chasse de Defense (French - Home Defence Fighter Squadron)
EFTS Elementary Flying Training School
ELD L'Escadrille Legers de Defense (French - Light Defence Squadron)
EMMA Etat-Major de l'Armee de l'Air (French - General Staff of the Army of the Air)
FAI Federation Aeronautique Internationale
FTS Flying Training School
GAO Le Groupe Aerien d'Observation (French Aerial Observation Wing)
GC Groupe de Chasse (French -Fighter Wing)
GMCP Le Groupe de Marche de Chasse Polonaise (French-Mobile Polish Fighter Wing)
HCU Heavy Conversion Unit
HQ Headquarters
IE Initial Equipment
IPMS Instytut Polski i Muzeum Sikorskiego (Polish Institute and Sikorski Museum), London
IR Initial Reserve
ITL Instytut Techniczny Lotnictwa (Technical Aviation Institute), before 1936 known as IBTL
ITW Initial Training Wing
KSUS Komitet do Spraw Uzbrojenia i Sprętu (Committee for Armament and Equipment Affairs)
KU Komenda Uzupełnień (Clearing Military Office)
LW Lotnictwo Wojskowe (Military Aviation), aviation force of the Polish Army, December 1918 - February 1940)
LWP Lotnictwo Wojsk Polskich (Aviation of Polish Forces), aviation of the Polish Army of the Polish Peoples' Republic, organized in the Soviet Union, 7 July 1943
LWS Lubelska Wytwórnia Samolotów (Lublinian Aeroplane Plant)
MAAF Mediterranean Allied Air Force
MDL Morski Dywizjon Lotniczy (Maritime Aviation Wing), aviation formation of the Polish Navy, March 1923-September 1939
MSU Motor Signals Unit
MTLRU Motor Transport Light Repair Unit
MU Maintenance Unit
nav. navigator
NCO Non-commissioned officer

NKVD Narodny Komissariat Vnutrennich Dyel (Russian -State Department of Internal Affairs)
obs. observer
O. de B. Ordre de Bataille (Order of Battle)
OKW Oberkommando der Wehrmacht (German-General Staff of the Armed Forces)
op., ops. operator, operation/operations
Op Gr Operational Group
ORB Operational Records Book
OTU Operational Training Unit
PAF Polish Air Force (the name officially applied to the Polish air arm of the Polish Armed Forces in Exile, March 1940 - January 1947)
PAFA Polish Air Force Association
pil. pilot
PLL LOT Polskie Linie Lotnicze LOT (Polish Airlines FLIGHT)
POW Polska Organizacja Wojskowa (Polish Military Organization), 1914-1918
PoW Prisoner of War
PRC Polish Resettlement Corps
PRU Photographic Reconnaissance Unit
PSP Polskie Siły Powietrzne (Polish Air Force)
PWS Podlaska Wytwórnia Samolotów (Podlasian Aeroplane Plant)
PZL Państwowe Zakłady Lotnicze (National Aviation Establishments)
RAAF Royal Australian Air Force
RAE Royal Aircraft Establishment
RAFVR Royal Air Force Volunteer Reserve
RCAF Royal Canadian Air Force
recce reconnaissance
RSU Repair and Salvage Unit
R/T Receiving/Transmitting
RWD Rogalski-Wigura-Drzewiecki design team
SAAF South African Air Force
SD Special Duties
SE Servicing Echelon
SFTS Service Flying Training School
SHAEF Supreme Headquarters Allied Expeditionary Force
SLP Stowarzyszenie Lotników Polskich (Polish Air Force Association)
SOE Special Operations Executive
sqdn squadron
TAF Tactical Air Force
tjn tajne (secret)
USAAF United States Army Air Force
WAAF Women's Auxiliary Air Force
VIP Very Important Person
ZWZ Związek Walki Zbrojnej (Union for Armed Struggle), formed in occupied Poland November 1939; reformed into AK.

GLOSSARY OF TERMS

RANKS

Before the war the Polish Armed Forces were headed by Inspector General of the Armed Forces in peacetime, or C.-in-C. Polish Armed Forces in wartime, directly responsible to the President of the Polish Republic. This position was held by Józef Piłsudski, succeeded after his death (12.05.1935) by Edward Śmigły-Rydz; both were made Marshals, the highest Polish military rank.

There were three General ranks in the Polish Armed Forces: generał broni (General of Arms), generał dywizji (Division General) and generał brygady (Brigade General). In the inter-war era Kazimierz Sosnkowski was Poland's only General of Arms. Władysław Sikorski, the war leader of Fighting Poland – the Prime Minister of the Polish Government in Exile and C.-in-C. of the Polish Armed Forces from 7 November 1939 – became the second. Generals of all three ranks were normally referred to in Polish documents, notes and historical narratives simply as Generals, and the author followed this practice except for a few specific instances where the use of a full title was preferable.

The Polish army and air force rank titles were the same. The LW was initially headed by a Brigade General, after Piłsudski's coup d'etat, when its role diminished, by a Colonel, and from the mid-1930s, again by a Brigade General.

Officer ranks

The Polish LW and PAF officer ranks and their RAF equivalents, in descending order were (abbreviations used in the text are highlighted in the table below in bold print):

Polish rank in full (abbreviations in brackets)	abbreviated translation of the Polish rank used in the text	RAF rank in full	abbreviation used in the text
no equivalent	—	Marshal of the RAF	**M/RAF**
generał broni (gen br)	(none in the PAF)	Air Chief Marshal	**ACM**
generał dywizji (gen dyw)	(none in the PAF)	Air Marshal	**AM**
generał brygady (gen bryg)	**Brig Gen**	Air Vice Marshal	**AVM**
no equivalent	—	Air Commodore	**A/Cdre**
pułkownik (płk)	**Col**	Group Captain	**G/Cpt**
podpułkownik (ppłk)	**S-Col***	Wing Commander	**W/Cdr**
major (mjr)	**Mjr**	Squadron Leader	**S/Ldr**
kapitan (kpt)	**Cpt**	Flight Lieutenant	**F/Lt**
porucznik (por)	**Lt**	Flying Office	**F/O**
podporucznik (ppor)	**S-Lt***	Pilot Officer	**P/O**

* the exact translations of these Polish titles are Sub-Colonel and Sub-Lieutenant, the British army equivalents of Lt Col and 2nd Lt respectively.

Other ranks

For the Polish LW and PAF NCO ranks the author has used in the text the same abbreviations as those applicable to the corresponding RAF ranks, to avoid difficulties arising from translation of more diverse titles. Pupils of the Polish Cadet-Officer schools, called podchorążówka, had the title podchorąży always added to their NCO ranks, and those who joined combat or operational training units before graduating to the officer rank enjoyed a special status, unique to the Polish Armed Forces.

Polish rank in full (abbr. in brackets)	abbreviation used in the text	RAF rank in full	abbreviation used in the text
podchorąży (pchor)	C-Off	(no equivalent)	—
chorąży (chor)	W/O	Warrant Officer	W/O
starszy sierżant (st sierż)	Sgt Mjr	Flight Sergeant	F/Sgt
sierżant (sierż)	Sgt	Sergeant	Sgt
plutonowy (plut)	Cpl	Corporal	Cpl
kapral (kpr)	LAC	Leading Aircraftman	LAC
starszy szeregowy (st szer)	AC 1	Aircraftman No. 1	AC 1
szeregowy (szer)	AC 2	Aircraftman No. 2	AC 2

BASIC UNITS

Structures of national air forces were generally similar, though because of differences of emphasis put on air power in various countries and their different strategic needs and tactical concepts, aviation units were not directly comparable in many ways. The RAF basic command unit was a squadron, which gradually increased in size as the war progressed. A precise meaning of this term is rather difficult to define since it was used to identify a variety of units which were generally smaller than wings and larger than flights. A squadron could be best described as the smallest self-contained air formation. Structures above this level were less explicit in the RAF than elsewhere. Wings came to prominence as basic command units only during the war. In contrast, the Luftwaffe structures were strictly formal and the size of its units remained largely unchanged throughout the war. In the LW, its peacetime command units varied considerably from the war ones introduced at the outbreak of war in 1939. The Polish air regiments were only the peacetime administrative structures and could be compared in some ways to the RAF stations. Eskadra was a basic command unit in the army cooperation force, while dywizjon took precedence over eskadra as a basic command unit in the fighter force.

Air force structures of the major European powers and the LW, generally recognized as comparable, are listed below and these are used in the text as translations of the relevant unit titles (usual numbers of aircraft per unit, as applicable at the outbreak of war, are given in brackets).

Unit	RAF	Armée de l'Air	Luftwaffe	LW
a small tactical formation of a few aircraft (a sub-division of the basic command unit)	flight (6 aircraft)	patrouille (3 aircraft)	Kette or Schwarm (3-4 aircraft)	klucz or pluton (3 aircraft)
the self-contained basic command unit	squadron (8-16 aircraft)	escadrille (6-16 aircraft)	Staffel (9-12 aircraft)	eskadra (7-11 aircraft)
a major command unit	wing (32-48 aircraft)	groupe (13-38 aircraft)	Gruppe (32-48 aircraft)	dywizjon (18-33 aircraft)
a large operational grouping	group (100-300 aircraft)	escadre (26-68 aircraft)	Geschwader (75-120 aircraft)	brygada (55-86 aircraft)

In 1940 the PAF in Great Britain adopted the Polish name dywizjon for a RAF squadron, and eskadra for a RAF flight. A RAF wing was called skrzydło, the new Polish term which never existed in the LW structures. This created a great confusion when relating the PAF wartime command units to their LW counterparts.

OPERATIONS' CODE NAMES*

Anti-Diver – Fighter operations against the German V1 flying bombs; Diver was the code name given to the V1

Circus – Daylight attacks by small bomber formations escorted by large fighter formations against short-range or "fringe" targets, intended to bring German fighters to battle (instituted at the beginning of 1941)

Flower – Fighter patrols over enemy night-fighter airfields, aimed to keep them on the ground or intercept any that took off

Instep – Fighter patrols over the Western Approaches and Bay of Biscay to intercept enemy aircraft interfering with the Coastal Command activities

Intruder – Offensive patrols over enemy territory to attack enemy aircraft and confuse enemy air defence systems (flown mostly at night)

Jim Crow – Defensive patrols of the English coastline to prevent enemy aircraft from crossing the coast, with spotting the Channel traffic as the additional task

Mosquito – Small-scale fighter ground-attack operations across the Channel, initiated in January 1941; in February renamed Rhubarb, to avoid any confusion with the Mosquito aircraft

Ramrod – Operations similar in content to a Circus, but with destruction of the target as the primary objective

Ranger – Free-lance penetrations of enemy territory, at first usually at the squadron or wing strength and later by even larger formations with the primary objective of wearing out the enemy fighter force

Rhubarb – Small-scale fighter or fighter-bomber attacks on ground targets of opportunity (initially called Mosquito)

Roadstead – Fighter-protected dive-bombing or low-level attacks on enemy ships at sea or in harbour

Rodeo – Sweeps over enemy territory by fighter aircraft only, to draw enemy fighters into the battle of attrition

Rover – Armed Reconnaissance operations against chance targets

*Main sustained air operations only

INTRODUCTION

1. The Origins of Polish Aviation

Poland, once a powerful Kingdom in Eastern Europe, was reborn as the IInd Polish Republic in November 1918, after 123 years of partition of its territory and enslavement of its people by Austria, Germany and Russia. The resurgent country had to be consolidated from three separate regions, each kept as a buffer zone on the frontier of a different occupying power and deliberately held back from development. Polish lands had to be reclaimed in spontaneous uprisings of the Polish people and the country's undefined borders had to be settled by force of arms. In those momentous days spontaneously formed Polish aviation units began to join the nation's fight for independence before the Polish Lotnictwo Wojskowe (LW or Military Aviation) could be formally established and organized.

During the dark period of partition Poles had to commit most of their intellectual and material resources to resisting the efforts of the occupying powers to destroy their national identity and culture. This prevented them from playing their due part in what was, for most of Europe, a period of intense industrial and technological advance. Nevertheless, a number of Poles were involved in the pioneering years of aviation flight and development at the turn of the 20th century.

Czesław Tański commenced trials with the "Lotnia" man-carrying glider in Janów Podlaski as early as 1896; Emanuel Wróbel made the first tentative flights in his monoplane in May 1910 in Opole, as did Stefan Kozłowski in his biplane a month later in Warsaw. The Borucki biplane, the Głowiński monoplane, Libański's "Jaskółka" ("Swallow") monoplane and the Zbierański and Cywiński biplane were the first aircraft designed and built in Poland to achieve sustained and fully controlled flights during 1911. In 1910 Prince Stanisław Lubomirski undertook to establish the Aviata aircraft factory, flying school and airfield at Mokotów in Warsaw. The factory began work in April and completed at least eight aeroplanes (five Farman-Aviata biplanes and three Bleriot monoplanes), before being forced into liquidation in March 1912. The entire Aviata establishment was taken over by the Russians and became a military base.

Poles who lived abroad or served in the armies of the occupying powers found greater opportunities to fulfill their aspirations and several of them made significant contributions to the advancement of flying in various countries.

Bronisław Matyjewicz-Maciejewicz and Grzegorz Piotrowski were among the first six officers from Russia selected for pilot training in France in the spring of 1910. The former graduated on 1 July, 1910 (Brevet No.152) and the latter, on 29 August, 1910 (Brevet No.195). Much to the embarrassment of the French and displeasure of the Russians, they both insisted that Polish, not Russian, nationality be entered on their pilot's licenses and in all the official documents. Piotrowski soon established the first international record to be officially credited to Russia in a spectacular 37 km (23 mile) flight with a passenger over the sea, from St. Petersburg to the island fortress of Kronshtadt on 22 September, 1910. Matyjewicz-Maciejewicz, who became one of the first military flying instructors in Russia, reached a height of 1,250 m (4,100 ft) at St. Petersburg on 3 October, 1910, establishing an all-Russian height record, duly recognized by FAI.

Other Poles played an important part in developing the Russian aircraft industry. Witold Jarkowski, Professor of the Petersburg Technological Institute, guided and assisted Igor Sikorski during the design and development of the giant four-engined Ruski Vitaz and Ilia Mourometz heavy bombers, which established Russian supremacy in the field of ultra-large aircraft. Adam Haber-Włyński, a brilliant aerobatic pilot and flying instructor, became the chief test pilot of the "Duks" factory in Moscow. Count Michał Scipio del Campo (Brevet No. 211), chief instructor at the Aviata factory's flying school in Warsaw, became later the head of a Moscow flying school and test pilot for Tereshtchenko aircraft. Jan Nagórski was the first man to fly an aircraft over the Arctic.

Zbierański and Cywiński Biplane, the most successful Polish design of the pioneering era, circled Mokotów aerodrome in Warsaw four times at a speed of some 45 mph on 25 September 1911.

The most notable among the Poles contributing to the birth of aviation in Austria was the Warchałowski family of industrialists and aviators, the brothers Adolf and Karol Warchałowski being particularly well known among the flying pioneers in Vienna. Adolf was the holder of Austrian Pilot's Certificate No. 1 and of several early Austrian air records and an international endurance record. He became the first military flying instructor in Austria and one of the first designers of aircraft to Austrian Army and Navy requirements (the Autoplan-Werke Warchałowski Military Type I and the Warchałowski Type XI floatplane, the second seaplane to go into service with the Austrian Navy).

Another outstanding Polish designer working in Austria was Henryk Brzeski, who in 1907 built a reaction-free bi-rotary engine, the rights to which were subsequently acquired by Gnome, Siemens and Daimler. The "Aquila" monoplane, developed by the Schindler brothers of Kraków and constructed with Brzeski's assistance in Vienna in 1910, was provided with Brzeski's unique powerplant, the 50 hp "Iskra" ("Spark") bi-rotary engine, driving contra-rotating airscrews.

France, the birthplace of European aviation, was a haven for several Polish inventors and pioneers. Stefan Drzewiecki, the first Polish aviation scientist of international repute, whose early works on the theory of airscrew and mechanical flight had been published in Russia in the 1880s, later advanced his airscrew theory in France. His revolutionary works on the subject, the Drzewiecki Elemental Blade Theory (1892) and the expanded Froude-Drzewiecki General Airscrew Theory, which provided the first rational method for calculating thrust of an airscrew, formed the basis of all subsequent airscrew designs. It was studied by the Wright brothers and exercised a profound influence on the early development of heavier-than-air flying machines. His wider research into the problem of aircraft stability led him to a design for an automatically-stable aeroplane, which materialized in 1912 in the form of the Tandem-Canard Monoplane.

Adam Ostoja-Ostaszewski, who began experimenting with various flying devices at his estate in Wzdów at the end of the 19th century, went to France in 1908 to widen his experience and built the Ost-1 biplane, which resembled the Wright 1908 Model A but with a wheeled undercarriage. Completed at Juvisy near Paris in the spring of 1909, the Ost-1 was successfully flown by various pilots, including the designer, who learned to fly on it. This is recognized as the first airworthy aeroplane to be developed by a Pole[1]).

Other noteworthy designs by Polish pioneers in France included the Bronisławski biplanes and the Wróblewski-Salvez monoplanes. Bolesław Bronisławski, a doctor of science, invented a new method of lateral control of aircraft, the Bronisławski stabilizing system, to replace ailerons. Two variants of the basic system were flight-tested with promising results at Juvisy in the years 1911 and 1912, but the idea did not find acceptance.

The brothers Piotr and Gabriel Wróblewski, later joined by the younger Edward, worked in France under the assumed name of Salvez. In 1909 Piotr designed a remarkable two-seat monoplane with steel-tubing framework, the Wróblewski Type W.I, which, constructed with the help of Gabriel, took to the air in April 1910 as the first successful aeroplane to be built in Lyon and made numerous flights. The more advanced Type W.II of 1911 attracted the attention of the French Army, and the brothers were invited by a French industrialist, Joseph Berthaud, to Ambérieu, to work in his Ambérieu-Aviation factory, where the Type W.III, better known as the Berthaud W (W for Wróblewski) was built. When, in 1913, the fledgling Service Aeronautique issued a requirement for a two-seat artwo-

[1] Januszewski, Rodowód polskich skrzydeł, p. 158.

mored machine, the brothers Wróblewski-Salvez developed the Type W.IV, with the forward fuselage protected by steel armour. Unfortunately, the aircraft crashed on its first flight on 1 March, 1914, killing Piotr and Gabriel Wróblewski.

However, because of the circumstances, these early activities by the Polish pioneers lacked continuity and had little bearing on the subsequent development of aviation in independent Poland.

Inspection of the Aviation of the 1st Polish Corps by Gen Dowbór-Muśnicki at Bobruysk on 2 April 1918.

2. The Birth of the Lotnictwo Wojskowe (Military Aviation)

The outbreak of World War I prompted open organization of Polish combat units in the partitioned Poland, ostensibly to fight on the side of one or other of the occupying powers. Formation of embryonic Polish military aviation became therefore a possibility, but proposals to organize a Polish squadron within the Polish Legion – formed by Commander Józef Piłsudski on the side of Austria and Germany – put forward in the spring of 1916 by Janusz de Beaurain came to nothing.

In Russia, however, the 1st Polish Aviation Unit, organized within Gen Dowbór-Muśnicki's 1st Polish Corps at Mińsk, came into being on 19 August, 1917, equipped with a single Nieuport fighter. When the Corps moved to Bobruysk a number of Russian aircraft were captured and the enlarged Aviation of the 1st Polish Corps, commanded by Col Abakanowicz, was ceremonially inspected by Gen Dowbór-Muśnicki and declared operational on 2 April, 1918. It comprised 12 Nieuport fighters, a few Morane Parasol two-seat monoplanes and a solitary Ilia Mourometz four-engined heavy bomber.

The sole Polish Ilia Mouromezt heavy bomber of the 1st Polish Corps at Bobruysk in April 1918. (T. Kopański)

Other Polish aviation units organized in Russia between the autumn of 1917 and the spring of 1918 included: Jakubowski's unit at Odessa, with nine aircraft, of which only one was serviceable; Cpt Grochowalski's 1st Polish Combat Aviation Unit at Dunayovice, with five Nieuport fighters (later supplemented by a few Spad VII fighters) and one Farman 30

A Nieuport 23, c/n 2731, of the 1st Polish Combat Aviation Unit, with its C.O. Cpt Tadeusz Grochowalski in the cockpit, at Kuzminna Hrebla, during the Battle of Kaniev in May 1918. The aircraft bears the original Russian markings, but the airmen have Polish Eagles on their caps.

seat biplane, which joined Gen Haller's 2nd Polish Corps at Kaniev; and the "Polonized" 14th Russian Aviation Unit at Iasi, which, combined with Cpt Niżewski's unit, formed the 2nd Polish Combat Aviation Unit of the 2nd Polish Corps, with four serviceable aircraft (three fighters and a Morane Parasol).

Early in 1918 the Polish aviation units in Russia had over 40 aircraft at their disposal, but they were all disarmed and interned by the advancing Germans in May 1918. Many of their members and the cadre of Polish military pilots formerly in Russian service managed to escape to France or returned home. They later played a crucial role in the formation of the Lotnictwo Wojskowe (LW, or Military Aviation).

As a result of persistent efforts of the Polish National Committee in Paris, an autonomous Polish Army began to form in France on 5 July, 1917, its command being taken over by Gen Józef Haller after his arrival from Russia. The Army was to be provided with its own substantial air force and M. Poincaré, President of the French Republic, presented it with a Polish Aviation standard on 22 June, 1918. Polish airmen from all parts of the world began to flock to France to enlist in the emerging force. At the end of July 1918 training began at Dijon and then at Istres and Pau, and Poles serving in French squadrons were permitted to join.

On Gen Haller's personal intervention, the Polish Camp was established at Pau under the command of Cpt Jerzy Kossowski. By 23 March, 1919, its personnel strength comprised 88 pilots and trainee-pilots, three observers, two technical officers and 110 mechanics. The French agreed to provide the Poles with seven fully equipped combat squadrons with 10-15 aircraft and a few reserve machines each, an aviation park and a flying school. This considerable force, largely manned to begin with by Frenchmen, progressively replaced by Poles, began to arrive in Poland at the beginning of May 1919.

By the autumn of 1918, the downfall of the Austrian and German empires was imminent and on 7 October the Polish Regency Council, formed in Warsaw, declared the establishment of the independent Polish State. The Declaration of Independence was swiftly followed by a spontaneous formation of improvised Polish military units in all parts of the resurgent country. These were unified under the command of Józef Piłsudski on 11 November; by April 1919 the strength had reached 250,000 men.

Disarmament of the Austrian troops in Southern Poland began at the end of October 1918 and by the 31st of that month Kraków had been liberated. An attempt at rebelling by Austrian airmen at Kraków's Rakowice airfield to escape by air was foiled and some 40 aeroplanes – about ten Brandenburg C.I and Oeffag recce. biplanes, several semi-derelict machines and a number of Brandenburg B.I trainers – were captured. The Ist and IIIrd Combat Squadrons began to form at Rakowice on 7 November, their Brandenburgs and Oeffags being immediately repainted in Polish colors: the red 'Z' insignia on a white square, on the wings, fuselage and vertical tail surfaces.

Cpt Mikijonek's Hansa Brandenburg B.I, serial 76.91, carrying the early Polish 'Z' markings of the Kraków province, after a crash on 26 January 1919.

During November other airfields were taken by the Poles by force. Hureczko airfield at Przemyśl twice changed hands before being finally secured by the Polish troops from Ukrainian insurgents, and six of its ten Hansa-Brandenburg biplanes, successfully evacuated to Rakowice, survived the fighting.

Further to the east, the large Ukrainian minority, supported by the Austrians, forestalled the liberation of Lwów and during the night of 31 October/1 November occupied all points of strategic importance in the town. The entire Polish population, including women and children, rose to oppose them. Polish airmen infiltrated the Lwów's Lewandówka air base at dawn on 2 November and took possession of hangars, workshops, stores and some 25 aircraft (18 Brandenburgs, a few Oeffags and Aviatiks, and a single Spad VII).

The first repaired machine – a Brandenburg C.I Type 27 – formed the initial equipment of the IInd Combat Squadron and went into action on 5 November. On the same day the well-appointed Lewandówka workshops commenced work on a permanent basis and during November 13 airframes were fully restored and 11 aeroengines overhauled.

With five Lloyd C.V (Type 46.) biplanes seized at Lublin airfield, Poles captured a total of approximately 83 aircraft from the Austrians, but only a few of these were in airworthy condition.

The Poles under German occupation began to shake off their oppressors in the early days of November 1918. In War-

S-Lt Jakubowski's Rumpler C.I, c/n C.4607, the first Polish aircraft to fly over the liberated Warsaw on 20 November 1918, with the early white-and-red shield markings used by the Warsaw squadrons.

Captured German aircraft at Ławica in Poznań in January 1919, repainted in the national white-and-red chessboard markings, but still without the opposite colour borders: Halberstadt C.L.II c/n Bay 208/18, with Albatros C.Ia c/n Bay 126/17 (facing the hangar door), LVG C.V and other reconnaissance biplanes behind. (L. Musiałkowski)

saw on 11 November Polish airmen demanded the surrender of Mokotów airfield. After an initial refusal, the Germans quietly abandoned the airfield four days later, leaving behind some 130 aircraft (most of them dismantled or damaged), an observer's school, 21 hangars, large workshops and well-stocked magazines, including buildings of the former Aviata factory, and much equipment. Restoration work on captured aircraft began immediately in the Mokotów workshops, the largest in Poland, which were reorganized as the Centralne Warsztaty Lotnicze (CWL or Central Aviation Workshops). On 20 November the first Polish military aircraft, a Rumpler C.I, piloted by S-Lt Stanisław Jakubowski, made a triumphal flight over Warsaw, sporting provisional national markings: the white-and-red shields.

At the end of December 1918 the Polish population rose in the Wielkopolska region and liberated Poznań from the Germans, except for the Ławica airfield, which refused to give in. On 6 January, 1919 the Poles presented the German airmen with an ultimatum to surrender immediately. Receiving no response, the Poles attacked and 20 minutes later Ławica capitulated. Its hangars, extensive workshop facilities and over 140 aircraft, of which 26 were reported to be ready for immediate use, provided a most welcome addition to the strength of the emerging Polish aviation.

In addition to Ławica, the Poles also captured a large Zeppelin hangar at nearby Winiary, which served as a reserve aircraft store and was packed with about 300 dismantled engineless airframes, including almost every German operational type except for the fighters, and vast stocks of bombs and ammunition. However, the airframes lacked engines and various components, delaying their restoration and in several cases making this impossible.

Aircraft from Ławica, bearing hastily painted Polish regulation white-and-red chessboards, quickly began operating in support of the Wielkopolan insurgents and the Wielkopolan Aviation Group, which was to consist of four combat squadrons, was formed at the base. Immediate help was also given to aviation units in other parts of the country. Thirty-two war-

Albatros D.III and D.V fighters of the Wielkopolan Aviation Group, Ławica, March 1919. In addition to the national chessboard marking, the aircraft carry also diagonal white-and-red stripes in individual positions. (M.Krzyżan)

planes and 12 engine-less airframes were dispatched to Warsaw on 14 January and nine fully restored aircraft were delivered to Kraków and squadrons on the Ukrainian front. According to the official Ławica report of 8 March 1919, there were 30 serviceable aircraft in the Wielkopolan Group, with a further 80 airframes, for which engines were available, capable of repair.

When, under the Versailles Treaty, the Wielkopolska and Pomorze regions were formally handed over to Poland at the beginning of 1919, the Polish military authorities took possession of the former German airfields in Toruń (on 18 January), Grudziądz (23 January), Bydgoszcz (24 January) and Puck (10 February). No aircraft were found on these airfields, apart from Puck, where there was one dismantled Friedrichshafen floatplane.

Official inventory records of 26 January, 1919, indicated the existence of 237 aircraft that had been captured in Poland, excluding Wielkopolska. Of these only 45 were serviceable. A detailed register of the captured aircraft (excluding Wielkopolska), compiled the following month, itemized 217 aircraft of 30 basic types and 42 different variants, and 232 aeroengines of 14 types. The Aviation Inspector's assessment, issued on 2nd May, 1919 (ref. L.328), listed 200 aircraft – nine single-seat fighters, 100 reconnaissance planes, 70 trainers and 21 condemned aircraft withdrawn from use – and added that only 12 of these were serviceable at the time of capture (scrapped airframes presumably accounting for the difference in numbers between various listings.)

With over 140 aircraft and about 300 engine-less airframes taken in Wielkopolska – which added 15 more different models to this motley collection of varied types – the total number of captured aircraft approached 700. Eventually about 560 were overhauled and made serviceable after additional engines and spare parts had been purchased, mainly from Germany.

Because of the spontaneous way in which various regions obtained their freedom, regional authorities emerged before an organized structure of central government. Military aviation followed the same pattern. Single aircraft went into action in support of local insurgent units before the first squadrons were fully formed or a national aviation organization established. The first steps were taken on 14 November, 1918, with the formation in Warsaw of the Aerial Navigation Section in the Technical Department of the Ministry of Military Affairs. This was headed by Col Hipolit Łossowski and subordinated to the Chief of the Polish General Staff.

On 1 December, 1918, uniform national markings, the distinctive white-and-red chessboards (originally used by Lt Stefan Stec since 1917 as his individual insignia), were adopted officially for all Polish military aircraft, replacing the different regional markings initially used in various provinces: a red 'Z' on a white square in Kraków, white-and-red stripes in Lwów and white-and-red shields in Warsaw.

During December 1918 orders were issued defining the structure and organization of aviation at all levels. Order No. 66 from 20 December created the Command of Military Aviation under Col Łossowski and established two Aviation Groups: Ist in Warsaw, with two squadrons, for Central Poland, and IInd in Kraków, with three squadrons, for Southern Poland. The order specified that each squadron – a mixed, multi-purpose unit – should possess six aircraft "for liaison, reconnaissance, artillery observation, destruction of enemy aeroplanes and co-operation purposes." Warsaw and Kraków became the bases for the 1st and 2nd Mobile Aviation Parks respectively; a military Aviation School was established in Warsaw and the Lower Pilots' School in Kraków. Within weeks the IIIrd Aviation Group completed its organization in Lwów and the independent Wielkopolan (later IVth) Aviation Group came into being in Poznań.

On 19 March, 1919, a new and rather unfortunate structure was introduced, which split control of aviation between the Aviation Inspectorate Office, the Aviation Command and the Section of Aerial Navigation, resulting in great practical difficulties and personal friction between various chiefs and commanders. The nominal strength of squadrons was increased to ten each, but the eight operational squadrons then in existence had only 32 serviceable aircraft between them (excluding two fully equipped operational squadrons of the independent Wielkopolan Aviation Group).

A further reorganization of military aviation, aimed at streamlining the system and the unification of all squadrons and units under the command of Gen Gustaw Macewicz, who became Head of the Aviation Inspectorate Office and Chief of the Aviation Command, took effect on 25 October, 1919.

3. The Fight for Poland's Independence (1918-1920)

The improvised Polish forces, reclaiming the former Polish lands, became involved in armed clashes with various minorities who resented the emerging Polish authority. Conflict with the Ukrainians developed into a bitter war. Fighting in besieged Lwów, the stronghold of the Polish resistance against the Ukrainians, marked the beginning of combat activities for the LW. Its first operational sortie was flown on 5 November 1918, in response to the Polish Command's request to bomb Lwów's Persenkówka railway station, which was used by the Ukrainian forces. Lt Stefan Bastyr, with Lt Janusz de Beaurain

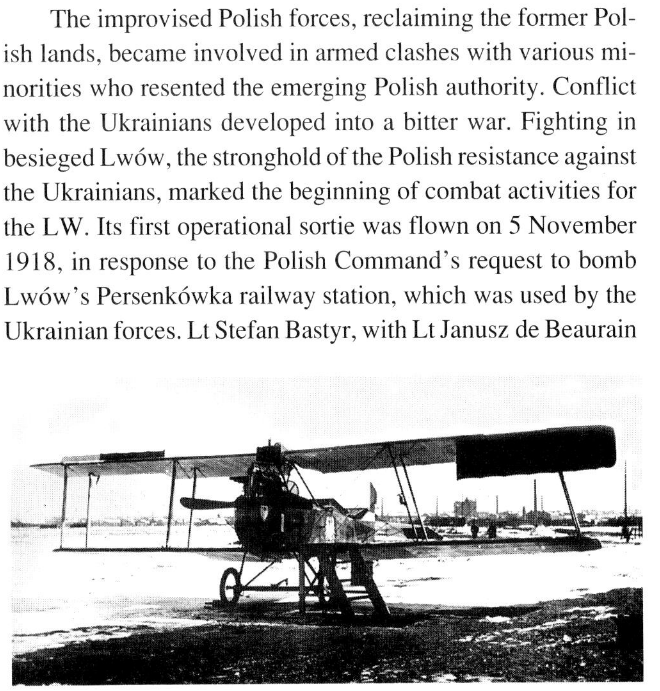

Brandenburg C.I, serial 27.10, newly restored by the Lewandówka workshop in Lwów in the middle of November 1918. The aircraft has white-and-red painted tips of the wing and white-and-red fin and rudder, used by the Lwów squadrons before introduction of the chessboards. (T. Żychiewicz)

Brandenburg B.I, serial 176.??, at the time of the fight with the Ukrainians in the spring of 1919, with an Albatros fighter overhead. (T. Żychiewicz)

Stec's Fokker E.V 001 at Lewandówka after the victory. The fighter bears Stec's individual markings, a horizontal S with white-and-red star in the middle.

in the observer's cockpit, immediately took-off from the Lewandówka airfield in a Brandenburg C.I Type 27 of the IInd Combat Squadron and accomplished this historic mission. Two more sorties followed on the same day.

Lwów was completely freed from the Ukrainians on 22 November, but remained cut off, and aircraft, apart from their vital aerial support, which in November amounted to 65 combat sorties in 78 flying hours, provided its only means of liaison with the rest of the liberated country. Even when the siege of Lwów was eventually lifted, heavy fighting spread along a much broader front, a situation further aggravated in January 1919 when the Czechs launched a surprise attack on the Cieszyn region.

In the middle of February the Ukrainians renewed their offensive, supported by Austrian-supplied aircraft. The hard-pressed IIIrd Aviation Group, consisting of the 6th and 7th Squadrons and the 5th Squadron transferred from the Cieszyn front, made a vital contribution towards keeping the Lwów rail link open and breaking up the enemy's attacks. During February the Group dropped 930 kg (1,990 lb) of bombs and had its first encounters with the enemy. Two Ukrainian aircraft were shot down by Polish reconnaissance biplanes, but one Polish machine was damaged by a Ukrainian fighter. Meanwhile, in the west, resistance by the well-organized German colonists, supported by the Grenzschutz (Border Guard), was quickly suppressed by Polish ground forces supported by the 2nd Wielkopolan Squadron, equipped with eight aircraft.

The Ukrainian offensive on the Lwów front continued in March with vastly superior forces, but after initial successes it was driven back by desperate Polish counter-attacks and better coordinated and more effective air strikes.

At the same time, intensive military activity engulfed the north-eastern frontier. Wilno surrendered to Russian Bolshevik troops and fighting flared up when combined Lithuanian-Belorussian and Polish force thrust north-east. The Ist Aviation Group composed of the 1st, 4th, 8th and 11th Squadrons, though with only 11 aircraft between them, was moved from Warsaw to the combat zone to support this action, which led to the recapture of Wilno on 19 April.

In the early spring of 1919 the Poles began to prepare for an all-out offensive in May against Ukrainian forces. The reinforced aviation of the Army Command "East" was regrouped into the IInd Aviation Group (1st Wielkopolan, 5th and 9th Squadrons, with 23 aircraft) and the IIIrd Aviation Group (6th and 7th Sqdns, with 15 aircraft including, for the first time, three Fokker E.V fighters received from Warsaw).

The Army's aviation began intensive air operations in April in preparation for the ground offensive. The IIIrd Aviation Group distinguished itself in numerous air strikes on Ukrainian railways, cavalry and reinforcements. On 19 and 20 April alone the Group carried out 90 sorties, forcing the withdrawal of a Ukrainian battery and armoured train. Its 27 sorties on 26 April, compelled the Ukrainians to abandon their field artillery which was subsequently captured by Polish troops. On 29 April, Lt Stefan Stec of the 7th Sqdn, flying his Fokker E.V, fought the LW's first victorious dog-fight, shooting down a Ukrainian Nieuport fighter which was escorting a Brandenburg reconnaissance biplane over the Sokolniki area.

The Polish offensive began at dawn on 14 May, 1919, with Gen Józef Haller in overall command, and made very good progress. The Aviation Groups, operating at full strength when weather permitted, softened up enemy positions and on occasion completely dispersed Ukrainian formations. Flying often at tree-top level, the squadrons suffered three aircraft lost and

Lt Stefan Stec, of the 7th Squadron, in the cockpit of his Fokker c/n E.V 187, LW serial 001, after his victory over a Ukrainian Nieuport on 29 April 1919. (T. Żychiewicz)

Gen Józef Haller welcomes one of the three Breguet 14 squadrons of the Polish Aviation of his 'Blue Army,' arriving from France at Mokotów aerodrome in May 1919. (T. Kopański)

Gen Gustaw Macewicz (with a beard), LW's first Aviation Commander, with an officer of the French Mission, in his staff car during inspection of the 3rd Wielkopolan Squadron at Góra near Jarocin in June 1919. Squadron's C.O. S-Col Marek Krzyczkowski is standing in front of the closed half of the door.

three damaged. By 27 May Polish forces reached their objectives, secured Lwów and the entire region and established contact with friendly Rumanian troops. The combat effort of the LW on the Ukrainian front in May 1919 amounted to about 160 sorties, during which 7,250 kg (16,000 lb) of bombs were dropped, 32,000 rounds fired and 100 photographs taken.

Mounting political pressure from France and Great Britain and rising tension on the German-Polish border over the imminent ratification of the Versailles Treaty, led to the suspension of further offensive activities against the Ukrainians and a concentration of Polish troops on the border with Germany favored by the Western Powers. The arrival of Haller's Army from France, in May-June 1919, also influenced these decisions, since its formidable air force – seven fully-equipped combat squadrons with reserves, totaling 101 warplanes – was partly manned by Frenchmen and its operational use on the eastern front had to be mitigated by political considerations. The 3rd and 1st Wielkopolan Sqdns, the latter pulled back from Ukraine, kept the German-Polish frontier regions under reconnaissance, and two of their aircraft were shot down by German anti-aircraft fire.

In July, the Ukrainians, trying to profit from the Polish preoccupation with the Germans, launched a surprise counter-offensive, which soon collapsed. At the same time fighting with the Russians in the North-east intensified, culminating in the Polish summer offensive, which ended with the capture of Mińsk and Bobruysk. The Polish Silesian uprising against the Germans in August received no support from Poland and crumpled quickly. By September 1919 all fronts became quiet and Poland emerged remarkably stable from its initial turmoil, but hopes for peace were soon to be dashed.

The autumn of 1919 provided time for respite and reflection: the Russian threat was still very real and the Ukrainian question not yet settled. Semen Petlura, a leading Ukrainian, proposed a solution to the latter problem. He offered to renounce all Ukrainian claims to the disputed Polish territories in exchange for Polish help to free central Ukraine from the Russians and promised a popular uprising within the Ukraine to assist the Polish forces. Piłsudski, whose ambition was to establish independent but friendly Ukrainian and Lithuanian-Belorussian states which would separate Poland from Bolshevik Russia, embraced these proposals with enthusiasm. The

The first aircraft to be constructed in Poland, a copy of the Brandenburg C.I Type 27, initially designated P 1, as completed by the IIIrd Aviation Park Lwów workshops at Lewandówka in April 1919. Ths aircraft was subsequently re-designated 27.III/1 – the III standing for the IIIrd Aviation Park. (T. Kopański)

Western Powers, suspicious of Piłsudski's intentions, refused to support the plan, which, because of its risky nature and Russian peace initiatives, also met with strong opposition inside Poland. However, Piłsudski's determination won the day and the agreement with Petlura was concluded in February 1920. Meanwhile, preparations for possible future military action in the East continued unabated and gathered urgency in the early months of 1920.

The problem of LW equipment became crucial in view of the anticipated offensive. A bewildering array of captured aircraft presented enormous maintenance difficulties and single-seat fighters were in short supply. During 1919 CWL in Warsaw managed to overhaul or rebuild 217 airframes and 169 aeroengines and the workshops at Ławica in Poznań restored about 145 aircraft. The Kraków and Lwów Workshops carried out similar work on a smaller scale. In addition, the Lwów Workshops undertook a limited production of the first new aeroplanes to be built in independent Poland, manufacturing small batches of Brandenburg C. I Type 27, 67, and 26 reconnaissance biplanes, the first of these flying in April, 1919. CWL planned production of the Hannover-Roland C.L.II, but dropped the idea after the first CWL-built example, serial 18.01, disintegrated in mid-air during its first public demonstration flight at Mokotów on 23 August, 1919. However, the attrition rate of the restored aircraft proved to be excessively high and the efforts of the workshops, gratifying as they were, could not make up for a crippling shortage of reliable new equipment.

Purchases from abroad provided the only solution. With fighters the top priority, Army Command "East" used its own initiative to buy 38 Oeffag-built Type 253 Albatros D.III biplanes from Austria, which reached Poland in June-August 1919. Attempts to obtain Fokker D.VII fighters and reconnaissance aircraft from Germany met with a very slow response, with delivery periods extending well into 1920. Purchase of 82 Austro-Daimler and 32 Mercedes engines posed fewer problems.

The unsuccessful Słowik Biplane, the CWL-built replica of the Hannover-Roland C.L.II, carrying the Polish serial 18.01 and the CWL White Eagle emblem on the fuselage, which was falsely claimed to be "the first aircraft to be built in independent Poland." (T. Żychiewicz)

The greatest injection of new equipment was provided by the Aviation of Haller's Army, which comprised: three Breguet (39th, 59th and 66th) with 26 Breguet 14A2 recce biplanes and 18 Breguet 14B2 bombers, three Salmson (580th, 581st and 582nd) with 39 Salmson 2A2 recce biplanes, and one Spad (162nd) with 18 Spad VII fighters – 101 warplanes in all, not counting 55 training aircraft of the flying school. Haller's Army Aviation, as well as the Wielkopolan Aviation, remained initially under independent command but with the replacement of the remaining French personnel by Poles and unification of all Polish aviation formations under the command of Gen Macewicz, the French Military Mission officially handed over the Haller Aviation to the Polish Government on 26 September, 1919.

Additional quantities of Breguets and Spads were ordered in France in the summer of 1919 to provide equipment reserves for existing and new squadrons. One more Breguet 14A2 was brought over by Lt Ludomił Rayski in June 1919 from Odessa by the way of Rumania. This aircraft originally belonged to the Aviation Squadron of the 4th Division of Polish Fusiliers, which was formed in Kuban in October 1916 and moved to Odessa when the town was freed by the White Russians and the French Expeditionary Force.

Breguet 14A2, c/n 7490, of the Haller Army's Aviation Escadrille Br 59, seen in August 1919 simultaneously bearing French and Polish markings on the wings and rudder. A month later the unit became the 17th Reconnaissance Squadron of the LW.

The LW's armament program for the year 1920, formulated in the autumn and finalized on 5 December, 1919, envisaged a force of 20 combat squadrons with 15 aircraft (ten operational and five reserve) each. Reappraising the equipment situation, the planners concluded that only LW's 60 Breguets (four sqdns) and 30 Albatros D.III fighters (two sqdns) could be incorporated in the 1920 program. New aircraft, expected to be delivered by the spring of 1920, included 45 more Breguet 14 recce biplanes (three sqdns) from France, 15 Fokker D.VII fighters (one sqdn) obtained privately by Poznań firms from Germany, and 15 Ansaldo Balilla fighters (one sqdn) and 60 SVA recce biplanes (four sqdns) from Italy. This represented a total of 225 warplanes for 15 squadrons.

Additional orders were placed for the five remaining squadrons and reserves. These comprised: 40 Spad XIII and 15 Spad VII fighters, as well as 20 ex-German Fokker D.VIIs from stocks captured by the French, 25 Breguets (an increase in the Breguet 14 order to 70 aircraft[2]) to be bought from France; a further 20 SVA recce biplanes and 20 Balilla fighters from Italy; and a quantity of war-surplus aeroplanes from Great Britain. About 20 ex-RAF aircraft were initially obtained for evaluation trials, including ten DH 9 recce biplanes and a few Bristol Fighter two-seat and SE 5a single-seat fighters. The trials resulted in an order for 105 Bristol F.2B Fighters. A further 32 warplanes – 12 DH 9s and 20 Sopwith F.1 Camel and 5F.1 Dolphin fighters – were given to Poland by King George V. With these increases the total number of combat aircraft to be delivered was expected to exceed 500.

The plan was never realized. Communist agitators prevented the handling of aircraft for Poland in Italian and British docks; unloading of Polish cargoes, directed to the Free City of Gdańsk (Danzig) because of lack of Polish port facilities, was sabotaged by the Germans, and unfriendly Czechs interfered with supplies in transit by rail to Poland. Only a small number of French aircraft and 15 Balilla fighters managed to get through in the spring of 1920; British planes began to reach Poland in July and nearly 40% of the machines on order arrived only after the cease fire in the autumn of 1920. The resulting shortages at a crucial period of the Russo-Polish war led to frantic efforts to obtain aircraft from across the border in Germany, but only 16 Albatros C.XV and a few DFW and LVG recce biplanes were delivered.

Advanced proposals for a formation of separate fighter and bomber groups, put forward during reorganization of the LW in the winter of 1919-20, were turned down by the Polish High Command. In April 1920, during final preparations for the Polish offensive against Ukraine, only 170 aircraft could be mustered, less than half of them serviceable, and these had to be parcelled out among the required 20 squadrons. The former Aviation Groups were renamed Dywizjons (or Dyons

[2] This order was later partly rescinded when the Poles found that the Breguets were from old stocks, with large numbers of flying hours behind them.

Gotha G.IV, c/n 100/17, the first twin-engined bomber of the 21st Bomber Squadron, supplementing its six single-engined AEG C.IV bombers, before the squadron's departure to the Ukrainian front at the end of April 1920. Note the White Eagle emblem on the bomber's nose. (T. Żychiewicz)

for short), three Dyons, composed of ten squadrons, being assigned to the Ukrainian front and three, also with ten squadrons, to the Lithuanian-Belorussian (Northern) front.

The LW Order of Battle on 25 April, 1920, was as follows:

Ukrainian front
- IIIrd Dyon: the 5th, 6th, 15th (former 4th Wielkopolan), 17th (former Breguet 59th) and 21st[3]) Sqdns, with 37 aircraft, attached to VIth Army (southern sector);
- IInd Dyon: the 2nd, 7th (Kościuszko) and 9th Sqdns, with 26 aircraft, attached to the IInd Army (central sector);
- Vth Dyon: the 3rd and 16th (former Breguet 39th), with 17 aircraft, attached to the IIIrd Army (northern sector).

The Lithuanian-Belorussian front
- VIIth (former Wielkopolan) Dyon: the 10th, 12th (former 1st Wielkopolan), 13th (former 2nd Wielkopolan), 14th (former 3rd Wielkopolan) and 19th (former Spad 162nd), with 49 aircraft, attached to the VIth Army (southern sector);
- IVth Dyon: the 1st (incorporating Salmson 581st), 4th (incorporating Breguet 66th), 11th and 18th (former Salmson 580th), with 31 aircraft, attached to the Ist Army (central sector);
- Ist Dyon: the 8th Sqdn only, with 10 aircraft, attached to the VIIth Army (northern sector).

The 7th Kościuszko Squadron, forming a part of the IInd Dyon at the Ukrainian Front, was at this time manned by Americans. In the summer of 1919 Cpt Merian C. Cooper, American Air Service (later Gen, USAAF) – whose great' great grandfather, Col John Cooper, was a personal friend of Gen Kazimierz Pułaski and in 1779 carried the mortally wounded Polish hero from the battlefield at Savannah to a ship where he died – took

[3] The 20th Squadron was omitted from the numeration sequence

the initiative to form a combat squadron composed of American volunteers to help Poland in her hour of need, as a gesture of gratitude for the participation of two great Polish Generals, Tadeusz Kościuszko and Kazimierz Pułaski in the American War of Independence.

His first recruit was Mjr Cedric E. Fauntleroy (of Rickenbacker's "Hat in Ring" Squadron fame) and American volunteers eventually numbered 17. They were posted to 7th Squadron, the first ten arriving on 16 October, 1919; Mjr Fauntleroy was appointed C.O. of the Squadron.

The unit's badge, designed by Lt Elliot W. Chess, displayed American stars and stripes with, superimposed, Kościuszko's cap and crossed scythes. The Americans fought with distinction until the end of the Russo-Polish war and were later succeeded by Polish pilots.[4]

[4] The Kościuszko Squadron traditions were continued by the 111th Sqdn of the LW, which defended Warsaw in September 1939, and by No. 303 Sqdn of the PAF in Great Britain.

Above: A Friedrichshafen G.IIIa, the second twin-engined bomber, joined the Squadron at the front in Lwów in May 1920. Note the white-and-red chessboard markings on the bomber's nose. (T. Żychiewicz)

Above: Mjr Cedric E. Fauntleroy (left), C.O. of the 7th 'Kościuszko' Squadron manned by American volunteers and Lt Ludomił Rayski (right), later LW's Aviation Commander, flanking the Kościuszko badge on an Albatros D.III fighter; 22nd November 1919. Below: Line-up of the OEFFAG-built Albatros D.III fighters of the 7th 'Kościuszko' Squadron on parade at Lewandówka aerodrome on 22 November 1919, the first anniversary of the liberation of Lwów. Mjr Fauntleroy's 'I' first in the line.

The Russians expected the main Polish attack in the north and the Polish offensive on the Ukraine, masterminded by Marshal Józef Piłsudski and launched on 25 April, 1920, caught them by surprise. The Polish armies, supported by Perlura's small Ukrainian force, achieved a lightning success, entering Kiev, 150 miles away, on 7 May, and establishing bridgeheads over the river Dnepr. The Polish operations were preceded by extensive aerial reconnaissance and benefited from close air support. The 3rd Sqdn began to operate from the Kiev airfield "Volhynian Post" as early as 8 May.

However, the promised Ukrainian uprising against the Russians never happened and the grossly overstretched and exposed Polish troops could move no further. This, combined with a Russian offensive in the north and strikes by Budyenny's highly mobile Cossack horse army from the south, resulted in the collapse of the Polish defences and a rapid Polish retreat along the entire front, Kiev being hurriedly evacuated on 10 June. In July, with the Russians resuming their offensive in the north with superior forces and Budyenny making headway towards Lwów, the situation grew increasingly chaotic. By 25 July great losses of equipment suffered by the LW left the 20 Polish combat squadrons with only 31 machines between them.

In the first half of August the Russian armies advanced into the heart of Poland, reaching the suburbs of Warsaw and the river Vistula further to the north. Retreating Polish army units regrouped in the south, between the rivers Vistula and Wieprz, for a most fateful and desperate battle, which was destined to decide the fate not only of Poland but also of Europe. It was later called by the Poles "the Miracle on Vistula" and by some Western historians, "the 18th decisive battle of the world," for it seemed certain that if Poland were overrun, the Bolsheviks would extend Communism to Germany and carry their revolution throughout Europe.

The Battle of Warsaw began on 14 August with a cunning strike by the Polish forces under the overall command of C.-in-C. Polish Armed Forces, Marshal Piłsudski to the north through the exposed Russian flank, which allowed the Polish forces to launch an attack on the Russian armies from behind. In the ensuing chaos and complete breakdown in communications, the northern wing of the Russian offensive continued to push on westward, with the northern armies not realizing for three days what was happening to the rest of the front. Their line of retreat was totally cut off and attempts to break through frustrated. By 25 August the Battle of Warsaw, which finally turned the tide of the war, was over. Long-awaited new aircraft, mainly Bristol Fighters, began to reach the Polish squadrons in quantity in the first half of August and the revitalized LW used all available aircraft in devastating strikes in support of the ground operations. At the height of the battle ten squadrons with some 80 serviceable warplanes played havoc among the fleeing Russian divisions.

Meanwhile, Lwów, threatened by Budyenny's Cossacks, became the second focal point of determined Polish resistance. As the weak and exhausted Polish troops in this sector could not be immediately reinforced, the IIIrd Dyon, then consisting of four squadrons, including the 7 Kościuszko Squadron, received an order to "stop Budyenny's cavalry army from advancing on Lwów." The order was fulfilled and during three days, 16 to 18 August, the Dyon managed to halt Budyenny's advance. Having on average 16 serviceable aircraft at its disposal, the Dyon flew a total of 190 sorties, during which about 8,000 kg (17,637 lb) of bombs were dropped and 27,000 rounds fired. By the 19th the Cossacks could stand it no more and began to retreat. This three-day ground attack by the IIIrd Dyon must be considered the most significant Polish air operation in the Russo-Polish war.

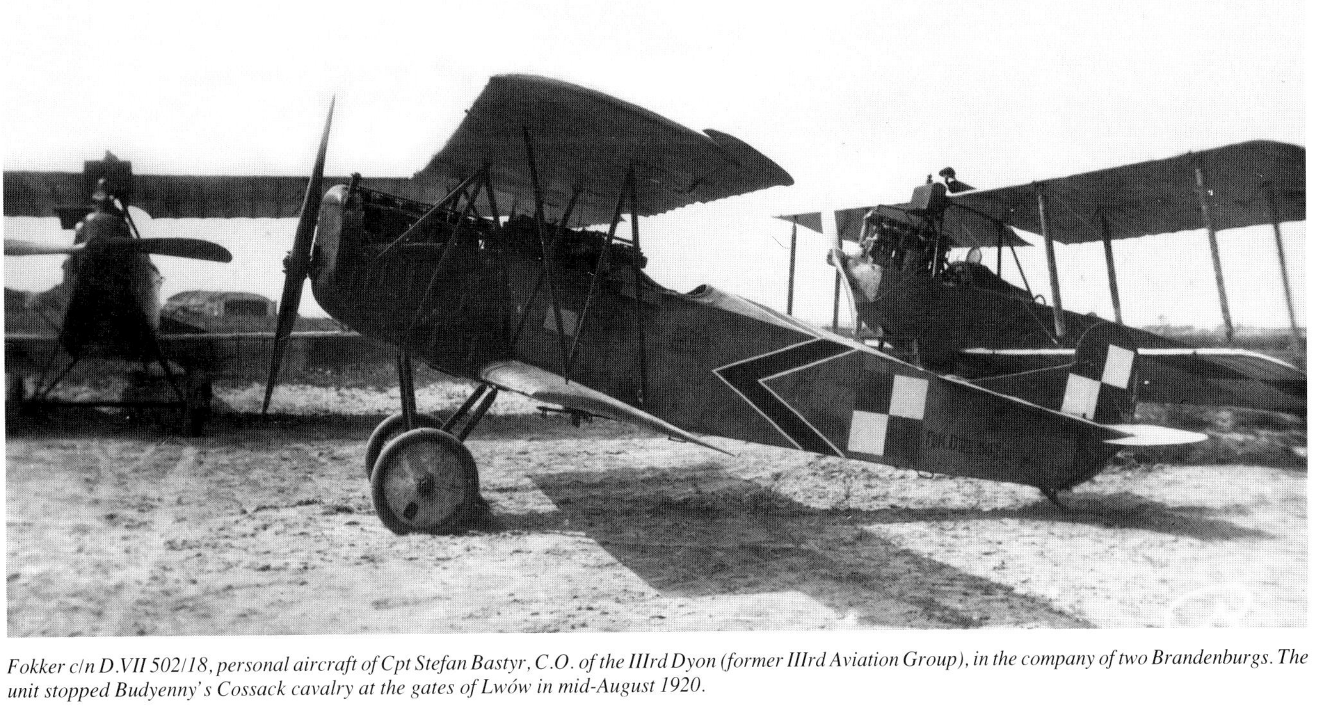

Fokker c/n D.VII 502/18, personal aircraft of Cpt Stefan Bastyr, C.O. of the IIIrd Dyon (former IIIrd Aviation Group), in the company of two Brandenburgs. The unit stopped Budyenny's Cossack cavalry at the gates of Lwów in mid-August 1920.

With the Polish forces facing north at the end of the Battle of Warsaw, the armies had to be regrouped to turn eastwards, to pursue the enemy. The September offensive over the river Niemen, combined with the Polish offensive in the south, met with outstanding success and forced the Russian retreat along the whole length of the eastern front. The cease-fire was ordered on 18 October 1920 at 24:00 hours. Poland was saved and the Russians defeated, but the political aims of the Polish spring offensive against the Ukraine were never realized.

The LW Order of Battle at the beginning of the Polish final offensive on 20 September, 1920, was as follows:

The northern front (51 aircraft)[5]
- 1st, 12th, 13th and 16th Sqdns, with 23 aircraft (Breguet, Bristol, LVC and Fokker D.VII), attached to the IInd Army;
- 3rd, 10th, 17th and 19th Sqdns, with 28 aircraft (Bristol, Albatros C.X, LVG and Spad), attached to the IVth Army.

The southern front (50 aircraft)
- 8th, 9th and 14th Sqdns, with 19 aircraft (AEG, Bristol and LVG), attached to the IIIrd Army;
- 6th, 7th and 15th Sqdns, with 31 aircraft (de Havilland, LVG, Balilla, Fokker D.VII and Spad), attached to the VIth Army.

Of the 101 aircraft at the front, 14 slightly damaged, were undergoing repairs at the Mobile Parks. The six remaining squadrons, with a total of 21 combat aircraft, constituting a reserve for the Polish High Command, were rearming and reorganizing at the rear.

The recovery of the LW strength from the crisis point on 25 July was partly due to deliveries of new aircraft from abroad, which at last began to reach the squadrons, and partly to remarkable efforts by the CWL, Ławica and – on a smaller scale – Kraków workshops, which extended the operational life of captured aircraft beyond expectations. In the peak three-month period, June to August 1920, Ławica alone provided the LW with 142 ex-German aircraft, out of a yearly total of about 220. The CWL total for 1920 amounted to repair or assembly of 327 airframes, including assembly of nearly 150 British aircraft in July and August, and the overhaul of 253 aeroengines. The Kraków Workshops dealt with a number of ex-Austrian machines and undertook almost complete reconstruction of the Albatros D.III fighters.

The Russo-Polish war was totally different from the World War I campaigns fought on the Western Front. It was characterized by the extreme mobility of the front, poor and primitive equipment on both sides and great deficiencies in command and training of the troops, the fledgling Polish army having no combat experience and the Red army being purged of its best "White" officers and crack regiments. The effect on air activity was particularly severe.

[5] All the figures come the technical or 'fifteen days' reports of the LW Command from September 1920.

Bristol Fighter, LW serial 20.23, and personnel of the 1st Squadron during the final Polish offensive over the river Niemen in September 1920.

With an average strength of between 60 and 70 serviceable aircraft available to the LW over a front of some 600 miles and very limited and desultory operations by the Red air force, there was limited scope for the use of air power. One bomber and a few fighter squadrons were formed from the initial force of army cooperation squadrons, but with little or no opposition from Russian fighters and little inclination on the part of army commands to make appropriate use of the bombers, even these were given army co-operation tasks. The LW retained therefore its restricted character of a single-purpose army ancillary service throughout the war, engaged mainly on reconnaissance and liaison. There were some exceptions, such as the bomber strikes on enemy reinforcements on the river Dnepr in the battle of Kiev and the ground attack sorties of the IIIrd Dyon in the defence of Lwów.

The combat record of the LW during the war of independence – quite impressive, bearing in mind its size – amounted to:

- 5 Nov., 1918, to 31 May, 1919,: 804 sorties in 1,191 flying hours, with 12 personnel losses[6];
- 1 June,1919, to 31 March 1920,: 716 sorties in 1,405 flying hours, with 17 personnel losses[6];
- 1 Apr., 1920 to 18 Oct. 1920: 3,652 sorties in 7,164 flying hours, with 53 personnel losses[6];

De Havilland DH 9, delivered to the 6th Squadron on the southern front in the final stages of the Polish offensive. (T. Żychiewicz)

– the war total: 5,172 sorties in 9,761 flying hours, with 82 personnel losses.[6]

Polish air victories included four Bolshevik and three Ukrainian aircraft shot down. Polish combat losses totaled 47 aircraft destroyed, of which three were shot down by Bolshevik fighters and 44 by enemy ground defenses (34 by the Bolsheviks, eight by the Ukrainians and two by the Germans); a further 20 aircraft were heavily damaged: four in air combat (three by the Bolsheviks and one by the Ukrainians) and 16 by enemy ground defenses (12 by the Bolsheviks and four by the Ukrainians).

Soviet historians admit the rather uncomfortable activities of the LW, but contend that their own aircraft exercised a "depressing effect" on the enemy and achieved "air superiority" over the North-eastern front between May and July 1920. In reality, the Polish forces experienced no significant interference from the air throughout the war. In the spring of 1920 the Russians grouped about 60 serviceable machines in the North, half of them fighters, to curb the activities of the Polish VIIth Dyon, but the Bolshevik pilots lacked the aggressive spirit and after a few encounters vanished from the sky. Polish fighters in the area were then used mainly for reconnaissance.

Polish army records indicate that Russian air attacks were infrequent and ineffective. On occasions Bolshevik aircraft hastily dropped a few bombs and quickly departed, but ground attacks on Polish troops were very rare and a few bursts of anti-aircraft fire were enough to chase off the enemy. The Red air force failed to keep pace with the mobile front. During the Battle of Warsaw not a single Bolshevik machine appeared over the city and the idle 19th Fighter Squadron, detailed for its defence, was assigned to ground-attack duties. The total absence of Russian air reconnaissance over the front enabled the Polish August offensive to succeed in disorganizing the Russian forces and cutting off their line of retreat.

According to official Soviet statistics, the Red air force made a total of over 2,000 sorties, dropped over 6,400 kg (14,100 lb) of bombs and fought about 35 air duels on the Polish front between April and October 1920. This was an average of about ten sorties a day, as compared with the average 18 sorties a day flown by the smaller LW, its IIIrd Dyon alone dropping more bombs in three days of August 1920 (16th to 18th) than the entire Bolshevik air force in seven months.

Unfortunately, these experiences led Marshal Piłsudski to dismiss the aeroplane as an effective instrument of war and exercised a major influence on Polish defence thinking in the interwar years.

4. The Years of Peacetime Expansion (1921-1935)

The Polish Nation paid dearly for its independence. Nearly 738,000 men were called up or volunteered to fight in 1918-1920 and of these about 47,000 perished, 15,000 remained missing and 115,000 were wounded. Although the cease-fire of 18 October, 1920, signified the end of major hostilities, the process of determining the borders of the Polish state was not fully completed until 1922.

LVG C.V "Halka Zosia" '5' of the 2nd Wielkopolan (later the 13th) Squadron. The unit recorded 547 sorties in 793 hours during the 1919-20 war, the highest among the Wielkopolan squadrons and was awarded an honorary plaque "For the defense of Eastern Frontiers" by the City of Poznań, carried on forward fuselages of all its aircraft. (A. Glass)

The problem of Wilno, the source of bitter conflict between Poland and Lithuania, was decided by force when the Polish troops under Gen Żeligowski took the town in October 1920. Some fighting on this front continued until the end of November in spite of an internationally supervised cease-fire and the 16th Sqdn remained on the scene until March 1921. The third Silesian uprising against the Germans, in the middle of 1921, conducted again without support of Polish regular forces, was the last action to be fought and led to the inclusion of a part of Upper Silesia within Poland. This was formalized in 1922.

The peacetime reorganization of the LW began on 18 January 1921, when its operational strength was reduced to 13 squadrons, nine reconnaissance and four fighter, with a nominal complement of ten aircraft each. Efforts were made to standardize equipment as far as circumstances permitted, but despite this the average number of serviceable machines was still only four to five per squadron.

The new LW structure envisaged the establishment of three air regiments composed of two dyons, one reconnaissance and one fighter, each with two to three squadrons per dyon. Six dyons were initially formed and the LW Order of Battle on 15 February 1921 was as follows (W. indicating a Reconnaissance Squadron, M, a Fighter Squadron):

- I Dyon, Warsaw; Squadrons: W.1 (Bristol Fighter), W.8 (LVG C.V and C.VI and DFW C.V), W.16 (Breguet 14);
- II Dyon, Łuck; Squadrons: W.6 (LVG C.VI);
- III Dyon, Lwów; Squadrons: W.5 (Bristol Fighter), M.7 Kościuszkowska (Ansaldo Balilla and Albatros D.III);
- IV Dyon, Brześć n/B.; Squadrons: W.3 (Breguet 14), W.12 (SVA 10), M.18 (Spad XIII);

[6] Flying crews only, killed, missing, and wounded, from all causes, including accidents.

- V Dyon, Bydgoszcz; Squadrons: M.13 (Fokker D.VII), M.15 (Fokker D.VII)
- VII Dyon, Grudziądz; Squadrons: W.14 (LVG C.V and C.VI and Albatros C.XV);
- Kraków; Independent Squadron W.10 (Bristol Fighter).

The 1st Air Regiment (Pułk Lotniczy) began to organize in Warsaw in May 1921 and the remaining two, the 2nd Air Regiment in Kraków and the 3rd Air Regiment in Poznań, in August.

Crippling equipment and personnel shortages, a restricted budget, a low level of industrialization and the inevitable administrative problems of the newly established state combined to frustrate expansion plans for aviation. In an attempt to expand the national production facilities a private Franco – Polish enterprise, the 'Francopol' Co Ltd came into being in May 1921 and received a provisional ten-year contract to build 300 airframes and 600 engines each year. The Company obtained substantial government advances towards the erection of a factory at Okęcie in Warsaw, but completely failed to fulfill its obligations.

First signs of the LW's revival came in 1923 with the arrival, on 20 January, of Gen Armand Leveque, a Frenchman, who replaced the sick Gen Macewicz as the Aviation Commander. Leveque's appointment coincided with the granting of a 400 million francs French armament loan to Poland, one quarter of which was allocated to the LW for purchases of aviation equipment in France. In 1923 Poland was on the verge of economic collapse with hyper-inflation reaching catastrophic proportions and without the French loan no improvement in the strength and equipment of the Polish Armed Forces in general and the aviation in particular would have been possible.

Leveque's plan envisaged a force of 50 squadrons with 572 warplanes, composed of 32 reconnaissance squadrons with ten aircraft each, 16 fighter squadrons with 15 aircraft each and two night-bomber squadrons with six aircraft each. This force was to be organized into eight four-squadron reconnaissance dyons and eight two-squadron fighter dyons, with two night-bomber squadrons remaining independent. Five army cooperation and five fighter dyons were intended for five proposed land armies. The remaining strength would be at the disposal of C.-in-C. Polish Armed Forces, but the LW was to retain its status as the army support service.

To meet the requirements of the plan, Gen Leveque ordered in France 150 Potez 15A2 reconnaissance biplanes, 70 Hanriot HD 14 trainers and eight flying boats (for the Polish Naval Aviation). In order to encourage the development of the Polish aviation industry, he also purchased licences for the Potez 15 and Hanriot HD 14 and H 19 models and 1,500 Lorraine engines for Polish-built airframes. Initial contracts for 175 Potez and Hanriot aircraft were placed with Plage & Laśkiewicz, 'Samolot' and PWS factories in the spring of 1924.

The E. Plage & T. Laśkiewicz Engineering Establishments in Lublin were contracted by the government in February 1920 to manufacture 200 Ansaldo A.300 army co-operation and 100 A.1 Balilla fighter biplanes per year, for the three years, from 1921 to 1923, with deliveries beginning at the end of 1920. Due to inexperience, only 96 aircraft (43 Balillas and 53 A.300s) had been completed by the end of 1923. The A.300 biplanes, more difficult to build than the Balillas, suffered severe structural defects which earned them the name of "flying coffins." In the spring of 1924 all production of the Italian models was stopped after delivery of a further 14 Balillas and 22 A.300s. The factory was completely reorganized and converted to manufacture of the Potez 15 with an order for 100 aircraft.

Prospects of production contracts for French aircraft encouraged the formation of two new private aviation factories

The Ansaldo A.1 Balilla fighter '1' of the 7th 'Kościuszko' Squadron at Mokotów in the immediate post-war period. (T. Żychiewicz)

Line-up of the French-built Potez 15A2 reconnaissance biplanes of the LW with the aircraft c/n 293 nearest. (M. Passingham)

One of the 75 Ansaldo A.300-2 reconnaissance biplanes delivered to the LW by the Plage & Laśkiewicz Lublin factory in 1923-24.(M. Krzyżan)

in 1923: WWS 'Samolot' (or Wielkopolan Aeroplane Plant 'Aeroplane') in Poznań and PWS (Podlaska Wytwórnia Samolotów or Podlasian Aeroplane Plant) in Biała Podlaska. 'Samolot' received an initial order for 25 Hanriot HD 14s and PWS for 50 Potez 15A2s, both plants commencing quantity production in 1924.

Gen Leveque's primary aim was to build up the LW strength to six air regiments, to establish adequate bases and to reorganize and expand training. In May 1924 he began to form the 4th Air Regiment at the Observers' School in Toruń, initially with only two squadrons. However, his adopted time scale for the LW's expansion to 50 squadrons by the end of 1926, which would require the creation of 35 new squadrons in the space of the remaining two and a half years was totally unrealistic.

On 17 February, 1924, Gen Władysław Sikorski became the Minister of Military Affairs in Grabski's Cabinet, whose most demanding and urgent task was the stabilization of the Polish economy (this was to be achieved by the introduction of a new Polish monetary system based on Złoty). Sikorski was a man of great vision, who shaped Polish defence policy with great foresight and vigor, believing that Germany and Russia were Poland's two equally dangerous enemies and posed a similar threat to Poland's security. Making good use of the French armament loan, he set himself the task of transforming the deficient and ill-equipped Polish Armed Forces into a progressive and well-armed fighting machine. Working together with Gen Stanisław Haller, Chief of General Staff, he prepared a comprehensive ten-year armament plan, which envisaged a land army of 60 infantry divisions, a substantial navy and a strong air force. Under his dynamic direction by the end of 1925 the Polish Army attained mobilization strength of 35 infantry divisions[7], its fire power was almost doubled and equipment standardized.

Sikorski was the only one among the Polish military leaders of the inter-war era to attach great importance to an air force and stress its decisive role in a future war. In 1926 he wrote: "Poland, with its open geographical borders and inland location, can rely only on a very strong air force to preserve her political independence and her status as a European power."[8] His views could be construed as the leading step to a formulation of Poland's own air force doctrine.

Sikorski did not consider it proper to have a Frenchman in charge of the LW and on 1 July, 1924, Gen Leveque was replaced by Col Włodzimierz Zagórski-Ostoja, who was promoted to General. In May 1925 Sikorski presented to the government a very ambitious twelve-year armament plan for the LW, which set the target of 200 combat squadrons with a total of 2,306 warplanes by the end of 1937. This force was to be

Gen Włodzimierz Zagórski-Ostoja, the Aviation Commander July 1924-March 1926, who began a dynamic expansion of the LW, increasing its strength from 13 squadrons with 86 aircraft to 26 squadrons with 236 aircraft in 21 months. (A. Glass)

composed of 110 army co-operation squadrons (1,100 aircraft), 46 single-seat fighter squadrons (690 aircraft), 28 two-seat fighter squadrons of a kind new to the LW (420 aircraft) and 16 multi-engined bomber squadrons (96 aircraft). About one-fifth of the LW strength, including all the bombers and a large proportion of the fighters, was to remain at the disposal of C.-in-C. Polish Armed Forces. Costs were estimated to require between 24% and 27% of the state's overall yearly armament budgets (similar percentages of the armament budgets were allocated to air forces in other larger countries in Europe).

The immediate aim was to achieve a mobilization strength of 50 squadrons with 560 warplanes by the end of 1927 and Gen. Zagórski set to work with tremendous energy to meet this target. Equipment stocks were an early area of his attention. He had at his disposal a large proportion of the French armament loan, which he used to finance orders for 582 combat aircraft (250 Spad 61 and 50 Spad 51 single-seat fighters, 250 Breguet 19 reconnaissance and light bomber biplanes and 32 Farman F.68 Goliath twin-engined night-bombers), and 101 trainers and seaplanes. He also purchased 150 Le Rhone rotary engines (for the Polish-built trainers) and licences for the Potez 27 and Spad 61. He revised the existing contracts with the Pol-

[7] In September 1939 Poland managed to mobilize 39 infantry divisions, or the same number as by the end of 1927 under the Sikorski program.

[8] "Memorial about the state of the army." He later expanded his views in a remarkable book "Przyszła wojna (Future War)", published in Warsaw in 1934 and subsequently translated into French, Russian, and English.

Potez 27A2 '5' of the 42nd Squadron, 4th Air Regiment, in flight. The Potez 27, combining the Potez 15 fuselage with the new Potez 25 wings, was manufactured only in Poland.

A Spad 61 fighter of the 121st Fighter Squadron, 2nd Air Regiment, sporting the geometric-style squadron markings introduced in the late 1920s. (J. Koniarek)

ish factories, replacing the Potez 15 in production with the Potez 27, and placed new orders for a further 405 aircraft with Plage & Laśkiewicz, PWS and 'Samolot', as well as with CWL, who were instructed to start manufacture of the Hanriot trainers. 'Francopol' received an order for 100 Spad 61 fighters. All this had a dramatic effect on the national aviation industry, the output of which shot up from less than 40 aircraft in 1924, to 182 aircraft in 1925 and 325 in 1926.

At the same time Zagórski also encouraged indigenous design and development. He initiated a design contest and authorized construction of five prototype and pre-production examples of the WZ X two-seat reconnaissance fighter, evolved by Władysław Zalewski as a possible equipment for the proposed two-seat fighter squadrons. These were built by CWL in Warsaw. The first prototype took to the air in August 1926; the first combat aircraft of national origin to fly in independent Poland.

Leveque's proposals for airfield construction were expanded. Zagórski authorized the construction of the Okęcie (Warsaw) and Sknilów (Lwów) air bases and expansion of the Ławica, Rakowice and Toruń facilities. Personnel training rapidly increased at Grudziądz, Poznań and Bydgoszcz, achieving an output in 1925 of some 250 pilots and 100 observers. Further improvements were expected in the years 1926-1927

The first indigenous combat aircraft to be constructed and flown in Poland, the Zalewski W.Z.X-1 first prototype, during its initial trials at Mokotów in August 1926, with Cpt Babiński, one of the test pilots, standing by it.

after opening of the Dęblin training centre, on which Zagórski spent 7 mil. złoty (about $1.4 mil.) by 1926.

Simultaneously, Gen Zagórski began to implement his plans to transform the LW from an ancillary army support service into an air force capable of independent operations. This entailed the eventual replacement of the existing mixed air regiments with separate, specialized army co-operation, fighter and bomber regiments, the latter two with a different structure and organization, intended to provide a powerful "combat brigade" at the direct disposal of C.-in-C. Polish Armed Forces. In May 1925 two new air regiments were formed: the 6th Air Regiment in Lwów and the first of the specialized units, the 11th Air Regiment, temporarily based at Mokotów in Warsaw and later to be transferred to Lida.

On 27 May, 1925, the LW combat elements were reorganized and all the units received a new designation system: dyons were given Roman sequence numbers beginning with I in each regiment, followed by a decimal figure denoting the air regiment to which they belonged; squadrons were identified by new decimal numbers, the first figure in the case of the army cooperation squadrons and the first two in the case of the fighter or bomber squadrons (starting with 1 in respect of the fighter regiments and with 2 in respect of the proposed bomber regiments) indicating the air regiment, and the last, the individual squadron sequence within the regiment (i.e. squadrons of the 2nd Air Regiment included Nos. 21, 22, 23, etc.; these of the 11th Fighter Regiment were to include Nos. 111, 112, 113, etc., and those of the proposed 21st Bomber Regiment, Nos. 211, 212, etc.). Except for minor amendments introduced in 1928, this basic system remained in use until the outbreak of war in 1939.

In the middle 'twenties Polish pilots made their mark internationally in a number of trail-blazing flights. In the second half of July 1925, 20 Potez 15s and Breguet 19s, led by Gen Włodzimierz Zagórski, made a formation flight from Paris to Warsaw via a circular route which twice crossed the Alps and passed over Italy, Austria and Czechoslovakia. Between 16 and 22 September of the same year Col Ludomił Rayski, flying a Breguet 19 and accompanied by Sgt Leon Kubiak, carried out a 7,850 km (4,900 mile) flight from Paris to Warsaw by the way of North Africa. However, by far the most spectacular and internationally acclaimed was Lt Bolesław Orliński's daring transcontinental flight from Warsaw to Tokyo and back, accomplished between 27 August and 25 September, 1926, on the same Breguet 19 as previously used by Rayski. Orliński also accompanied by Kubiak, flew across the Soviet Union, Manchuria and Mongolia, a distance of 22,600 km (14,050 miles) in 121 flying hours.

Early in 1926, in consequence of a press campaign aimed at preparing the ground for Piłsudski's military coup, Zagórski was discredited politically. On 18 March, 1926, he was temporarily suspended from duty pending investigation of the press allegations and Col Ludomił Rayski became the acting Aviation Commander. After Piłsudski's coup d' etat in May 1926 Rayski's appointment was formally confirmed, and Zagórski was arrested for alleged financial irregularities, but because of lack of evidence, he was ostensibly released on 6 August, 1927 without ever being charged. Taken under escort from the Wilno Military Prison to be set free in Warsaw, he vanished without trace from the moment of his arrival in the capital.[9]

Gen. Zagórski left behind 26 squadrons, with a total of 236 operational aircraft (with the two-squadron MDL, or Naval Aviation Dyon, temporarily subordinated to the LW, 28 squadrons), with eight more squadrons in organization, consistent with the programmed expansion to 34 squadrons by the end of 1926 (with a further 16 to be formed in 1927). These units still suffered serious deficiencies in personnel strength and accommodation which measures planned by Zagórski were intended to overcome. Unfortunately, in consequence of the political upheavals of May 1926 the rapid expansion of the LW was stopped in its tracks and its progress towards independent status was reversed.

[9] Circumstantial evidence pointing to his abduction and assassination is presented in Zbigniaw Cieślikowski's book "Tajemnica śledztwa K0-1042/27 ("Mystery of the inquest K0-1042/27"), MON, Warszawa, 1979.

Haunted by experiences of the Russo-Polish war, Marshal Piłsudski refused to recognize the importance of air power in modern warfare and changed the direction of armament priorities. Rayski's extremely modest plan for the LW from August 1926, proposing a mobilization strength of 30 squadrons by the end of 1928, with a progressive build-up to 52 squadrons by 1935, was rejected as too expensive and his aviation budget was cut by a massive 73 percent as compared with Zagórski's last budget.

Piłsudski's Directives for aviation, dated 13 October, 1926, requested, among other things, no expansion, for the time being, beyond the existing 28 squadrons. In Section III "Use of aviation at the time of war", he defined the LW's combat role in only one brief sentence: "The Aviation at the time of war shall be used principally for the purpose of reconnaissance and liaison." A special study group of the General Staff concluded from this directive that the LW needed a new kind of 'light' or 'companion' aviation, specially adapted to provide infantry, cavalry and artillery formations with observation and liaison capability.[10] Cheap, low-powered and comparatively light aircraft with short-field performance were requested for these purposes. Recommendations for future LW expansion centered upon the formation of such 'light' or 'companion' units.

These retrograde trends included a complete ban, ordered at the end of 1926, on training of pilots. After two and a half years of rigorous enforcement a restricted training programme was allowed, lasting until 1933. LW advancement was also hampered by a new regrouping, ordered on 14 July 1928, which dispensed with the specialized 11 Fighter Regiment in Lida,[11] and replaced it with a mixed army cooperation unit, the 6th Air Regiment.

[10] The first such unit, the Observation Squadron, was created by Gen Zagórski towards the end of 1925 at the 2nd Air Regiment on Morane Saulnier AR-35 high-wing monoplanes for evaluation purposes.

[11] The rejected concept of specialized, independent single-purpose formations was brought back to life 11 years later by Gen Zając with the establishment of the Pursuit and Bomber Brigades.

The huge Farman F.68BN4 Goliath twin-engined night bombers, delivered after Zagórski's departure, were never formed into operational units. In the late 1920s and early 1930s they were used mainly for parachute training.

In his memorandum of 12 March 1928, which also included proposals for a force of 37 squadrons with 347 combat aircraft incorporating five bomber squadrons of six aircraft each, by the end of 1929, Rayski set out his own objectives. These were: to link future LW growth with development of the Polish aircraft industry under state ownership; to use the LW "sparingly" because of its small size; and to make it as universal as possible. Large stocks of aircraft of foreign origins, which became surplus to requirement when the formation of new squadrons was suddenly halted, and availability of the first fully qualified graduates from Polish technical universities, both helped the prospects of a smooth conversion from foreign models to indigenous designs. Creation of a bomber force was also one of Rayski's greatest personal ambitions[12] and although by the end of 1929 the LW had reached a strength of only 31 squadrons with 295 aircraft, this embodied three bomber squadrons (18 aircraft) of the three-engined Fokkers F.VII/3m.

In the first years after taking office Rayski awarded Plage & Laśkiewicz, PWS and 'Samolot' development contracts for original designs. He began moves to expand the state aviation industry by transforming the CWL workshops, which first undertook production of new aircraft under Gen Zagórski's orders, into PZL (Państwowe Zakłady Lotnitze or National Aviation Establishments), a major airframe manufacturing plant, specializing in the development of all-metal structures. With emphasis dramatically shifting from manufacture to development, the output of the aviation factories sharply fell from 325 aircraft in 1926 to 124 in 1927, 198 in 1928 and 216 in 1929, only to drop again to 148 in 1930 and in 1931. However, the fledgling Polish design offices exploded into a flurry of activity and numerous prototypes began to emerge in quick succession.

The most notable were the Plage & Laśkiewicz's huge Lublin R-VIII single-engined recce.-bomber, designed by Jerzy

Rudlicki's impressive Lublin R-VIII recce.-bomber first prototype of 1928, powered by the 550 hp Farman 12W engine.

Rudlicki and first tested in March 1928. Two prototypes and four pre-production examples were ordered. 'Samolot' works' designs by Ryszard Bartel, the BM 4 primary and BM 5 intermediate trainers, were the first indigenous aircraft to go into quantity production with deliveries of 22 BM 4 and 60 BM 5 trainers beginning early in 1929.

The first Polish aeroengine plant, the Polish Skoda Works funded with the help of Czech capital, was installed in the premises of the dissolved 'Francopol', which had failed to fulfill its obligations; by the middle of 1926 it had completed only two Spad 61 fuselages. Skoda began to supply the industry with home-produced high-quality power units, starting with the 450 hp Lorraine Dietrich "W"-type water-cooled engine, followed by the 220 hp Wright Whirlwind radial and various high-powered Bristol radial engines. Wright and Bristol engines became the most widely used powerplants in Poland during the 'thirties.

PZL, formed on 1 January 1928, was entrusted with the design and development of five basic types of aircraft: the P.1 all-metal fighter monoplane, the Ł.2 liaison/'companion' monoplane, the PZL 3 four-engined heavy night bomber monoplane (the largest aircraft ever to be projected in Poland but never built because of its high cost), the PZL 4 three-engined pas-

[12] During World War I Rayski flew operationally in the defence of Dardanelles in Turkey and witnessed the poor effectiveness of fighters against attacking bombers. The LW's 21st Bomber Squadron, the only one in existence in 1920, was formed largely on his own initiative.

First batch of Le Rhone-powered Bartel BM 4a primary trainers after completion by 'Samolot' factory in Poznań in February 1929. These were the first quantity-production aircraft of Polish design to enter service with the LW.

The PZL P.1/II, the second prototype of Puławski's revolutionary all-metal fighter of 1929. The aircraft, registered SP-ADO for the Bucharest International Fighter Contest in June 1930, outperformed all its competitors.

PWS 10 '9' of the 131st Fighter Squadron, 3rd Air Regiment, showing off the squadron's new "Raven" badge at the flying display at Mokotów in June 1932. (A. Glass)

senger transport monoplane and the PZL 5 trainer/tourer biplane.

The French Spad fighters, although of a modern design with good overall performance, suffered a fatal structural weakness, and development of a satisfactory Polish replacement became an urgent priority. The revolutionary P.1, designed by Zygmunt Puławski, needed further development and adaptation for the selected Polish-built Bristol Jupiter radial engine, this resulting in the P.6 and P.7 fighters. In the meantime PZL manufactured under licence 25 Vibault 7 all-metal fighters. The Vibault gave the factory good experience in metal structures but proved difficult to handle in flight and PWS was awarded a contract for 50 improved Czech Avia BH-33 fighter biplanes known as PWS A. These were followed in production by PWS's own design, the PWS 10 high-wing fighter monoplane, with 80 delivered by the middle of 1932.

In 1931 the P.7 was selected as the standard single-seat fighter of the LW and production orders for 150 machines were placed with PZL. During 1933 all first-line fighter elements were re-equipped with the P.7, and the Polish fighter force became the first in the world to be entirely armed with all-metal monoplanes. A progressive, Bristol Mercury-powered development, the P.11, succeeded the P.7. Fifty P.11a and 175 P.11c were ordered and began to enter service in 1934. However, with the subsequent formation of additional squadrons, supplies of the P.11 proved insufficient and three fighter squadrons were still equipped with the P.7 in September 1939, by which time the aircraft was totally outclassed as a fighter.

'Towarzyszące' or 'companion' units, initially in the form of three-aircraft platoons, later increased to seven-aircraft squadrons, came into existence in 1932, equipped with the PZL Ł.2 and Lublin R-XIII monoplanes. The R-XIII, designed by Jerzy Rudlicki, became the standard equipment of the 'companion' units and 273 examples of various versions, including a twin-float seaplane, were eventually completed by 1937; 59, including 10 seaplanes, were still in operational use in September, 1939.

Development of a Polish replacement for the 'liniowe' or army cooperation biplanes was less urgent because of large stocks of the Potez 25s and Breguet 19s. This eventually materialized as the PZL P.23 Karaś, which did not reach operational status until 1937. Initial P.23 Karaś production was included in Rayski's new, three-year aviation development plan, covering the period from 1 April 1933 to 31 March 1936. The plan also envisaged design studies and the construction of prototypes for a new generation of Polish warplanes: a fighter, a medium bomber and a 'companion' class aircraft. Because of uncertainty over future trends in military aviation in general,

The very formidable PZL P.7a all-metal fighters in the colors of the 142nd Fighter Squadron, 4th Air Reg., with the "Wild Duck" badge on the fuselages in 1933. Unfortunately, because of insufficient budgets and wrong armament priorities, three squadrons flew the P.7a still in September 1939.

A Lublin R-XIIIter/hydro, serial 712, the 1933 seaplane version of Rudlicki's ubiquitous 'companion' monoplane, at the Puck naval base in December 1934.

The PWS-manufactured Potez 25A2 biplanes of the 35th Army Cooperation Squadron, 3rd Air Reg., which retained its geometric-style markings until the rearmament with the P.23 Karaś.

The French-produced Breguet 19A2 biplanes of the 12th Army Cooperation Squadron, 1st Air Reg., in July 1936. (A. Glass)

development of these new Polish designs was delayed and extended into the next period, during which more comprehensive and precise programmes were worked out.

The LW Order of Battle for 1 July 1936 comprised: 17 army cooperation squadrons (170 Potez 25 and Breguet 19 biplanes); 13 fighter squadrons (100 P.11 and 30 P. 7 fighters); 3 bomber squadrons (18 Fokker F.VII/3m bombers); and 33 'companion' platoons (99 R-XIII monoplanes – partly grouped into 8 squadrons). This amounted to a first-line strength of 417 aircraft, but, leaving out the 'companion' platoons which had no effective combat potential, represented a true combat force of 33 squadrons with 318 aircraft and pointed to a stagnation in the LW's growth over the last decade. Only about one-third of this force, 100 P.11 fighters, could be regarded as up-to-date. Imminent rearmament of the army cooperation squadrons with the P.23 Karaś offered prospects for early improvement. On the other hand, the P.11 had reached the peak of its capability and, with no successor in sight, a rapid deterioration of fighter strength was to be expected. With rearmament gathering pace in Europe, and in Germany in particular, the balance of air power began to swing against Poland at an ever-increasing rate.

This unfortunate state of affairs arose from the negation of the LW's importance and wrong armament priorities after the 1926 coup d' etat and not from failures by the young Polish aviation industry; on the contrary, it had succeeded in producing some remarkable and advanced aviation designs.

The line of Puławski fighters, beginning with the P.1, which, in the International Fighter Contest in Rumania in June 1930 outperformed other contemporary fighters, put Poland in the forefront of international development. Puławski selected a gull-wing monoplane configuration which was characterized by inboard panels tapering sharply in chord and thickness towards the root, with their thinnest sections sloping down to the fuselage attachment points. This unique arrangement, which became known the world over as the Polish- or Puławski-wing and was subsequently copied in many countries, including France, Germany and the Soviet Union, eliminated the obstructed vision imposed by the wing in all other fixed-wing aircraft. The patented landing gear, of so-called 'scissors-type', in which extension levers acted on shock-absorbers placed on opposing sides inside the fuselage, was equally ingenious and offered exceptional aerodynamic cleanness. The circular-section fuselage of semi-stressed-skin metal construction, which was introduced on successive P.1 developments (the P.6, P.7, P.11, and P.24) also represented a unique technological achievement. The P.6 created a sensation at the 1930 Paris Show and repeated its Paris success a year later at the National Air Races in Cleveland, Ohio.

Completely obsolete Lublin-built Fokker F.VII/3m bombers remained in equipment of the LW bomber squadrons until replaced by the PZL 37 Łoś bombers in 1938. (K. Stembrowicz)

The Lublin R-XIIID monoplanes, newly delivered to the 43rd 'Companion' Squadron, 4th Air Reg., in 1935. The R-XIIID was the most numerous variant (95 built) and was still used by seven squadrons in September 1939.

Early-production P.11c Puławski-wing fighters during a visit to Sweden in May 1936. The P.11c '1', serial 8.14, of the 112th Fighter Squadron, III/1 Dyon, 1st Air Reg., bearing the "Fighting Cock" badge, nearest.

Foreign interest in the Puławski-wing fighters exceeded all expectations and sales inquiries were received from several European countries, but a number of prospective sales came to nothing for lack of credits and financial backing. However, Rumania purchased 50 P.11b fighters and licence rights to the aircraft. The ultimate Puławski-wing development was the Gnome-Rhone-powered P.24, the first fighter outside the Soviet Union to embody wing-mounted twin cannon. On 28 June, 1934, the P.24 achieved a maximum speed of 414 km/h (257.2 mph) establishing a new FAI-recognized international speed record for radial-powered fighters. Exhibited at the 1934 Paris Show, the aircraft was acclaimed as the world's best armed and fastest fighter. Despite this it was never ordered by the LW, who began to favor the re-equipment of its fighter force with a multi-purpose twin-engined two-seat aircraft.[13] The P.24 was manufactured in quantity until 1938 for export only, and in 1939 constituted standard equipment of the Bulgarian, Greek, Rumanian and Turkish fighting forces.

At the same time PZL originated development of a sparless wing, the central element of which was a torsion-box made of corrugated heavy-gauge duraluminum sheet, covered with stress-bearing smooth skin. Inspired by Piotr Tułacz's patent of April 1927 for a 'sandwich' wing, this method of construction was developed by Franciszek Misztal and introduced on the PZL 19 Challenge tourer of 1932. Misztal's revolutionary wing design was subsequently used on the P.23 Karaś and most other pre-war PZL types, years ahead of similar structural techniques adopted elsewhere in the 1940s and later.

In the late 1920s, Jerzy Rudlicki, the chief designer of the Lublin factory, conceived a revolutionary tail unit, consisting of two inclined surfaces in place of a conventional tail. The Rudlicki Vee-tail offered considerable aerodynamic advantages and improved the field of fire for rear gunners. A specially modified Hanriot HD 14 trainer became the first aircraft in the world to fly with the Rudlicki Vee-tail and its trials, in the middle of 1931, proved completely successful and trouble free. The Rudlicki Vee-tail development of the R-XIII, designated the R-XIX, was built in prototype form, but the invention was

[13] Later, when this decision was severely criticized, the excuse was offered that the P.24 was powered by the French Gnome-Rhone engine which was not manufactured in Poland and could not be bought in quantity because of scarcity of the foreign currency. This, however, lacks validity, since the French offered Poland a large armament loan (equivalent to about $60 million) and half of it, to be spent in France, remained unused for some time.

Left: The P.6/I prototype, with Cpt Bolesław Orliński at the controls, as flown in the 1931 National Air Races at Cleveland, Ohio. (Underwood, Collinge & Asso.) Right: The last of the long line of Puławski-wing fighters, the 970 hp Gnome-Rhone-powered PZL P.24G, at Okęcie in the spring of 1938, before delivery to Greece. The P.24s of the Royal Hellenic Air Force proved very successful in combat with the Regia Aeronautica during the Greco-Italian war of 1940-41.

P.23B Karaś II Army Cooperation aircraft of the 42nd Squadron, in the first line, and 41st, behind, on parade at the base of the 4th Air Regiment in Toruń on 19 June 1938.

totally rejected by the Aviation Department and the company was fined for doing "unauthorized work."[14]

The Polish aviation industry also established an early lead on the use of high-lift devices and in research into STOL (Short Take-Off-and-Landing) aircraft, embracing the slotted-wing concept in advance of other aircraft-manufacturing countries. Handley-Page slots along the entire leading-edge of the wings and flaps on the trailing-edge were employed on the PZL Ł.2 in 1929. The Ł.2 gained international fame at the beginning of 1931, when its first production example, flown by Cpt Stanisław Skarżyński, made a spectacular tour around Africa, covering 25,770 km (16,012 miles) in 147 flying hours.

All Polish aircraft developed for the 1932 and 1934 Challenge Trophy Championships featured slotted wings. In 1932 Franciszek Żwirko and Stanisław Wigura won the Challenge on the RWD 6, and in 1934 the RWD 9 competition tourers, powered by the remarkable Nowkuński's GR 760 radial won the first two places in the final classification, with Cpt Jerzy Bajan in the lead. During the Championships the RWD 9 flown by Bajan remained airborne at 54 km/h (33.6 mph), the lowest recorded flying speed for powered fixed-wing aircraft. The Challenge experiences were fully applied in the design of the quantity-produced RWD 13 tourer, which flew for the first time in February 1935, becoming the world's first practical STOL aircraft for popular use.

The Instytut Badań Technicznych Lotnictwa (IBTL, or Institute of Aviation Technical Research) played a crucial role in assisting the Polish designers and the aviation industry in research and development of new aircraft and equipment. Established on 1 August, 1926, from a reorganization of the earlier WCBL, or Military Centre for Aviation Research, whose functions included quality-control, the Institute gradually assumed the character of a major research center with increasingly widening interests and responsibilities.

Its tasks, as initially envisaged, were: first, to conduct studies in research and scientific and technical experiments and, secondly, to determine the most suitable flying equipment and aviation materials for Polish needs. Its specialist teams included airframe, aero-engine, armament, wireless, photographic, flight instrumentation, and technological sections. From the late 'twenties the ITBL conducted static tests for airframes of all new Polish designs, evaluated Polish prototypes while recommending improvements and issuing certificates of airworthiness, and tested engines, airscrews and various items of aviation equipment.

Under the guidance of its Scientific Council, composed of distinguished Professors of the Warsaw Technical University, and following a programme of Prof Gustaw Mokrzycki, the Institute was to become "the creative center for the Polish aviation of tomorrow and the research facility to find solutions to its problems of today."

Reorganized and renamed the Instytut Techniczny Lotnictwa (ITL, or Technical Aviation Institute) on 1 August, 1936, the establishment was given responsibilities for preparing detailed technical specifications for new designs to requirements of the Aviation Command, for preparing rules on construction and testing of Polish aircraft and for coordinating works of other research centers, such as the Aerodynamical Institute (established in September 1927 in Warsaw under the leading Polish aerodynamicist, Prof Czasław Witoszyński). The Institute's staff rose from 85 in 1929 to over 900 in 1939.

In view of the Institute's heavy involvement in flight-testing and certification of prototypes and acceptance trials of early production aircraft, a special Experimental Dyon (Wing) was established at the IBTL in 1931. Later the enlarged unit became known as the Independent Experimental Dyon and comprised the Prototype Aircraft Squadron and the Production Aircraft Squadron.

[14] Experimental work on the Rudlicki Vee-tail, rejected by the Department for no reason, was subsequently undertaken in Great Britain, France, Germany, Italy, and Russia and the formula was adopted by the Beech Aircraft Corporation in the later 'forties, under the name of "butterfly tail," for several of its own designs, the company claiming in a publicity campaign the credit for the invention!

The revolutionary Rudlicki Vee-tail, air-tested in Lublin on a Polish-built Hanriot HD 14, serial 30.83, for the first time in 1931. The Rudlicki Vee-tail, later copied in many countries, including the USA, became popularly called the 'Butterfly' tail.

In the first half of the 1930s, Rayski, promoted to Brig Gen on 1 January, 1934, finally succeeded in nationalizing the Polish aviation industry. The 'Samolot' factory, damaged by fire in September 1929, was refused Government help and went into voluntary liquidation following year, its machine-shop equipment, tools and other assets being taken over by PZL. PWS, which in the years 1924-1930 delivered about 390 licence-built aircraft, achieving the highest average yearly output among the Polish aviation factories, went bankrupt in 1932. Between 1926 and 1932 the company constructed prototypes of 19 different designs, but only one of these, the PWS 10 fighter, was ordered in quantity and two others were built in pre-production series of five aircraft each. Government advances for provisional orders, which were never confirmed, were used – not always in a proper or efficient manner – to keep the company going, but mounting debts led to its collapse. Towards the end of 1932 the Government took possession of the factory against outstanding liabilities and organized it as a State enterprise trading under the same name. Collapse of the Plage & Laśkiewicz Lublin company was engineered in the autumn of 1935, when the Aviation Department rejected delivery of the improved R-XIIIF series for trivial reasons and canceled the entire production order, the only one which the factory then had. The Lublin factory was taken over by the State and reactivated as LWS (Lubelska Wytwórnia Samolotów or Lublinian Aeroplane Plant). The order for the R-XIIIF was duly reinstated, the aircraft becoming suddenly acceptable again.

The Lublin R-XIX of 1932, the rebuilt R-XIII prototype serial 56.1, fitted with the Rudlicki Vee-tail.

Only DWL (Doświadczalne Warsztaty Lotnicze or Experimental Aviation Workshops), a small factory established at Okęcie in 1930 to produce light touring and training aircraft to the design of the RWD team (Rogalski, Wigura and Drzewiecki), successfully forestalled all attempts by the Aviation Department to nationalize the company. In addition to the famous RWD 6 and RWD 9 competition tourers, DWL turned out a number of outstanding designs. The RWD 7, with Jerzy Drzewiecki at the controls, established two FAI-recognized International Records for Light Aircraft in Category 3: the Speed Record of 178.75 km/h (111.05 mph) over a 100-km course on 12 August, 1931, and the Height Record of 6,023 m (19,760 ft) on 30 September, 1931. The RWD 5bis, flown by Cpt Stanisław Skarżyński, made a spectacular crossing of the South Atlantic from St. Louis, Senegal to Maceio, Brazil in 20 h 30 min on 7-8 May, 1933, and established the new International Distance Record of 3,582 km (2,224 miles) for Light Aircraft in Category 2.

The official attitude towards foreign capital was totally different. Towards the end of 1935 the Polish Skoda Works engine factory at Okęcie, founded by the Czechs, was bought out by the State at a highly inflated price. The establishment was renamed the PZL WS 1 (Wytwórnia Silników or Engine Plant No. 1), which had an average yearly output of about 300 high-powered engines.

Towards the middle of the 1930s aircraft production began to recover, from 148 aircraft in 1931, to 247 in 1932, and, after a drop to 184 aircraft in 1933, to 271 aircraft in 1934 and 369 in 1935, the latter figure exceeding for the first time the 1926 output.

5. Plans and Realities of Polish Aviation Rearmament (1936-1939)

The second comprehensive Polish rearmament plan began to take shape immediately after Marshal Piłsudski's death in May 1935. The plan, formulated when the arms race in Europe had already started in earnest, was intended to make good large deficiencies in every field of armament which arose over the nine years since the rejection of Sikorski's programme. However, the proposals lacked clear definition as to direction and priorities and proved too ambitious for what was still an inadequately industrialized country. Furthermore, the plan did nothing to reform the national command structure. This had been split in 1926 into two unrelated 'peacetime' and 'wartime' channels and left the LW without a unified command for its direction and control. Its subordinate status as an ancillary army service remained unaltered.

The LW's expansion proposals were evolved by Gen Rayski in 1935-36, building on the basis of his earlier three-year plan from 1933. Rayski, as the Aviation Commander in the 'peacetime' structure (and de facto only a chief of an aviation service), prepared his new plan in a vacuum, without knowledge of any specific defence aims stipulated for the air arm or of overall strategic concepts. These remained the secret prerogatives of the Central Staff (the 'wartime' command structure) and could not be divulged to him. In truth no such aims existed. The Central Staff, composed of army officers who had no understanding of air warfare, failed to define a clear-cut role for the air arm in a future war. Paradoxically, whilst the army chiefs were completely engrossed in producing detailed plans for war with the Soviet Union, the proposed expansion of the air arm was being shaped by the hypothesis of war with Germany!

Failure to define the real operational needs was reflected in the proposed composition and tactical characteristics of the future LW which were more appropriate to the German concept of a fast all-out attack than to providing defensive countermeasures to such a threat.

At that time Gen Guivil Douhet's prophetic theory, which proclaimed air power as a new and totally independent component of armed strength, overshadowing land and naval forces as the decisive weapon in any future war, was a subject of intensive study and heated debate the world over. With Douhet's emphasis on the power of the bomber, against which – according to him – there was no effective defence, some foreign experts forecast the demise of single-seat fighter aircraft. Rayski, lacking guidance from the Central Staff as to the future requirements of the LW and probably influenced by Douhet's views and Germany's example, embarked upon a vast built-up of the Polish bomber force. He subordinated all other aviation needs to that objective, with catastrophic consequences in the fighter-interceptor field.

On the basis of Rayski's calculations, submitted to a special study-group of the Central Staff in the spring of 1936, it was estimated that 63 bomber squadrons with 378 twin-engined bombers, supplemented by 360 dive-bomber/attack fighters, would be needed to destroy selected military and industrial targets in Germany. The programme was dismissed as financially unacceptable, and its various elements were reassessed, resulting in Rayski's reduced expansion proposal of June 1936. The plan, to be implemented over a five-year period starting in April 1937, at a cost of 1,153 mil. złoty ($240 mil.), envisaged a force of 30 bomber squadrons (180 aircraft) and 32 new-class dive-bomb/attack fighter squadrons (320 aircraft)[15] of the Dispositional Aviation, plus 18 single-seat fighter squadrons (180 aircraft), 8 reconnaissance squadrons (80 aircraft)[15] and 18 'companion' squadrons (126 aircraft) of the Armies' Aviation; a total of 106 squadrons with 886 first-line aircraft. 100% equipment reserve (30% to replace aircraft temporarily unserviceable, 40% to replace aircraft undergoing maintenance and 30% true reserve) was to be provided.

[15] These squadrons were to succeed the 'liniowe' (army cooperation) squadrons, which were to be split into light bombing/attack and reconnaissance squadrons.

Brig Gen Ludomił Rayski, Aviation Commander, LW, May 1926-March 1939.

Further expansion of the state-owned aviation industry was authorized, including two new large PZL factories: the WP 2 (Airframe Plant 2) in Mielec with an estimated yearly output of 450 all-metal aircraft and the WS 2 (Engine Plant 2) in Rzeszów with an estimated yearly output of 1,000 engines.

Pressures on the aviation budget led to further reduction in proposed LW strength. In July 1936 Rayski resisted a further substantial reduction of the bomber force suggested by the Central Staff and eventually agreed to dispense instead with 16 new-class dive-bomber/attack fighter squadrons and three single-seat fighter squadrons. This reduced the planned size of the LW to 87 squadrons with 708 first-line aircraft and lowered the cost of the program to 1,002 mil złoty ($209 mil).

Although on paper the plan assumed a progressive improvement in equipment strength from 1937 onwards, the reverse happened in reality. Production rose from 396 aircraft in 1936 to the record level of 511 in 1937, but then dropped to 469 aircraft in 1938, falling still further in 1939. The whole program, because of wrong emphasis, policy changes and technical difficulties, became a study in failure, with no chance of being implemented within the stipulated time table.

To meet the equipment requirements of his plan, Rayski authorized the construction of prototypes of six new designs, two each – one interim and one principal service model – for bomber, army support and 'companion' squadrons. A prototype order for a single-seat fighter was conspicuously absent.

The P.30 Żubr (European Bison) and the P.37 Łoś (Elk) were to be developed for the bomber force. The P.30 had been conceived as a passenger transport, but after the design had been rejected by the State airline PLL 'LOT', the prototype was converted to a bomber configuration during the construction stage. The bomber, an ungainly composite structure of metal, wood, and fabric, powered by two 400 hp Pratt & Whitney Wasp Junior radials, commenced tests in March 1936. Performance proved quite inadequate and the aircraft was hastily re-engined with two 680 hp Pegasus radials with the minimum of modifications, to serve as insurance against a possible failure of the very advanced P.37 Łoś.

One of the unsuccessful LWS 4 Żubr bombers at the Dęblin Aviation Training Center in April 1939. (R. Łopacki)

An initial production order for 16 Żubr bombers was placed with LWS under the designation LWS 6, which was later changed to LWS 4, to provide interim equipment for two squadrons by the end of 1937, with an option for the establishment of four more Żubr squadrons in 1938. The Rumanians were also considering purchase of 24 aircraft, but the re-engined P.30 prototype disintegrated in mid-air on 7 November, 1936, killing members of the Rumanian mission who were evaluating it.

Despite unresolved structural problems, which required endless modifications, ever increasing empty weight, which rendered the aircraft useless as a bomber and serious delays in deliveries, the production of 16 aircraft went ahead. The Żubrs eventually became the most expensive aircraft on the LW's order book and could be utilized for training purposes only.

In contrast, the P.37 Łoś, designed by Jerzy Dąbrowski, became one of the best aircraft developed in Poland and proved a leader among international designs of its class. Of exceptionally small overall dimensions, the bomber had a remarkably high ratio of useful load to basic weight – the aircraft could lift a load almost equal to its own empty weight – and was characterized by excellent performance and harmonious lines. Its wings, of laminar-flow section, employed an ingenious torsion-box structure, open at the bottom to permit a bombload of up to 2,595 kg (5,720 lb) to be hung in a series of double tiers across the wing center-section and fuselage, between the engine nacelles. Revolutionary landing gear, invented by Piotr Kubicki, which comprised twin wheels hinged on articulated joints to the oleo-legs of each unit, was adopted for the production model.

The P.37/I prototype flew for the first time on 13 December 1936 and PZL was asked to prepare for the manufacture of at least 180 production aircraft but eventually orders for only 124 Łoś bombers were confirmed. The first production Łoś A bombers began to reach the LW's Experimental Dyon in the early spring of 1938 and the major production model, the Łoś B, went into service with Łoś conversion units the following autumn. Various "teething" troubles and a series of mysterious accidents, which were eventually traced to aerodynamic overbalancing of the twin rudders, prevented the bomber from achieving operational status before April 1939 and delayed completion of the series. Prototypes of a more powerful development, the P.49 Miś (Teddy Bear), powered by two 1,300 hp Hercules engines and provided with 20 mm dorsal cannon, were under construction in 1939.

The P.37 Łoś attracted great international attention and several countries expressed interest in its purchase, but many of the export inquiries had to be turned down because of limited production capacity.

To satisfy expanded army support requirements, which were not yet clearly defined, a development of the P.23 Karaś with dive-bombing capability was briefly considered as the interim equipment. The fourth Karaś prototype, the P.23/IV, later designated P.42, was adapted for this purpose and tested in 1936, but it was found not entirely suitable.

The main equipment of the army support formations was to be provided by a multi-purpose twin-engined two-seat divebomber/attack fighter, the P.38 Wilk (Wolf). This design, conceived by Jerzy Dąbrowski and Franciszek Misztal, was fundamental to Rayski's rearmament proposals, for in addition to the army support role, the P.38 was intended to supplement the LW's bomber potential, provide bombers with fighter protection and contribute to national air defence as a fighter, thereby reducing the need for a conventional single-seat fighter. Such a solution was aimed at keeping rearmament costs down

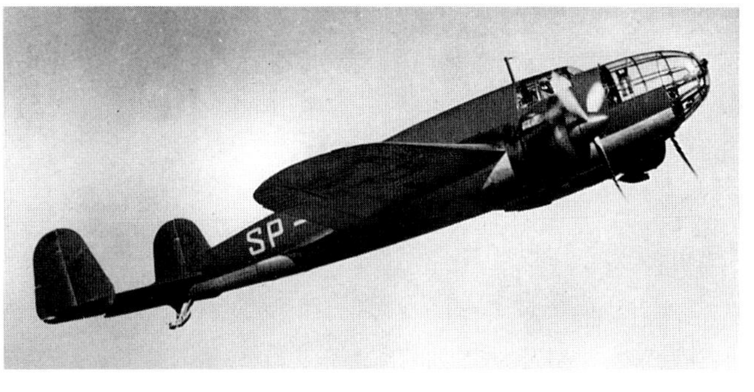

The P.37Abis Łoś Abis demonstrator, serial 72.14, with the civil registration SP-BNL for the flight to the Belgrade Aeroshow in May 1938.

and to achieve the planned expansion of LW strength within a restricted budget.

The aircraft, to be armed with two fixed 20 mm cannon[16] and two guns in the nose, plus twin machine guns in the rear gunner's cockpit, capable of carrying a 300 kg (660 lb) bomb under the fuselage and with an estimated maximum speed of 520 km/hr (323 mph), was developed around a new Polish eight-cylinder inverted Vee engine, the Foka (Seal). The engine, designed by Stanisław Nowkuński, was packed with unconventional and untried ideas and for its estimated maximum output of 600 hp had uncommonly low weight and small overall dimensions.

The Wilk was a very advanced aircraft, both technologically and tactically, predating such warplanes as the Messerschmitt Bf 110, de Havilland Mosquito and Bristol Beaufighter. Because of its comprehensive specification to meet a variety of conflicting requirements, a reasonable period should have allowed for its full development. Yet, despite this and the novel nature of its powerplant, PZL received a provisional order for 500 Wilks as early as the spring of 1936 (to be confirmed before the end of the year) with the delivery of the first 90 scheduled for the end of 1937. Envisaged as the replacement for the P.23 Karaś, the Wilk was to provide equipment for 32 new-class support/attack squadrons and eight army reconnaissance squadrons by 1942, this amounting to 400 machines (45%) of the LW's entire proposed strength of 886 first-line aircraft.

The P.38/I prototype, eventually completed at the end of 1937, turned out to be 20% heavier than anticipated and the Foke prototype engines manifested many shortcomings and initially developed a maximum output of less than 400 hp, which made them unacceptable for flight testing. The second airframe, the P.38/II, fitted with the much heavier 420 hp Ranger SGV-770B engines, began flight trials in the April 1938, from which it was clear that the aircraft in its existing form was

The Ranger-powered PZL P.38 Wilk prototype at Okęcie in February 1939. (T. Kopański)

The RWD 14b Czepla production prototype, manufactured by the LWS under the DWL licence in 1938. The white painted rudder indicated the aircraft was undergoing the official ITL trials.

unfit for operational use.[17] With no suitable substitute engine available, the whole Wilk programme collapsed. An extensive redesign was undertaken and prototypes of the new version, the P.48 Lampart (Leopard), powered by two 700 hp Gnome-Rhone Mars radial engines, were well advanced by the sum-

[16] Only one 20 mm cannon was installed in the Wilk prototype.

[17] The Wilk prototypes were then renumbered: the Ranger-powered aircraft which flew first becoming the P.38/I, and the Foka-powered one, the P.38/II.

The LWS 3/III, the third Mewa prototype, during tests in the spring of 1939. First production examples were ready at the beginning of September 1939.

mer of 1939. A more powerful development, the P.54 Ryś (Lynx) was on the drawing board.

The 'companion' squadrons, later renamed observation squadrons, were to receive the RWD 14 Czapla (Heron) high-wing monoplane with short take-off and landing capability and powered by the Polish 430 hp PZL Mors II (Walrus) radial engine. The Czapla was expected to start replacing the Lublin R-XIIIs in 1937 as the interim type, to be followed by the much superior LWS 3 Mewa (Mew) in 1939. Unfortunately, because of protracted development, the Czapla began to enter service only in the spring of 1939 and its production was limited to 65 aircraft.

Rayski's lack of appreciation for a single-seat fighter was reflected in his curious proposal that the sturdy, all-metal P.11 should be replaced by the P.39/LWS 4, a cheap, wire-braced light-weight monoplane of composite construction, powered by either PZL Foka or Ranger V-770 engines. Under his plan this aircraft was to remain in first-line service with fighter squadrons until April, 1942, and was to serve also as an intermediate fighter trainer. It could not operate at night or in winter conditions, its estimated top speed was 380 km/hr (205 mph) and its practical ceiling was limited to only 1,500 m (4,900 ft). The design contrasted sharply with other Polish projects, all technically advanced, and when, in June, 1936, Rayski asked the Government for permission to build the P.39/LWS 4 prototypes, he was refused. In the meantime, the specialist PZL fighter team responsible for the P.7, P.11 and P.24 was dissolved and Wsiewołod Jakimiuk, who led the team after Puławski's death, was entrusted with the design and development of a twin-engined passenger transport, the PZL 44 Wicher (Gale).

Rayski's 1936 plan was not officially endorsed, either in its original or revised form, but served the Central Staff only as the basis for the preparation of their own plan. This was the work of the Infantry Col Jan Jagmin-Sadowski and was ratified at the XVIIth Session of the KSUS (Committee for Armament and Equipment Affairs) on 13 October 1936. The KSUS plan, originally to be completed by April 1941, set a target of only 78 combat squadrons with a total of 642 aircraft and 100% equipment reserve, to be composed of five categories:

- 21 bomber squadrons (126 medium bombers), for destruction of fixed targets at the far rear of the enemy;
- 14 army support squadrons (140 aircraft), for armed reconnaissance and attack on mobile forces and fixed targets in the battle area;
- 18 'comparion' squadrons (126 aircraft), for battlefield reconnaissance and close cooperation with major army formations in the field;
- 15 single-seat fighter squadrons (150 aircraft), armed with highly manoeuverable airplanes, for close defence of the battlefield;
- 10 pursuit squadrons (100 aircraft), armed with fast aeroplanes, for defence of the country's air space in depth.

The PZL P.46/I Sum I prototype at Okęcie in February 1939.

The KSUS plan was based on a more conventional tactical concept, which called for new designs for army support and single-seat fighter squadrons and adopted a cautious approach to the new class of P.38 Wilk two-seat fighter squadrons, reducing their planned first-line strength to 100 aircraft. To meet the requirement for army support aircraft, Stanisław Prauss evolved a new single-engined three-seat monoplane, the P.46 Sum (Sheat-fish), which generally followed the P.23 Karaś pattern. But the gap arising from interrupted development of a classic single-seat fighter to succeed the P.11 proved difficult to close.

Feasibility studies in the light-fighter field were to be expanded and in addition to some further work on the improved P.39/LWS 4 project, a superior PZL design, the P.45 Sokół (Hawk) proposed by Kazimierz Korsak, was investigated in detail. Later, two more competitive designs, the PWS 42 and the RWD 25 entered the Sokół light-fighter programme and although the future of such an aircraft was not yet decided, all three Sokół contenders, powered by the 700 hp Gnome-Rhone Mars radial engine, were awarded prototype contracts.

Jakimiuk, still working on the PZL 44 Wicher transport, did not begin studies for an interceptor-fighter, the P.50 Jastrząb (Hawk), until the winter of 1936-37. His creation was a modern all-metal low-wing monoplane, employing a thin aerofoil-section wing and other advanced aerodynamic features, with built-in development potential. However, the P.50 specification, drafted to Rayski's directives, represented a subtle departure from the KSUS requirement in calling for a fast aeroplane, with manoeuverability taking second place to speed, capable of carrying a 300 kg (660 lb) bomb. These were the characteristics of the 'pursuit' fighter, affording insurance against the looming failure of the P.38 Wilk, rather than those of the classic fighter sought by the KSUS. An ill-matched powerplant, the Mercury VIII radial with a nominal output of only 730 hp and a maximum rating of 840 hp, which was forced upon the designer to save time and money, offered little prospects for the fighter's success.[18] The British Dowty company was to supply an inwardly-retracting landing gear for the prototype.

The P.50 was hastily ordered off the drawing board into mass order for production. PZL received an initial order for

[18] The P.24 then manufactured in Poland for export was only fitted with the 970 hp Gnome-Rhone radial, yet the 25% heavier P.50 airframe was to be provided with a 13% less powerful engine!

Left: Though the PZL P.50/I Jastrząb I was surrounded by great secrecy and no photographs of the whole prototype were ever released, the aircraft was shown to Count Ciano, the Italian Foreign Minister, during his visit to Warsaw at the end of February 1939. Here Ciano examines the Jastrząb's cockpit. Right: Author's artist impression of the P.50/I, based on four photographs of small fragments of the aircraft taken during Ciano's visit to the PZL.

150 Mercury VIII-powered examples, which was soon increased to 500, with deliveries of the first 50 fighters to commence in September, 1939. The P.50/I prototype was to be ready to fly in October, 1938, but late delivery of the Dowty gear delayed its flight tests until February, 1939. The P.50/II second prototype airframe, adapted for a more powerful radial of up to 1,600 hp and armed with two 20 mm cannon in addition to four guns, stood idle awaiting a powerplant, which remained unspecified until June, 1939, when the 925-1,145 hp Bristol Taurus II was ordered for it.

Re-assessment of Poland's budgetary and industrial capabilities in 1937 led to postponement of the completion of the plan until April, 1942. A the same time the original KSUS plan was re-adjusted to take into account a new, more adequate organization of the aviation units entailing additional aircraft for the squadron and dyon commands which increased the LW's first- line strength to 689 warplanes. The failure of the P.38 Wilk programme was also reflected. According to the Aviation Command's chart ref. 393/mob.37, from 23 October, 1937, details of the revised KSUS plan were as follows:

Schedules		**Armies Aviation**		**Dispositional Aviation**		**Totals**
	'Companion' Sqdns (a/c number – type)	Army Support Sqdns (a/c number – type)	Fighter Sqdns (a/c number – type)	'Pursuit' Sqdns (a/c number – type)	Bomber Sqdns (a/c number – type)	Sqdns (a/c number)
Actual Strength, 01.04.37	8[1] (126 - R-XIII)	17 (170 - Potez 25/P.23)	13 (130 - P.7/P.11)	–	3 (18 - Fokker)	41 (444)
01.04.38	18 (126 - R-XIII)	17 (178[2] - P.23)	15 (166[3]) - P.11	–	5 (18 - Fokker/ 12 - P.30)	55 (500)
01.04.39	18 (132 - R-XIII/ RWD 14/LWS 3)	17 (178 - P.23)	15 (166 - P.11)	–	8 (54 - P.30/P.37)	58 (530)
01.04.40	18 (132 - RWD 14/LWS 3)	17 (178 - P.23/P.46)	15 (166 - P.11/P.50)	2 (21 - not specified[4])	15 (99 - P.30/P.37)	67 (596)
01.04.41	18 (132 - RWD 14/LWS 3)	14 (146 - P.46)	15 (166 - P.50)	6 (63 - not specified[4])	21 (140 - P.37/ new type[5])	74 (647)
01.04.42	18 (132 - LWS 3 or new type)	14 (146 - P.46)	15 (166 - P.50)	10 (105 - not specified[4])	21 (140 - P.37/ new type[5])	78 (689)

[1] In addition to 8 organized squadrons, a number of independent platoons
[2] 17 squadrons @ 10 aircraft, plus 8 dyon commands @ 1 aircraft each
[3] 15 squadrons @ 10 aircraft, plus 15 squadron commands and 1 group command @ 1 aircraft each
[4] Possibly the P.48 Lampart, not yet ordered in quantity
[5] Probably the P 49 Miś

A small adjustment to this plan was introduced in January 1939 with the reorganization of the bomber force into nine-aircraft squadrons and a corresponding decrease in the number of bomber squadrons to 14.

In the meantime, personnel training was rapidly expanded to meet the anticipated growth in aircraft strength. After a complete stoppage in training of new pilots, which resulted from Piłsudski's decision in the autumn of 1926 to halt further expansion of the air arm, several observer officers were allowed to retrain as pilots in the early 1930s. Pilot training of new cadet-officers began at the Szkoła Podchorążych Lotnictwa (SPL, or Aviation Cadet-Officer School) in the Dęblin Aviation Officers' Training Centre – later renamed the Centrum Wyszkolenia Lotnictwa Nr.1 (CWL 1, or the Aviation Training Centre No.1) – only in the autumn of 1934, with the first 30 pilots graduating in the IXth Passing out from Dęblin in 1936. This was hardly enough to cover even a natural wastage in existing squadrons.

During the 1930s aviation training underwent frequent reorganization and in the second half of the decade entered a period of vast intensification and improvement. Hitherto largely neglected engineering personnel training received an important boost with the formation of the Technical Training Centre at Bydgoszcz – later renamed the Centrum Wyszkolenia Lotnictwa Nr.2 (CWL 2, or Aviation Training Centre No.2) where the Technical Cadet-officer School was established in the autumn of 1933. As the result of a complete revision of curricula at the Dęblin and Bydgoszcz Aviation Cadet-officer Schools in the years 1935-1937, training programmes were extended from two to three years, with training reaching exceptionally high standards.

At the CWL 1 the number of squadrons was increased from three to six and these were formed into three Dyons (Wings), new satellite airfields were established at Podlodów, Ułęż, Borowina and Zajezierze, and in 1937 the number of pupils increased to about 450, of which 200 were in the first year of training, 170 in the second and 80 in the third. In 1939 321 cadet-officers, over 65% or them pilots, graduated from Dęblin; 173 at the XIIth Passing out and 148 at the XIIIth Passing out, which was speeded as war loomed.[19]

The expanded Technical NCOs' School of the CWL 2 in Bydgoszcz had in 1936-37 about 400 pupils, but the Technical Cadet-officer school was detached from the CWL 2 in 1936 and moved to the Warsaw-Mokotów airfield as the Aviation Cadet-officer School-Technical Group, being at the same time subordinated to the CWL 1. By September 1939, 186 Technical Officers, and 34 Reserve Officers with full Engineering Degrees, passed out from Mokotów. This provided squadrons with valuable technical expertise.

In the middle of 1939 about 800 cadet-officer pupils were in training at the CWL 1, which then comprised:

An RWD 8a of the CWL 1 in flight near Dęblin. The RWD 8 provided standard "ab initio" equipment at all the LW training centers. The type was the most produced aircraft in Poland, with almost 600 delivered between 1933 and September 1939 to the LW, flying clubs and private owners. It was manufactured under the DWL licence in a few other countries. (R. Łopacki)

- Aviation Cadet-officer School – Combat Group, at Dęblin and its satellite airfields, having two Training Dyons (Wings) of three squadrons each, and one Exercise Dyon (Wing) of three fighter squadrons and observers' squadron;
- Aviation Cadet-officer Reserve School, at Sadków, consisting of one Training Dyon (Wing) of three squadrons;
- Higher Fighter School, at Ułęż, transferred from Grudziądz and the successor to the Gunnery and Bombing School;
- Aviation Cadet-officer School – Technical Group, at Warsaw-Mokotów, transferred from the Bydgoszcz CWL 2.

The CWL 2, moved from Bydgoszcz to Krosno and Moderówka, also had about 800 pupils under training in the summer of 1939 and comprised:

- NCO School for the Youth, at Krosno and Moderówka, composed of three Training Dyons (Wings);
- A ground course for Engineer Chiefs;
- Ground courses for Aircraft Engineers and for Wireless Engineers.

Almost 1,000 other airmen were undergoing primary and advance training at the Pilot's School in Grudziądz, the Training Battalion for ground maintenance crews at Świecie and the

The PWS 26, a progressive development of a long line of PWS trainers, served with the Dęblin and other LW training centers as a basic trainer from 1937 onwards. Some 300 were delivered. (R. Łopacki)

[19] "Czyż mogli dać więcej" ("Could they have given more"), by the Colleague Circle of the XIIIth Passing out from the Dęblin Aviation Cadet-officer School, p. 468-471, Veritas Foundation Press, London 1989.

Aircraft Engineers Course in Lwów, as well as at special proficiency training courses run by all Air Regiments at their bases. The overall number of the LW officers and men under training in 1939 was probably in excess of 2,500.

At the beginning of July 1936, in a belated move to entrust someone within the GISZ-controlled 'wartime' command channel with specific responsibilities for air defence the Inspektorat Obrony Powietrznej Państwa (IOPP, or Inspectorate of Air Defence of the State) was created, with Gen Orlicz-Dreszer, a cavalry officer, appointed as its head. Dreszer was killed in an air crash twelve days later and was-succeeded by Gen Józef Zając, an infantry officer. In the autumn of 1937 a minuscule Air Staff was formed at the Central Staff. This was a new body, completely unconnected with the Air Staff of Rayski's Aviation Command in the 'peacetime' channel (subordinate to the Ministry of Military Affairs). The two Air Staffs did not even communicate with each other.

Gen Zając was a novice to aviation, but after a few months in office, he saw himself as qualified to pronounce judgment on intricate questions of operational and equipment policy. Charged with the country's air defence, he began to promote the view that the top priority should be given to development of the fighter force. He completely rejected the proposed build-up of the bomber force, maintaining that Poland could not afford twin-engined bombers and did not need such aircraft in general, and the P.37 Łoś in particular. He contended that the bomber capability should be provided by the French or British air forces as and when necessary. He began to voice increasingly strong criticism of Rayski's decisions and opposed the ethos of the KSUS plan, which bore the hallmark of Rayski's influence.

In 1936 Zając unveiled his own plan. This envisaged 102 combat squadrons with 900 warplanes by April, 1942; 30 single-seat fighter squadrons (300 aircraft), 15 'pursuit' fighter squadrons (150 aircraft), 22 army support squadrons (220 aircraft), 20 observation squadrons (140 aircraft) and 15 bomber squadrons (90 aircraft) – a large proportion of the bombers to be provided by French or British units stationed in Poland. The 'pursuit' fighters were to be organized in five dyons and strategically place in three large groupings with their own early-warning networks.

This coincided with a deepening equipment crisis. Polish troop concentrations in connection with the Lithuanian and the Czech emergencies in 1938 revealed alarming weaknesses in the combat readiness of the LW squadrons. Although the fighter force was increased to 15 squadrons by the formation of two new squadrons, shortages of P.11 aircraft necessitated retention of the totally obsolete P.7 in three squadrons. What was even worse, early in 1938 the P.11 steel engine frames began to crack due to metal fatigue, this severely restricting their use.

After the Czech crisis a shortage of the P.23 Karaś aircraft reduced the number of army support squadrons from 15 to 12 and enforced disbandment of the two remaining Potez 25 squadrons. The R-XIII was declared unsuitable for combat duties

Line-up of the Łoś Ibis bombers of the 210 Bomber Dyon (211th and 212th Squadrons) at Okęcie on 19 March 1939, when the Dyon was officially declared operational, with aircraft serials 72.43 nearest and 72.17 behind. (A. Glass)

and all the 18 'companion' squadrons were temporarily withdrawn from the first-line strength pending rearmament with the RWD 14 Czapla and the LWS 3 Mewa in 1939. Earlier on the existing three Fokker bomber squadrons were transferred to reserve and the first two P.37 Łoś squadrons in the process of conversion training did not reach operational status until March, 1939. Thus, at the end of 1938, the LW's first-line strength fell alarmingly to a mere 27 (15 fighter and 12 army support) squadrons with a total of only 290 aircraft, of which 130 (P.11 fighters) were affected by flying restrictions. Under the KSUS plan the LW should have had at that time at least 55 combat squadrons with 500 aircraft.

At the end of November, 1938, Zając prepared a secret defence report on the LW's state of readiness, which pinpointed the devastating deficiencies in equipment and stressed that apart from the bombers, the situation would worsen in 1939 and would not significantly improve before 1941. He warned that in case of war in 1939 or 1940 he LW would not last longer than two to four weeks. On 14 January, 1939, Gen Rayski tendered his resignation – for the third time (he had done so twice before: first, verbally, in 1937, and then in the spring of 1938) – and on 19 March it was accepted. He was succeeded by Gen Władysław Kalkus as the Aviation Commander – in the 'peacetime' channel – whilst Gen Zając was nominated to the post of C.-in-C. of the LW in case of war – in the 'wartime' channel. To bridge the nonsensical isolation of the two channels, which under the existing statutory government and military structures had to remain separated, Kalkus was made personally (and not because of his function) responsible to Gen Zając.

With the German seizure of Czechoslovakia in March 1939 adding to the gravity of the situation, Zając at once set about the revision of all production programmes in line with his own ideas. Purchases of aircraft from abroad were to ease the equipment crisis. Although his tactical concepts had some merit, practical steps taken in haste to resolve the crisis were rather confusing and haphazard and seriously disrupted production schedules of the Polish aviation industry.

In the following weeks the KSUS plan was considerably

reshaped. The 1st Air Regiment was to be greatly enlarged to embody the Dispositional Aviation, a combat formation with a strength of 330 aircraft at the disposal of the C.-in-C. Polish Armed Forces, comprising six single-purpose Brigades, each consisting of two to three two-squadron dyons, as follows:

- one Bomber Brigade: three dyons with 19 aircraft each, a total of 57 P.37 Łoś bombers;
- two Army Support (Light Bomber) Brigades: each of three dyons with 21 aircraft each, a total 126 P.46 Sum and Fairey Battle light reconnaissance bombers;
- three Pursuit Brigades: one of three dyons and two of two dyons, each dyon with 21 aircraft, a total of 147 aircraft, possibly the P.48 Lampart.

One of the P.43B recce.-bombers, still in Bulgarian markings, being reassembled at PZL WP 1 at the beginning of September 1939.

The remaining five mixed Air Regiments, supplemented by a new 7th Air Regiment to be established at Wilno the following autumn, were to provide the Armies Aviation with the nominal strength of 362 aircraft, as follows:

- two two-squadron reconnaissance dyons and four independent reconnaissance squadrons, a total of 82 P.46 Sum aircraft;
- six two-squadron observation dyons, a total of 126 LWS 3 Mewa aircraft;
- seven two-squadron fighter dyons, a total of 154 single-seat fighters of an unspecified type.

This combat force of 66 squadrons with 692 aircraft was to be augmented by an operational reserve of 33 squadrons with some 340 aircraft, providing 50% backing for all categories of squadron.

Adjusting production orders to his requirements, Zając wanted to halt the manufacture of the P.37 Łoś at 104 aircraft, but since the assembly of an additional 20 airframes had already been transferred to the WP 2 at Mielec, giving this new factory its initial work, they were allowed to be completed.

PZL's firm order for 160 P.46 Sums was increased to 300, with the first deliveries expected at the end of 1939. Meanwhile, in the early months of 1939, the main production effort of the WP 1 the at Okęcie centered on completion of the 42 P.43B recce. bombers (the Gnome-Rhone-powered export version of the P.23 Karaś) ordered by Bulgaria. In March 1939, at the height of the Czechoslovak crisis, it was briefly intended to retain these warplanes for the LW,[20] but this decision was soon reversed and in spite of crippling shortages of aircraft in the Karaś squadrons, 33 P.43Bs were delivered to Bulgaria in the summer. The last nine, already crated for dispatch, were unpacked and reassembled at the WP 1 in the first days of September. Amid all this, Poland decided to purchase 100 Fairey Battles from Great Britain, aircraft which were generally inferior to the P.43s and less suited to the specific operational conditions prevailing in Poland. Five combat squadrons were to be equipped with the Battles.

In the all-important fighter field the situation became equally chaotic. Disappointing results of the Mercury VIII-powered P.50/I Jastrząb prototype trials, which proved the aircraft manifestly underpowered, forced Zając to defer large scale manufacture of the fighter, limiting this to the first 30 examples which were already on the production line. Production was intended to be resumed later, after improvement of the basic design, which included development of a new wing and employment of a more powerful engine. As a stop-gap measure, limited production of new variants of the P.24 or the P.11 was considered and eventually the P.11g Kobuz (Hobby), which utilized the Mercury VIII radial originally intended for the P.50 Jastrząb, was selected. An urgent order for 90 P.11g Kobuz fighters was placed with the PWS factory, which had hitherto specialized in all-wood structures and had no experience in building all-metal airframes. Earlier PWS orders for 130 PWS 33 Wyżeł (Pointer) and 150 PWS 35 Ogar (Hound) trainers, were reduced to 25 and 50 aircraft respectively. At the same time a new-generation PZL fighter, evolved by Jerzy Dąbrowski around a 1,400-1,600 hp Hispano-Suiza 12Z upright-Vee liquid-cooled engine and believed to be designated P.62, progressed to the mock-up stage.

Meanwhile, towards the middle of 1939, Poland, disturbed by high-altitude flights by German reconnaissance aircraft op-

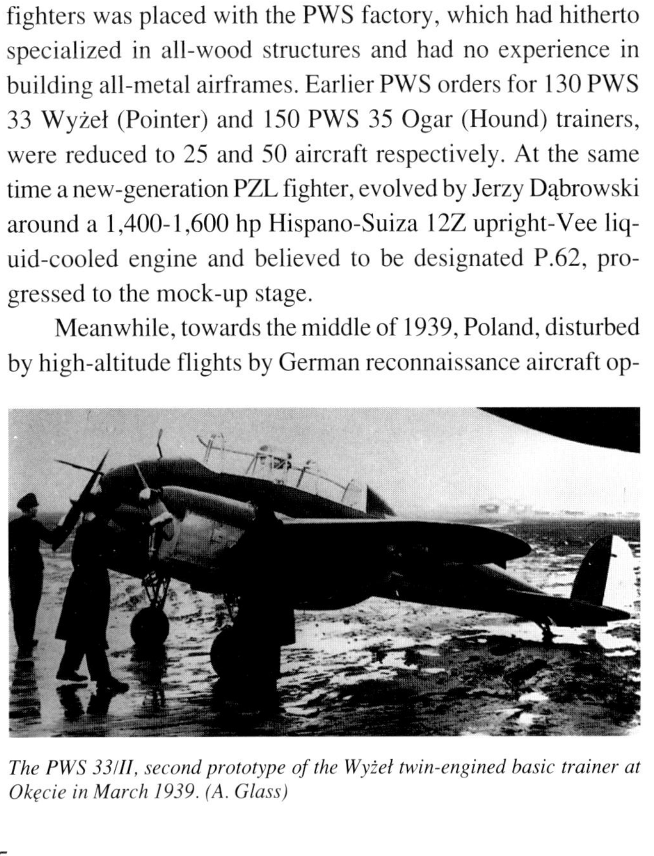

The PWS 33/II, second prototype of the Wyżeł twin-engined basic trainer at Okęcie in March 1939. (A. Glass)

[20] Communication from the Army Administration Office, dated 28 March, 1939, which also advocated mass production of the P.50 Jastrząb and the P.46 Sum.

erating with impunity over its territory, began frantic attempts to secure speedy delivery of foreign fighters. Several American designs were briefly considered, with the Curtiss Hawk 75A (export version of the P.36) and Seversky EP-1 (P.35) receiving particular attention, but a French armament loan and attractive delivery schedules led to selection of the Morane-Saulnier MS 406, of which 120 were ordered. The first 20 were ready for dispatch at the end of August and the next 50 were to be shipped in two batches by 1 October, 1939. Great Britain also agreed to supply 14 Hurricanes and one Spitfire without delay.

In the spring of 1939 12 of the 18 'companion' squadrons, renamed observation squadrons, were reinstated in the first-line. Five of them were re-equipped with the RWD 14 Czapla and the rest were to have their obsolete R-XIIIs replaced in the autumn with the LWS 3 Mewa, 200 of which were in full production at the LWS factory. Work on a progressive development of the LWS 3 was in hand and a number of new projects, mainly for army support aircraft, including a dive-bomber, were at various stages of design development at PZL, PWS and DWL.

At defence conferences held in Paris and London by French, British and Polish aviation missions at the end of May, the French promised to send to Poland a bomber formation of five Amiot 143M squadrons (62 aircraft) at the outbreak of hostilities. In August, they decided to supplement this force by a reconnaissance wing of Breguet 690s or Potez 63s. The British did not commit themselves, but during later discussions began to explore the possibility of 'pendulum' bombing operations by RAF Wellingtons.

Taken together, these measures represented a belated effort to develop the LW consistently with an overall defence strategy for the first time since 1926, but they could not have come at a worst time, for the war did not wait. The immediate effect of the production changes was a dramatic drop in the output of the Polish aviation factories. In the first eight months of 1939 they managed to turn out only about 210 aircraft, of which some 45 were built for export and only about 115 were warplanes for the LW – the lowest number for the past five years.

According to a secret report of the Air Staff, at the beginning of May, 1939, the LW had on charge a total of 1,863 aircraft, of which:

- 383: 18 P.37, 124 P.23, 127 P.11, 30 P.7 and 84 RWD 14 and R-XIII, constituted combat equipment of 41 operational squadrons;
- 300: 48 P.37 (including the second Łoś dyon in the process of formation), 71 P.23, 52 P.11, 34 P.7 and 95 R-XIII – most of them undergoing maintenance or repairs – represented operational reserves;

The Seversky EP-1 fighter NX2587 '68' in front of the ITL hangar during demonstration to the LW at the Okęcie airfield in June 1939. Standing in front, right to left, are George Burrell, Seversky's pilot, Stanisław Riess, the ITL's test pilot, and Stanisław K. Prauss, the ITL's technical interpreter. (J. Hopla/ J. Konierek)

P.37 Łoś and P.11 aircraft of the 1st Air Regiment on the Okęcie airfield early in 1939. First behind the Łoś is a P.lla '3' of the 111th 'Kościuszko' Squadron, with P.11c and P.11a fighters of the 111th and 112th Squadrons further away.

- 1160: various training types, more than half of them RWD 8 and PWS 26 and about 400 obsolete ex-operational types (mainly Potez and Breguet), provided training equipment.

By September the LW's combat strength was augmented only by 18 P.37 bombers (the second Łoś dyon).

The failure to make a proper use of home production resources was highlighted by the Aviation Command's own appraisal of the Polish aviation industry in the summer of 1939. This assessed its maximum peace-time production potential at 62 combat aircraft and 35 high-powered engines per month, or 745 aircraft and 420 power units per year. On a full war-time footing and two-shift working this was estimated to rise to 1,140 aircraft and 540 engines per year.

PART I

First Encounters with the Luftwaffe

(1939-1940)

Chapter I

THE LONELY FIGHT WITH THE LUFTWAFFE

September 1939

1. Political Background to the War

Powerful interests in the defeated Germany had never accepted the provisions of the Versailles Treaty, especially those affecting Eastern Europe. Separation of East Prussia from the rest of Germany by the so-called Polish 'Corridor,' the detachment of Gdańsk (Danzig) and the return to Poland of parts of the Silesia and Poznań provinces, which had been forcibly incorporated into Germany by the partition of Poland and for over a century systematically colonized by German settlers, were not regarded by German statesmen as permanent. Some leading German politicians went even further, expressing the conviction that a German-Russian collaboration aimed at total destruction of Poland was a historical necessity. Gen Hans von Seeckt, the architect of the German military revival and arbiter of foreign policy of the Weimar Republic, was the staunchest promoter of this view. During the biggest Bolshevik offensive against the Poles in the summer of 1920 he considered schemes for an invasion of Poland which would permit a direct link with the Red Army and subsequent common action against France and Great Britain.

The USSR was a willing partner waiting for an opportunity to destroy the Versailles order. In October 1920, at the end of the Polish-Soviet war, Lenin declared publicly:

"If Poland had become Soviet, the Versailles Treaty would have been shattered and the entire international system built up by the victors would have been destroyed."

These common interests led to a formal understanding between the Weimar Republic and the Soviet Union, Germany's overwhelming desire to weaken the dominance of the Western powers and to destroy Poland taking precedence over its fear of communism. The resulting Rapallo accord was concluded between the two countries on 16 April, 1922. Gen. von Seeckt, encouraged by its implications, wrote the following in September:

"Poland's existence is intolerable and incompatible with Germany's vital interests. She must disappear and will disappear through her own weakness and through Russia with our aid ... (With it) will fall one of the strongest pillars of the Versailles Peace, the hegemony of France ... The attainment of this goal must be one of the firmest guiding principles of German policy, as it is capable of achievement – but only through Russia or with her help. A return to the frontiers of 1914 should be the basis of agreement between Russia and Germany."

Lenin expressed similar sentiments:

> "An independent Poland is very dangerous to the Soviet Russia; it is an evil, which, however, at the present time has also its redeeming features; for while it exists, we may safely count on Germany, as the Germans hate Poland and will at any time make common cause with us in order to strangle her ... Germany wants revenge and we want revolution. For the moment our aims are the same."[1]

The Rapallo Agreement alarmed the Poles, reviving fears of a new partition of their country. The German and Russian threats were regarded by them as equally menacing to their newly restored independence and Polish defence preparations, conducted under the leadership of Gen Władysław Sikorski and the Chief of General Staff, Gen Stanisław Haller, assumed the possibility of attack by either of the two enemies, or both of them at once.

The Franco-Polish alliance of 19 February, 1921, which formed the basis for close economic, political and military cooperation between the two countries, and – more importantly – provided for mutual assistance in case of unprovoked aggression, became a cornerstone of Polish defence policy. However, the Polish trust in the alliance was soon to be grossly undermined.

The British, anxious to contain French influence in Europe, began to oppose anti-German solutions and even to sup-

[1] All quotations from Jan Karski, "The Great Powers and Poland 1919-1945," University Press of America, TM, Inc., Lanham, 1985 (Lenin's quotations, pp. 62 & 124, von Seeckt's quotation, p. 84).

port Germany in order to achieve a desirable balance of power. A multinational conference at Locarno in October 1925 was intended to enhance political stability in Europe and guarantee the existing borders. In fact, because of the British stand, it had the opposite effect. French attempts to include Poland and Czechoslovakia in the Locarno package were vetoed by Great Britain siding with Italy and Germany. Territorial guarantees were limited only to the western borders. In effect, the Locarno Pact subverted the permanence of the Versailles order, indirectly reaffirmed British recognition of the temporary character of Germany's eastern frontiers and gave a tremendous boost to German revisionists. Arguably, the Locarno Pact lent legitimacy to Hitler's claims to Czechoslovakia in 1938 and to Gdańsk and the Polish western lands in 1939 and led him to believe that the Western powers would not go to war to oppose his "legitimate" demands.

Gen Sikorski denounced Locarno as being specifically aimed against Poland and "provoking war." The pact gave rise to Poland's suspicions of the French commitment to the Franco-Polish alliance, which increased with the admission of Germany to the League of Nations in 1926 with the privileged status of a "major power."

French efforts to build up friendly relations with Russia, France's traditional ally from Tsarist times, were also worrying Poles, who interpreted their alliance with France in the widest of terms, insisting upon its full application to a threat from the East as well as from the West.

With Piłsudski's successful coup d'etat in May 1926, Polish foreign policy and strategy entered a new era. Skeptical of the French alliance, he thought it more probable that Poland would have to fight alone in a war with Russia. He believed the German threat to be exaggerated and less imminent and always regarded Russia as the more dangerous and permanent enemy. He persuaded the general staff to adopt a strategy based on the assumption of Soviet rather than German aggression.[2] With Poland placed between two menacing powers, this led him to seek an accommodation with Germany. His initial approaches to Berlin in December 1926 met with an uncompromising German stand on Gdańsk and the border issue. Germany's hostility towards Poland grew alarmingly during the 1932 election campaign and the unyielding German attitude prevailed until the spring of 1933.

Soviet politicians quickly realized that Piłsudski might seek some understanding with Germany. Worried about this their policy moved towards normalization of relations with Poland, including a non-aggression pact. In the changing political situation in Europe, both sides saw advantages in this and after difficult and protracted negotiations the Soviet-Polish non-aggression pact was eventually signed on 25 July, 1932. Improved relations between the two countries even included exchange visits by military aircraft. In the spring of 1933 Soviet TB-3 heavy bombers arrived in Warsaw and the LW reciprocated by sending five Lublin R-XIIID monoplanes, led by Col. Rayski, on a visit to the Soviet Union the following November.

When Hitler came to power, the course of German policy underwent a substantial change, particularly in relations with the Soviet Union. His political vision – the Thousand-year Reich, master of the European Continent – was irreconcilable with the Rapallo spirit. To assure real "Lebensraum" for Germany, he needed vast expanses of Russia and Gdańsk, and revision of the Polish border became a side issue, not his main objective. He wanted agreement with Poland and hoped for her cooperation.[3]

However, the beginning of Hitler's chancellorship was marked by a serious confrontation over Gdańsk when the city's German-controlled Senate decided to replace the special harbour police with its own force in February 1933. On 6 March Poland reacted by reinforcing the Polish garrison at Westerplatte in Gdańsk and mobilizing the border division, demonstrating to Hitler and to the world Poland's resolve to defend her rights in Gdańsk and her western borders to the point of war if necessary.

Anxious to settle this problem while the Germans were still weak and to concentrate on his eastern defences, Piłsudski is said to have approached France with the idea of a preventive war with Germany, but this story lacks corroboration.

The French were not in the mood to support Poland. On the contrary, the Western Powers, who were then working on the ill-famed Four-Power Pact (which was eventually signed between France, Great Britain, Italy, and Germany on 15 July, 1933, in Rome, but never ratified), condemned the reinforcement of the Polish garrison at Westerplatte, and the League of Nations ordered restoration of the "status quo" in Gdańsk as of 15 February 1933 by both sides. Unbeknown to Poland, the Western Powers secretly agreed with the German contention that the Polish border needed substantial revision, especially in respect of Gdańsk and the "Corridor," which were to be ceded to Germany "in Poland's own interest" to ensure a "durable peace." Germany was also granted equal rights in armaments.[4]

The Gdańsk incident provided a lesson to the Germans, not to provoke Poland until they were ready for war, and to Piłsudski, not to expect the support of the Western Powers in a dispute with Germany. This led him to renew attempts to seek an accommodation with the Germans. In April 1933, Alfred Wysocki, the Polish envoy in Berlin, was instructed to find out if the Germans would be willing to negotiate. For Hitler this was an opportunity not to be missed and Wysocki was granted an audience with him on 2 May.

To achieve his political and strategic aims in Europe, Hitler had to separate Poland from France. Dispensing with the League of Nations and its conception of "collective security" and taking up Poland's proposal to conduct direct negotiations, of-

[2] Ibid, P. 94, 172, and 181.

[3] Herbert von Dirksen; "Moscow, Tokyo, London – Twenty years of German foreign policy," p. 110; Norman, University of Oklahoma Press, 1952.

[4] Jan Karski: "The Great Powers and Poland 1919-1945," p. 152-156.

fered him the best means to this end.[5] In October 1933 Germany withdrew from the League of Nations and from the Disarmament Conference. During the same month Józef Lipski replaced Wysocki as the head of the Polish Legation in Berlin. On 15 November, 1933, a joint communiqué was issued, which informed the astonished world that the Polish and German governments agreed "to deal with questions concerning both countries by means of direct negotiations and to renounce all application of force in their relations with each other for the consolidation of European peace."

On 26 January, 1934, a ten-year non-aggression pact was signed between the two countries, changing the political balance in Europe. However, Poland firmly repudiated all German hints and suggestions regarding any military cooperation against the Soviet Union, and the Polish government also reaffirmed its adherence to the obligations arising from the Franco-Polish alliance of 1921. Both signatories to this alliance wanted to preserve it intact, but on their own terms and with different political objectives in mind. This proved incompatible in practice and in effect the Polish-German pact had a disturbing impact on Poland's relations with the Western Powers, and with France in particular.[6] Exchanges between Paris and Warsaw grew more and more acrimonious and the Western Powers – who themselves acted with complete disregard to Polish interests at Locarno and Rome (the Four-Power Pact), encouraging German revisionism in the east – began to condemn Beck's policy as pro-German and unfriendly towards France, with French diplomats hinting that France might eventually repudiate the alliance with Poland altogether.[7]

France, with her eastern alliance in disarray decided to restore it by signing a mutual assistance pact with Russia. The Germans used this as an excuse to denounce the Locarno Treaty and marched into the demilitarized Rhineland to the sound of fresh reassurances from Hitler about his desire for peace. With only three German battalions crossing the Rhine, the French army of 13 divisions on the German frontier could have forced a German retreat with a negligible engagement of its strength. This almost certainly would have spelled the end of Hitler, but the French, paralyzed by internal wranglings, did nothing. From then on, Hitler's quest for the dominance of Europe, by combination of deceit and force, was unstoppable.

After Piłsudski's death on 12 May, 1935, his policies and strategies were ardently adhered to. The Polish Central Staff continued to give top priority to preparing of a detailed defence plan to meet the military threat from the East. Work on this came to an end only in February 1939. Accordingly, the whole military set-up was arranged to face eastwards and army engineering units and military building resources were concentrated on the construction of fortifications in the East. Because of Soviet air surveillance the fortification works were provided with fighter protection and in the late summer of 1936 a Russian R-5 reconnaissance biplane was shot down by Lt Witold Urbanowicz of 111th Fighter Squadron, flying a P.11, who led a two-aircraft patrol in the Sarny region. However, some work on a general evaluation of the German threat was undertaken by the end of 1933 and in 1935 Gen Śmigły-Rydz issued instructions to initiate the study "Germany." This led to a general conclusion that Germany would be fully prepared for a major war in the years 1942-43, but the possibility of its earlier readiness from 1940 onwards should not be ruled out. An assessment of Germany's war potentialities, presented by Gen Tadeusz Kutrzeba in 1936, was greatly expanded under Kutrzeba's direction by Col Stefan Mossor, his Chief Staff Officer, on request from Śmigły-Rydz, resulting in a preliminary "Study for Poland's strategic plan against Germany."

This, the first and the only such plan then in existence, was presented to Śmigły-Rydz towards the end of January, 1938. The plan indicated that Poland would not be able to conduct a war with Germany on equal terms and pointed to the necessity of outside help. Its recommendations included alliance with France and material and industrial help – in armour, aircraft and raw materials – from the Soviet Union. Because of this latter suggestion and Śmigły's continued preoccupation with war in the East, the plan was shelved.

Hitler's preoccupation with the Spanish Civil War and Austria, and negotiating the Anti-Comintern Pact with Japan and Italy, brought about a relatively quiet time on the Polish scene. However, Polish public opinion became increasingly disturbed by the scale and speed of German rearmament and by anti-Polish outbursts from various German leaders and officials, especially with regard to Gdańsk and the Pomorze ("Corridor") and Silesia regions.

The seizure of Austria provided Germany with tremendous strategic advantage. Encouraged by Chamberlain's incredibly foolish appeasement policy, Hitler proceeded with preparations for "Fall Grün" ("Case Green") – the conquest of Czechoslovakia by force. This led to a flurry of diplomatic activity and first warnings from France, Russia and Britain to Germany (the "May crisis" of 1938), with Poland remaining on the sidelines, and eventually brought Europe to the brink of war in September, 1938, which was only averted – or rather temporarily postponed and made even more certain – by the sacrifice of Czechoslovakia and the humiliating capitulation of the British and the French at Munich on 29 September, 1938. With this, Hitler's greatest bloodless victory, the German conspirators in the Army High Command, who, led by Gen Franz Halder, were prepared to overthrow Hitler the moment he gave the order to attack Czechoslovakia if Britain and France declared publicly their determination to stand by that country,[8] lost heart. Thus, the second, and the last, chance to get rid of

[5] William L. Shirer: "The Rise and Fall of the Third Reich," p. 266.
[6] Jan Karski: "The Great Powers and Poland 1919-1945," p. 177.
[7] Ibid, p. 183.

[8] Gen Ewald von Kleist, the first emissary of the plotters, arrived in London on a secret mission on 18 August, 1938, but, apart from Winston Churchill, no British politicians were inclined to commit themselves. On 2 September he was followed by Halder's own emissary, Col Hans Boehm-Tettelbach, who was also largely ignored.

Hitler was spoiled by the Franco-British surrender at Munich, at a time when the balance of military power was still greatly in favour of the combined Western and Czechoslovakian forces.

During the "Anschluss" and "Fall Grün" operations, Germany made a calculated effort to improve their relations with Poland. The Germans in Gdańsk, regarded as the barometer of relations between two countries, were on their best behavior. Hitler also adopted a policy of encouraging those countries which had territorial claims upon Czechoslovakia to share in the spoils. Both Hungary and Poland had large minorities living in Czechoslovakia. Poland was interested in the recovery of the small Cieszyn (Teschen) region, with a Polish population of about 135,000, which had been overrun by the Czechs in January 1919 and Hungary in the Carpathian Ruthenia province, inhabited by about 500,000 Hungarians. Poland made it clear that she would not allow Russian troops or aircraft to cross her territory to help Czechoslovakia and Polish forces entered the Cieszyn region on 2 October, 1938, with a short-sightedness matching that of the British and the French which in the end would prove catastrophic for Europe.

By the end of October, 1938, Hitler was preparing to invade the rest of Czechoslovakia and to occupy the Lithuanian Memel district. Looking still further ahead, he also began diplomatic efforts to detach France from Great Britain, which resulted in a pact of friendship signed by the German and French foreign ministers in Paris on 6 December. This was wrongly interpreted in Germany as meaning that after Munich the French were no longer interested in Eastern Europe and would not go to war over Poland. The stage for "Fall Weiss" ("Case White"), a surprise attack on Poland, began to be set.

2. The German "Fall Weiss" and the Polish "Z" Plan

Captured German documents do not reveal when exactly the plan code-named "Fall Weiss" had first been drawn. After the bloodless conquest of the Sudeten territory the Gdańsk issue was a prominent item on Hitler's agenda. On 24 October, 1938, Ribbentrop raised the subject "in strict confidence" in conversation with Józef Lipski, the Polish ambassador in Berlin, adding that in addition to the "return" of Gdańsk, Germany desired to build an extra-territorial motorway and a double-track railroad across the Polish "Corridor" to connect Germany with Danzig (Gdańsk) and East Prussia and that Hitler wished Poland to join the Anti-Comintern Pact against Russia. In return, Germany offered to extend the Polish-German non-aggression treaty by a further ten years and to guarantee Poland's existing borders. The Polish reply, delivered on 19 November, was negative, but as a gesture of good will the Poles expressed willingness to replace the League of Nations guarantee of the Free City status of Gdańsk with a similar, but direct Polish-German agreement, warning at the same time that "any attempt to incorporate the Free City into the Reich must inevitably lead to conflict."

P.11a and 11c fighters of the 111th 'Kościuszko' Squadron on the Aleksandrowice airfield in Bielsko, taking part in the Polish recapture of the Cieszyn region at the beginning of October 1938. (T. Kopański)

Karaś aircrews of the 22nd Squadron, 2nd Air Regiment, during the Cieszyn operation, October 1938. (T. Kopański)

Hitler's reaction was reflected in a new directive to the armed forces of 24 November. This supplemented the October orders covering Czechoslovakia and Memel by adding that "preparations are also to be made to enable the Free City of Danzig to be occupied by German troops by surprise" in "quasi-revolutionary" circumstances, "not a war against Poland."

On 5 January, 1939, during a meeting with Józef Beck, the Polish Foreign Minister, Hitler declared, "Danzig is German, will always remain German and will sooner or later become part of Germany." Beck responded that he saw "no possibility whatever of agreement" on the Gdańsk issue. Since the first oblique reference to the "Fall Weiss" appeared in the Luftwaffe document "Planstudie 1939" of 7 February, 1939, it is almost certain that the initial directive for this plan was prepared in utmost secrecy in January, following Hitler's talk with Beck.

By the end of February Beck became aware of the dangers of his foreign policy and, as German pressure increased, he began to consider closer links with the Western Powers. Hitler's ruthless annihilation of Czechoslovakia on 15 March outraged public opinion in Britain and France, but in spite of the solemn guarantees given to the Czechs at Munich neither country made any move to oppose it. Nevertheless, their policies towards Germany were swiftly redirected. On 21 March Chamberlain suggested that Britain, France, Poland and the Soviet Union should consult immediately about steps to stop further German aggression. This echoed Litvinov's earlier and wider proposals, which had been rejected by Chamberlain because of his deep distrust of Russia. However, to Poland, which had good reason for even more profound distrust of the Soviets, inclusion of Russia was not acceptable. Ironically enough, on the same day Ribbentrop, dispensing with the earlier gentle approach, turned the heat on Poland, requesting agreement to the demands of October 1938. He warned that, "Poland could not take a middle course between Russia and Germany" and that her "only salvation was a good relationship with Germany and the Führer", and, "a joint anti-Soviet policy."

The occupation of Memel by the Germans on 23 March shook Poland because of parallels between the Lithuanian port and the situation of Gdańsk. On the same day a partial "alarm" mobilization was ordered and Polish troops began to concentrate around Gdańsk. Polite but firm rejection of the German demands was delivered by Lipski to Berlin on 26 March, the only concession by the Poles being an offer to discuss the German road and rail link across the "Corridor", but without giving it extra-territorial rights. On 31 March Chamberlain declared in the House of Commons:

> "In the event of any action which clearly threatened Polish independence and which the Polish Government ... considered it vital to resist with their national forces, His Majesty's Government would feel themselves bound at once to lend the Polish Government all support in their power. They have given the Polish Government an assurance to this effect. ... the French Government have authorized me to make plain that they stand in the same position in this matter."

By then Hitler was convinced that Poland would fight if he moved against Gdańsk and the Chamberlain guarantee threw him into one of his rages, but he still believed that a war with Poland would be a short, isolated affair and would be over before the West could act.

The directive "Fall Weiss", a lengthy document with numerous enclosures, was issued to the armed forces in the greatest secrecy on 3 April, with only five copies being released. It read: "The present attitude of Poland requires ... the commencement of military preparations to remove, if necessary, any threat from this direction forever." The aims were "to destroy Polish military strength ... (by) ... a surprise German attack." To achieve this, the German policy would seek "to isolate Poland if possible ... to limit the war to Poland only" and the war was to start "with sudden, heavy blows" and gain "rapid successes." Preparations were, "to be made in such a way that the operation can be carried out at any time from 1 September, 1939, onward." An assessment of the military situation contained a significant and ominous sentence "Intervention by Russia ... cannot be expected to be of any use to Poland."

On 6 April Beck signed a provisional mutual-assistance agreement with the British. On 28 April Hitler delivered a thundering speech to the Reichstag, which ridiculed President Roosevelt's plea for peace, dismissed all suggestions that Germany was preparing an attack on Poland and, in the same breath, denounced the Polish-German non-aggression pact as "no longer in existence" because of Poland's mutual-assistance treaty with Great Britain.

Germany had begun overtures to Moscow on the possibility of improving relations soon after Munich. Stalin was unhappy to be left out of the Czechoslovakian deal and according to Litvinov's "Notes for a Journal" told him towards the end of 1938: "we are prepared to come to an agreement with the Germans ... and also to render Poland harmless." At the beginning of May, 1939, Litvinov – a Jew and therefore an inconvenient person to deal with the Nazis – who had worked for a collective security alliance of the Soviet Union with Britain and France with little success, was suddenly dismissed from the post of Foreign Commissar. He was replaced by Vyacheslav Molotov, pointing unmistakably to a turning point in Soviet foreign policy. On 7 May the French Foreign Minister received an intelligence report from Germany informing him that Hitler was looking for an understanding with Russia which included a fourth partition of Poland.

On 23 May, the day after the signing of the "Pact of Steel" between Germany and Italy, Hitler bluntly told his military chiefs that war was unavoidable, explaining: "Danzig is not subject of the dispute at all. It is a question of expanding our living space in the East ... " Non-German lands in the East were to provide Germany with a source of labour (a hint of the slave labor programme). Poland was to be attacked "at the first suitable opportunity,"[9] but it was essential to isolate her. Almost all of Germany's military strength was to be concentrated on "Fall Weiss." Hitler also referred to the possibility of a war "of life and death" with England and France, which would include occupation of neutral Holland and Belgium. This was "no longer to be a question of right or wrong but of to be or not to be for eighty million (German) people."

The operational plan "Fall Weiss," prepared by Gen Walther von Brauchitsch, Commander in Chief of the Army, was accepted by Hitler on 15 June, 1939. Its aim was to prevent a full mobilization of the Polish Armed Forces by a surprise invasion and to destroy the main body of the Polish Army, which was expected to be west of the Vistula-Narew line, by a concentric strike from Silesia on the one side and from Pomerania-East Prussia on the other.

Two army groups were set up to achieve these objectives: Army Group "South," under the command of Gen Gerd von Rundstedt, made up of the 8th, 10th and 14th Armies, which was to thrust from Silesia in the general direction of Warsaw and secure the banks of the Vistula both south and north of the capital city; and Army Group "North," under the command of Gen Fedor von Bock, consisting of the 3rd and 4th Armies, which was to push across the "Corridor" to establish a link between the Reich and East Prussia and to swing south towards-Warsaw. Gdańsk was to be declared German territory on the first day of war. The order of deployment for the offensive was to be put into operation on 20 August, by which time all preparations were to be concluded. For protection against a possible Anglo-French intervention from the West, Germany was to rely mainly on fortifications of the West Wall (the Siegfried Line), which was to be brought to a state of maximum readiness by 25 August at the latest.

Initial, phase "Z" mobilization started on 15 August. On the evening of 23 August Hitler set "Y" Day, commencement of the onslaught on Poland, for the 26th day (Saturday) of that month at 04:30 hours. The secret general call-up (Fall "X") would begin on the same day. The "Y" Day date was confirmed in the "Fall Weiss" executive order, issued to the armed forces on 25 August at 15:02 hours. By then 37 infantry and 15 armored, motorized and fast divisions were concentrated on the Polish border, ready to attack.

First directives for the Polish "Z" Plan ("Z" for "Zachód" or the "West"), the case of defensive war with Germany, were issued by Marshal Śmigły-Rydz to the Central Staff only in the second half of February 1939. At a conference on 4 March the Central Staff established mobilization, operational, and quartermaster terms and divided the work on the Plan between them. The annexation of Czechoslovakia by Germany in March worsened the strategic situation of Poland and increased the likelihood of a main German attack from Silesia. This led to a feverish revision of the defence proposals. In addition to the five Armies of the original Plan – "Modlin," "Pomorze," "Poznań," "Łódź," and "Kraków" – a new Army "Karpaty" was to be formed and deployed in southern Poland to counter the new threat from Czechoslovakia. The seizure of Memel by the Germans also necessitated reinforcement of the northern flank and this was to be provided by a new Independent Operational Group "Narew."

In assessing German military strength and intentions the "Z" Plan drew on the earlier Kutrzeba-Mossor's studies. In operational terms it considered a number of variants of German aggression, from an all-out attack along the whole 850-mile length of the Polish border aimed at the total destruction of Poland, to a limited local action against Gdańsk or the "Corridor", or possibly also against the Wielkopolska region, aimed at retaking the lands returned to Poland under the Versailles Treaty. For this reason it was considered essential to defend at all cost and despite immense practical difficulties, the "Corridor" and the Western provinces, particularly in view of uncertainty about the timing and scale of Western help.

Recognizing German superiority in men and equipment, the "Z" Plan set out the main operational aims as follows: to defend territories regarded as essential to the conduct of war and inflict the heaviest possible losses on the enemy in the process; to counter-attack in favorable situations in order to hold the front until the Western Allies began their offensive operations; and to explore tactical opportunities resulting from an Allied offensive in the West and the relief of enemy pressure on the Polish front.

In contrast to the "W" Plan (the case of war with Russia), which took years to prepare, the hurriedly improvised "Z" Plan defined operational objectives only in general terms. It was not worked out in detail, never put to the test in war games, nor did it extend beyond the first phase of operations, i.e. the border battle of the main defence position. Tasks for the reserves of the Commander-in-Chief remained largely unspecified.

As work on the Plan progressed, it became imperative to establish as quickly as possible the extent of military action by the Western Allies and of financial and material help from abroad. Intensive talks between the Polish and the French defence chiefs, conducted in Paris in May 1939, pressured the French into signing a provisional bilateral agreement.[10] This obliged France to provide Poland with five bomber squadrons and to begin immediate air operations if Germany attacked in the East, to start probing strikes across the border on the third day of general mobilization, and to launch an offensive on the fifteenth day. Poland undertook to begin operations immediately in the case of German aggression against France. However, the Franco-Polish agreement, as it stood, was of little practical value, as the French evaded the fundamental question of the scale of their anticipated military commitment. Simi-

[9] From minutes of the meeting by Hitler's Adjutant, 2nd Lt Rudolf Schmundt, as quoted by William L. Shirer in *The Rise and Fall of the Third Reich*, p. 590.

[10] The agreement was conditional upon validation of the comprehensive Franco-Polish political accord which was not signed until 4 September, 1939.

lar talks with the British proved even less promising and only a visit by Gen William Ironside to Warsaw in July encouraged greater Polish optimism. Whilst the French and the British would give no firm undertaking for offensive operations in the West, they were more accommodating in providing war credits and some material help. These were to be too late to be of use.

In the meantime, the Polish Armed Forces were increasing their defence readiness. A new, imaginative Mobilization Plan "W" ("W" from Gen Józef Wiatr, its originator), which came into force on 30 April 1938, was flexible and comprehensive. It allowed for selective call-up of men, horses, equipment and material reserves by quantity and territory, according to emerging tactical needs. The mobilization process was divided into three stages: a preliminary one, which prepared the ground for a wider mobilization, secret "alarm" mobilization (partial or a full one) of troops required to reinforce threatened border areas, and an open general mobilization.

The preliminary and partial "alarm" mobilization, which began on 23 March, 1939, was stepped-up in April to improve combat readiness in the "Corridor" and along the German border and alert anti-aircraft defences. A much wider partial "alarm" mobilization was ordered on 13 August in response to provocative German excesses in Gdańsk, this being followed by the full "alarm" mobilization in all western provinces on 23 August and in the remainder of Poland three days later. Marshal Śmigły-Rydz ordered the strategic deployment and defensive grouping of the Polish forces on the German border on 29 August, but, bowing to great diplomatic pressure from Britain and France, which were asking him "not to annoy the Germans," he postponed public proclamation of the general call-up by one day, until 30 August. By 1 September the Polish Army comprised 26 infantry divisions, one armoured, eight cavalry and three mountain brigades, and 56 National Defense battalions.

3. The Russian Treachery and Political Agitation at the Eleventh Hour

Moscow moved very carefully in the summer months of 1939, exploring the political positions of both sides: the Germans, who temporarily adopted a patient attitude, and the British and French, whose discussions with the Russians were marked by dither and delay. In August, with a date for the "Fall Weiss" operation – set for the 26th day of that month – fast approaching and the growing possibility of Western involvement in the conflict, it became crucial for Hitler to reach a speedy settlement with the Soviets. On 14 August Germany pressed for talks about "the possibility of restoring German-Russian friendship and clarifying jointly territorial questions in Eastern Europe," including "the Baltic States, Poland and the southeastern area." Ribbentrop offered to go immediately to Moscow for discussions.

The Russians, who were at that very moment negotiating with France and Great Britain in Moscow the terms of a possible anti-German accord, warmly welcomed the German initiative and in response proposed a Nazi-Soviet non-aggression pact. In an act of unsurpassed political double-cross, matched only by Hitler's own deceitfulness, Stalin was hypocritically playing both fields simultaneously, to win either way. His price for allying himself with the Western Powers was a demand that Soviet troops be permitted to enter Poland – through the Vilno and Galicia provinces – to be deployed throughout the country to "protect" it from the Germans and that similar arrangements should also be made with Rumania. His terms for "friendship" with Germany were the incorporation of the eastern half of Poland in the Soviet Union and a free hand in respect of the Baltic States and Rumania. Whilst the British and French negotiators, aware of Poland's uncompromising stand on the subject of Russian troops, could not provide the Soviets with satisfactory answers, the Germans were enthusiastically obliging and on the afternoon of 19 August Stalin made up his mind to opt for the German deal. At the same time, the Russians allowed discussions with the Western Powers to drift for six more days.

A large German delegation, headed by Ribbentrop, arrived in Moscow on board two Fw 200 Condor four-engined transports at noon on 23 August. The Germans, extremely anxious to reach an immediate agreement, went straight to business and the non-aggression treaty, together with the "Secret Additional Protocol"[11] which was the crucial part of the agreement, was signed the same night. The secret protocol defined "the delimitation of both countries respective spheres of interest in Eastern Europe" as follows:

> "1. In the case of a territorial and political transformation in the lands belonging to the Baltic States (Finland, Estonia, Latvia, Lithuania), the northern border of Lithuania shall represent the frontier of the spheres of interest both of Germany and the U.S.S.R.
>
> 2. In the case of a territorial and political transformation of the lands belonging to the Polish State, the spheres of interest of both Germany and U.S.S.R. shall be bounded approximately by the line of the rivers Narew, Vistula and San.
>
> The question whether the interest of both sides make the maintenance of an independent Polish State appear advisable and how the borders of such a State should be drawn can be definitely determined only in the course of further political developments. In any instance, both Governments will resolve this question by means of friendly understanding."

The Russian interest in Rumanian Bessarabia was also recognized by the Germans.

[11] Both parties treated this protocol as most secret and its contents became known only after the war, with the capture of the secret German archives.

Thus, a new partition of Poland was agreed to by Germany and Russia, as three times before during the 18th century. With his eastern flank now secure against intervention by the Soviet Union, Hitler confidently informed his generals that "Fall Weiss" would go ahead as planned and that Poland would be destroyed within a week or two before anyone could come to her aid.

When the announcement of Ribbentrop's visit to Moscow was made in Berlin, the British cabinet stated that a Nazi-Soviet pact "would in no way affect their obligation to Poland ... which they are determined to fulfill." This was followed by a transformation of the earlier British unilateral guarantee of Poland into the much more meaningful pact of mutual assistance against German aggression, signed on 25 August just after 17:30 hours. Hitler, elated by his triumph in Moscow, was deeply shaken by the news from London and hesitated. His sudden dilemma was heightened at the same time by a communication from Mussolini, who warned him that should the German attack on Poland not be a localized affair and lead to a counter-offensive by the Western Powers, Italy would not be able to participate in the war, because "of her present state of war preparations."

The latest deadline for postponement of "Fall Weiss" had been set at 15:00 hours on 25 August, yet the news from London and Rome was grave enough to prompt Hitler to halt the opening of hostilities, set for 4:30 hours the next morning. Urgent orders were flashed to all units to stop all "Fall Weiss" movements at 20:30 hours, creating confusion among the troops, some of which were already marching towards the frontier.

Hitler regained his composure in less than 24 hours, and on 26 August decided to attack Poland on 1 September. The intervening time was to be used for political moves intended to drive a wedge between Poland and Britain. His new proposals were designed to persuade the British that "peace" negotiations were still possible and purported to present "moderate" demands, which would appear reasonable to the British but would probably not be acceptable to the Poles. After an exchange of diplomatic notes, the British advised the Polish government to enter into direct negotiations with Germany. Hitler's agreement to this, expressed on 29 August, was simply a ploy to keep Britain out of the conflict.

Hitler had his "justification" for the invasion of Poland already well prepared, regardless of what happened on the diplomatic front. The immediate cause of war was to be provided by a grotesque border incident, code-named "Operation Himmler," for which 150 Polish uniforms and small arms had been obtained some weeks earlier.

This was to be a faked "Polish" attack on the German radio station at Gleiwitz (Gliwice) in Silesia, which was arranged by the SS and Gestapo using condemned German concentration camp prisoners and the Polish uniforms, so that "Polish soldiers" could be blamed for an attack on Germany. A selected prisoner – referred to as "canned goods" – was first to be injected with drugs to make him unconscious and then shot by SS men wearing Polish uniforms and left dead on the ground for the benefit of the press and propaganda officials. The "captured" Gleiwitz radio station was to broadcast a speech in Polish calling for war with Germany. The operation, originally set up for the night of 25 August, was eventually carried out on 31 August at 20:00 hours.

In the early hours of 31 August, while Hitler still claimed he was waiting for a Polish emissary to negotiate peace, the German Army received orders to attack Poland on 1 September. At 12:40 hours the formal "most secret" Directive No.1 "for the conduct of the war" was issued. Its first two points read:

> "1. Now that all the political possibilities of disposing by peaceful means of a situation on the Eastern Frontier ... are exhausted, I have determined on a solution by force.
> 2. The attack on Poland is to be carried out in accordance with preparations made for "Fall Weiss". ... Allotment of tasks and the operational target remain unchanged.
> Date of attack: 1 September, 1939.
> Time of attack: 4.45 hours."

The remaining two points dealt with war in the West and stressed that "the responsibility for the opening of hostilities should rest squarely on England and France. For the time being minor frontier violations should be met by purely local action." Orders to go over to the attack Hitler expressly reserved for himself.

4. Strength and Deployment of the Polish and German Air Forces in September 1939

Under a general operational plan prepared by the embryonic Air Staff in the spring of 1939, which outlined the wartime organization of the LW, first-line aviation elements were to be regrouped to meet anticipated needs of the Army High Command to:

- the Dispositional Aviation, an independent tactical formation under the direct orders of C.-in-C. Polish Armed Forces, composed of a Pursuit Brigade (five fighter squadrons, detailed for the defence of Warsaw) and a Bomber Brigade (four P.37 Łoś medium bomber squadrons and five P.23 Karaś light bomber squadrons);
- the Armies' Aviation, composed of seven tactical aviation formations under the orders of army commanders in the field, with six consisting of one fighter unit (either a dyon or a squadron), one P.23 Karaś reconnaissance squadron and two RWD 14 Czapla or R-XIII observation squadrons each, and one with only reconnaissance and observation squadrons. Originally five such formations were envisaged for the five proposed armies, but when the land forces were expanded by the addition of Independent Operational Group "Narew" in the north and the Army

"Karpaty" in the south, the available aviation had to be stretched to cover them. No fighters were available for the last named Army.

War tasks for the LW were set out in the "General directives for the use of aviation" appended to mobilization instructions from 15 April 1939. Part I, "The aviation at the disposal of Armies' Commands," defined the air reconnaissance limits for each of the armies and permitted use of reconnaissance squadrons for bombing purposes in extreme emergency. The fighters were to impede enemy air activities – and his reconnaissance in particular – over the respective armies territories, set up "ambush" ("look-out") posts, afford protection to reconnaissance aircraft and provide air-strike support to ground troops on the battlefield in critical situations. The directives required fighters to operate "seldom, but in strength, rather than frequently in weak numbers." The observation aircraft were to operate on the Polish side of the front line.

Tasks for the Dispositional Aviation, detailed in Part II, entailed reconnaissance on behalf of the C.-in-C. Polish Armed Forces, particularly on the flanks of the front. The bombers were to "intervene on the battlefield and at the immediate rear against live forces of the enemy," attack the enemy's airfields and his rail and road transport. The fighters were to provide air defence for specified cities or areas of the country.

In July 1939 Gen Zając introduced a number of changes to the air arm's tasks and defined its war objectives more precisely, limiting their scope to tactical operations in the narrowest sense of the word. Reconnaissance was to be limited to shallow tactical operations on armies' behalf, because in Zając's opinion "the Germans, resorting to war, were bound to complete all troop concentrations before the commencement of hostilities" and there was "the overwhelming need to preserve own aircraft for immediate demands of the battle." He also emphasized that the bombers should give absolute priority to intervention over the battlefield and should operate at intervals but always in force, so that they could recover their strength after periods of intensive activity.

Cpt Adam Kowalczyk, C.O. of the IV/1 Fighter Dyon, dispatches pilots of the 113th Squadron on an operational training task in February 1939, a few weeks before incorporation of the unit in the Pursuit Brigade. Squadron's C.O., Lt Wieńczysław Barański, first on the left.

At the outbreak of war the Bomber Brigade received an even more restrictive directive, indicating a further change of mind; it was to be exclusively used against the enemy's armour in the field.

The LW's combat objectives, as specified, indicated a considerable inability on the part of army and aviation chiefs to make sensible use of the extremely limited Polish air arm. The fragmentation of the meager fighter force into insignificant units allocated to army commands proved most unfortunate. The request that such units – mostly devoid of any means of early warning and ground control – should operate "in strength" and at the same time set up "ambushes" with two or three aircraft per ambush and even participate in the ground battle, was totally unrealistic and contradictory. This dispersion of fighters precluded their coordinated use for defence of Polish airspace as a whole and enabled enemy bombers to operate unmolested over large parts of the country.

Confining air reconnaissance requirements to close-range ignored the mobility potential of modern land forces and could prevent the C.-in-C. Polish Armed Forces from receiving vital strategic information on enemy movements behind the fast changing front. To restrict the bomber force, which was specifically designed and trained to destroy large reinforced land

P.37Abis and 37B Łoś Ibis and II bombers of the 210th Bomber Dyon at the Okęcie base of 1st Air Regiment, just before becoming operational in March 1939.

P.23B Karaś II aircraft, carrying the 'Wawel Dragon' badges of the 22nd Squadron, on the Rakowice base of 2nd Air Regiment in 1938. In 1939 the unit became the 2nd Squadron of the Bomber Brigade. Karaś '4', serial 44.68 nearest, with '3', 44.67, close behind.

Line-up of Lublin R-XIIID 'companion' monoplanes of the 36th Squadron, 3rd Air Regiment, in 1938. In 1939 the Squadron was incorporated in the Army "Poznań" Aviation. (A. Glass)

objects, to attacks on widely dispersed armour and troops moving over open areas was an especially ineffectual decision. It meant that the Luftwaffe enjoyed a freedom to operate from its forward airfields with absolute impunity and that no coherent plan on how to deal with the enemy's air threat was made.

The LW, which had progressively mobilized since 23 March, completed the process when the full "alarm" mobilization was ordered on 23 August. Its mobilized strength approached 16,000 men and included 1,181 fully trained pilots, 497 observers and 219 air gunners. Of these about 1,300 aircrew members and 7,200 ground personnel served with the combat units, which began to move from peacetime bases to the intermediate "alarm" airfields on 27 August and then proceeded to their assigned combat airfields.

This double move was intended to confuse the enemy's intelligence and conceal the location of the airfields. Almost all combat aircraft reached their war destinations by the morning of 1 September, but three P.11 fighters and four P.23 Karaś recce-bombers were lost due to landing accidents. Airfields for the anticipated arrival of the five French Amiot squadrons were also in readiness.

The LW's outgoing operational strength on 1 September at dawn comprised 392 serviceable warplanes (for the LW Order of Battle see Appendix 1):

- 158 (128 P.11 and 30 P.7) single-seat fighters, of which 53 in the Pursuit Brigade and 105 in the Armies' Aviation;
- 114 P.23 Karaś recce-bombers, of which 50 were in the Bomber Brigade and 64 in the Armies' Aviation;
- 36 P.37 Łoś medium bombers, all in the Bomber Brigade;
- 84 (35 RWD 14 Czapla and 49 R-XIII) observation monoplanes, of which seven were at the disposal of the Aviation Command and 77 in the Armies' Aviation.

In addition the Naval Aviation Dyon had on strength 12 (ten R-XIII and two R-VIII) floatplanes. These were totally obsolete and unsuitable for combat use.

Discounting the 84 observation planes, the effective LW's combat strength amounted to only 308 aircraft, of which 150 could carry bombs. The fighters had between four and five years squadron service behind them in the case of P.11 and up to seven years in the case of P.7, which was now too slow to be effective. The P.23 Karaś light recce-bombers, which had on average about two years service behind them, were approaching obsolescence and only the 36 Łoś bombers, the only twin-engined warplanes on the LW's inventory, could be regarded as modern, up-to-date equipment (for data of the Polish operational aircraft see Appendix 2).

This flimsy force was supplemented by 12 Fokker F.VII/3m transports, 94 unarmed RWD 8 trainers attached to various aviation units for liaison and 28 miscellaneous liaison aircraft of the Staff Squadron at the disposal of the Aviation Command. The Naval Aviation Dyon also had four (three RWD 8 and one R-XIII) liaison monoplanes, this bringing the overall number of aircraft in the Polish operational units to 510.

All serviceable warplanes were engaged on the German front, leaving the eastern border completely undefended from the air. Apart from 47 newly produced P.37 Łoś bombers, which were not yet fully armed and equipped, equipment reserves were insignificant. They consisted of some 30 RWD 14 Czapla, 50 P.11 and 70 P.23 Karaś aircraft, all of which were undergoing repairs or overhaul. About 30 reserve P.7 fighters and 130 R-XIII observation monoplanes could not be regarded as useful operational types.

German intelligence reports of May 1939 estimated the combat strength of Polish military aviation at about 900 aircraft: 150 bombers, 315 fighters, 325 reconnaissance aircraft,

Karaś IIs of the 41st Reconnaissance Squadron escorted by P.11 fighters, during exercises in 1939. (M.Krzyżan)

50 naval aircraft and the rest transports. These exaggerated figures continue to be quoted in many German publications.

Air power was to provide the mainstay of Germany's military might and its expansion received the highest priority in German economic planning and allocation of national resources. Aircraft production programmes of hitherto unheard-of magnitudes were put in hand and the Luftwaffe rapidly became a political as well as military weapon which Hitler exploited to intimidate the rest of Europe into abject submission to his demands.

In the early 1930s the German aviation industry was of a size comparable to that of Poland and in 1933 its average monthly output amounted to 31 aircraft. In January 1934, Gen Erhard Milch had formulated the "Rhineland Programme," which in its ultimate form called for the delivery of 4,021 aircraft, including 822 multi-engined bombers, by the end of September 1935. Production rocketed to an average of 164 aircraft per month in 1934, and to 265 per month in 1935. With this increasing availability of military aeroplanes and the clandestine formation of aviation units, Germany decided on 26 February 1935 to proclaim the formation of the Luftwaffe, to be officially established on 1 March. Its initial strength amounted to about 20,000 men and 1,822 aircraft of all types. In the autumn of 1935 Milch prepared even more ambitious plans.

These called for delivery of 11,158 aircraft, of which 3,820 were combat planes, including 1,849 bombers by 1 April, 1938. Subsequent running revisions of the production programs (to 18,000 aircraft) resulted in increases in the yearly output figures to 5,112 aircraft in 1936 (426 per month) and to 5,606 aircraft in 1937 (467 per month), but despite Göring's promises of "unlimited resources," output could not be further expanded or even sustained. Drastic steps were taken to improve

P.11c fighters of the 142nd Fighter Squadron, 4th Air Regiment, incorporated in September into the Army "Pomorze" Aviation (a prewar photo; two of the aircraft have white painted tips of fins and rudders for exercises).

Karaś IIs of the 41st Reconnaissance Squadron, 4th Air Regiment, which formed the bomber/reconnaissance element of the Army "Pomorze" Aviation, on the way to the Karaś overhaul base at Okęcie just before the outbreak of war.

Below: The Łoś bombers of the 210th Dyon (renumbered Xth during the mobilization) of the Bomber Brigade rolling off to take-off position in March 1939. (T. Kopański)

the efficiency of the aircraft industry and it was decided to concentrate on the production of only five basic types of the "new generation" warplanes: the Heinkel He 111, Dornier Do 17, Junkers Ju 87, and Messerschmitt Bf 109 and Bf 110. With re-tooling for the new types, production dropped in the early months of 1938 and then recovered, bringing the yearly total to 5,235 aircraft, or 436 per month, instead of 1,180 envisaged in Göring's industrial mobilization plan.

At the time of Munich in September 1938, the Luftwaffe's first-line strength amounted to 2,928 aircraft, of which 1,669 were serviceable. Faced with the possibility of general war, Hitler regarded this as totally inadequate and ordered an accelerated expansion of the Luftwaffe to 18,500 warplanes, of which 10,300 were to provide first-line equipment for 98 regiments and reconnaissance formations, with 8,200 in operational reserves. This force, to be achieved by April 1942, was to be composed of 58 bomber (Ju 88 and He 177), eight dive-bomber (Ju 87), 16 heavy fighter (Bf 110 and Me 210[12]) and 16 fighter (Bf 109) regiments, at a cost of 60 billion marks.

Production Plan No.9, formulated in November 1938 to meet this demand, envisaged an overall production totaling 45,700 aircraft (from 1933) and included orders for: 703 He 177, 7,327 Ju 88, 2,000 He 111, 900 Do 17 (production of these last two types being extended in view of delays in development of the He 177 and Ju 88), 2,002 Ju 87, 3,320 Bf 110 and Me 210, and 4,331 Bf 109. In 1939, because of continuing development difficulties with the He 177, Ju 88 and Me 210, various amendments were introduced, including expanded production of the He 111 and Do 17. In the summer of 1939 the monthly output of the German aviation industry, which comprised 51 airframe and 17 aeroengine factories and employed over 170,000 people, began to exceed 700 aircraft and its total production for 1939 reached 8,295 aircraft. Despite this, German aviation rearmament was then far from being completed and the Luftwaffe was not considered ready for a major war, which had not been expected before 1942.

At the beginning of 1939 the Luftwaffe underwent a major reorganization of its command structure aimed at achieving improved operational flexibility. The office of Inspector-General of the Luftwaffe was subordinated to that of Secretary of State for Air under the overall command of Gen Erhard Milch. Gen Hans Jeschonnek was appointed Chief of the Air Staff.

On 1 February, 1939, the former regional Luftwaffe Group Commands were transformed into three Luftflotten (Air Fleets):

- Luftflotte 1 (Ost), under the command of Gen Albert Kesselring, with headquarters in Berlin, covering North-east Germany including East Prussia;
- Luftflotte 2 (West), under the command of Gen Helmut Felmy, with headquarters at Brunswick, covering North-west Germany;
- Luftflotte 3 (Süd), under the command of Gen Hugo Sperrle, with headquarters at Munich, covering South-west Germany.

Luftflotte 4, commanded by Gen Alexander Löhr, with headquarters in Vienna, was established on 18 March, to cover South-east Germany, Austria and occupied Czechoslovakia.

At the same time the Naval Air Command was dissolved and its function taken over by the Luftwaffe General at the Naval High Command (OKM) and the newly-organized Armies' Aviation was subordinated to the Luftwaffe General at the Army High Command (OKW).

The Air Fleets differed in strength depending on their operational tasks and comprised one to three Fliegerdivisionen (Air Divisions – in 1940 renamed Air Corps). These were mixed formations, normally of 200 to well over 1,000 aircraft each, consisting of single purpose regiments (Geschwadern), except for the Lehrdivision, a combat-proficiency development unit, which was composed of regiments for a variety of purposes. Regiments were normally composed of three or four wings (Gruppen) and a command squadron or flight, the fourth wing usually being a replacement and operational training unit. The wings, with their own airfields, constituted the basic tactical units of the Luftwaffe. They were composed of three squadrons (Staffeln), each of nine to 12 aircraft (sometimes up to 16 aircraft), and a command flight[13]).

With the formation of Air Fleets the numeration of regiments was changed from the earlier three-digit system to a two-digit one. Under this Air Fleet 1 was allotted regiment numbers 1 to 25 and wing numbers ending in 0 and 1, Air Fleet 2, regiment numbers 26 to 50 and wing numbers ending in 2, Air Fleet 3, regiment numbers 51 to 75 and wing numbers ending in 3, and Air Fleet 4, regiment numbers 76 to 99 and wing numbers ending in 4. The independent Lehrdivision comprised two regiments individually numbered LG 1 and LG 2. The naval units retained their earlier three-digit numbers ending in 6.

The Arabic regiment numbers carried letter prefixes, the letters preceding G (standing for regiment) indicating a kind of aviation (K for bombers, St. for dive-bombers, Schl. for attack aircraft, J for fighters, Z for "destroyers" or heavy fighters and L for the Lehrdivision regiments). Roman numerals in front identified wings, i.e. III/KG 76 denoted IIIrd Wing of 76th Bomber Regiment. The independent reconnaissance wings had the designation letters AGr followed by their individual Arabic numbers, with squadron numbers – to which the "kind of unit" letter was added in brackets ((F) for long-range and (H) for army reconnaissance) – in front, thus 3(F)/AGr 121 denoted 3rd Squadron of 121 Reconnaissance Wing, frequently written in a simplified form as 3(F)/121.

[12] With the change of BFW name to Messerschmitt in July 1938, the company's new projects adopted the designation Me instead of Bf.

[13] Wing I had squadrons numbered 1 to 3, Wing II, squadrons 4 to 6, Wing III, squadrons 7 to 9 and Wing IV, squadrons 10 to 12. The reconnaissance aviation was divided into wings and its basic tactical units were squadrons, which frequently operated independently.

The Luftwaffe's "Fall Weiss" war objectives for Operation "Ostmarkflug," the air assault on Poland, were defined in the middle of May as follows:

"- to prevent the Polish air arm from effective participation in operations and from attacking the Reich by its destruction on the ground, as well as its bases, installations and supply sources;
- to carry out concentrated air strikes on the near and far rear of the fighting Polish armies, preventing mobilization and deployment of forces, paralyzing command and liaison networks and frustrating evacuation attempts (destruction of transport systems); also to provide close support for ground troops by attacks on the forward positions of the Polish forces and especially by strikes supporting armoured thrusts, to isolate battlefields in selected areas of advance and to prevent the enemy from establishing new lines of defence;
- to carry out a massive air strike on military installations, government establishments and armament industries in Warsaw (Operation "Wasserkante")."

Preparatory air reconnaissance activities began in April 1939, with clandestine high-altitude flights by two special squadrons of the Luftwaffe High Command, 1(F) and 2(F)/Ob.d.L., equipped with specialized variants of Do 17, the Do 17S and 17U.[14] These aircraft were immune to interception by Polish fighters because of their high operating ceiling and great speed.

The Luftwaffe started mobilizing in March 1939, under the initial "increased tension" stage of the German multi-stage mobilization plan. On 25 August the "X-Fall," or secret general mobilization, was ordered, this being followed by the "Mob-fall", or public general call-up announced on 1 September. Its personnel strength rose from about 373,000 to 1,300,000 men. The strength of anti-aircraft defence units, which formed an integral part of Luftwaffe organization and were incorporated in these figures, increased from about 107,000 to 900,000 men.

According to the records of the Luftwaffe Quartermaster General 6th Department, the strength of "Operative Luftwaffe" – comprising the four Air Fleets and the Lehrdivision – on 1 September 1939, amounted to a total of 3,374 aircraft, of which 2,956 were serviceable, as follows (numbers of serviceable aircraft given in brackets):

- bombers (K): 787 (685) He 111, 371 (319) Do 17 and 18 (0) Ju 88 – total 1,176 (1,004);
- dive bombers (St.): 339 (292) Ju 87 and 27 (26) Do 17[15] - total 366 (318);
- attack aircraft (Schl.): 40 (36) Hs 123;
- fighters (J): 743 (651) Bf 109 and 28 (25) Ar 68 – total 771 (676);
- "destroyers" (heavy fighters, Z): 95 (82) Bf 110 and 313 (295) Bf 109[16] – total 408 (377);
- long-range reconnaissance (F): 257 (235) Do 17;
- army reconnaissance (H): 275 (251) He 126, 67 (51) He 46 and 14 (8) He 45 – total 356 (310).

Naval aviation totaled 224 (206) warplanes, of which 57 (56) – 21 (20) He 111, 12 (12) Ju 87 and 24 (24) Bf 109 – were put at the Luftwaffe's disposal, and 167 (150) – all seaplanes – remained under orders of the Naval High Command:

- marine aircraft (M): 8 (8) He 115 and 58 (46) He 60 – total 66 (54);
- maritime reconnaissance (F): 62 (55) Do 18;
- general purpose (Mz): 31 (30) He 59;
- ship-deck aircraft (BFl.): 8 (8) Ar 196.

This air strength, totaling 3,598 (3,162) warplanes, was supplemented by 552 (540) multi-engined transports, of which 547 (535) were Ju 52, thus increasing the number of aircraft at the disposal of the German Armed Forces to 4,150 (3,702). In addition, over 300 machines of various types, comprising about 20 special-purpose bombers at the disposal of Göring's HQ, special reconnaissance units formed for the duration of the Polish campaign, the Fi 156 Storch observation units and miscellaneous liaison and courier aircraft, supplemented the first-line formations.

Equipment reserves included almost 1,000 combat planes in the replacement and operational training units, about half of them of older types, such as the Ju 86 bombers. Apart from this, some 2,500 earlier "first generation" aircraft were in the Luftwaffe's training establishments, which trained about 15,000 pilots annually.

At that time the Luftwaffe was the most powerful air force in the world. Against its 3,598 warplanes the Allies could master the following air strength:

Britain: 536 bombers, 608 fighters, 312 recce a/c, 304 marine a/c – total 1,760;
France: 463 bombers, 634 fighters, 444 recce a/c, 194 marine a/c – total 1,735:
Poland: 86 bombers, 158 fighters, 148 recce a/c, 12 marine a/c – total 404;
overall:1,085 bombers, 1,400 fighters, 904 recce a/c, 510 marine a/c – total 3,899.

However, apart from some 530 British Spitfire and Hurricane fighters and about 440 French MS 406 and Hawk 75A fighters, most of the Allied equipment was totally obsolete and

[14] Spy flights over Poland had begun as early as 1937 by the secret Rowehl's Kommando, which used three Lufthansa He 111 passenger transports, specially adapted for clandestine photographic reconnaissance work.

[15] Aircraft of unit commands.

[16] Pending rearmament with the Bf 110, many "destroyers" units still operated the Bf 109C and 109D variants.

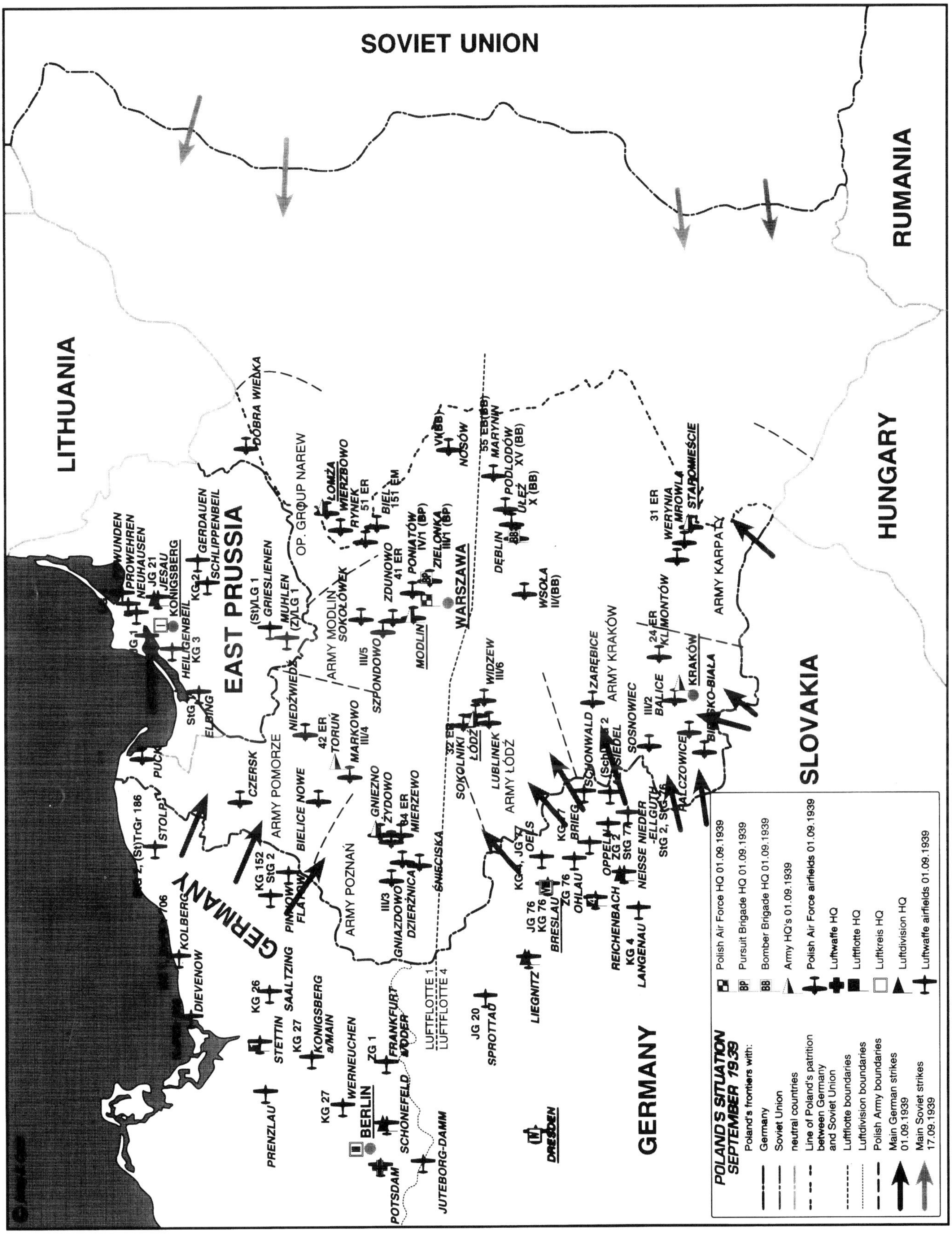
SOVIET UNION
LITHUANIA
EAST PRUSSIA
GERMANY
HUNGARY
SLOVAKIA
RUMANIA
WARSZAWA
POLAND'S SITUATION SEPTEMBER 1939
Poland's frontiers with:
Germany
Soviet Union
neutral countries
Line of Poland's patrition between Germany and Soviet Union
Luftflotte boundaries
Luftdivision boundaries
Polish Army boundaries
Main German strikes 01.09.1939
Main Soviet strikes 17.09.1939
Polish Air Force HQ 01.09.1939
Pursuit Brigade HQ 01.09.1939
Bomber Brigade HQ 01.09.1939
Army HQ's 01.09.1939
Polish Air Force airfields 01.09.1939
Luftwaffe HQ
Luftflotte HQ
Luftkreis HQ
Luftdivision HQ
Luftwaffe airfields 01.09.1939

its offensive potential extremely limited, whereas the Luftwaffe was armed with modern aircraft, about 70 percent of which had been delivered within the last eight months.

Göring's signal, "'Ostmarkflug', 26 August, 04:40 hours," sent from his Headquarters at Wildpark Werder near Potsdam the previous day at 18:30 hours, was canceled 90 minutes later. The order was re-issued on 31 August in the afternoon, setting the date of "Ostmarkflug" at 1 September, 04:45 hours.

The 1st and 4th Air Fleets, detailed for the Operation "Ostmarkflug," were reinforced with the Lehrdivision and units from the two remaining Air Fleets. The units assigned to the Polish front moved from their peacetime bases to operational airfields on 24 August in the afternoon, with the notable exception of the KG 27 (three He 111 wings), which was to go over to its war airfield after bombing Warsaw as part of Operation "Wasserkante."

The larger, 1st Air Fleet, ordered to operate over the Army Group "North" war theatre, was concentrated on airfields in Pomerania and East Prussia and comprised 23 1/2 wings: 13 bomber, 4 1/2 dive-bomber, 3 heavy fighter and 3 fighter wings. The 4th Air Fleet, operating with the Army Group "South" was concentrated in Silesia and had in its strength 19 wings: 8 bomber, 4 dive-bomber, 1 attack, 2 heavy fighter and 4 fighter wings. The "Operative Luftwaffe" on the Polish front consisted of the following (for the Luftwaffe Order of Battle see Appendix 3):

Marshal Göring's Headquarters:
– 2 bomber squadrons Ln.Abt.100[17]: 10 Ju 52, 9 He 111 and 2 Do 17U;
– 2 long-range recce squadrons: 9 Do 17 and 8 He 111; – total: 38 aircraft

Air Fleets:	*1st Air Fleet*	*4th Air Fleet*	*totals*
– bombers:	274 He 111	81 He 111	355 He 111
	142 Do 17	177 Do 17	319 Do 17
– dive-bombers:	143 Ju 87	127 Ju 87	270 Ju 87
	12 Do 17	11 Do 17	23 Do 17
– attack:	–	37 Hs 123	37 Hs 123
– fighters:	119 Bf 109	144 Bf 109	263 Bf 109
– heavy fighters:	36 Bf 109	40 Bf 109	76 Bf 109
	53 Bf 110	29 Bf 110	82 Bf 110
– long-range recce:	45 Do 17	30 Do 17	75 Do 17
– overall:	824 aircraft	676 aircraft	1,500 aircraft

Reich defense fighter units in the East	– total: 102 Bf 109
Army reconnaissance units:	
– long-range reconnaissance:	– total: 83 Do 17
– close-range reconnaissance (He 126 and He 46[18]):	– total: 218 aircraft
	overall total: 1,941 aircraft

The Luftwaffe deployed about two-thirds of its entire serviceable strength against Poland – 1,941 aircraft[19] out of 2,956 – and included 75% of its strike capability: 695 twin-engined bombers, 270 dive-bombers and 37 attack aircraft, or 1,002 aircraft out of 1,333. The remaining third, which kept watch in the West, was largely stripped of bombers and consisted mainly of fighter and reconnaissance aircraft. The Luftwaffe's crack 7 (Parachute) Fliegerdivision, under Gen Kurt Student, was also posted to the Polish front, to stand by at immediate readiness for any special tasks that might be required.

The Luftwaffe's superiority over the LW was overwhelming: in twin-engined bombers it represented a ratio of 19:1 (695 aircraft to 36), in single-engined bombers, a ratio of 5.4:1 (270 aircraft to 50); and in fighters (single-seat and heavy), a ratio of 3.3:1 (523 aircraft to 158). On 3 September that superiority was further increased by the transfer to the Polish front of two wings of KG 55 and two squadrons of KG 54, a total of 77 serviceable He 111 bombers.

5. Air War over Poland – September, 1939

On 1 September 1939 at 04:30 hours, 15 minutes before "X-hour," the time set for a start of the massive assault which was to open the undeclared war on Poland and trigger off World War II, three Ju 87B dive-bombers of 3./St.G 1, commanded by Lt Bruno Dilley, crossed the Polish border to demolish detonation points and leads at Tczew (Dirschau) leading to mined bridges over the Vistula river, to prevent Poles from destroying them. The Stukas did not succeed nor did the Do 17Z bomb-

[17] An experimental bomber unit, specializing in blind and night bombing techniques, later developed into the KG 100.

[18] A few He 45s were still retained by some of the units, but presumably none were serviceable on 1 September.

[19] On 22nd June, 1941, the Luftwaffe deployed 1,945 warplanes – including 510 twin-engined bombers, 290 dive-bombers and 480 single-seat and heavy fighters against the Soviet Air Force, which was ten times stronger than the LW.

Three Junkers Ju 87B dive-bombers of 3./St.G 1 on the Luftwaffe's first combat mission in World War II, on 1 September 1939 at dawn.

ers of III/KG 3, which bombed Tczew an hour later. The bridges, providing a link with East Prussia of immense strategic importance to the Germans, were blown as intended. The Luftwaffe's first attack did not achieve its purpose.

Early fog and low clouds over Northern, Western and Central Poland severely curtailed the morning activities of the Luftwaffe, whose principal objective was to obliterate the Polish air arm by a surprise all-out strike on its bases, installations and supply sources. Uncertain weather prospects also compelled Göring to suspend Operation "Wasserkante," a concentrated raid by combined bomber forces of both Air Fleets on Warsaw.

Luftflotte 1, most affected by the weather, managed to launch only a few of its bomber wings and these attacked various coastal and naval aviation targets, indiscriminately bombed Puck, and severely damaged the Małaszewicze air base. Before 07:00 hours about 60 He 111s of LG 1, flying with Bf 110 escort, approached Warsaw in a probing attack, with Okącie airfield as their main target. The Pursuit Brigade, in full readiness since dawn on combat airfields at Zielonka and Poniatów, was alerted by observation posts and its four squadrons, the 111th, 112th, 113th and 114th, were ordered off at about 06:50 hours, with the Brigade's fifth squadron, the 123rd, which had just arrived from Kraków, remaining on the ground. Climbing through clouds, the fighters came into contact with scattered enemy formations a few minutes later. Lt Aleksander Gabszewicz scored the Brigade's first victory, destroying a He 111. In vicious individual fights, which lasted over an hour and prevented the bombers from dropping any bombs on Warsaw, six He 111s were shot down and a Bf 110 of Mjr Walter Grabmann, the C.O. of I/(Z)LG 1 and veteran of the Legion Condor, was damaged and he himself wounded. Germans credited their crews with four P.11s, but in fact only three were brought down. Widespread reconnaissance activity around noon resulted in the spectacular destruction of one Do 17 over the Warsaw suburbs by Lt Stefan Okrzeja of the 112th Squadron.

As the weather improved, Luftflotte 1 rapidly intensified its operations, directing its main effort against the Polish peacetime airfields and Warsaw, which was attacked in waves by some 200 aircraft: He 111s, Do 17s, and Ju 87s, protected by Bf 110s and Bf 109s. Over 30 P.11s and P.7s of the Pursuit Brigade took off in defence, joined by a few fighters already in the air. Determined Polish attacks prevented many of the bombers from hitting their targets: airfields at Okęcie, Mokotów, and Gocław and the Vistula bridges. In the fierce air battle the Germans lost two Heinkels, one Do 17, one Ju 87 and five Bf 109s, but the defenders also suffered heavy casualties: six Polish fighters were destroyed, three heavily damaged and one, although only moderately damaged, never returned to its squadron. The escorting Bf 110s shot down four P.7s, which proved easy prey and damaged another P.7 and one P.11c. Polish pilots were repeatedly fired at as they parachuted down and two were severely wounded. One of the Polish pilots reported that Germans under attack formed a defensive circle.[20] On this first day the Pursuit Brigade, which flew about 105 sorties, lost 13 fighters and, with 19 more damaged and temporarily unserviceable, it was left with only 21 ready for immediate action. In the evening its C.O., Col Stefan Pawlikowski, was ordered to use the Brigade "more sparingly."

Better weather conditions over Southern Poland enabled Luftflotte 4 to operate more freely at first light, with the main blow falling on Kraków's Rakowice air base. This was first bombed by 60 He 111s of KG 4, then by the Stukas of I/St.G 2 and finally, from low level, by the Do 17s of KG 77. The returning Ju 87s surprised Polish fighters just beginning to take off from Balice airfield and Sgt Frank Neubert, who flew Ju 87B T6+GK, shot down the P.11c of Cpt Mieczysław Medwecki, the C.O. of III/2 Fighter Dyon, who was killed – the Luftwaffe's first air victory. Minutes later S-Lt Władysław Gnyś, flying P.11c '5' who was taking off with Medwecki but managed to evade the Ju 87s, in a solitary attack on two returning Do 17E bombers of KG 77, shot down both (one of the bombers was coded 3Z+FR), achieving the distinction of destroying the first German aircraft in World War II. Bombers of

[20] Later a familiar sight in British skies during the Battle of Britain.

A close-up of the P.11c of Cpt Tadeusz Opulski, C.O. of the 112th Squadron, Pursuit Brigade, one of the fighters defending Warsaw, which came to grips with Luftwaffe on the morning of 1 September (a prewar photo, taken earlier in 1939).

Luftflotte 4 raided several other targets, including airfields at Krosno, Moderówka, Sadków, and Masłów, and Heinkels of KG 4 flew across the Carpathian Mountains to make a long-range attack on Lwów's Skniłów air base. Numerous direct support missions were also flown, with the heaviest attacks directed against the Polish cavalry and troop concentrations at Wieluń.

On the first day of war Luftflotte 1 flew about 1,200 aircraft sorties, including 30 bomber missions at wing strength, 17 of which were directed against Polish air arm targets, eight in support of the Army and five against naval facilities and, according to its own records, achieved nine air victories for a loss of 14 aircraft. Luftflotte 4 carried out about 900 sorties and army reconnaissance formations about 600, bringing the total to some 2,700 sorties. The Luftwaffe strength returns showed a total loss of 37 aircraft with four more damaged beyond repair.

At least 30 Polish peacetime airfields and various aviation factories were hit, resulting in the destruction of about 180 aircraft, but only 24 of these could be counted as operational reserves, the rest being trainers and obsolete and civil types. The most serious damage was inflicted on the Małaszewicze base, where nine reserve Łoś bombers were wrecked in four Luftwaffe strikes. The Polish combat squadrons, dispersed on well camouflaged airfields, did not suffer in these attacks. Luftflotte 1 reported: "Whereabouts of the enemy air arm largely unknown."

Over 100 air raids were recorded by various Polish cities and towns, large and small, many of which suffered indiscriminate bombardment of the civilian population. Ominously, on a request of Gen Hans Jeschonnek, the Luftwaffe's Chief of Staff, made in the morning, the Soviet Government agreed to extend

Heinkel He 111H bombers of KG 1 'Hindenburg' on the way to a Polish target.

the broadcasting hours of the Minsk radio station and include a special identification of the station in its programmes, to provide a beam bombing aid for German bombers operating at night or in bad visibility.

The official Polish war communiqué also put the number of German aircraft destroyed at 37. Of these 24 were credited to Polish fighters, but while Waraw was well defended by the Pursuit Brigade, which brought down 15 enemy raiders, Polish air space remained largely unprotected. Two-thirds of the Polish fighter force, split into insignificant groupings between various armies in the field, flew 102 sorties and fought a number of isolated skirmishes, but accounted for only nine German aircraft. Devoid of efficient command and early-warning systems and having to operate mainly from "ambush" ("look-out") posts on accidental landing strips in flight strength or even in pairs or singly, the Armies' fighters were rendered largely ineffective. The Karaś reconnaissance squadrons flew

S-Lt Hieronim Dudwał of the 113th Fighter Squadron taking off from Okęcie in the middle of March 1939 in his P.11c '10', 8.70 (underwing call-number 170-N). (T. Kopański)

Dudwał's '10' severely damaged in the battle of the Pursuit Brigade with Bf 110s over Warsaw in the afternoon of 1 September.

20 sorties and the observation units about 55, bringing the total number of LW sorties to some 280. Losses amounted to 29 aircraft, of which 19 were fighters.

On 2 September the Luftwaffe continued to concentrate its efforts on the Polish air arm, industrial and air defence targets, including radio stations working for the early-warning system. In the morning 88 He 111P bombers of KG 4 with Bf 110 escort attacked CWL Dęblin and its satellite airfields. The raids, repeated in the afternoon, caused great devastation. Other airfields and the LWS, PWS and PZL WP 2 factories were also targeted and over 50 training and unserviceable machines, as well as more than 90 disassembled RWD 8 airframes in storage at Sadków, were destroyed. However, exhaustive searches by German reconnaissance aircraft for the Polish combat airfields drew a blank and this, combined with the absence of Polish warplanes in large formations, led the mystified German generals to the mistaken conclusion that the LW had been totally destroyed by the first strikes on their known bases. The official German communiqué for 2 September stated: "All existing enemy aircraft in hangars or in the open were set on fire. From this it can be assumed that the Polish air arm has received a mortal blow." German war propaganda made the most of this fallacy.

With a proportion of the German bombers kept on the ground in readiness for an immediate strike on Polish airfields when located by air reconnaissance, only a small formation was despatched against Warsaw, the bombers splitting into sections and approaching the capital from various directions. These diversionary tactics confused the observation posts and about 20 fighters of the Pursuit Brigade which took off failed to make contact with the enemy.

The Armies' fighters on this second day had more success. In the afternoon eight fighters from the 161st and 162nd Squadrons, intercepting a German raid on Łódź, came to grips with 12 Bf 110s of I/ZG 76 and in a violent dog-fight destroyed two Messerschmitts at a cost of two of their own number. The fighters of the Army "Pomorze" were engaged in numerous encounters and brought down eight Germans, including one Bf 110, but the 141st Squadron was misused in the afternoon for a futile low-level strafing attack on the advancing German troops and suffered grave losses: three pilots killed, three aircraft destroyed, one damaged and five slightly damaged. The Army "Poznań" fighters were also very active and shot down four enemy raiders. Altogether the Polish fighters flew 171 sorties and accounted for 17 German machines for a loss of

S-Lt Władysław Gnyś, of the 121st Squadron, Army "Kraków," who destroyed the first two Luftwaffe aircraft in World War II on 1 September early in the morning. (W.Gnyś/J.F. Lansdale)

Above: The fuselage of Dornier Do 17E bomber 3Z+FR of KG 77 shot down by Gnyś. Below: The fuselage of his other victim. Both Dorniers crashed in the village of Żurada, south of Olkusz.

Left: Some of the nine Łoś bombers destroyed and damaged at Małaszewicze on 1 September 1939. The most fascinating feature of this photograph are squadron markings of the dissolved 12th (Karaś) Squadron on Łoś II '3' of the XXth (220th) Dyon, the non-operational Łoś reserve unit. (T. Kopański) Right: Newly received Dornier Do 17Z-2 bombers of KG 3 "Blitz," preparing on the Heiligenbeil airfield, East Prussia, for their next strike on Poland.

nine P.11s (three of them in the ground attack) and three P.7s. The Luftwaffe recorded 14 aircraft totally lost and four damaged beyond repair.

The Bomber Brigade went into action for the first time and flew nine reconnaissance and 18 bomber sorties, the latter by Karaś light-bombers of the 64th and 65th Squadrons, which successfully attacked a developing thrust by the German armour on Częstochowa but lost seven aircraft. The Karaś reconnaissance squadrons made 30 sorties, including six by the 24th Squadron, which involved a bombing attack against German armour, and the observation units about 50 sorties. All told the LW made about 280 sorties and lost 25 aircraft.

From 3 September onwards, the Luftwaffe, satisfied that air supremacy had been fully achieved, began to turn its main effort to direct support of the advancing German forces, which were striving to break out from the frontier regions in a great pincer movement designed to trap the Polish armies west of the Vistula river. Luftflotte 4 was reinforced by about 100 (77 serviceable) He 111s of KG 54 and KG 55, arriving at the Polish front from their peace-time airfields.

Nevertheless, air and naval bases and industrial establishments continued to be heavily bombed. Two of the largest Polish warships, the destroyer "Wicher" and the minelayer "Gryf," were sunk. Warsaw was the target for multiple strikes by He 111s of LG 1 and KG 27 and the Stukas. The Germans again managed to confuse the air defences by probing Bf 110 raids, which were intended to draw Polish fighters into the air by frequent course changes as they approached the city. The Pursuit Brigade, which after repairs of damaged aircraft had at its disposal 40 fighters, flew 75 sorties but managed to engage only the Bf 110s, shooting down two and losing four fighters in the battle and one due to pilot error. The unmolested bombers caused considerable damage and hit the PZL WP 1 and WS 1 plants and hangars of the 1st Air Regiment at Okęcie. During the night bombers of the specialist Ln.Abt.100, helped by the extended Minsk radio broadcast, dropped their bombs on Palmiry ammunition depot. This was the first time that VHF radio beams were used for guidance and bomb release (the X system). During the following days several such night raids were made by the Ju 52s and He 111s of Ln.Abt.100.

The Armies' fighters put up 120 sorties, with the Army "Poznań," Army "Kraków" and Army "Pomorze" fighters – the last named flying standing patrols above the Vistula river crossings – being particularly busy. In the course of desperate

Reconnaissance Do 17P, 7A+EM of 2.(F)7/21, which worked for the Command of the 1st Air Division, Luftflotte 1, and was involved in a desperate search for the secret Polish combat airfields.

dog-fights 11 enemy aircraft were destroyed and eight fighters lost. Together with the Pursuit Brigade, the Polish fighters brought down 13 raiders and lost 13 of their own aircraft, including one in a take-off accident. Total German losses for the day were 24 aircraft: 21 destroyed and three damaged above 60%.

The Bomber Brigade's Karaś light bombers were heavily engaged with considerable success in repeated strikes on the most menacing spearheads of the German armoured forces in the Częstochowa-Radomsko region in South-western Poland. In addition to four reconnaissance flights, the Brigade carried out 43 bomber sorties and lost nine aircraft. The Armies' Karaś squadrons logged 25 reconnaissance and 18 bomber sorties also directed against German armour, and the observation squadrons, about 45 sorties. With the fighter force still in good shape and the bomber force widening its activities, the LW flew 330 sorties, probably the greatest number of the war in a single day, and lost 33 aircraft.

During the period 4 to 6 September the war entered a new phase, with the Germans breaking through Polish defences in the north and south. The Luftwaffe intensified and extended its ground attacks, neutralizing the Polish fortified positions, breaking up attempted counter-attacks and helping armoured divisions in their thrusts towards Warsaw. By the nightfall of 7 September Luftwaffe 4 had flown 1,634 sorties, dropping 743 tons of bombs and firing 36,650 gun rounds, in direct support of the XVIth Panzer Corps alone. Rail and road networks were attacked with increasing vigour in order to isolate Polish front-line forces, and Polish communications were destroyed to such an extent that effective military command became impossible. Attacks on aviation and industrial targets and on Warsaw also continued. The Luftwaffe flew over 2,000 sorties each day.

On 4 September Warsaw suffered one of its most damaging bombardments. At Okęcie PZL WP 1 buildings were hit, several aircraft demolished and the WS 1 engine plant set on

S-Lts Jan Dzwonek and Kazimierz Rębalski of the 161st Fighter Squadron, Army "Łódź," studying a map by Dzwonek's P.11c at the Basiówka airfield near Lwów on 31 August 1939 before a departure to the Widzew combat airfield. Mechanic Suwiński is sitting in the cockpit.

Below: 3rd September was a day of the maximum effort for the Karaś light bombers of the Bomber Brigade, attacking the German armour spearheads which pierced the Polish defence lines in the south. The Brigade's 2nd Squadron alone lost three aircraft, including C.O. Cpt Kazimierz Słowiński's '1' – he perished with his crew. The photograph shows Cpt Słowiński entering his Karaś II at the Rakowice base on 24 April 1939.

fire. The Pursuit Brigade flew 74 sorties and destroyed four raiders, but was unable to stop the bombers. The Brigade lost two fighters in combat and one due to excessive wear. The Armies' fighters logged 91 sorties, and shot down seven of the enemy. Their own losses amounted to ten fighters. Five of these, one P.11 and four P.7s, were destroyed on the ground (three) and in the air (two) at Widzew – the first Polish combat airfield to be detected and successfully attacked whilst in use – by Bf 109Ds of I/ZG 2, whose pilots claimed 11 PZL fighters shot down. During the day the Luftwaffe lost 20 of its machines and a similar number sustained damage between 10% and 60%.

For the Bomber Brigade this was the day of maximum effort. The Łoś bombers, which had earlier flown only single reconnaissance missions, went into action in force for the first time. In five bombing raids, totaling 56 sorties, 50 of them by Łoś bombers, and concentrating on German armour in the Radomsko-Piotrków area, forward elements of the XVIth Panzer Corps suffered crippling casualties, which the Germans described as "considerably successful tactical action by the enemy"[21] The Brigade, which also made 12 reconnaissance sorties, or 68 in all, lost nine of its 36 Łoś bombers. The Armies' Karaś reconnaissance squadrons launched 32 sorties, half of them bomber missions. The LW flew a total of about 300 sorties, during which 30 machines were lost.

On 5 September, in view of the successes of the enemy's diversionary tactics over Warsaw and increasingly unreliable reports from the rapidly disintegrating early-warning network, the Pursuit Brigade adopted new operational methods. Instead of taking-off in strength in response to incoming alerts, it began to fly standing patrols by small sections of aircraft. This resulted in more frequent encounters and in 40 sorties five victories were achieved; three fighters were lost. The Armies' fighters flew 51 sorties, accounting for two raiders for no loss to themselves. The German records listed seven aircraft lost and a further ten damaged in various degrees.

The Bomber Brigade, after its great exertions the previous day and because of the need to change airfields, was unable to maintain the pressure on the German armour and mounted only 22 aircraft sorties: nine combined recce-bomber missions by Łoś and 13 by Karaś aircraft. Its effort was further weakened by the need to switch some of its bombers to attack enemy armour which was breaking through in the north, in the Ciechanów Area. Three aircraft were lost. The number of sorties flown by the LW dropped to 180 and 15 of its aircraft were lost.

6 September was the most dramatic day in the German-Polish war of 1939. Warsaw came under threat from the advancing German spearheads and by nightfall the Polish government had decided to leave the capital. The Pursuit Brigade, with 29 serviceable aircraft, was ordered to carry out reconnaissance of the Łódź-Kielce region, the first mission not directly connected with the air defence of Warsaw, followed by a fighter sweep of the Kutno-Koło battle area. Its dismayed pilots, who until then had not been aware of the gravity of the ground situation, attacked enemy formations with grim desperation in an all-out attempt to keep the Luftwaffe at bay, and fought over the Kutno front what was to be the last air battle of the war. In the evening the Brigade was ordered to move immediately from the Warsaw area to airfields in the Lublin area, thus the most purposeful and distinguished period of its opera-

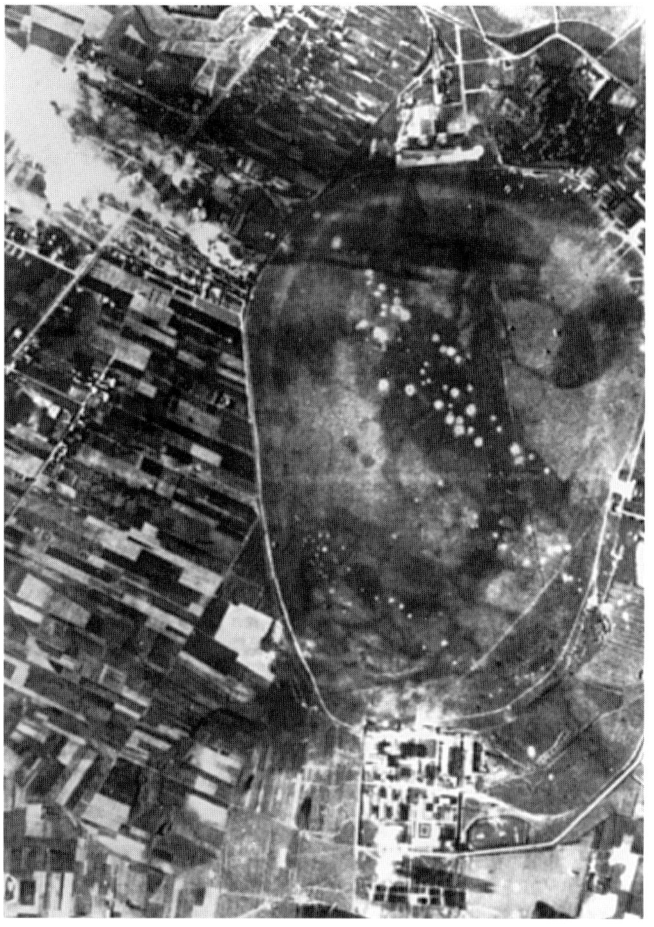

A Luftwaffe reconnaissance photograph of the Okęcie airfield after the bombardment on 4 September, with the burning PZL WS 1 aero-engine plant, top left, eight large hangars of the 1st Air Regiment, opposite on the right, and the bombed PZL WP 1 at the bottom.

P.11c of the 161st Fighter Squadron, Army "Łódź," destroyed by Bf 109s at Widzew on 4 September.

[21] "Luftwaffe in Polen," Ob.d.L., 8 Abt., a secret study, dated 11 July, 1944 (copy of Air Ministry's translation, AHB 6, VII/33, in the author's archives).

Left: P.7a '10', 6.133, of the 162nd Fighter Squadron, Army "Łódź" shot down by Bf 109s in the woods near Widzew on 4 September. (T.E. Willis) Right: Łoś bombers of the Xth Bomber Dyon, Bomber Brigade, soon after arrival at the Ułęż combat airfield. The Dyon moved to Kuciny on 3 September, from where it began to operate in force against German armor the following day. (M. Krzyżan)

tions came to an end. During the day the Brigade had flown 75 sorties, scored eight victories and lost seven fighters in combat and five because of airfield changes, only 19 of its aircraft being able to fly to Lublin.

This was also the last day of intensive activity for the Armies' fighters operating in their original groupings. Sixty-seven serviceable Armies' fighters flew 89 sorties – with the fighters of Army "Pomorze" and Army "Poznań" accounting for 40 sorties each – and destroyed five enemy aircraft for a loss of two. An order of the C.-in-C. Polish Armed Forces, requesting withdrawal of the fighter units from the armies to reinforce the Fighter Brigade in the Lublin area, began to reach them in the evening.

The Luftwaffe's strength returns for the day showed a loss of 16 machines. About 20 more were damaged. During the six days of bitter combat, marking a heroic stand of the obsolete and overwhelmingly outnumbered Polish fighters against the mighty Luftwaffe, the Germans according to their own records,[22] lost 126 aircraft totally destroyed or critically damaged (more than 60%), with a roughly similar number sustaining repairable damage. This amounted to almost half of all their losses in the September campaign. The flying personnel losses for the same period amounted to 315 killed, missing and wounded. 42 of the aircraft – 21 He 111 and Do 17 bombers, ten Ju 87s, four Bf 110s, six Bf 109s and one Hs 126 – were destroyed during raids on Warsaw, with 31 falling to the guns of Polish fighters and 11 to the anti-aircraft defences.

The Pursuit Brigade, with an initial strength of 53 fighters, flew during this period 408 sorties, shot down 34 enemy planes (including those destroyed outside the Warsaw area during the fighter sweep) and lost 36 aircraft, averaging one loss per 11 sorties. With three P.11s received as replacements and one taken over by the Armies' aviation, only 19 (16 serviceable) fighters remained. The Armies' fighters, with an initial strength of 105 machines, flew 585 sorties, achieved 51 victories and lost 38 aircraft, averaging one loss per 20 sorties. The Armies' aviation, which received two replacement P.11s, was left with 69 (64 serviceable) fighters. In six days of fighting the Polish fighter force was almost halved. Personnel losses totaled 13 pilots killed, four missing and 19 wounded.

These figures point to the tremendous effort and combat achievement of the meager Polish fighter force and reflect its excellent morale. However, because of the immense disproportion between the two sides in numbers of aircraft employed and sorties flown, the Luftwaffe, which sustained probably no

[22] "Luftwaffe in Polen," Ob.d.L., 8 Abt., dated 11 July, 1944, and Meldungs der Luftwaffe an das O.K.W., Gen.Qu. 6 Abt.

Left: An unserviceable P-11c '4' (underwing call-number 804-N), of the Pursuit Brigade, abandoned at the Okęcie airfield, with a German sitting in the cockpit. (T. Kopański). Right: S-Lt Dudwał's P.11c '10', 8.70, with part of the pilot's headrest fairing removed and the fuselage top recovered with unpainted metal, was left behind partly repaired on the Brigade's Poniatów landing ground and captured by the Germans.

more than one loss for 100 sorties from all causes and one for over 140 sorties from the opposing fighters, cannot be said to have been seriously challenged.

Activities of the Bomber Brigade on 6 September were very limited. They amounted to six bomber sorties by the Łoś bombers against the enemy's northern armoured force and six reconnaissance flights. No bombers were lost. During the first six days of war the Brigade flew 176 sorties: 76, including 65 bomber sorties by Łoś and 100, including 77 bomber sorties by Karaś bombers. About 100 tons of bombs were dropped on German armour. Its losses totaled 12 Łoś and 16 Karaś aircraft. One Łoś and one Karaś were received as replacements.

The Armies' Karaś reconnaissance units flew 26 sorties, ten of them bombing missions. In common with the Armies' fighters, they were removed from Army command to join the Bomber Brigade. The observation squadrons, the only ones to remain under Army orders, carried out about 40 operational flights. The LW's total effort on that day numbered about 240 sorties, with a loss of 23 aircraft. During the six days of strenuous fighting the LW logged about 1,610 sorties and lost 155 aircraft, or 40% of its initial strength.

During the period 7 to 17 September the Luftwaffe's main task was to prevent a Polish retreat over the Vistula and to assist in neutralizing encircled Polish troops on the western banks of the river. The main West-East roads were under continuous air attacks, bridges over the Vistula were bombed and rail tracks systematically destroyed, bringing traffic to a standstill in many parts of the country.

Remnants of a Ju 87B dive-bomber which was shot down while attacking the Vistula bridges and crashed in the Saski Garden in Warsaw.

Karaś II '5' of the 21st Squadron, which became the 1st Squadron of the Bomber Brigade, left partly camouflaged with tree branches at the Rakowice airfield in Kraków. Most Karaś aircraft operated from combat airfields with the wheel spats removed.

Karaś II recce.-bombers of the 41st Squadron, Army "Modlin," very active on 6 September, reconnoitered and bombed a large German armoured force approaching Warsaw from the north (a prewar photo, showing Karaś '6', 44.215, of the 41st Sqdn earlier in 1939).

Henschel He 126s, working for German Army Commands (Kolufts), were mainly used for reconnaissance.

The rapidly changing ground situation necessitated a major re-grouping of the Luftwaffe, which was fully effective by 10 September. Luftflotte 4 was considerably reinforced by units from Luftflotte 1 and by the only Stuka wing not until then deployed on the Polish front, III/St.G 51. This unit arrived from Luftflotte 3 on 8 September to strengthen the ground-support Aviation for Special Tasks, which began to move to forward airfields inside Poland. Combat elements of Luftflotte 1 were formed into the Strike Group "West" and Strike Group "East," and the "East Prussia" Air Command was dissolved. Hitler's War Directive No.3 of 9 September reflected his growing anxiety over the possibility of an Allied offensive in the West and urged transfer to the West of land and air formations no longer essential to keep the Poles from establishing new defence lines. As a result, a limited withdrawal of Luftwaffe units from the Polish front began on 12 September and reached considerable proportions between 19 and 22 September, after the Russians invaded Poland from the East.

On 8 September spearheads of the 4th Armoured Division raced towards Warsaw and in the afternoon entered Okęcie airfield. At the same time a heavy raid was made on the city, inflicting widespread damage and starting large fires. The fol-

Ju 87B dive-bombers of I./St.G 2 'Immelmann,' of the Aviation for Special Tasks, after arriving at a combat airfield near Częstochowa, inside Poland, on 8 September.

lowing morning an assault by 120 tanks was repelled by the Polish defenders and Warsaw defied the aggressors for the next 19 days. During the period of 9 to 12 September 135 aircraft of the Aviation for Special Tasks, reinforced by bombers of KG 55 and KG 77, incessantly pounded the encircled Polish armies in the Battle of Iłża, preventing all attempts to break out to the eastern bank of the Vistula.

On 9 September at 17:00 hours Gen. Kutrzreba's Army "Poznań", which had been conducting an orderly withdrawal eastwards towards Warsaw during hours of darkness completely undetected by German air reconnaissance, and had incorporated remnants of the Army "Pomorze," launched a surprise counter-offensive over the Bzura river. The German line was pierced in a number of places and the Germans, who at first failed to appreciate the magnitude of the Polish attack, had to stop the dash for Warsaw and turn their troops to repair the Bzura front. On 11 September – for the first time since the beginning of the war – Army Group "South" called urgently for the commitment of maximum air strength against Kutrzeba's Army. The Luftwaffe accordingly switched its main effort from the Battle of Iłża, which was then effectively over, to the Bzura front. Even the crack airborne troops of the 7th Fliegerdivision forming the High Command Reserve, were air-lifted by Ju 52s to reinforce the breaking Bzura battle line. The Luftwaffe's devastating attacks, which reached their peak on 16 and 17 September in uninterrupted low-level strikes by hundreds of aircraft, led to the collapse of the Polish counter-attack and destruction of the Army "Poznań."

On 13th September the Operation "Wasserkante" appeared again in Luftwaffe orders and Warsaw became the target of a

A German column enters the Okęcie airfield, passing partly completed Łoś bombers.

devastating terror raid by 183 aircraft, which caused raging fires and heavy casualties among the civilian population. Further massive attacks on the city had to be temporarily suspended because of the commitment of all the Luftwaffe's forces to the Bzura front.

According to the Luftwaffe's operational records, Luftflotte 4 flew during the period 1 to 15 September a total of 9,029 combat sorties, of which 483 were directed against Polish aviation targets, 4,806 were made in support of land operations in general, and 3,740 (including 1,693 in the Battle of the Bzura), in direct support of troops in the frontier and final battles. Luftflotte 1, for which no detailed data are available, flew probably between 11,000 and 12,000 sorties and the army reconnaissance units over 5,000, bringing the total number of sorties in the first 18 days of war to be about 26,000. Recorded German casualties for this period amounted to 205 aircraft counted as totally lost – destroyed or damaged beyond repair (60% or more) – and 482 aircrew killed, missing and wounded. This represented the average loss of one aircraft for about 127 sorties and one such loss for about 250 sorties due to interception by Polish fighters.

The major regrouping of the LW commenced on 6 September. During the first days of fighting it soon became apparent that the aviation units attached to the armies were not operating effectively and that the developing situation would make this even less likely. The armies' commands seem to have been unable to use their aviation formations in general and fighters in particular in accordance with specified war directives. The P.11s of the army "Modlin" were rapidly expended chasing enemy aircraft to the outskirts of Warsaw and on occasion fighting alongside the Pursuit Brigade. The fighters of the Army "Łódź" were ordered to carry out futile sweeps over the front and those of the army "Pomorze" were engaged in very costly and largely ineffective ground attacks. The fighter and Karaś squadrons of the Army "Kraków" had to leave their designated operational area as early as 3 September because of lack of prepared landing fields at the rear of the retreating Army. Fail-

S-Lt Erwin Kawnik's P.7a '5' of the 123rd Squadron, Pursuit Brigade, damaged in the air battle over Warsaw on the afternoon of 1 September, was captured by the Germans at Okęcie.

A Hs 126 of an unidentified reconnaissance unit, shot down by Polish fighters.

A Luftwaffe photograph of Okęcie and the WP 1 plant, taken around 20 September, shows a few bomb craters and semi-finished Łoś bombers dispersed on the airfield.

ure of the early-warning networks in the armies' regions forced the fighters to operate blindly and further reduced their efficiency. The Karaś reconnaissance squadrons were increasingly used for bombing, but because of the small numbers of aircraft involved and a resulting lack of bombing concentration, they could not arrest the enemy's advance.

This, combined with an absence of equipment reserves and the need to bring the Pursuit and the Bomber Brigades to their full strength, led to the decision of 6 September to withdraw the fighter and some of the Karaś squadrons from the armies in the field so as to reinforce the Brigades. Individual orders were dispatched to the respective army commands the same evening. The Army "Poznań" was the only one to protest and was allowed to keep its P.11 and Karaś units, partly because of remarkably good results achieved – in six days its fighters flew 175 sorties out of 585 logged by all the Armies' fighters and shot down 11 aircraft – and partly because its aviation was essential to the success of the contemplated counter-offensive over the Bzura river. The complete regrouping of the LW was outlined in the all-embracing "Reorganizational Directives for the Aviation Combat Units" of 10 September, the issue of which coincided with a total regrouping of the Polish fighting forces and creation of three fronts: the Northern front, along the lower Bug river and extending to the East over the Polesie Marshes; the Central front, along the middle Vistula; and the Southern front, along the San river. The all-important Southern front, defending the gateway to Rumania, was the only one to be given fighter protection and the 152nd Squadron was to leave the Army "Modlin" to provide it.

This change of heart, reversing the disastrous dispersal of the greater part of combat aviation between the armies, came too late. The squadrons were by then preoccupied with incessant changes of airfields, short of most essential supplies and no longer able to function effectively. This apart, the complete breakdown of communications and general chaos resulted in orders arriving late or not reaching the relevant commands at all. The assignment of the 152nd Squadron to the Southern front was never effected and the squadron was eventually incorporated in the Pursuit Brigade.

During 7 September the Pursuit Brigade was installing itself on airfields in the Lublin area. The fighter units withdrawn from the armies began to arrive on the same day. The Brigade, left to its own devices, found itself without petrol, ground personnel and means of communication and received no operational orders. Its new combat tasks were never specified. After desperate searches, small amounts of fuel and some other supplies were brought over from the bombed CWL Dęblin and from the Lublin Supply Depot. During 8 and 9 September the Brigade, working on its own initiative, resumed operations on a limited scale and shot down three Germans over the Lublin area.

Alarming information from the front and contradicting orders prompted the Brigade to withdraw to airfields in Volhynia, far to the Southeast, on the evening of 9 September.

This photograph of the Pursuit Brigade on Młynów combat airfield in Volhynia, was taken around 15 September. Hay-camouflaged P.11a and 11c fighters of the 113th Squadron by a haystack.

On 10 September its units were ordered to return to the Lublin area. They managed to fly to their allotted airfields the next day, only to be told the following evening to go back to Volhynia. The ground personnel and transport columns, scattered along congested and bombed roads, and traveling at a snail's pace in opposite directions, were unable to link up with the aircraft. The Brigade reassembled and sorted itself out only on 15 September. On the 16th, after the final incorporation of all ex-armies' fighters, the Brigade's operational strength rose to 46 P.11s, which were organized into the Warsaw Dyon and the Kraków Dyon and eight P.7s, which were formed into the Reconnaissance Squadron.

A He 126, destroyed on that day by S-Lt Hieronim Dudwał, became the last German to be brought down by the Brigade, which by then constituted the entire Polish fighter force. On 17 September the Brigade shot down four Soviet aircraft, but the Soviet invasion necessitated an immediate evacuation of the fighters to Rumania.

In these exceedingly adverse circumstances, during the period 7 to 17 September, the Brigade and its subordinate fighter units still managed to fly 194 sorties and shot down five German and four Soviet machines, this bringing the Brigade's overall combat effort to 602 sorties and its total score to 43 aircraft destroyed for certain.

On 7 and 8 September the remaining fighters of the Armies' Aviation flew 18 sorties before being transferred to the Pursuit Brigade. The fighters of the Army "Poznań" – the only ones to stay with an army to the end – continued to fight until 16 September, considerably contributing to the army's resilience and operational initiative. During the period 7 to 9 September they carried out 65 sorties and scored two victories at a cost of three machines lost in combat. Its further three P.11s had to be sent to the rear because of excessive wear.

At the beginning of the Bzura counter-attack only 13 P.11s were left and the Army's attempts to obtain reinforcements (one or two squadrons) from the Pursuit Brigade proved unrealistic. During the seven days – 10 to 16 September – moving from one landing field to another, the Poznań P.11s heroically defied the Luftwaffe and managed to fly just over 100 sorties, during which they destroyed at least nine Germans and lost ten machines. On 17 September the last three fighters left the rap-

A P.11a with the underwing call-number 935-N and painted-over LW chessboards from the same group at Młynów. The aircraft has a substitute spoke wheel (presumably from a servicing trolley) on its port side.

idly contracting Army's territory, but two were lost in the process and only one succeeded in reaching Rumania. Skillfully led by Mjr Mieczysław Mümler, the Poznań P.11s accounted for 343 sorties out of the 771 fighter sorties flown by the Armies' Aviation during the 17 days of operations and, with 22 certain victories to their credit out of the 62 achieved by the Armies' fighters, became the Armies' highest scoring unit. Their outstanding bravery and determination provided one of the finest chapters in the LW's lonely fight against the overwhelming power of the Luftwaffe.

Altogether the fighters of the Pursuit Brigade and the Armies' Aviation logged 1,373 combat sorties, destroyed 105 enemy aircraft and lost 120 (including nine during evacuation) of their own. Their personnel losses were 20 killed , five missing and 25 wounded.

A hastily organized fighter unit of the Higher Pilots' School, operating from 14 to 16 September in the defence of combat airfields in South-eastern Poland, added two more Germans to the LW's overall score, bringing it to 107. Both aircraft were shot down by Lt Henryk Szczęsny, flying the P.11g Kobuz prototype.[23]

On 7 September the Bomber Brigade substantially increased its activity. Łoś and Karaś dyons flew 41 bomber sorties against German armour in the North (Rożan area) and on the central front (Łódź area) and five reconnaissance sorties. The Łoś bombers were attacked by Bf 109s and suffered heavy casualties, a whole flight of three aircraft of the 16th Squadron (XVth Bobmer Dyon) being destroyed. The Brigade lost on that day five Łoś and four Karaś bombers. One Łoś was received as replacement. On the next day small Łoś and Karaś formations continued to bomb the Germans in the North, flying 26 sorties, but in view of the rapidly deteriorating ground situation the Brigade was ordered to withdraw to airfields in South-eastern Poland and only nine sorties were flown on 9 September. By then the Brigade was left with 21 Łoś and 31 Karaś aircraft.

The "Reorganization Directives" of 10 September requested the Bomber Brigade to bring the two Łoś dyons to full strength by collecting 20 replacement bombers from reserves evacuated from the Małaszewicze base, and to bring the VIth Dyon to full strength with the remaining Karaś machines from the Brigade's II Dyon and those from the 32nd, 34th and 42nd Squadrons withdrawn from the armies. The newly established fronts were allocated the following Karaś units: the Northern Front, the 41st Squadron reinforced by the 51st Squadron; the Central Front, the 24th Squadron; and the Southern Front, the

Pilots of the Brigade breakfasting in the open near haystacks at Młynów.

[23] This aircraft, with the enclosed cockpit, was frequently wrongly referred to as the P.24 in various Polish publications.

31st Squadron, reinforced by the 55th Squadron from the Bomber Brigade. However, the Directives were unrealistic and confusing. None of the reserve Łoś bombers was fully armed and fitted out, and eventually only eight could be collected (in addition to the two already received on 6 and 7 September). The 34th Squadron was to stay with the Army "Poznań" and the 41st and 51st Squadrons had no serviceable Karaś aircraft left by 12 September.

On 10 and 11 September the Bomber Brigade was organizing itself on the Volhynian airfields. The next day, during which 23 sorties – 18 of them by Łoś bombers – were recorded, was the last of the Brigade's more significant operations. Bf 109s intercepted one of the Łoś raids and shot down one bomber. An acute shortage of fuel and bad weather severely restricted further Łoś operations. On 14 September, when 12 sorties were flown, the returning Karaś bombers of the VIth Dyon were followed by a German aircraft to Hutniki airfield and towards evening the Luftwaffe launched its first really successful attack of the war on an active Polish airfield, then repeated it the next morning. This wiped out the Brigade's Karaś force. At least 17 aircraft were totally destroyed and the rest badly damaged. In the afternoon of 14 September the Łoś dyons received orders to move near the Rumanian border, or, if no suitable airfields could be found there, to cross over to Rumania. The dyons' C.O.s were dismayed at the prospect and tried to resist the execution of this order. On 16 September the Łoś bombers still managed to fly eight combat sorties, the last of the campaign.

Between 7 and 17 September the Bomber Brigade flew 142 sorties: 59 by Łoś and 53 by Karaś bombers. Apart from a few Karaś reconnaissance flights, all these were bombing, or combined recce-bombing missions. About 80 tons of bombs were dropped and losses from all causes amounted to 15 Łoś and 34 Karaś aircraft. The Brigade's overall war record totaled almost 320 sorties: 135 by Łoś and 183 by Karaś bombers, during which about 180 tons of bombs were dropped and 27 Łoś and 50 Karaś machines were lost; 69 airmen were killed and 56 missing or wounded.

The Armies' Karaś reconnaissance aircraft, depleted by heavy losses which were aggravated by the need to operate from constantly changing and badly prepared airfields, flew 72 sorties between 7 and 17 September, only two of them being regarded as specifically bombing missions. This brought their overall 17-day war effort to 245 sorties; 54, during which about 22 tons of bombs were dropped, were bombing missions. Total casualties amounted to 56 aircraft lost, 39 airmen killed and 22 missing or wounded.

The observation units, which remained with the armies to the end and whose combat activities are inadequately documented, carried out an estimated total of 250 operational sorties between 7 and 17 September, and for the whole 17-day period, about 520 sorties. Eighty aircraft were lost, 21 airmen killed and 49 missing or wounded.

Appendix 5 provides a summary of the LW's war effort and losses in September 1939.

6. The Soviet Invasion and the Final Act of the Polish Drama

Hitler attached tremendous strategic and political importance to active participation of the Soviet Union in the annihilation of Poland. Germany's Western frontier was drained of troops to provide a maximum concentration of forces on the Polish front in an attempt to achieve a swift fait accompli before the Western powers might intervene. But such a risk could not be sustained for long and a Soviet invasion from the East would release many German divisions to return to face France. Politically, having an accomplice for such a blatant act of aggression would make Hitler's imaginary "grievances" appear more convincing and ease Germany's relations with neutral countries.

The Western Allies' failure to fulfill their Treaty obligations towards Poland and launch the promised offensive was a missed opportunity – never to be repeated – to direct the developing war along a totally different course, which, if the Allies had taken the initiative as Hitler feared, could have led to a prompt defeat of Germany. The French, in the words of the British General J.F.C. Fuller, with "the strongest army in the world," had then at their disposal 86 divisions[24] against 34 German divisions, several of them reserve, in the West. German Generals testified at Nuremberg:

> - "The success against Poland was only possible by almost completely baring our western border. The French ... would have been able to cross the Rhine without our being able to prevent it and would have threatened the Ruhr area, which was the most decisive factor of the German conduct of war." (Gen Halder).
> - "We soldiers had always expected an attack by France during the Polish campaign and were very surprised that nothing happened ... A French attack would have encountered only a German military screen, not a real defence." (Gen Keitel).

But the defeatist French launched only a token attack with a few divisions (Operation "Saar"), which, starting on 7 September, ended as a fiasco. Five squadrons of the Amiot 143M bombers promised to Poland never arrived and the specially prepared airfields waited for them in vain. Only one bomber raid was mounted by the Allies during the Polish campaign, when 24 RAF bombers attacked German warships at Wilhelmshaven on the night of 6 September and lost seven aircraft. However, even this limited activity was abandoned under French pressure, the French fearing German retaliatory strikes, and the RAF bombs were exchanged for leaflet-drop-

---

[24] German Generals believed that the French possessed over 100 divisions at the beginning of the war.

ping! Meanwhile, Polish cities, towns and hamlets were turning to dust under the ceaseless and savage bombardment of the Luftwaffe and Poland was left to fight and bleed to death alone, with the Allies offering sympathy instead of the promised military action.

Despite Allied inactivity, the Germans regarded their position in the West as precarious and as early as 4 September began to prompt the Russians to enter Poland, increasing diplomatic pressure a few days later. The Soviets responded that "the unexpectedly rapid German successes" place them in "a difficult situation," since they needed to find a plausible excuse for their intervention, which would look "reasonable to the masses and avoid giving the Soviet Union the appearance of an aggressor." After frantic military preparations by the Russians, the Germans were informed on 14 September that the Red Army would march sooner than had been expected, but in order to justify this Warsaw must fall first.

To the Germans, who were trying to convince the Russians that the war with Poland was practically over, this was an embarrassing demand. It came at the time of the Polish counterattack over the Bzura and the increasingly stubborn resistance of Warsaw.[25]

The next day a "most urgent" and "top secret" message was received by Molotov, assuring him that Warsaw would be captured "in the next few days" and pressing him to name "the day and the hour" of invasion. On the evening of 16 September the Germans were informed that the Russian intervention was imminent and that the public justification for it would be formulated as follows: "The Polish state had disintegrated and no longer existed ... ; the Soviet Government considered itself obliged to intervene to protect its Ukrainian and Belorussian brothers and make it possible for these unfortunate people to work in peace." In the early hours of 17 September the German ambassador in Moscow was informed: "The Red Army would cross the Soviet border at 06:00 hours ... Soviet aircraft would commence today to bomb the region east of Lwów." Operation "Wasserkante," which the Luftwaffe was to launch on that day against Warsaw to enforce its immediate surrender, was postponed once again at the last moment by Hitler's order committing all available air strength to crush the Polish forces on the Bzura front. Strategic considerations took precedence over political tactics that became less important.

On 17 September at dawn Soviet forces violated the Polish border on its entire 600-mile length and quickly swept westwards, meeting patchy resistance from weak and totally confused units of the KOP (Border Protection Corps) and the Polish army. The invading Red Army was supported by the Aviation of the Belorussian and Ukrainian Fronts, comprising about 2,000 warplanes: the I-16 and I-15bis fighters (about 60% of the total), SB-2 twin-engined bombers (about 30%) and R-5 reconnaissance aircraft (about 10%). This force was supplemented by a small number of the heavy four-engined TB-3

[25] The German radio announcement on the evening of 8 September implied that Warsaw was already captured.

Hitler's partners in aggression – the Red Army soldier stands guard over a Polish PWS 26 trainer wrecked by the Luftwaffe on the Wilno's Porubanek airfield.

bombers used mainly for transport and the special police squadrons of the NKVD. The main thrust of the Soviet invasion was directed against the extreme South to cut off the retreating Polish armies from the Rumanian border.

Polish commands first learned of the Soviet intervention from various messages received in the morning and were not sure how to react to it. Air reconnaissance confirmed the reports and the hostile intentions of the Russians soon became apparent. The Aviation High Command's orders requiring immediate evacuation of serviceable aircraft from South-eastern Poland to Rumania and from the North-east, where a large number of training and flying club machines were sheltering, to Latvia, began to reach unit commanders in the afternoon. Evacuation of the aircraft started on 17 September before nightfall and continued the next day. During the night the Polish Government and the C.-in-C. Polish Armed Forces crossed the Rumanian border.

During 17 September Polish fighters flying reconnaissance and patrol sorties along the Eastern border were involved in several clashes with Soviet aircraft. In the morning S-Lt Stanisław Zatorski, attacked by a flight of three fighters – I-16 and two biplanes – over Rokitno, brought down one biplane and damaged the other, but was himself hit and died on crash-landing at Sarny. At about the same time S-Lt Tadeusz Koc destroyed a reconnaissance plane, which he thought was an Hs 126 but turned out to be a Soviet R-5. In the late afternoon a solitary P.11 attacked three SB-2s bombing railway lines near Nadwórna and shot down two; the identity of the pilot remains unknown. Soviet aircraft carried out various limited strikes on Polish troops, military installations and depots, armoured trains and railway lines, in particular those leading to the Rumanian border. The SB-2s tried to bomb an airfield of the XVth Łoś Dyon in the afternoon, but dropped their loads on a neighborhood field. The Red Army High Command's communiqué for 17 September reported the shooting-down of seven Polish fighters and three bombers.

During the last phase of war, 18 to 30 September, the scope of Luftwaffe activities over Poland diminished and a substan-

A Ju 52 which took part in the massive attack on Warsaw on 25 September, carrying the inscription "Uwaga" in Polish. This was undoubtedly inspired by the broadcasted Polish coded air-raid warnings beginning with the words: "Uwaga, uwaga, nadchodzi ..." ("Attention, attention, approaching ...").

tial proportion of its force, including all the He 111 bomber units, returned to the West. By 20 September fighting at the Bzura had died down and the neutralization of the Polish armies East of the Bug river was approaching its final stages. Warsaw and Modlin became the main focus of German attention, as well as the Hel Peninsula, where a handful of Polish defenders successfully resisted an overwhelming German force. A large scale leaflet-dropping action began, calling on Warsaw to surrender and warning about consequences of further resistance.

Intensified air strikes on Warsaw resumed on 20 September. Involving a large Ju 87 force, these were specifically aimed at the destruction of power stations, gas works and water mains supplying the city. German land attacks on Polish positions launched at the same time were repelled. Bad weather on 22 and 23 interrupted the German onslaught. On 24 September Göring ordered the Luftwaffe to launch the long positioned Operation "Wasserkante" and to break Polish resistance at all costs. Since there were no He 111s left on the Polish front, some 100 Do 17 bombers of KG 77 were assisted in the attack by 30 Ju 52s loaded with incendiary bombs, all the eight Stuka wings and the Hs 123 assault aircraft of I (Schl.)/LG 2. Over 400 machines in all were to participate in this biggest air assault of the war.

On 25 September at dawn a thousand guns opened an avalanche of fire and at 08:00 hours the Luftwaffe commenced an all-day bombardment of the city. In 1,176 sorties 486 tons of high-explosive bombs and 72 tons of incendiaries were dropped. All hospitals were hit and raging fires could no longer be extinguished. A still-burning Warsaw was attacked again the next morning by a smaller force and in view of the hopelessness of the situation and heavy civilian casualties, the Warsaw Defense Command decided to open negotiations for surrender with the enemy. By then nearly 10% of all buildings were destroyed and a further 41% damaged.

On the afternoon of 26 September the P.46/II second Sum prototype which had been interned in Bucharest, Rumania, was secretly prepared by the Poles for a daring mission and, piloted by Cpt Stanisław Riess, escaped to Poland. The aircraft landed in darkness under German fire at Mokotów in Warsaw and delivered final orders to the defenders of the capital, including a call for the formation of an underground army, which was destined to become one of the most powerful clandestine military forces ever to exist. The machine successfully took-off from Mokotów later the same night and reached Kowno in Lithuania, its crew eventually escaping to England.

A cease-fire on the Warsaw front came into force on 27 September at 14:00 hours and the city surrendered the next day. Meanwhile, the Luftwaffe switched its attacks to the Modlin fortress and in two days flew 800 sorties against this target, dropping 318 tons of bombs. The fortress capitulated on 28 September, but Hel Peninsula continued to defy the Germans until 1 October. The last German aircraft to be shot down in the Polish war fell to the Hel defenders at 13:55 hours on that day. The last Polish troops surrendered to the Germans on 6 October, after the Battle of Kock in the Lublin area.

According to official German records, the Wehrmacht's losses in the Polish Campaign of September 1939 amounted to 16,343 killed, 320 missing and 27,640 wounded. The Polish losses are estimated at over 70,000 killed (67,177 war graves were officially recorded within the borders of the post-war Polish state, which did not embody Poland's lost eastern provinces) and about 140,000 wounded. The German army records indicate that 588,304 Polish soldiers (including 11,452 officers) were taken prisoner.

Armed opposition to the Red Army also continued and Soviet aircraft were engaged in intensive reconnaissance activities and occasional ground support operations until the end of September. On 20 September two enemy aircraft were shot down in an attack on the Polish "Głowacki" armoured train, and during the next ten days various KOP units and troops of Gen Kleeberg's "Polesie" force were strafed and bombed. The last major battle with the Red Army, which entailed a substantial air support, was fought in the Kowel area on 24 September. As late as 30 September Soviet fighters and about 20 SB-2 bombers were engaged in breaking the resistance of the "Kobryń" Division, scattered cavalry units and the remnants of the "Polesie" Operational Group.

The fourth partition of Poland was now finally sealed. Article 2 of the "Secret Additional Protocol" to the German-Soviet Non-aggression Treaty of 23 August 1939, provided for the possible retention of a small Polish State, but on 19 September Molotov intimated to the Germans that Stalin was against this. Six days later Stalin proposed that all Polish territories should be partitioned between Germany and Russia along the factual demarcation line of the Bug river (further to the East than originally stipulated) and that, in return, the Germans should waive their claim to Lithuania. He added that if the Germans consented, the Soviet Union would immediately resolve the problem of the Baltic countries – Estonia, Latvia and Lithuania – and expected in this matter "the unstinting support of the German Government."

The Germans agreed and the "German-Soviet Boundary and Friendship Treaty" of 28 September incorporated all Stalin's proposals in secret additional protocols, adding one new and sinister clause: "Both parties will tolerate in their territories no Polish agitation which affects the territories of the other party. They will suppress ... all beginnings of such agitation and inform each other concerning suitable measures for this purpose." Thus Germany and Russia obliterated Poland from the map of Europe once more and their dictators agreed to institute in Poland regimes of unprecedented terror, with such consequences as the Auschwitz extermination camp and the mass-murder of Polish officers at Katyń and other places.

A substantial number of partly finished new aircraft and others undergoing overhaul and repairs were captured by the Germans at the WP 1 at Okęcie which stopped work on 4 September. These included some 20 Łoś bombers, over 20 Karaś and P.43 recce.-bombers, a few fighters and the Wilk and Wyżeł prototypes. This photo, taken from a Ju 52 on 22nd September, shows a proportion of this haul.

A damaged P.43 at Okęcie, originally intended for Bulgaria.

7. Assessments and Conclusions of the Air War over Poland

During the war with Poland the Luftwaffe enjoyed greater superiority over an opposing air force than either during the campaign in France, the Battle of Britain, or the war with Russia. In terms of quantity, this expressed itself in the overall ratio 5:1 with regard to warplanes employed by both sides, and 12:1 when comparing the German attacking force with the defending fighters. In terms of quality, the disparity was even greater. The committal of such overwhelming power assured the Luftwaffe a total air supremacy, which was exploited to wage total war with unprecedented savagery. Hitler had told his generals " ... it is not right that matters, but victory. Close your hearts to pity. Act brutally. Be harsh and remorseless! Be steeled against all signs of compassion."[26] No distinction was made between military targets and the civilian population and no concessions were made to humanitarian sentiments.

The Luftwaffe largely achieved its objectives. Although its attempts to destroy the combat elements of the Polish air arm on the ground in the opening hours of the conflict proved a failure, air strikes almost annihilated reserve and training aircraft and all ground installations, and brought the Polish aviation industry to a complete standstill in the first days of war.

Squadrons on secret combat airfields were thus isolated from their supply sources and support maintenance facilities. The Luftwaffe's support of ground forces, both direct and indirect, made a decisive contribution to the collapse of the Polish defences and the speedy conclusion of the campaign. Its most significant achievement was the total dislocation of Polish communications which made effective military command virtually impossible and shattered all Polish attempts to establish new defence lines. The Ju 87 Stuka bombers, the Hs 123 assault aircraft and the Bf 110 "destroyers" proved to be the most effective weapons at the Luftwaffe's disposal.

However, the Polish campaign, notwithstanding the Luftwaffe's unquestionable success, also revealed certain shortcomings, such as limitations of conventional bombers in locating targets, persistently overenthusiastic – and therefore misleading – reports of the German air crews on results of their missions, the serious failure of aerial reconnaissance in allowing Gen Kutrzeba to launch a powerful and well coordinated counter-attack to the total surprise of the German command, and also failure to prevent determined attacks by Polish aircraft on German forward troops. What is more, the spectacular usefulness of the Luftwaffe for concentrated short-term tactical operations as manifested during the Polish campaign, was to become the source of its main weakness in what were to be the changing requirements of a prolonged air war, demanding a sustained effort of strategic dimensions. This lesson was completely lost in the euphoria of the Luftwaffe's victory in September 1939.

German assessments of the September air war – both those of wartime and more recently – express surprise that, contrary to all expectations, the Polish air arm had never appeared in force, except for a single instance on the afternoon of 1 September over Warsaw, and never attempted to raid the Reich's territory and never undertook any concentrated air action. They admit to substantial losses over Poland, which they tend to attribute to accurate Polish anti-aircraft fire rather than to the Polish fighter operations. At the same time they also stress the remarkably high morale of the Polish airmen, which was never

[26] From records of the Military Conference of 22 August 1939, at Obersalzberg, as presented at the Nuremberg trials.

Some of the partly completed Łoś II bombers captured at Okęcie were still unpainted. Fortunately, all were cut swiftly to pieces and turned to scrap, when a careless German officer ordered Polish workers to "clear the airfield of Polish aircraft."

broken by the reverses of the war, the exceptional determination of Polish bomber crews, who continued to fly isolated missions until the time of the Soviet invasion and the reckless courage of the Polish fighter pilots, who came to be feared by the Luftwaffe as daring and deadly opponents even though encounters were infrequent.

The respective effectiveness of the Luftwaffe and the Polish air arm should not be measured by the same yardstick. The answer as to why Polish aircraft failed to appear in strength or to bomb Germany lies in the diametrically different war doctrines of the two sides and in constraints put upon the LW not only by its size and status, but also by unreasonable restrictions imposed on air operations over German territories in response to heavy diplomatic pressure from France and Great Britain. Under the very limited and confusing objectives set out by the Polish Chiefs of Staff in September 1939 the LW was not expected to defend the entire Polish air space nor to make a major contribution to land operations. Its effectiveness should be assessed from this perspective.

The Luftwaffe's combat effort on the Polish front amounted to about 33,000 sorties, including operations of reconnaissance units attached to the armies. About 12,000 of these were flown in the first six days of the war, about 14,000 in the next 12 days and a further 7,000 in the concluding phase of the campaign. About 25,000 were bombing sorties, during which about 19,600 tons of bombs were dropped on Polish targets.

The Abschluss-Gesamt-Meldung (Closing Total Report) of the Luftwaffe's Quartermaster General 6th Department, embracing total aircraft losses from all causes during the month of September 1939 on both fronts, Polish and Western, which was issued on 5 October 1939, listed: 78 bombers (He 111 and Do 17), 67 fighters (Bf 109), 12 "destroyers" (Bf 110), 31 dive bombers (Ju 87),[27] 28 long-range reconnaissance aircraft (Do 17), 35 close-reconnaissance aircraft (He 126 and Heinkels), 12 transports, 13 seaplanes and nine liaison aircraft, totaling 285 aircraft, of which 258 were lost over Poland and 27 in the West. In addition, damaged aircraft, capable of repair[28], were as follows: 69 bombers, 67 fighters, 27 "destroyers", 22 dive-bombers, ten long-range reconnaissance aircraft, 45 close-reconnaissance aircraft, 22 transports, five seaplanes and 12 liaison and other aircraft; totaling 279 aircraft, of which 263 were damaged over Poland and 16 in the West.[29] Total personnel casualties, including anti-aircraft artillery units, amounted to 279 killed, 221 wounded and 234 missing, or 734 in all. Aircrew losses alone comprised 189 killed, 126 wounded and 224 missing, or 539 in all, of which about 20 were sustained in the West. These figures represent an average total loss of one aircraft for about 128 combat sorties and one aircrew casualty for 63 sorties on the Polish front.

The document does not identify types and quantities of the German machines destroyed and damaged on the Western front, but the author's own researches indicate the following break down:[30]

	Polish Front			Western Front			Overall totals		
	Lost	**Damaged**	**Total**	**Lost**	**Damaged**	**Total**	**Lost**	**Damaged**	**Total**
Bombers	73	66	139	5	3	8	78	69	147
Fighters	56	59	115	11	8	19	67	67	134
"Destroyers"	12	27	39	–	–	–	12	27	39
Dive-bombers	31	22	53	–	–	–	31	22	53
Long-range Recce.	25	9	34	3	1	4	28	10	38
Close Recce.	32	43	75	3	2	5	35	45	80
Transports	12	32	44	–	–	–	12	32	44
Seaplanes	8	3	11	5	2	7	13	5	18
Liaison and others	9	2	11	–	–	–	9	2	11
Totals	**258**	**263**	**521**	**27**	**16**	**43**	**285**	**279**	**564**

[27] The Hs 123 assault biplanes were almost certainly grouped in this listing together with the Ju 87s.

[28] Damage assessed at between 59% and 10%; damage below 10% was not recorded.

[29] The loss figures of 285 aircraft destroyed and 279 damaged in the Abschluss-Gasamt-Meldung were wrongly interpreted in most German and foreign publications as being applicable to the Polish front only (not to both fronts). This misinterpretation in turn encouraged aviation writers in Poland to believe that the score of 126 German aircraft previously credited to the Polish fighters was too low by some 20 aircraft, whilst in fact the opposite was true, as the author's scrutiny of the German losses and the Polish fighter victories will later show.

[30] Compiled from the British, French, and German communiqués.

Since some 60% of the damaged aircraft never returned to service, the Luftwaffe suffered the outright loss of about 415 machines, averaging one for about 80 sorties, over the Polish front from all causes. Taking into account an average 10% loss from causes other than enemy action, it should be assumed that about 230 German aircraft, or one for 143.5 sorties, were destroyed outright by the Polish fighter and ground defences, and with the damaged machines eventually written off, this total rose to 370-375 aircraft – with the Polish fighter share amounting to about 165-170 – or one for 89 sorties.

During September scrutiny of Polish fighter claims by the LW Command was very rigorous, but due to the loss of most of the original documents, a list of victories was carefully compiled during the war years by the Polish Aviation Historical Office in Great Britain from surviving unit records and pilots' statements. After final screening and readjustments in 1945 by a special Fighter Claims Commission led by Col Jerzy Bajan, the official list of Polish victories comprised 126 aircraft destroyed,[31] 10 probables and 14 damaged, 105 (80%) of the confirmed victories occurring during the first six days of war. Bajan's Commission did not have access to the captured Luftwaffe records which became available to researchers in Great Britain early in 1947. These revealed that during the first six days of war the Luftwaffe lost 126 aircraft in Poland from all causes and that the Polish victories confirmed by the Commission exceeded the actual German daily losses for 2, 5, 6 and 8 September (see Appendix 6, Luftwaffe operational losses in Poland). The author's own study of available Polish and German war records indicates a loss of only 85 German aircraft in the period 1 to 6 September due to the Polish fighters. A further 18 German and four Soviet machines were shot down in the period 7 to 17 September, bringing the total fighter score over the 17 days of air fighting to 103 German and four Soviet aircraft destroyed. Another ten enemy aircraft were claimed by the Łoś and Karaś aircrews. This leaves about 115-120 Germans shot down over the period of 30 days by the Polish ground defences.

Bearing in mind the circumstances in which the fighters had to work, the results achieved can be regarded as truly outstanding, especially because Poland had no integrated air defence system. The theoretically sound observation-post/early-warning network, which was painstakingly built up on a regional basis, to be developed eventually into a centrally-controlled and coordinated air defence system for the whole country, proved virtually useless in war conditions because of totally inadequate communications which relied entirely on vulnerable and overloaded telephone lines. Except for the Pursuit Brigade, the Armies' fighters, operating from widely scattered and frequently changing combat airfields, had only very erratic and short-lived links with their regional early-warning centers, limited to the first day or two of war, or none at all.

[31] Mjr Olszewski's researches put the number of Polish "certain" victories at 105, whilst Gen Zając assessed them as "between 110 and 130."

The best part of the slim fighter force was unwisely fragmented into insignificant and isolated units attached to the various armies to provide localized defence. Contrary to conflicting war directives, they could not possibly operate "in strength." Working blindly, these units were forced to disperse even further, to set up "ambushes" or "lookout" patrols at section strength or even less. Fighter cover was therefore confined by design only to Warsaw itself and to some limited areas of the country. Their deployment inhibited concentrated action in strength in the sectors of the enemy's most intensive aerial activities. These crippling drawbacks were further aggravated by a lack of repair facilities and any stores or supplies on combat airfields and by the low performance of the old and weary PZL fighters with a maximum speed inferior to almost all of their opponents.

The only asset the Polish pilots had was their unsurpassed moral strength, which enabled them to fly and defy the enemy for 17 days and achieve surprising results. They managed to launch 1,373 combat sorties and score 107 air victories – including four against Soviet aircraft – at a cost of 120 fighters and 50 pilots killed, missing and wounded, averaging one victory for 13 sorties on the credit side, and one fighter lost for 11.5 sorties and pilot casualties for 27.5 sorties on the debit side. This was the true measure of their effort and combat efficiency.

According to the official findings of Bajan's Fighter Claims Commission, the top Polish fighter pilots and their scores in September 1939 were as follows: S-Lt Skalski Stanisław, 4 1/4[32], S-Lt Dudwał Hieronim, 4; S-Lt Główczyński Czesław, 3 1/2; S-Lt Gedymin Włodzimierz, 3 1/2; Cpt Leśniewski Mirosław, 3.

The Polish bombers were also seriously handicapped by their improper tactical deployment and operational difficulties. Neither the Łoś bombers nor their crews were adequately equipped to attack small formations of moving and widely scattered tanks and troops which became their sole targets. The only bombs at the disposal of the Łoś squadrons – 100 kg (220 lb) high-explosives – were singularly unsuitable for the purpose.

Because of the nature of the targets, the adopted method of maximum harassment by small numbers of aircraft was likely to be more effective than single strikes by larger formations. However, instead of sustained pressure on a particular enemy force or area, the bombing effort was dispersed among various sectors of the front, and although numerous German tanks and vehicles were immobilized, the scattered casualties did not hinder the enemy's ability to push forward.

Ill-prepared, grass combat airfields forced the Łoś and Karaś bombers to operate with greatly reduced bomb-loads. The maximum-load bomb-salvo of the Bomber Brigade's 86 aircraft – discounting 10 replacement bombers – amounted to

[32] After the war, Skalski, who returned to Poland and was imprisoned by the Communist regime, had two more of his September claims officially recognized as "certain" by the Communist authorities after his release and rehabilitation.

almost 128 tons. Yet, in spite of the determination of the aircrews, who kept on flying and harassing the enemy until 17 September, the total weight of bombs dropped by the Brigade came to about 180 tons, or only 1.5 times its potential bomb-salvo. Nevertheless, out of 2,511 German tanks deployed in Poland, 217 were totally destroyed and 457 damaged, a significant proportion of these losses being attributed to the Polish bombers. During its 318 sorties the Brigade lost 86 machines, including all its Karaś bombers, the last of which were destroyed by the Luftwaffe at Hutniki airfield, and suffered crew casualties of 125 killed, missing and wounded, averaging one aircraft lost for 3.7 sorties. The price the bombers had to pay for their courage was extremely high.

Although during the interwar era the LW was oriented predominantly towards reconnaissance, which was regarded by the Polish High Command as its prime function, war directives restricted such operations to shallow reconnaissance only. The techniques employed were very primitive. Only a few Karaś aircraft were provided with cameras of adequate quality. Photo reconnaissance sorties were few in numbers and almost all reconnaissance work relied on visual observation, hand-drawn sketches and written reports. These were generally imprecise and lacked details of the composition and strength of enemy columns, and sometimes even confused these with Polish troops. The Karaś reconnaissance units flew about 245 sorties, over a quarter on bombing missions, and 22 tons of bombs dropped by these aircraft brought the total bomb-load dropped by the LW in September to just over 200 tons – almost 100 times less than the Luftwaffe. With 56 aircraft lost and 61 airmen killed, missing or wounded, they suffered an average loss of one Karaś for 4.4 sorties.

The hard-pressed observation units which, contrary to expectations, had to venture beyond the front line and on occasions even to undertake low-level attacks on enemy troops, also lost nearly all of their aircraft, suffering their biggest casualties from indiscriminate fire from Polish troops during low-level approaches to drop or deliver reconnaissance reports or orders. Their losses of 80 machines and 70 crew members during some 520 operational sorties averaged one aircraft lost for 6.5 sorties.

All the LW equipment loss figures refer to losses from all causes and are somewhat inflated by the deliberate destruction of combat weary or unserviceable aircraft during evacuation. In addition, during the 17 days of fighting many machines, approaching a quarter of all losses, had to be abandoned for trivial reasons – even such as flat tires or shortage of fuel – because retrieval or repair facilities were unavailable. Out of its initial strength of 392 warplanes and about 40 replacement machines, the LW, which flew over 2,450 combat sorties, lost 333 aircraft, or one for every 7.4 sorties, from the following causes:

Only a small number of the LW aircraft in usable condition were captured by the Germans. The P.7a fighters taken at Dęblin were overhauled by the Luftwaffe and utilized by its A/B 124 Pilot School, established at the former Polish CWL 1 Centre. (V. Koos)

- about 245-250 (one for every 10 sorties) due to enemy action; of these about 100 were destroyed in air combat or by enemy ground defences; 25 destroyed by the Luftwaffe on the ground; and 120-125 lost in forced landings due to battle damage or crew injury (more than half of these presenting the preventable loss);
- about 35 shot down by own ground defences;
- about 20 destroyed during evacuation;
- the rest lost due to wear, mechanical failure or pilot's error.

Available records indicate that 99 warplanes of the LW combat units were evacuated: 98 (44 P.11 and P.7 fighters, 19 Łoś bombers, 17 Karaś recce aircraft and 18 Czapla and R-XIII observation monoplanes) reached Rumania and one (P.11) crash-landed in the Soviet Union.

Almost 50 other combat aircraft, from reserve units and training centers, arrived in Rumania, bringing the total to 146, as follows: 44 P.11 and 14 P.7 fighters, 27 P.37 Łoś bombers, 19 P.23 Karaś recce. bombers, 17 RWD 14 Czapla and 23 R-XIII observation machines, one Potez 25 and one Fokker F.VII/3m. In addition, about 150 more aircraft: military trainers, PLL "LOT" passenger transports and auxiliary aircraft, and private and club lightplanes, also crossed over to Rumania. These included 61 RWD 8s, 11 RWD 17s, 26 PWS 26s, five Lockheed 10s, four Lockheed 14s, three Fokker F.VII/3ms, one DC 2,

A damaged LWS 4 Żubr examined by the Germans. Though useless as bombers, the aircraft were well suited for training, and a few captured examples were refurbished and used by the Luftwaffe blind-flying school at Schleissheim.

one Ju 52, one PWS 24, 28 RWD 13 cabin tourers including five RWD 13S ambulances, and about ten more lightplanes of miscellaneous types. Over 10,100 airmen and 2,100 highly qualified aviation designers, engineers and technicians also found their way to Rumania.

A special Polish-Rumanian commission was to compile a detailed inventory of all Polish aircraft in Rumania and establish their capital value, but since the Rumanians decided to requisition all this equipment for their own use, the commission was never instituted. One-sided Polish attempts to put together such a register were obstructed by the Rumanian authorities and the Polish lists never presented a complete record.

About 85 reserve and civil aircraft from North-eastern Poland reached Latvia, and these included one newly repaired P.11 fighter, eight R-XIII and a few Czapla observation monoplanes, about 20 PWS trainers, 40 RWD 8s and some lightplanes of other types.[33] Individual aircraft flew to Hungary, Lithuania and Sweden.

The final conclusions can be drawn as follows:

- Poland was a poor and badly industrialized country. Her economy could not sustain the levels of armament necessary to permit her to resist military threats from her two mighty neighbors, Germany and Russia. Apart from that, since 1926 the distribution of the country's slim armament budget between various arms severely disadvantaged the LW, whose subordinate status as an army support service without a clearly defined combat role, combined with ignorance of aviation matters by the Polish Army chiefs, was the root cause of its weakness. Consistent underfunding had a crippling effect on its organization, size and operational strength. Belated and confused attempts to define the LW's place in Poland's overall defence strategy brought about a dramatic change of direction in aviation armament priorities and catastrophically deepened the equipment crisis on the eve of war.
- The war directives and tactical deployment of the LW units reflected indecision on the part of the Polish High command. The command's inability to make good use of the fighting potential of the air arm impeded its effectiveness and precluded its operations in strength.
- Whilst the concept of moving all combat units from their peace time bases to improvised landing grounds to save them from destruction was the right one, it was not put into practical effect in the detail necessary for the mobility of the squadrons to be a workable proposition. Apart from poor preparation of airfield surfaces which severely reduced aircraft performance, the lack of adequate communications, transport, maintenance and supply facilities grounded a large proportion of the force for several days in the later stages of the campaign, almost nullifying potential operational benefits of airfield changes.

[33] A German register of captured aircraft on Latvian and Estonian territories, dated 6 July 1941, listed 74 machines of PWS and RWD types.

One of the probably two P.37 Łoś bombers to carry the Luftwaffe markings, as seen at the Victory Exhibition in Vienna in the autumn of 1940. The other crashed during tests at the Luftwaffe test centre at Rechlin in May 1940.

An ex-Polish Łoś II, impressed into service with the FARR (Royal Rumanian Air Force) and carrying the Rumanian number 211, in flight over Transylvania in 1940. The Łoś bombers equipped 76th and 77th Bomber Squadrons of the FARR's 4th Bomber Group. (M. Moisescu/M. Passingham)

One of the ex-Polish RWD 14 Czapla observation monoplanes in service with the FARR in 1940. (M. Moisescu/M. Passingham)

- Symptomatic of the dismissive attitude towards air power was the lack of an integrated system of air defence for the country as a whole and comprehensive methods of combating the Luftwaffe. Not surprisingly, the Polish air operations bore the characteristics of guerrilla warfare rather than a coordinated campaign.
- Exceptionally high standards of training, combined with patriotic and moral values and generally excellent and purposeful line command at unit level, helped the Polish airmen to overcome their disadvantages, wrong tactical decisions and enormous equipment deficiencies. Because of these qualities they were able to withstand the Luftwaffe and inflict casualties upon the enemy for 17 days, until the Soviet invasion made further operations impossible.
- Although encounters with the Polish fighters were infrequent, the Luftwaffe aircrews quickly learned to recog-

The Polish PWS 26 trainers interned in Latvia. (T. Kopański)

nize the Polish pilots as dreaded opponents, whose superior flying skills and utter disregard for their own safety turned clashes even with solitary fighters into deadly contests. Attacks from very close quarters and accurate marksmanship largely compensated for the poor fire power of the PZL fighters and usually led to destruction of the quarry, unless the speed advantage allowed it to escape. The Polish bomber and reconnaissance crews displayed similar determination in attacks on German troops and armour.

- Despite the magnitude of Poland's defeat and terrible losses suffered by the Polish armies and air arm, the airmen's resolve to fight on was never broken and the painful evacuation to Rumania was accepted by them as a necessary step to continue the struggle for freedom from foreign soil.

- Poland, fighting in isolation, would not have been able to defend herself against Germany's overwhelming strength and the Soviet invasion sped up the collapse of Polish resistance. Better military preparations could have changed the course of the September campaign and made it more prolonged and costly to the enemy, but its final outcome was inevitable. Only the promised major Allied offensive in the West, which, as Poles trustfully believed, was to provide the cornerstone of a common defence strategy, might have changed that. But the Allies did nothing and Poland was left to bleed to death alone.

- Poland's lonely fight against an overwhelmingly superior enemy lasted for 35 days and whilst her armies were defeated, the nation never surrendered. Eight months later, in the campaign in the West, the numerically stronger armed forces of four countries – France, Great Britain, Belgium and Holland – resisted the Germans for 38 days and three of them surrendered, with only Great Britain, saved from a land assault by the sea, surviving. The unyielding stubbornness of the Polish resistance was summed up by the German war chiefs, who later estimated that destruction of one ill-equipped Polish division required on average three times the combat effort to shatter one Allied division in the West in 1940.

Chapter II

THE POLISH AIR FORCE REBORN IN FRANCE

1. Formation of the Polish Government in Exile and the Exodus to France

The September defeat caused great bitterness among the Polish, but at the same time strengthened the nation's determination to fight on and its belief in ultimate victory.

With the internment of President Ignacy Mościcki and leading members of the Polish Government in Rumania, Polish ambassadors and statesmen gathered in Paris towards the end of September 1939 to resolve the government crisis. Władysław Raczkiewicz, Chairman of the World Federation of Poles, emerged as the most acceptable statesman to become the new President and Mościcki formally transferred his powers to him. Raczkiewicz was sworn into office on 30 September and his first presidential act was to dissolve formally the pre-war government and appoint Gen Władysław Sikorski[1] to the post of Prime Minister and Minister of Military Affairs. At the beginning of October 1939 Gen Sikorski formed the Polish Government in Exile, which embodied prominent figures from all major Polish political parties and carried the hallmark of a government of national unity. By a presidential decree of 7 November, he also became the C.-in-C. Polish Armed Forces.

The new government, in a major declaration of its intentions and policies of 18 December 1939, stated that its uppermost objectives were: "by means of a most effective participation of Poland and her armed forces in the war, to free lands of the Republic from foreign occupation and ensure for Poland borders ... which would provide lasting security" and "reconstruction of the Republic, rich not only in land masses and large population, but also in spiritual values."[2]

Immediate rebuilding of Polish fighting capability became the task of paramount urgency and already by the middle of October plans were drawn for the formation of an independent Polish force to be established as a separate entity under orders of the Polish High Command. This was to be composed in the first instance of an army with two infantry corps and an armoured division based in France, and an air arm with 20 combat squadrons, based in Great Britain or in France. The human resources were to be provided by the Polish military personnel evacuated to Rumania or sheltering in other neighboring neutral countries and by recruitment among Poles living in France and other Allied countries and volunteers of Polish descent from America. It was thought that up to 180,000 men would be available for service with Polish formations in this way. While consultations regarding these proposals were in progress with the French and British authorities, energetic steps were taken to facilitate a quick transfer of Poles from the Balkan and Baltic states to France.

The Polish air arm found itself in a particularly favorable position. About 80 percent of its mobilized strength had managed to cross the borders of the overrun homeland to continue the fight, and its crucial role in modern warfare was fully appreciated by Gen Sikorski, who himself was an ardent advocate of air power. In the new situation, Sikorski recognized in an air force a tool of predominant importance for political as well as military reasons and the only means of direct link with the occupied country. Gen Zając's proposals of the end of October, recommending establishment of a separate Aviation Command, were well received.

High-ranking Polish officers in Rumania, who endeavored to preserve military order and organization among the mass of evacuees, declared their loyalty to the Polish Government in France, though a few irresponsible supporters of the former government agitated against it.

Gen Zając forecast the rebirth of the Polish air arm at a meeting with evacuated LW personnel at the sports arena in Radauti (Rumania) on 19 September. He departed for Paris on 3 October to begin the task of renaissance. Ten days later he became the acting Chief of Aviation Command, the Command itself being officially instituted on 30 October. He left Gen Stanisław Ujejski in charge of the evacuation of Polish airmen from Rumania and issued directives defining personnel priorities. Highest on the list were mechanics and equipment main-

[1] Although Gen Sikorski remained strictly neutral during Piłsudski's coup d' etat of May 1926, in March 1928 he was removed from all his duties and his attempts to return to active service in September 1939 were ignored.

[2] "Monitor Polski" No.277/284, 19 December 1939, Angers, France.

tenance specialists, cadet officers and younger pilot/aircrew officers, followed by technical officers. Older, high-ranking officers were to be recalled individually.[3]

Meanwhile, the situation of the Polish personnel in Rumania grew more difficult. By the beginning of October Polish airmen had been moved by the Rumanian authorities from initial assembly points around Tulcea to internment camps in the malaria region of Babadag in Dobrudja. During the first few weeks of chaos and confusion some 300 managed to escape to the West on their own initiative. In addition, 562 left by early October in small groups or individually before full records, indicating ranks and specializations, could be prepared, their escape routes leading via Yugoslavia to Greece or Italy.

A comprehensive register of Polish aviation personnel, including also anti-aircraft artillery, dated 25 November 1939.[4] comprised 1,491 officers, 3,361 NCOs and 4,424 airmen, or 9,276 in all.

Of these 1,027 officers, 1,023 NCOs and 369 airmen, or 2,419 in all, had already left for France in organized transports. With about 860 men mentioned earlier and not covered by the register, the total number of military aviation personnel evacuated to Rumania exceeded 10,100. However, this figure included a small percentage of men unfit for duty and a few who turned out not to belong to any aviation formations. In addition over 1,000 Polish airmen found temporary refuge in Hungary; 207, including 57 officers, in Latvia; and 121, including 38 officers, in Lithuania.

Mass evacuation posed tremendous problems. Apart from complicated formalities and logistics, civilian clothes had to be provided for thousands of travelers. Large sums were needed to finance the operation and bribe various officials. Private aircraft, cars and personal belongings had to be sold in Rumania at desperately low prices to supplement the evacuation funds. Communications between internment camps and discipline were difficult to maintain and secrecy was impossible to achieve. The main evacuation route led from the Black Sea ports, mostly from Balchik, to Beirut in French-controlled Syria, where the airmen, usually arriving in hired Greek ships, were transferred to French vessels to be taken to Marseille. The German Ambassador in Bucharest made furious protests, charging Rumanians with breaches of neutrality and hostility towards the Reich, and German agents and money were sent to Rumania to hinder the evacuation process, their activities curtailing the flow of refugees in the early months of 1940. By 1 May 1940, 3,003 airmen[5] – 102 officers and 2,901 other ranks – were still in Rumania, but despite great obstacles a large proportion of them eventually managed to leave and the last transport from the Balkans, numbering 930 people, arrived in England via Palestine in the autumn of 1940.

In Hungary the Polish refugees met with a very friendly reception and understanding. Although outwardly the Hungarian authorities preserved the appearance of strictness and neutrality, the assembly posts and internment camps were well supplied and help was secretly given to evacuation plans. The largest contingent of aviation personnel, 505 men including 40 officers, was directed to Nagy-Kata from where 405 men were evacuated to the West in less than three weeks, starting at the end of October. Another group, about 70 strong, "escaped" from Jolsva camp, driving away in hired coaches.

Gen Władysław Sikorski, the Prime Minister of the Polish Government in exile and C.-in-C. Polish Armed Forces, with Gen Józef Zając, Chief of Aviation Command next to him on the right, and Col Stefan Pawlikowski on the extreme right, in company of high-ranking French and Polish officers. (IPMS)

German protests resulted in more stringent control of the camps and by 15 February, 1940, 463 airmen (no officers)[6] were still in Hungary. Most of them departed in the following few weeks to join Polish embarkation transports in Rumania. This evacuation ended only when Admiral Horthy, the Hungarian Head of State, was interned by the Nazis. By that time 903 Polish airmen had left Hungary, or about 90 percent of the total.

By contrast, evacuation from Latvia and Lithuania proved almost impossible, as both of these countries yielded to Soviet pressure and detained all Polish military personnel. An attempted mass escape of 57 aviation and three naval officers from the fortress at Libau (Latvia) on 19 October, 1939, did not succeed and according to the Polish Military Attaché in Stockholm, only 51 airmen (43 officers and eight other ranks) managed to get out of the two countries via Sweden, by January 1940.

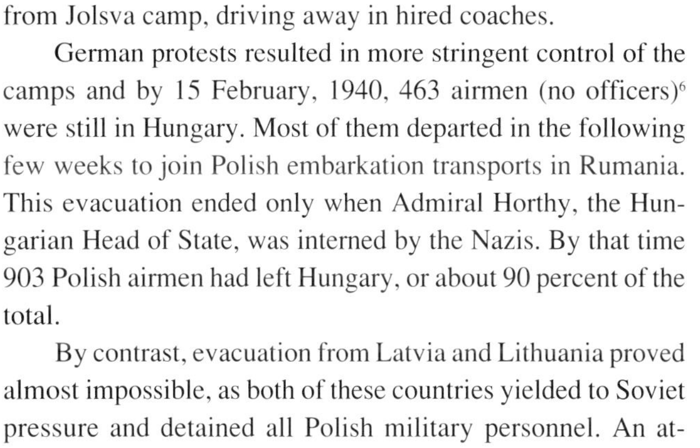

The Greek ship "Patris," which evacuated Polish airmen from Rumania to France and Syria. (T. Kopański)

[3] Note ref. Ldz.44/tj. 39/Lot., 17 October 1939.
[4] Gen Ujejski's report ref. Ldz.49/39/Lot., Bucharest, 26 November 1939.
[5] Polish Embassy note ref. Ldz.2025/Lot./tj., Bucharest, 2 May 1940.
[6] MSWojsk, note ref. Ldz.796/tj./Ewak./40, 27 May 1940.

Many young men in Poland, hearing about the formation of the Polish Armed Forces abroad, embarked upon hazardous individual escapes to France. The escape movement assumed considerable proportions and during the autumn of 1939, in the aftermath of the September war, escapes were generally successful. The most popular routes led South, through Rumania and Hungary. Some men reached France via Lithuania and Sweden and others slipped across Germany to Belgium, Holland or Denmark, or even went East to Russian ports and traveled as stowaways on ships. Early in 1940 border crossings became far more dangerous and by March all routes were guarded so closely that most escape attempts ended in death.

To assist the transit of military personnel as well as the volunteers from Poland, a network of evacuation centers and forwarding points, official and clandestine, was built up throughout Europe, including major centers at Bucharest, Constance, Budapest, Stockholm, Kaunas, Riga and Helsinki, transit centers at Belgrade and Rome, smaller ones at Athens, Salonica, Istanbul and Zagreb, and a few, which operated only briefly, in other towns. The main reception center for all Polish sea-going transports was established at Marseille. Local attitudes towards a seemingly unending stream of Polish escapees varied from hostile or indifferent to friendly, but the warm welcome and assistance given to those who traveled through Yugoslavia deserves special recognition. Despite Italy's official alliance with Germany, considerable help was also given individually by the Italian people.

About 38,000 men, evacuees and escapees, eventually reached France, but the early Polish estimates of the possible size of the Polish Armed Forces in France proved far too optimistic. Although almost 125,000 men went before the Polish Recruitment Commissions when mobilization of the Poles living in France was proclaimed by the Polish Government in Exile and 103,000 of them were found fit for combat duties, with a further 10,000 suitable for second-line service, about 47,000 (largely miners) were reclaimed by the French authorities as essential work force. Ultimately only about 46,500 were enlisted in the Polish Armed Forces, which by June 1940 numbered about 84,500 men. About 11 percent of this total were airmen.

Precise numbers of the available aviation personnel are difficult to establish because various official registers listed different groupings of men: some embracing only those evacuated to Rumania, others, all on the Rumanian territory (including airman in transit from Hungary, escapees from Poland, etc.), and still others referred to airmen serving with the Polish Air Force in France, or in France and Great Britain. The "Report on state of organization and training of the Polish Air Force formations in France and Great Britain" prepared by Gen Zając for Gen Sikorski under the date of 14 May 1940,[7] which should be regarded as the most authoritative and comprehensive source, comprises the following personnel strength totals:

"I. Personnel in France:

A. Aviation – Officers: 1,427; Cadet-Officers: 259; Other ranks: 5,285;

B. A.A. Artillery – Officers: 187; Cadet-Officers: 41; Others ranks: 712;

C. Signals – Officers: 29; Cadet-Officers: 13; Other ranks: 327;

II. Aviation personnel in Great Britain:
Officers: 233; Cadet-Officers: 70; Others: 1943.

III. Aviation personnel in Rumania (approximate strength, presently in Rumania):
Officers: 124; NCOs: 836; Others: 2,143[8]
Of these only about 800 airmen to be evacuated as suitable for service."

The above figures represented totals of 8,280 men in France, of which 6,971 were serving with the aviation units and 2,246 in Great Britain, all in the aviation units, or overall 10,526 men under PAF command.

Attempts were also made to retrieve some Polish aircraft evacuated to Rumania. Of the greatest interest to the Polish Government were two RWD 15 tourers – in particular the long-range RWD 15 SP-KAT – and the PLL "LOT" airliners, all of which were considered suitable for the contemplated courier flights to occupied Poland, as well as the precious Łoś bombers.

In November 1939 Mjr Zygmunt Trebinka, acting on his own initiative, took steps to recover the Polish P.37 Łoś bombers requisitioned by the Rumanian authorities. These aircraft would have been useful to the Allies in view of a lamentable shortage of modern medium bombers in France. Gen Zając did not need much persuasion to give his full backing to the idea. Mjr Trebinka went to Rumania, where he held talks with the Rumanian Minister of Armament and the Minister of Aviation. Later the Polish Military Attaché in Rumania, S-Col Tadeusz Zakrzewski, presented official Polish proposals to the Aide of King Carol II. These suggested that for each aircraft returned, one Łoś bomber would formally become the legal property of the Rumanian Government as part of the deal. The released bombers were to be flown by Rumanian pilots to Baba-Eski airfield at Istanbul and there handed over to Polish pilots to be ferried to France. The Polish authorities also offered to cover delivery costs. Rumanian reluctance to return the bombers induced Gen Zając to write to the French Minister of Air on 22 May 1940, and urge him to press the Rumanians through French diplomatic channels to accept the proposal.[9] However, these moves were overtaken by the capitulation of France and

[7] Ref. Ldz.1505/tjn/40, Paris, 14 May 1940 (IPMS, Lot.A IV,1/1a).

[8] These figures showed personnel strength as of 1 April 1940 and included 100 airmen (22 officers and 78 other ranks) removed from the Rumanian register by 1 May: 8 erroneously listed as airmen and 92 in transit to France.

[9] Gen Zając's letter ref. 1648/tj., Paris, 22 May 1940.

all the Łoś bombers stayed in Rumania, being later used by the Rumanian Air Force on the Eastern Front.

The British also showed great interest in the Polish civil transports of the PLL "LOT" in Rumania, which were valued at two million dollars. Imperial Airways was particularly anxious to get hold of the four Lockheed L14H "Super Electra" airliners. In order to obtain release of the aircraft from Rumania, Imperial Airways agreed to acquire the entire "LOT"'s inventory evacuated to Rumania and to offer employment to the "LOT" personnel. The notification of purchase was communicated to the Rumanian Government by the British Embassy on 20 October 1939. The Rumanians did not protest, but put forward certain conditions and obstacles preventing departure of the aircraft, and despite the British request refused to let them go.

2. Establishment of the Independent Polish Air Force in France – Proposals, Agreements and Initial Setups.

Preliminary proposals, put forward by Gen Zając in the middle of October 1939, considered three options for the rebuilding of the Polish Military Aviation. The first, and the most desirable, was the formation of the entire air force at the side of only one Allied power, either Great Britain or France, with Great Britain being Zając's first choice. The second was the formation of fighter and bomber units, for independent air operations, in Great Britain, and of army cooperation units in France. The third was to be accepted only as a last resort, and that was the formation of bomber units in Great Britain and fighter and army cooperation units in France. During secret talks with Gen Sikorski it was agreed that the Polish airmen should at first be used sparingly in order to avoid excessive casualties, the intention being to preserve a strong and well trained air force cadre until the decisive battles and then, after final victory, bring it back to Poland to provide the basis for the future Polish air force.

Guided by provisional personnel returns from the Balkan and Baltic states, Gen Zając proposed the formation of 20 combat squadrons with 200 aircraft, as follows:

Independent Air Force:
- one Fighter Brigade: eight squadrons with 10 aircraft each,
- one Bomber Brigade: eight squadrons with 10 aircraft each;

Armies' Air Force:
- one Army Cooperation Wing: two squadrons with 10 aircraft each
- one Close Reconnaissance Wing: two squadrons with 10 aircraft each

The brigades were to consist of two regiments each, and each regiment would comprise two two-squadron wings. He also proposed the formation of a Polish Air Force Command separate from the Army Command.

The Allies seem to have been taken by surprise by the numbers of Polish aviation personnel available and by the size of the proposed air arm. Neither was prepared to provide facilities for such a large force, explaining their inability to meet the Polish request by shortage of combat aircraft and suitable accommodations.

Preliminary consultations with the Armee de l'Air and the Royal Air Force led to a tripartite conference at the HQ of the Armee de l'Air in Paris on 25 October 1939, Gen Romatet leading the French side, Air Vice-Marshal Evill the British, and Gen Zając, the Polish. It appears that the French and the British agreed beforehand on a common approach to the Polish proposals and when Gen Zając held unofficial talks with AVM Evill a day earlier, he had learned in advance that the British would accept only a proportion of the Polish aviation personnel and offer to form two Polish bomber squadrons, with two more as reserve.

At the conference Gen Zając expressed the strong desire of the Polish Government to have the whole of the Polish aviation concentrated in one country and explained why Great Britain was preferred. This would facilitate the formation in Canada of units composed of Polish volunteers from the United States. Moreover, Polish personnel were familiar with British aeroengines, which powered the aircraft used in Poland before the war. He added that the Polish aviation personnel whose arrival was expected amounted to between 5,000 and 6,000 men, including 650 pilots, 200-300 observers and air gunners, 250 technical officers, 1,500 riggers and fitters and 2,500 auxiliary ground personnel. Of these, 200 pilots, 50 observers and gunners, 40 technical officers, 50 riggers and fitters and 300 ground and factory personnel were already in France. In addition, the anti-aircraft artillery personnel still in Rumania and Hungary numbered about 140 officers and 600 of other ranks.

The French side suggested that half the Polish personnel should be sent to Great Britain and half stay in France. The British delegation supported this view in principle, offering to take a total of 2,300 Polish airmen (300 pilots and 2,000 others). It was argued that such a solution would ease the strain on equipment supplies and accommodations, and at the same time expedite formation of Polish units and reinforce the combat potential of both Allies. The French declared their willingness to organize within a few months one Polish groupe de chasse (fighter wing) of 38 aircraft (27 combat establishment, two unit command's aircraft and nine operational reserve), and one groupe de bombardement (bomber wing), of 26 aircraft, which would operate jointly with the Polish Army in France.

The conference of 25 October clarified several basic points and established guidelines which paved the way for the rebuilding of the Polish air arm along realistic lines. However, the understandings reached still had to be ratified by the governments involved and many intricate legal problems had to be worked out. A Franco-Polish working party was set up to resolve these issues and prepare the ground for a comprehensive Franco-Polish agreement.

On 28 October 1939, the Polish Cabinet speedily passed a nine-point motion, moved by the Minister of Military Affairs, which, while endorsing the conclusions of the conference, recapitulated the Polish standpoint. It read:[10]

> "1. The Polish air arm constitutes a part of the Polish Armed forces forming at the Allied powers, the highest authority over which rests with the President of the Polish Republic and the highest command with the C.-in-C. Polish Armed Forces, and the organization of which is directed by the Ministry of Military Affairs.
> 2. The Polish Government accepts from necessity the proposals to form the Polish air arm in England and France. The Government agrees to the commencement of formation of fighter and reconnaissance units (the latter for cooperation with the Polish land army) in France and bomber units in England.
> 3. The Polish Government accepts that as soon as technical circumstances would permit it, the Polish air arm shall be developed on the territory of only one of the Allied powers, except for Canada, where its units can be formed from the Poles permanently living in that country.
> 4. The Polish Government accepts that the whole of the Polish air arm, after completion of its formation, shall be used on one section of the front, in tactical cooperation with the Polish Army and under one overall Polish command."

The remaining five points urged rapid approval of the above proposals by the Allies; dealt with the establishment of the Polish anti-aircraft, signal and air observation units; accepted the French proposal to appoint Lyon as the main assembly and segregation centre for all Polish aviation personnel arriving in France; suggested that the Polish civil technical personnel should be given employment in the Allied aviation industries appropriate to their professional qualifications; and implied that the cost of organization and maintenance of the Polish air arm on territories of the Allied powers should be for the time being born by the countries concerned, to be reimbursed by the Polish Government after the war.

It is interesting to note that in Article 2 the Polish Government strictly followed the third of Zając's original alternatives and omitted any reference to the French offer of 25 October to form a Polish bomber unit. Acceptance of this would have meant a further dispersal of the Polish air arm between France and Great Britain, which the Polish authorities were anxious to avoid. During subsequent talks the French asked for clarification of this point and were officially informed that the Poles did not intend to form bomber units on French soil, because all available Polish bomber personnel would be needed in Great Britain. They proposed to form as the first step a Fighter Regiment in France, composed of two two-squadron fighter wings, to be followed by one or two army cooperation squadrons. In response, the French, while accepting the above program in principle, flatly turned down the Polish idea of a Fighter Regiment, stating that the war structure of the Armee de l'Air did not accommodate such a formation and therefore no Polish command of this order could be created in France.

The French Government moved slowly and with considerable reluctance. Gen Denain, Chief of the French Military Mission at the Polish Government, raised one difficulty after another, which delayed ratification of the Franco-Polish agreement. The French wanted the Poles to integrate with French air units singly or in small groups and Polish insistence upon formation of an independent, national air force met with the French counter-proposal that Polish three-aircraft flights should be incorporated in French squadrons. This was rejected, but the Poles agreed that Polish fighter squadrons would initially serve with French wings to gain operational experience before all-Polish fighter wings were formed.

Two other possibilities were considered by the Polish Government at this time. One was the suggestion of Anthony Drexel Biddle, the American Ambassador to the Polish Government, that the Poles should obtain aircraft directly from the USA and form their own, totally separate air force, fighting under the Polish national colours. The other was the formation of additional Polish air arm units in Syria, in anticipation of a possible use of that country as a base for attack against the Axis powers. While the practicability of the first of these proposals was rather doubtful, the second received further attention and re-emerged in some of the Polish Air Force expansion plans in May 1940.

On 4 January 1940, the long awaited Franco-Polish military agreements were signed by the Prime Minister Edouard Daladier and Gen Władysław Sikorski. The main treaty, the "Military Accords concluded between Poland and France," provided legal basis for the formation of the Polish Armed Forces in France in general; a second treaty, the "Agreement relating to the Polish Air Force," constituted the annex to the former. Article 1. of this Agreement confirmed in principle Article 1. of the Polish Government's resolution of 28 October 1939, but added that the Polish units, while "maintaining the character of the Allied Air Force, shall be used by the French High Command on the same basis as the French units," and "whenever possible ... shall be engaged on those sections of the front on which the Polish land army is operating."[11] This wording restricted the independence of the Polish Air Force and contradicted Article 4 of the said resolution.

Other points of the Agreement established the right of the Polish High Command to inspect the Polish aviation units and to recruit Polish volunteers from other countries, and added that the Agreement would be supplemented by a special detailed accord to be worked out between the French Air Ministry and the Polish Ministry of Military Affairs.

[10] IPMS, Lot.A IV,1/12h-5.

[11] IPMS, Lot.A IV, 1/12i-4

Such a comprehensive document, the "Procedural accord relating to the establishment of the Polish Air Force in France,"[12] was concluded between the French Minister of Air Guy La Chambre and Gen Sikorski on 17 February, 1940. Its opening phrase linked it to the agreements signed on 4 January. Article 1 of the new document established the following priorities for the formation of the Polish Air Force units in France:

1st – one groupe de chasse (fighter wing) of two squadrons
2nd – second groupe de chasse (fighter wing) of similar composition
3rd – one (or two) escadrille de cooperation (cooperation squadron)
4th – unit establishments, pending availability of personnel and armament.

Gen Sikorski welcomed by Gen Denain, Chief of the French Military Mission at the Polish Government, on his arrival at the Polish Air Base at Lyon-Bron on 27 March 1940. (K. Chołoniewski)

Under Article 2 French rates of pay were to be applied to the Polish units and maintenance and provisions provided by the Armee de l'Air as for French units.

Article 3 read: "Before the formation of the exclusively Polish wings, Polish squadrons will each be initially attached to a French fighter wing for the purpose of familiarization of their personnel with methods of combat on the Western front. They will be subordinate to the French Wing Commanders and used under the same conditions as the French squadrons and in association with them. ... the Polish officers assigned to the Polish wing commands will undergo practice in French wing commands. The Polish wings, when formed, will be subordinate to the French Aviation High Command under the conditions specified in Article 1 of the 'Agreement relating to the Polish Air Force' of 4 January 1940."

Article 4 referred to the possibility of employment of the Polish design, technical and airline staff (as proposed in Article 8 of the Polish resolution of 28 October 1939), which "could be individually put at the disposal of the French Command, to be used in accordance with appropriate qualifications."

Article 5 dealt with training, saying that "A Training Centre for the Polish Military Aviation shall be established at Lyon-Bron Air Base and initially staffed by Polish and French cadre, but as soon as circumstances permit shall become exclusively Polish. The Polish command of the Centre shall be subordinate to the French command of the Air Base." Detachments of the Polish flying and maintenance personnel under Polish officers and NCOs were to be posted for practice at French schools and training centres and then to rejoin the Lyon-Bron Training Centre to be distributed as required in line with Articles 3 and 4 above.

Articles 6, 7 and 8 stipulated respectively that: the Polish personnel will temporarily be wearing French uniforms with Polish insignia, until Polish Air Force uniforms become available; the Polish flag will fly over the bases occupied by Polish units; the Polish aircraft will carry the Polish white and red chessboard markings on both sides of the fuselage beside the French regulation markings. The remaining Articles, 9, 10 and 11, covered additional matters, including medical and religious care and application of Polish Army Regulations (except for the Polish airmen serving or training at the French establishments), and added that all questions still requiring clarification would be resolved either by complementary agreements or by correspondence between the French and the Polish Air Chiefs.

The legal basis, defining the status and organization of the Polish Air Force in France, had at long last been determined. However, this followed rather than induced the process of rebuilding of the Polish military aviation which had been energetically pursued, both practically and legally, since the first arrivals in France the previous autumn.

The Polish Aviation Command, formed within the framework of the Ministry of Military Affairs, had begun work on 30 October 1939. Initially the Command had to set itself up as a working team, which would assume basic command functions over Polish airmen as they arrived, assess personnel needs and initiate training. Its first tasks were to establish an efficient machinery for the receiving, assembling and segregating of airmen, and for subsequent personnel management consistent, as far as possible, with French and British establishments and organization.

The first expression of Command's activities was an order of 30 October, 1939, regulating establishment of the Aviation Receiving Point at Marseille and the Aviation Assembly Stations at Lyon, Istres, Salon, Paris-Le Bourget (for officers) and at the Polish Army Assembly Center in Coëtquidan (for other ranks).[13] Formation of the Polish training camp began at the Lyon-Bron base with an initial approval of the French Air Ministry given on 22 November – ahead of the Franco-Polish Accords – and the Lyon Assembly Station was incorporated in the Centrum Wyszkolenia Lotnictwa (CWL or Aviation Training Centre). The existing assembly network was reorganized

[12] IPMS, Lot.A IV, 1/12j.

[13] Le Bourget and Coëtquidan, organized in the middle of October, were the first two.

on 7 February 1940; the Assembly Stations at Istres, Salon and Coëtquidan were closed, a new Temporary Aviation Assembly Station was set up at Stepfonds, and the Lyon CWL Training Centre was further expanded with a view to absorbing the airmen from Stepfonds at a later date. Le Bourget became the selection center for Polish airmen departing for Great Britain.

The second phase of the Aviation Command's work started on 22 February 1940 with the historic decision of Gen Sikorski that the Polish Military Aviation, until then incorporated with Army's organization as one of its services, was to become an independent member of the Polish Armed Forces, to be known as the Polskie Siły Powietrzne (PSP), or the Polish Air Force (PAF). Sikorski's order to the airmen read "... My ambition is to create this year a Polish Air Force numerically at least equal and technically superior to that with which the Polish Armed Forces entered the fight in the autumn of last year. ... "[14] With effect from 1 March, 1940, the Aviation Command became the Polish Air Force Command, and within its framework separate PAF Commands were to be formed for France and Great Britain, as well as the Command of Anti-Aircraft Defences. A few days later Gen Józef Zając was confirmed in office as the Commander of the Polish Air Force, with Gen Stanisław Ujejski as his Deputy in France and Gen Władysław Kalkus, the Deputy Commander in Great Britain. Col Izydor Modelski took the command of the Anti-Aircraft Defenses.

Although it would be true to say that the separation of the PAF from the Polish Army was at this stage largely a symbolic one, with Gen Zając having in practice the status of an army commander,[15] it should not be dismissed as mere cosmetics, as some high-ranking army officers grudgingly suggested. The prevailing conditions and small size of the Polish Armed Forces, with slim material and human resources, made it necessary to direct the military organization and operations from one nerve centre, but Sikorski's decision signified a basic change in the Polish attitude towards the air force and pointed to the shape of things to come in Polish military organization.

Gen Zając immediately set to work on detailed proposals to be put before the Polish High Command, which were to advance autonomy of the PAF in practical terms. A great effort was made to develop the PAF Command in such a manner as to enable it to deal not only with matters of the moment, but to prepare it for returning to and functioning in a liberated Poland. With this in mind, various commissions and working parties were set up at the Command and the Counter-Espionage Office was formed at Lyon-Bron. During this period, apart from adjusting itself to the demands of its new status, the Command organized the first Polish combat units and detached the first Polish fighter flights to the front.

However, subsequent developments seemed to indicate that the PAF Command might have chosen the wrong priorities and overstretched itself. During the third phase of the Command's activities, which commenced in April 1940, while the work of various commissions progressed well, the actual direction, shape and running of the emerging PAF units showed complete disregard for the letter and spirit of the "Procedural Accord" of 17 February 1940.

In March 1940 the Command gave its consent for six Polish three-aircraft flights to be posted one each to different French wings at the front as an exceptional arrangement – the Accord had emphasized that Polish teams at squadron strength only would be attached to French wings – on the understanding that after a few weeks of combat experience the flights would be recalled to be incorporated into Polish squadrons or wings. Contrary to this, the flights were retained by the French wings and the practice, instead of being the exception, began to form a pattern. As French casualties mounted, the French persisted in using Poles in small groups or even singly as replacements. This cut right across prospects of forming the anticipated PAF units, as experienced Polish personnel were progressively drained away.

Dispersion of the PAF strength in France, the very issue the Accord was to guard against, was further heightened by Polish approval for the formation of a Polish fighter wing for Finland, to fight on the Finnish front against the Soviet Union. Its departure from France was prevented only by the end of the Russo-Finnish war. The Polish decision to allow formation of Polish bomber units in France after all, which reversed the earlier Polish stand, was another example of this.

It could be argued that these moves were justified by desperately slow progress with the organization of PAF combat units in France and the need to give the increasingly restless airmen in the barracks something to do, but they sent the wrong signals to the French and provided them with excuses – if they needed any – to ignore the Franco-Polish agreements or interpret them to suit French convenience. This led to the complete erosion of the PAF's autonomy in practical terms, with the French Air Ministry usurping the command of PAF formations and establishments. Administrative decisions affecting Polish personnel began to be taken by the French without any reference to the PAF Command, sometimes countermanding Command's directives in respect of movement of individual Polish airmen. Attempts were made to classify some high-ranking Polish officers as "non-active," certain restrictions were imposed on Poles in training, and even the execution of the right of the PAF Command to inspection of the Polish units met with a distinctly unhelpful attitude by the French Air Ministry.

The French Commander of the Lyon-Bron air base, Col Hug, regarded himself as the sole Commandant of the CWL Training Centre, interfered with its internal running at every level and involved himself with the matters specifically reserved – under Article 5 of the "Procedural Accord" – for the Polish Command. Instructions issued by the Polish Command of the Centre were ignored by the French and its authority was reduced almost to a farcical level. Extremely strained relations

[14] The Polish Armed Forces were then more than ten times larger.

[15] Zając postulated creation of the post of Undersecretary of State for Air.

with Col Hug led Col Pawlikowski, the Polish C.O. of the CWL, to tender his resignation and to the appointment of Col Wacław Iwaszkiewicz in his place on 20 May 1940.

Contrary to French promises, apart from the solitary Polish Fighter Wing (I/145), originally intended for Finland, no new Polish fighter squadrons or wings, nor any reconnaissance units, were formed. The position of the Polish higher-ranking aviation officers became desperate, as they could not find suitable positions within the modest framework of the PAF, deprived as it was of the units that had been expected to be formed.

Except for a very few cases, the promised postings of Polish officers to French combat unit commands for practice did not materialize. At the beginning of June only one Polish officer was serving with the command of a French fighter wing at the front, and Polish operational command structures did not exist even at squadron levels.

Attempts to resolve these serious problems by means of correspondence proved totally futile, as the lower-echelon French officers appointed by the French Air Ministry to deal with the Polish grievances had neither the rank nor authority to take effective action.

During May 1940 the PAF Command commenced strenuous efforts to find solutions and regain effective control. New proposals for quicker expansion of the PAF's combat strength were drawn and presented to Gen Sikorski for his appraisal.[16] The proposals evaluated the progress of training at the Lyon CWL Training Centre and noted that retraining courses for ground personnel lasted between three and four weeks and were conducted in groups of up to 400 men; courses for fighter and reconnaissance pilots lasted five weeks and, for bomber pilots, eight weeks, in groups of 80 to 90 pilots. Based on this, two alternative propositions for formation of the Polish combat units were put forward:

- 1st: four fighter wings, three close-reconnaissance wings and three bomber wings, of which one close-reconnaissance and one bomber wing would be established in Syria, if the French Air Ministry agreed;
- 2nd: in the event of no Polish units being formed in Syria: six fighter wings, four close-reconnaissance wings and two bomber wings, with priority to be given to four fighter wings, two close-reconnaissance wings and two bomber wings.

Men were already available for these units except for air gunners. A further 25 percent, to maintain initial personnel reserves, could be supplied almost immediately. The remaining 175 percent – to meet 200 percent reserves envisaged by the agreement – were to be obtained by training younger personnel. At the same time comprehensive plans were formulated for establishment of maintenance parks, support services and signal and anti-aircraft units, which were to comprise four heavy artillery squadrons (75-90 mm caliber) and eight light artillery squadrons (25-40 mm caliber).

To improve working relations between the French and Polish commanding officers and authorities and define clearly their respective areas of responsibility, sets of detailed instructions were prepared in French and Polish. A draft of a new order, transforming the CWL Training Centre at Lyon-Bron into an autonomous and exclusively Polish establishment, was also drawn up.

At the beginning of June 1940, the PAF Command sent a protest note to Gen Picard, the Chief of Staff of the Armée de l'Air, which outlined Polish complaints[17] together with new proposals for solutions and additional agreements. At the same time the Command turned down a fresh French demand for a further 100 Polish fighter pilots to be posted to the front in individual three-aircraft flights and for 50 bomber pilots as individuals, not as complete crews.

The new Polish proposals requested instead:

- the immediate formation of six fighter squadrons, to be achieved by expanding the six existing, original Polish three-aircraft flights at the front to squadron strength, and these squadrons to be regarded as the elements of three future wings; this, together with the existing I/145 Polish Fighter Wing would bring the Polish fighter force to four wings;
- the immediate formation of four bomber and two reconnaissance squadrons with the Polish personnel who had already completed training on French equipment, as the first step towards establishment of appropriate wings;
- the immediate posting of the Polish officers appointed to command the future wings to suitable French commands for operational practice, with two posts being additionally requested for Polish officers at the level of bomber brigade command.

Unfortunately, by this time with the collapse of French defences and chaos and confusion before the impending surrender of France, these belated moves were of no more than academic interest. Other activities by the PAF Command included plans for the re-establishment of the PAF in Poland and its future status and structure after the war. The PAF Regulations were approved and about to be issued and a comprehensive order defining autonomy of the PAF was in the final stages of preparation when France capitulated. Work on these projects was later resumed in Great Britain in changed circumstances.

3. Organization and Training of the PAF Units in France

Polish airmen left Poland driven by an overwhelming desire to get even with the Germans and to continue the fighting from Allied territory. In that respect their morale was of the highest order, with no signs of defeatism. However, the mood

[16] Ref. Ldz.1505/tj./40, Paris, 14 May 1940 (IPMS, Lot.A IV,1/1a)

[17] Ref. Ldz.1916/I/tj./40, 6 June 1940 (IPMS, Lot.A IV,1/12a-40).

of the evacuees caused the Polish military command considerable concern.

The shock of the defeat, shattering deficiencies in the strength and readiness of the Polish air arm and incompetence in the Polish Army and Aviation Commands revealed during the September campaign led to a crisis of confidence among the younger officers and men in their leadership. This, combined with the unworthy behavior of some high-ranking officers during the evacuation and in Rumania, impaired military authority and affected the discipline. Prolonged disputes in barracks and perhaps excessively harsh criticism of recent events, which was not yet tempered by an understanding of French and British weaknesses, deepened discontent. The main blame for the state of pre-war Polish aviation was attached to Gen Rayski, but Generals Zając, Kalkus and Ujejski, who retained their positions in the new PAF structure, enjoyed little confidence.

It was not appreciated that Gen Sikorski's hesitant decision to retain Zając in command of the PAF was more a matter of expediency than a choice and was regarded by him as a temporary solution. Like-minded groups of ideological or political character came into being in France with the aim of "purifying" the Army and Air Force officer cadres. Each of them promoted different individuals.

The most significant among them was the "Chain of good friends," inspired by Gen Sikorski himself. The "Chain" attracted a substantial number of younger aviation officers, as well as NCOs, led by Cpt Roman Niewiarowski. Several members of this group belonged before the war to the ideologically similar, secret "Cross and Sword Order." At least two more groupings, one of them with a nationalistic right-wing orientation, made their presence felt. Sikorski's concern was to appoint an experienced and "neutral" Air Force Chief who could rebuild the Service quickly, until a more reputable officer, who had earned general trust and respect in future air operations, emerged.

Towards the end of 1939 the patience of the impetuous young men, waiting idly in the camps for some sort of action was approaching the breaking point. The process of organizing the planned units showed no signs of progress, stifled by the cumbersome French bureaucracy. Poor facilities and rough accommodations, lacking heating and bedding, which were a far cry from the decent quarters Polish officers were accustomed to in Poland, and the patronizing attitude of the French, who generally did not want to fight, blamed the war on Poland and largely dismissed the Polish September resistance as ineffectual, fueled resentment.

By the middle of December 1939 the Assembly Station of the Base Aérienne Mixte de Lyon (Mixed Aviation base of Lyon), initially commanded by S-Col Bronisław Wojtarowicz, became the biggest concentration of Polish airmen in France. Within three weeks of the French approval of 22 November the Polish Aviation Command had set up a framework for the CWL Training Centre at Lyon. This was soon to be recognized by the French as Le Depot d'Instruction de l'Aviation Polonaise (DIAP).

Col Stefan Pawlikowski, the former Commander of the Warsaw Pursuit Brigade, was appointed C.O. of the CWL and arrived in Lyon on 16 December to implement its organization. On 22 December the Lyon Assembly Station was subordinated to him to become a part of the newly established CWL and at the same time the Station's command passed from

A mass said on 27 March 1940 in one of the CWL's hangars at Lyon-Bron, with an altar erected behind the tail of the Centre's Potez 63.11 '1', c/n 138, which bears the PAF chessboards on the fuselage (partly obscured by the left rudder).

Wojtarowicz to Col Edward Karaś. Apart from administrative and support units, and the Assembly Station, the Centre comprised: Ground Training Command under S-Col Stanisław Nazarkiewicz; Air Training Command under S-Col Piotr Dudziński; Fighter Wing under S-Col Leopold Pamuła with Mjr Eugeniusz Wyrwicki as his Deputy and composed of 1st Fighter Squadron under Zdzisław Krasnodębski and 2nd Fighter Squadron under Cpt Adam Kowalczyk; and Observation (Close Reconnaissance) Wing under Mjr Edward Młynarski, composed of 1st Observation Squadron under Cpt Czesław Kujawa and 2nd Observation Squadron under Cpt Aleksander Łukiński. The skeleton Fighter and Close Reconnaissance Wings were a pointer to the intended future composition of the PAF in France, to be specified later in the "Procedural Accord" of 17 February 1940.

Polish pilots of the so-called "Montpellier Squadron' showing off their new uniforms, with the PAF Eagles on their caps and Polish distinctions, on parade at the Lyon-Bron base on 27 March 1940. The three Captains in the first row are: Łaszkiewicz, Pentz and Sulerzycki. (IPMS)

In practical terms, the establishment of the CWL meant, to begin with, only the formal allocation of the Lyon personnel resources to its various formations. It could do little to ease the frustration of officers and airmen deprived, as they still were, of training and equipment. Persistent complaints, addressed directly to Gen Sikorski, and calls to bring to trial those responsible for the September catastrophe, were deeply worrying to Gen Zając. On 19 December, three days after arriving in Lyon, Pawlikowski received a request from him to take strong "no nonsense" measures to quash the "rebellion of the young," as the protests became to be called. Pawlikowski, who better understood the men, was reluctant to do this and, after an investigation of their grievances, a delegation of three representatives of "the young" was allowed to see Gen Sikorski in Paris. On 5 January 1940, during a meeting at which Gen Zając was also present, Sikorski listened to their complaints and promised personal intervention.

At that time, as the result of Polish pressure which preceded the imminent signing of the Franco-Polish Military Accord (on 4 January 1940), some improvement in the situation began to take place. At a conference at the French Air Ministry on 29 December it was agreed that training of Polish pilots on the Morane-Saulnier MS 406 fighters would commence at the Fighter Aviation Training Center at Montpellier by the middle of January. The French proposed a four-month course – in accordance with French practice – while the Poles suggested that, bearing in mind the experience of the Polish pilots, a two-month course would be sufficient. It was agreed that the duration of the course would depend on the progress of the Poles. Col Pawlikowski was immediately informed and asked to select the first group of 19 pilots to go to Montpellier.

The preliminary moves towards organization of the first Polish fighter wing already taken in December with the formation of the CWL, gathered momentum in January 1940, resulting in expanded plans and a rapid re-distribution of the most experienced Lyon personnel among the two proposed fighter wings: 1st Warsaw Fighter Wing under S-Col Leopold Pamuła, staffed with pilots of the former Pursuit Brigade, and 2nd Poznań Fighter Wing under Mjr Mieczysław Mümler, staffed mostly with pilots of the former 3rd (Poznań) and 2nd (Kraków) Air Regiments. Preparation of the cadre for two more wings – 3rd Fighter Wing under Mjr Józef Kępiński and 4th Fighter Wing under Mjr Eugeniusz Wyrwicki – was also initiated.

The selection of a handful of pilots (for one squadron only) from among many for the Montpellier course was a very delicate matter. Pawlikowski called a special meeting of the wing commanders and explained his proposal that the whole group should be made up of pilots from the same wing, expressing the hope that the rest of the wing's pilots (for the second squadron) and all ground personnel would be trained at the same time at the Lyon CWL Training Centre, so that the first complete Polish fighter wing could be ready for action in two months time. Mümler strongly supported Pawlikowski's plan and pleaded that his 2nd Poznań Fighter Wing should be chosen. Mümler commanded great respect because of the part he had played in the September fighting and his proposal was accepted. The 19 pilots from his wing, with Cpt Stefan Łaszkiewicz in command and 18 others were as follows: Cpts Mieczysław Wiórkiewicz, Mieczysław Sulerzycki, Jan Pentz; Lts Władysław Goettel, Kazimierz Bursztyn, Stefan Zantara, Józef Brzeziński; S-Lts Erwin Kawnik, Włodzimierz Karwowski, Bolesław Rychlicki, Stanisław Chałupa, Władysław Chciuk, Bohdan Anders, Władysław Gnyś, Wacław Król; NCOs Leopold Flanek, Antoni Beda, Eugeniusz Nowakiewicz.[18] Lt Józef Sokołowski, a Frenchman of Polish origin, joined the group as liaison officer and interpreter.

All men received brand new uniforms, made of the regulation French Air Arm navy-blue fabric but with the Polish Eagle emblems and distinctions, and departed for Montpellier on 7 January, 1940. After two days' familiarization with the establishment, on 10 January the Poles were incorporated into 1st Fighter Wing, forming its first squadron (the second squadron remaining French). The Polish Aviation Workshop, manned by some 40 technicians and mechanics under Cpt Eng Stefan

[18] The ranks and order of names are quoted after the French Air Ministry documentation.

Wagner, was also organized to service the Polish unit. A short theoretical course was followed by a few refresher flights on MS 230 high-wing monoplane trainers, from which the pilots graduated to the MS 406 monoplane fighters, commencing training on these aircraft on 3 February. The Poles mastered the modern Morane fighters immediately and after a week or so, when each pilot put in about 20 hours flying without the slightest accident, the French, somewhat surprised, concluded that their conversion training should be regarded as completed. The whole group returned to Lyon-Bron on 16 February and became known as the "Montpellier Squadron," although after completion of further operational training at the Bron base, the pilots were split into six three-aircraft flights and never operated together as one unit. At that time various organizational changes were taking place at CWL and the Montpellier pilots, whose early return was not fully anticipated, were first given two weeks' special leave, which they spent skiing at Mégève.

On 20 January 1940, Gen Sikorski arrived at Lyon at the start of a two-days' inspection in order to sound out the situation at the CWL Centre for himself and disperse the discontent. He was greeted at the railway station by a keen and well-drilled detachment of Polish airmen in brand new uniforms and soon found out that the "rebellion of the young" was far less serious than initially feared by some high-ranking officers. Soon after his visit the Training Centre underwent further expansion and the command of its Assembly Station passed on 1 February from Col Karaś to S-Col Stefan Berezowski. The status of CWL was reaffirmed by the "Procedural Accord" of 17 February, and at long last serious training of the Polish airmen began at the Centre. A limited number of training aircraft, mainly the Romano Ro 82 biplane trainers, MS 230 high-wing monoplane trainers, Caudron C.635 Simoun cabin monoplanes and a few obsolete Dewoitine D.501 single-seat fighters were put at the disposal of the Poles. Some 20 Polish officers were dispatched to various French commands for practice, and training of ground personnel and mechanics of various trades also started.

On 3 March the Fighter Wing of the PAF Training Division, comprising two squadrons: the Fighter Training Squadron, which moved to the Lyon-Mions airfield, and the Fighter Exercise Squadron, at Bron, was formed at the Lyon CWL Training Centre under the command of a Frenchmen, Cpt Baville Rougevin and Cpt Mieczysław Wiórkiewicz (from the "Montpellier Squadron"). Wiórkiewicz also became the first C.O. of the Fighter Exercise Squadron, with ten experienced fighter pilots beside him: Mjr Eugeniusz Wyrwicki (who later briefly commanded the Squadron), Cpt Feliks Gazda, Lts Wieńczysław Barański, Jan Borowski, Aleksander Gabszewicz, and S-Lts Rajmund Kalpas, Bogusław Mierzwa, Włodzimierz Miksa, Jerzy Poniatowski and Jerzy Radomski. The Squadron was equipped with 11 MS 406 fighters and entrusted with the air defence of the Lyon base. It became the first Polish unit in France to be ready for action, although in all probability it flew no operational sorties before May 1940.

Two Polish pilots (marked with arrows) of the 1st Caen Polish Training Squadron, with French instructors and ground crew, during training on a LeO 20-BN5 bomber in May 1940 (P. Riviere/ K. Chołoniewski).

Bloch MB 210-BN5 bomber with a Polish crew, preparing for a take-off at Caen. (P. Riviere/K. Chołoniewski)

At the same time, ostensibly to facilitate wider training for less experienced personnel, a large number of Poles were dispersed from the CWL in March and April to various French establishments. In reality this move appears to have been a half-hearted attempt to appease the Poles and give them something to do rather than a serious commitment to meaningful training. The tuition provided was very slow, erratic and of a very basic nature, generally worthless even for semi-qualified personnel. In many instances qualified Polish pilots, instead of being trained, were simply used to carry French pupils on various training flights or as reliefs for other exercises.

On 1 March 583 Polish airmen – 200 trainees, 71 pilots, 217 mechanics and 95 other ground personnel – left for Africa to be distributed among four training establishments: Rabat (Bomber and Observers' School), Marrakech (Flying School), Blida (Gunnery and Bombardment School) and Maison Blanche (Maintenance School). Later a number of Poles were moved to Fes (Bomber School). Early in March several centers in France received Polish airmen and Polish training units were established where numbers warranted. By the middle of April these training facilities for flying personnel were as follows:

- Avord: a group of 12 Polish fighter pilots training on MS 230 trainers, D.501 and MS 406 fighters, and practicing formation flying on North American A.57 Yale basic trainers.
- Bordeaux, Merignac: three groups of Polish pilots: one fighter group on D.500/501s practicing mock combat with French bombers; one flying Potez XXVs with French observers, practicing reconnaissance and blind flying; one

practicing multi-engined flying on LeO 20 and Potez 54 bombers.

- Caen: three groups of pilots: the 1st Caen Polish Training Squadron commanded by Mjr Czesław Krajewski,[19] comprising: 11 bomber pilots training on LeO 20, Potez 54, Amiot 143 and Bloch MB 210 bombers, one group of nine pilots allocated to target-towing and flying Potez XXV tugs, one group of 16 pilots allocated to air-gunnery school and flying Potez XXVs with French trainee gunners for firing practice. A fighter flight of four Polish pilots, which was temporarily transferred to Etampes for training on MS 230s (from 1 to 24 April) was to continue training at Caen after its return.[20]

- Cazaux: a group of Polish bomber pilots training on Potez XXVs, Amiot 143 and Bloch MB 210 bombers, prior to graduating to Potez 63s; Polish fighter pilots training on Loire 45 (prototype) and 46 fighters at the Cazaux Air Firing School.

- Chateauroux: two groups of pilots: one fighter group of ten, training on Potez XXVs and Nieuport-Delage NiD 622 sesquiplane and Loire Lo 46 Puławski-wing fighters; one bomber group of nine pilots training on Potez 56 general-purpose monoplanes.

- Clermont Ferrand: four groups: one of four fighter pilots, training mainly on D.501 fighters; one of 11 bomber pilots, allocated to air-gunnery school and flying Potez XXVs with French trainee gunners; one of 32 observers, training on Potez XXVs, Bloch MB 210s and Potez 63s in reconnaissance and army cooperation work which included direction of artillery fire (most of the observers had only one flight each by the middle of April); one of 13 radio navigators, still in the middle of April awaiting training on Bloch MB 131s and Potez 63s.

- Etampes: a fighter group of three instructors and ten pilots (including four transferred from Caen) undergoing intensive training on MS 230 trainers and D.500/501 and MS 406 fighters, and also practicing formation flying on North American A.57 Yale basic trainers.

- Istres: a group of bomber pilots, which, after completion of training on Caudron C.445 Goelands, LeO 20s and Potez 54s, flew Potez 63s.

- Montpellier: two groups of pilots: one, allocated to target-towing work; one of five fighter pilots transferred from Toulouse on 13 April, where they were given no training.

- Pau: two groups of pilots, one fighter and one bomber; pilots of both groups were used to carry French trainee observers/gunners on Potez XXVs; only four pilots from the fighter group underwent training on D.501 fighters, to which was to include mock attacks on the school's Potez XXVs.

- Rennes: four groups of bomber personnel and a Polish fighter training flight – the largest concentration of Polish airmen within French training centres in France proper, under overall command of Mjr Jan Biały: one of four instructors and 30 bomber pilots, training on Caudron C.445 Goelands and LeO 20 and Potez 54 bombers; one of 30 observers/gunners, training on Potez XXVs; one of 15 radio navigators, undergoing theoretical training in navigation, bombardment and gunnery on the ground; one of six pilots, flying Potez XXVs with Polish and French trainee observers; a Polish fighter flight of five pilots under command of S-Lt Antoni Kolubiński, initially training on MS 230 trainers and converting to D.501 fighters (in May formed into the "Kol" Chimney Flight)

- Tours: three groups of pilots: one of four fighter pilots training on one available D.501 fighter and also used as relief pilots to fly French pupils on Potez XXVs; one of six pilots flying Potez XXVs with French observers; one of bomber pilots, training on Potez 54 bombers with French crews.

- St. Jean d'Angely: a group of bomber pilots, practicing on LeO 20 bombers.

- Versailles: a group of fighter pilots carrying French observers/gunners on Potez XXVs, about to convert to D.500/501 fighters and start mock attacks on Potez XXVs.

A Polish pilot sitting on top of the Loire 45-01 fighter prototype at the Cazaux Air Firing School (note the four-blade airscrew fitted to the fighter). The Loire 45 and 46 fighters, featuring the Puławski-wing layout copied from the PZL fighters, brought nostalgic memories to the Poles. (P. Riviere/K. Chołoniewski)

Loire 46-C1 fighters of the Cazaux firing school allocated to the Polish fighter-pilot group, with the aircraft '1', c/n 36, in the full view. (P. Riviere/K. Chołoniewski)

[19] Formation of this Squadron was initiated in January 1940.

[20] This unit, composed of S-Lt Stanisław Andrzejewski, its C.O., S-Lt Wilhelm Śniechowski, C-Off Romuald Gadus and NCO Bronisław Kościk, is misleadingly referred to in some current publications as "The Caen Chimney Flight." Such a Flight was not included in any of the PAF listings of the Polish combat units in France and available Polish documents (The Caen Training Center file, IPMS, Lot.A IV, 1/2a-10; Krajewski's report, A IV, 1/2b-6) indicate that it remained in training until the capitulation of France.

Polish mechanics servicing a Potez 631 Ins (trainer), c/n 58, used by Polish crews training at Istres, April 1940. (T. Kopański)

Training in French Africa were the following:

- Blida: two groups: one of six fighter pilots training on NiD 622 fighters and also carrying French pupils on Potez XXVs on various exercises; one of ten bomber pilots flying LeO 20s prior to conversion to Bloch MB 200 bombers – all pilots to be employed in flying trainee navigators, bomb-aimers and gunners.
- Fes: groups of bomber pilots undergoing dual-control and solo flights on Caudrons C.635 Simoun and C.445 Goeland trainers and Potez 54 bombers, including instrument flying, prior to graduation to Potez 63s.
- Marrakech: groups of pilots undergoing initial training, starting with Morane MS 315 trainers, prior to transfer to Rabat for bomber training.
- Rabat: groups of bomber pilots undergoing refresher flights on Potez XXVs; one group of nine pilots training on LeO 20 bombers.

Polish aircraft and aeroengine maintenance personnel began training at various courses at Clermont Ferrand, Montpellier, Rennes and at Maison Blanche in Africa, and groups of Polish technical and engineering personnel went to industrial establishments at Aulnat Sud, Bordeaux, Couzon, Limoges and Toulouse. Training of the PAF signal units' members was undertaken by the Aviation Communications Centre at St. Jean d'Angely and at Versailles, and of meteorologists, at Saint Cyr. The Anti-Aircraft Artillery Training Centre at St. Nazaire accepted the first Polish recruits of the 500 expected and PAF cadets were sent for basic military training at the Cadets' School at Coëtquidan. A Polish Staff Flight (Caudron C.635 Simoun aircraft) was also established at Orly to provide liaison/courier services for the PAF Command.

Although the Poles were cordially welcomed at the French centres, their training was regarded by the PAF Command as totally inadequate, particularly that of fighter pilots, who were largely misused. Most of them were confined to routine flights on the obsolete D.500/510 fighters and were given no opportunity to improve their air combat and tactical skills. In view of this efforts were made to concentrate all training of flying personnel at the Lyon CWL Centre and no further qualified Polish pilots were sent to the Montpellier fighter school.

Training facilities at the Lyon base were expanded. According to the CWL report of 12 April 1940, the in-flight training for less advanced pilots was then conducted in the Training Squadron, which had an average availability of 6 Romano Ro 82 biplanes and 2 Caudron C.635 Simoun cabin monoplanes. The course usually consisted of six 1- to 1.5-hour flights on the Ro 82 and eight 1-hour flights on the C.635 Simoun per pilot, with 109 pilots already trained on the Ro 82s and 87 on the Simouns, and a further 12 in the course of training on the Ro 82s and 14 on the Simouns. From the Training Squadron the pilots graduated either to the Fighter Training Squadron or to the Observation (Close Reconnaissance) Training Squadron.

The Fighter Training Squadron had on average 9 serviceable Morane MS 406 fighters and 3 unarmed Caudron CR 714

Polish mechanics and French instructors grouped in front of a LeO 20-BN5 bomber trainer at the Rennes training centre at St. Jacques de la Laude in April 1940. (IPMS)

A damaged Caudron C.635M Simoun cabin monoplane. The Simouns were used by the PAF in France as liaison and training aircraft. (P. Riviere/K. Chołoniewski)

Cyclone lightweight fighters at its disposal. Pilots were required to make eight 2-hour flights on the Caudrons and 20 to 25 flights, totaling about 12 hours, on the Moranes. On 12 April, 59 were undergoing training in the Squadron: 24 on Caudrons and 35 in the final stages of training on Moranes. With the completion of this course the pilots were considered ready for service either with the Fighter Exercise Squadron or with any other semi-operational or operational unit.

The Observation Training Squadron had on average the use of two Potez 39 high-wing monoplanes, one Caudron C.445 Goeland twin-engined trainer and one Potez 63.11. The training program comprised 20 flights, totaling about five hours, on the Potez 39, a similar amount of flying on the Goeland, and 20 to 25 flights, totaling about 10 hours, on the Potez 63.11. On 12 April 34 were in training, with several already flying the Potez 63.11.

However, Pawlikowski's hopes that the remaining pilots (for one squadron) needed to complete the establishment of Mümler's 2nd Poznań Fighter Wing – which already had six separate flights (the "Montpellier Squadron") at the front – would be quickly trained, were not realized because of different priorities imposed upon the CWL by the French.

In December 1939 France decided to send an expeditionary force to Finland, to assist in its defence against the Soviet invasion, and appealed for volunteers. The Polish Government in Exile briefly considered the prospects of organizing a Polish fighter unit in Finland, possibly armed with British aircraft, as one means of more quickly reviving the PAF's strength. Over 100 Polish airmen volunteered to go, but investigations showed that neither sufficient combat equipment nor training and support facilities would be available in Finland, and the idea – for which there was not much enthusiasm in general – was soon abandoned.[21] However, towards the end of January 1940, after discussions with the French, the Polish Government eventually agreed that one Polish two-squadron fighter wing, fully trained and equipped in France, would be sent to Finland as part of the Polish contribution to the French expeditionary force, which was to also include one Polish infantry brigade. The 3rd Fighter Wing under Mjr Józef Kępiński, was chosen and in addition to its C.O. and deputy C.O., Cpt Piotr Łaguna, 33 pilots from the former Pursuit Brigade and the Dęblin Aviation Training Centre, one technical officer and 29 mechanics and ground personnel were selected to provide the cadre of the unit, which became known as the "Finnish Wing." It was hoped that additional personnel, bringing the wing to its full strength, would be provided by Polish airmen still in Sweden and possibly also by Polish escapees from Lithuania and Latvia. Col Klemens Rudnicki in Stockholm was asked to implement these proposals and to open negotiations with the Finnish Government, to ensure that the Polish wing be regarded as part of the Polish Armed Forces and would preserve its national character, although its tactical use would be decided by the Finnish High Command.[22]

The French attached great importance to the formation of the "Finnish Wing." Mjr Lionel de Marmier became its French supervisor and liaison officer and the Lyon CWL Training Centre was requested to give top priority to its organization and training. From the outset the French wished to equip the Wing with the Caudron CR 714 Cyclone lightweight fighters, although the imported Dutch Koolhoven FK 58 fighters were also briefly considered as a possible alternative. In the meantime, the Wing began training on the available MS 406s.

The Caudron CR 714 Cyclone lightweight fighter, built mainly of wood and armed with four wing-mounted 7.5-mm MAC M.39 guns, was powered by the 450 hp Renault 12R 01 or 03 twelve-cylinder aircooled inverted Vee engine. The aircraft entered production in the summer of 1939, the initial order for 20[23] being followed by two orders for 100 each. The main advantage of the CR 714 was its cheapness and ease of manufacture, but its overall performance was generally unimpressive, and very inadequate rate of climb and poor manoeuverability eventually led to its rejection by the Armee de l'Air, which decided to retain only 20 machines for development, evaluation and training purposes. The second order for 100 was canceled and of the remaining 100 aircraft 80 were to be sent to Finland and 20 exported to Yugoslavia. The first production CR 714s were ready in January 1940 and the following month six were despatched to Finland and five, not yet armed and fully equipped, were delivered to the CWL at Lyon-Bron for evaluation trials, where the Poles began to familiarize themselves with the aircraft. Later three more CR 714s were delivered to Lyon.

On 12 March, 1940, the Finnish-Soviet war came to an end, as did all Allied plans to help the Finns. The "Finnish Wing" was no longer needed and the French wanted to disband it. The Polish High Command strongly resisted and proposed that the unit should provide the basis for a Polish fighter wing, which – as the first such PAF unit to come into being – should be renamed the Ist Polish Fighter Wing of Warsaw. After a personal intervention by Gen Sikorski, the French Chief

[21] Gen Zając's letter to Col Rudnicki on "Organization of the Polish Aviation in Finland" of 12 February 1940, Ldz.326/Tj./Org./40 (IPMS, Lot.A IV,2/2a.

[22] Ibid.

[23] Serial numbers I-191 to I-210 (c/ns 8533 to 8552 and CR 714 airframes numbers 1 to 20 respectively). The first batch of 44 aircraft from the first 100 order carried serials I-211 to I-255 (cns 8553 to 8597, airframe numbers 21 to 65).

of Air Staff agreed, giving the unit the French number 145, and his order (DSP Nr.15/4) of 3 April, followed by the French Air Ministry order (EMAA Nr. 1.115/I/IS) of 6 April, authorized formation of the I/145 Groupe de Chasse Polonaise de Varsovie (Warsaw Fighter Wing).

The Wing was promised a full complement of the Bloch MB 152 fighters, but, as a temporary measure, received 20 MS 406s for training. The French Air Ministry order of 6 April stipulated that the Wing's two squadrons, when ready, should be separately attached to the French MB 152 fighter wings as third squadrons to gain combat experience; the Wing's command, ground personnel and its MS 406 fighter trainers were to remain at the Lyon-Bron Base. This was opposed by the PAF Command, anxious to preserve the unity of the Polish Wing and resulted in an impasse, because French units had priority in deliveries of the MB 152 fighters.

On 10 May, at the start of the German offensive, the Armee de l'Air took over 38 surplus, non-operational Caudron CR 714 fighters then in stock[24] as an emergency measure and two days later announced that these aircraft would be allotted to the I/145 Warsaw Fighter Wing – to cool Polish impatience to get into the battle – until MB 152s became available.

Earlier, in connection with the preparation of Polish troops for the Norwegian expeditionary force, the formation of a "Norwegian Wing" was briefly considered, but soon abandoned as the situation developed. This also affected PAF planning and the CWL organization. With the establishment of the I/145 Warsaw Fighter Wing (originally the Polish 3rd Fighter Wing), Pamuła's Fighter Wing (originally the Polish 1st Fighter Wing) became the 3rd Fighter Wing. At mid-April 1940 the PAF combat units in process of formation were as follows:

- 2nd Poznań Fighter Wing (Mümler), also sometimes referred to by the names of Kraków-Poznań Fighter Wing or Ławica Fighter Wing, with the 1st "Ravens" Squadron (intended to embody pilots from the so-called "Montpellier Squadron") and the 2nd (unnamed) Squadron;
- 3rd (unnamed) Fighter Wing (Pamuła), with the 1st "T.Kościuszko's" Squadron and the 2nd "Salamanders" Squadron;
- 4th Okęcie Fighter Wing (Wyrwicki), with the 1st "Eagle-owls" Squadron and the 2nd "Swallows" Squadron;
- Observation (Close Reconnaissance) Wing (Młynarski) with two unnamed squadrons.

[24] With eight CR 714s used for training at Lyon-Bron, the Armee de l'Air had then 46 Caudron fighters on charge. In all about 80 CR 714s had been completed when France collapsed and most of the remaining 40 incomplete Caudrons were captured by the Germans at the Guyancourt factory.

One of the first Caudron CR 714-C1 Cyclone lightweight fighters, delivered to the CWL at Lyon-Bron in February 1940, being examined by a Polish airman. (T.E. Willis)

4. The PAF in Combat over France

With the end of the Polish September campaign in the East and with no further operations by the Western Powers after the midget Operation "Saar," Hitler was certain that peace in Europe would be negotiated within a few months, if not weeks. Apart from action at sea and skirmishes in the air over the Western front, hardly a shot had been fired on the ground and the Germans began to call this the Sitzkrieg ("sitting war"), the French, the dróle de guerre ("funny war") and the British, the "phoney war." Hitler's peace overtures, combined with the effect of a very hard winter on air activity, led the French to the naive belief that the Germans were reluctant to attack for fear of the French strength. Even the unexpected German in-

Some of the 18 Morane-Saulnier MS 406-C1 fighters delivered to the Lyon-Bron base for the 1st "Ravens" Squadron (familiarly known as the "Montpellier Squadron") of the 2nd (II/8) Poznań Fighter Wing. (IPMS)

Gen Sikorski, with Gen Denain on his left, begins an inspection of the Polish Lyon-Bron base on 27 March 1940.

The French welcoming flights from six fighter wings, which came to Lyon-Bron on 27 March 1940 to be joined by Polish flights assigned to wings, reviewed by Gen Sikorski. (IPMS)

vasion of Denmark and Norway on 9 April 1940 was taken by many Frenchmen as diversionary tactics, confirming this. Notwithstanding the Polish September experience, the French refused to take effective measures to protect their country in general and the air force in particular against a surprise air strike, cherishing the notion that the Germans would not dare to launch such an attack against them. The black-out was hardly observed. French aircraft stood in the open on their peacetime airfields, with no attempt at dispersal or camouflage. Most pilots were quartered in private accommodations in towns and – except for the duty stand-by flights – turned up at their bases only at breakfast time; no anti-aircraft gun emplacements were manned before then.

The impatience of Polish airmen to go to action was not understood and the French showed no urgency in helping the PAF units to achieve operational status. The first to be sent to the front were the six three-aircraft flights (the so-called "Montpellier Squadron") composed of the pilots who had completed training at the Montpellier Fighter Aviation Training Centre and who had been waiting for weeks at the Lyon-Bron base for their operational aircraft. The flights, each to be attached separately to French Fighter Wings for about two months to gain experience in French tactical methods, were to provide the first half – 1st "Ravens" Squadron – of the 2nd (II/8) Poznań Fighter Wing.

In the middle of March 1940 18 Marane-Saulaier MS 406 fighters arrived at the Polish Lyon-Bron base. There the French cockades on the aircraft fuselage sides were overpainted with the PAF regulation white-and-red chessboards – the rest of the French markings left intact – and each of the pilots completed 15 hours of familiarization flying on his machine. The flights were fully established, each having Polish ground crews of 11 to 12 men .

On 27 March Gen Sikorski arrived at the Lyon-Bron base to declare the Polish CWL Training Centre officially open, and to inspect the six Polish combat flights and pronounce them "operational." Welcoming flights from each of the six French fighter wings to which the Poles were allocated also visited the base and lined up opposite the Polish Moranes to take part in the parade and then to escort their Polish comrades in arms to their combat airfields. The Poles and the Frenchmen exchanged their pilot badges and the day ended with an impressive flypast by the fighters. However, because of poor weather conditions, departure of the flights to their appointed airfields was not fully completed until 29 March.

The Polish flights, numbered 1 to 6 and designated by the first letters of their C.O.s' names, were allocated to French fighter units as follows:

- No.1 "Ła" Flight: Cpt Stefan Łaszkiewicz, Lt Stefan Zantara, Sgt Mjr Leopold Flanek, to 1st Squadron GC III/2 at Cambrai;
- No.2 "Pe" Flight: Cpt Jan Pentz, S-Lt Bogdan Anders, S-Lt Włodzimierz Karwowski, to 4th "Jeanne d' Arc"

Left: Gen Sikorski hoists the Polish Air Force flag at Lyon-Bron during the ceremonial opening of the CWL Training Centre on 27 March 1940. The Centre's Potez 63.11 '2', c/n 132, of the organizing Polish reconnaissance wing, is standing next to the flag-staff. (IPMS) Right: Polish airmen on parade at the CWL Centre on 27 March 1940, with the Potez 63.11 '2', flanked by MS 406 and CR 714 fighters, in the background.

Morane fighters of the six Polish front-line flights, facing Moranes in the welcoming French flights at Lyon-Bron on 27 March 1940. The nearest aircraft, c/n 1031, assigned to No.4 "Bu" Flight, has the Bronzavia flame-damper shrouds attached to its exhaust ports (these were later removed).

Squadron GC II/6 Anglure-Vornarce;
- No.3 "Su" Flight: Cpt Mieczysław Sulerzycki, S-Lt Erwin Kawnik, S-Lt Bolesław Rychlicki, to 2nd Squadron GC III/6 at Vez-Thuizy near Reims;
- No.4 "Bu" Flight: Lt Kazimierz Bursztyn, S-Lt Władysław Chciuk, S-Lt Władysław Gnyś, to GC III/1 at Toul Croix de Metz;
- No.5 "Br" Flight: Lt Józef Brzeziński, S-Lt Stanisław Chałupá, NCO Antoni Beda, to 1st Squadron GC I/2 at Xaffevilley;
- No.6 "Go" Flight: Lt Władysław Goettel, S-Lt Wacław Król, NCO Eugeniusz Nowakiewicz (MS 406 c/ns 953, 956, 959) to 3rd Squadron GC II/7 at Luxeuil.

The Polish front-line flights had initially a rather uneventful life. Their activities consisted mainly of routine patrols over the border sectors, emergency scrambles to chase away German reconnaissance aircraft operating in the locality of the airfields and escort missions to protect French recce. or bomber aircraft making shallow excursions into Germany. Some skirmishes were fought with enemy fighters and bombers, but these were infrequent and generally inconclusive. The records of No.6 "Go" Flight give an indication of the modest activity during this period: in 24 days, from 7 to 30 April, the Flight logged 50 hours 55 min on operational sorties.

S-Lt Wacław Król, of No.6 "Go" Flight, receives French "wings" from a French lieutenant of GC II/7 during the exchange of badges between French and Polish pilots at Lyon-Bron on 27 March 1940. (IPMS)

Apart from the six front-line flights, one more Polish unit was considered by the Poles to have achieved operational readiness in March. This was the Fighter Exercise Squadron, with 11 MS 406 fighters, which was to protect the Lyon airfields. It was not engaged in combat until the German offensive began.

The "funny war" came to an abrupt end at dawn on 10 May 1940 with the surprise German offensive against France and the Low Countries. Heavy air strikes on unsuspecting Allied air bases were aimed at assuring the Luftwaffe immediate air supremacy.

The Luftwaffe used for the attack (Operation "Fall Gelb") Luftflotten 2 and 3, with an available first-line strength of 1,120 Do 17, He 111 and Ju 88 bombers, 342 Ju 87 dive-bombers, 42 Hs 123 attack aircraft, a 1,016 Bf 109 fighters and 248 Bf 110 heavy fighters – a total of 2,768 serviceable warplanes. These were supplemented by 591 Do 17 long-range and Hs 126 close reconnaissance aircraft and 154 naval aircraft, bringing the total to 3,513. Paratroops, detailed to seize fortifications, bridges and other objects of strategic importance had 401 Ju 52 transports and 45 DFS 230 troop gliders at their disposal.

This mighty force was opposed by 177 Belgian and 124 Dutch combat aircraft, most of them totally obsolete, about 1,500 French first-line aircraft, largely obsolete, and 25 squadrons (eight Battle, six Blenheim, five Lysander and six Hurricane) with 408 aircraft of the British Air Force in France, a total of about 2,210 machines, of which only about 1,700 were serviceable, and only 760 of them fighters. Within hours of the German offensive the Belgian and Dutch air forces virtually ceased to exist. Only a handful of the Dutch Fokker D.XXI fighters survived to the next day and continued to fight the Germans with great courage until 14 May.

On 10 May the French fighter force in Metropolitan France comprised 24 combat-ready Groupes de Chasse (GC) or fighter wings – 12 Morane MS 406, seven Bloch MB 151 & 152, four Curtiss Hawk 75A and one Dewoitine D.520 – with a total of 552 serviceable aircraft. Of these 19 were distributed among the five (the 22nd, 23rd, 24th and 25th) Fighter Groupings, four at the front, covering the entire Belgian and German frontier area and one (the 21st) defending Paris and the lower Seine region; one remained in Southern France (at Néant) and four were allocated to the Alps Command, to guard against a possible Italian attack.

France, like Poland, had not defined a clear role for its air force. Its fighters were widely dispersed, their combat efficiency being affected by a confusing dual command system – comprising the land armies' commands and front aviation commands – which resulted in contradictory orders, a fragmented and largely ineffective early warning system which collapsed in the first days of the German offensive, and a lack of reserve combat airfields. These factors made concentration of the fighter effort in specific combat areas difficult to achieve and prevented coordinated fighter defence of the country as a whole.

At the beginning of the Luftwaffe's onslaught the PAF in France had 6,971 airmen on active service, 3,304 of them at the Lyon-Bron Aviation Training Centre. This total included 2,205 flying personnel, of whom 1,599 were pilots and trainee-pilots. But only 18 Polish pilots and 69 ground-staff were then serving with the combat squadrons at the front (manning the six MS 406 flights of the "Montpellier Squadron"), and 23 Polish officers were attached to the French operational units for command experience. In addition 16 Polish pilots of the Lyon Fighter Exercise Squadron (on MS 406s) flew operationally on their own initiative, as did the I/145 Warsaw Fighter Wing, which set up stand by flights after its airfield had been bombed on the opening day of the offensive.

The PAF state of readiness was then as follows:[25]

already formed:
- six separate fighter flights (the so-called "Montpellier Squadron"), gaining combat experience at the front with French units and intended to provide the backbone of the PAF's 2nd Fighter Wing;
- I/145 Warsaw Fighter Wing, under Mjr Kępiński, fully organized, but not yet equipped with its specified combat aircraft, awaiting full formation and equipment;
- 2nd Kraków-Poznań Fighter Wing, which received the designation II/8, under Mjr Mümler, with training on MS 406 fighters completed, awaiting final organization and combat equipment at the Lyon CWL Training Centre;
- 1st Close Reconnaissance Wing, under Mjr Młynarski, with training on Potez 63 aircraft completed, awaiting final organization and combat equipment at the CWL,

in process of training at the CWL;
- 3rd and 4th Fighter Wings, respectively under S-Col Pamuła and Mjr Wyrwicki undergoing in-flight training;
- 2nd Close Reconnaissance Wing, in the early organization stages, undergoing ground training.

MS 406-C1 '3' of the I/145 Warsaw Fighter Wing at Lyon-Bron in May 1940, with one of the Centre's three Bloch MB131-RB4 recce-bombers taking off in the background.

With a rapid increase in the intensity of the battle and mounting losses, the Armée di l'Air grew desperately short of pilots, but still resisted the formation of all-Polish fighter wings. It asked instead for more separate Polish fighter flights to reinforce the French units, either as individual front-line flights strengthening the French Wings (GC) at the front, or as local defence flights with a new category of auxiliary units: the Home Defence Fighter Squadrons (L'Escadrilles de Chasse de Defense, or ECD), usually with nine aircraft each, and the rather smaller Light Defence Squadrons (L'Escadrilles Legeres de Defense, or ELD). The ELDs, also frequently referred to as the Patrouilles de Protection (Protection Patrols), had three to seven aircraft each and were entrusted with the protection of specific targets, usually factories, hence their colloquial name "Chimney Flight."

The PAF Command reluctantly met these requests as a temporary measure, enabling more Polish pilots to get combat experience, and the necessary steps to form the required units were rapidly put in hand at the Lyon CWL Training Centre. With these units envisaged as the foundation for new all-Polish squadrons and wings, the PAF formed them as six-pilot double flights whenever the circumstances permitted. Attempts were also made – with some success – to obtain postings for the new units to the French wings which already incorporated Polish flights.

The forming units were initially given provisional team numbers.[26] Once they were established, their provisional for-

[25] Based on PAF Command's report for Gen Sikorski; L.dz.1505/tjn/40, dated 14 May 1940 (IPMS, Lot.A IV,1/1a).

[26] Available PAF records indicate that this practice had ended at the end of May 1940 with Team No.10.

PAF OPERATIONAL UNITS IN FRANCE, MARCH-JUNE 1940
(initial operational airfields only)

mation team numbers were replaced by new designations. The new front-line flights were identified by consecutive numbers starting with No.7 (since the Nos. 1 to 6 were already allocated to the six "Montpellier" Flights at the front) and by the first letters of their C.O.s' names, but the situation soon became confused when one home defence two-pilot team (No.9 "Ko" Flight) was also included in the numerical sequence, and after No.10 "Ga" Flight the numbering of the front-line flights seemed to have been abandoned. Only the identification by letters of the C.O.s' names was retained.

The Polish "Chimney Flights" were initially given consecutive Roman numbers beginning with I. and the customary designation letters from the names of their Commanders, but later the Flights became known mainly by the names of the places which they were protecting. Records indicate that at least two Polish fighter pilots served singly with French fighter units.

Details of combat activities of all PAF units in France until hostilities ceased are given below, with a summary of their war effort in Appendix 7. For the record, the provisional team numbers are added in brackets after the units' designations where applicable.

A. The All-Polish Fighter Units

(1.) I/145 Warsaw Fighter Wing[27]

On 10 May the Wing, organized on the lines of French fighter wings as a two-squadron unit, had in its strength 32 pilots – 22 officers and 10 NCOs (the proportion of officers to NCOs being higher than in the corresponding French wings) – with supporting ground staff of five officers and 184 men under the command of Cpt Adam Jaworski. A liaison staff of two officers and three NCOs was included in the above figures. In addition, Mjr Lionel de Marmier was attached to the Wing as its French supervisory and advisory officer.

The flying personnel comprised:

- Wing's C.O.: Mjr Józef Kępiński, Deputy C.O.: Cpt Piotr Łaguna;
- Squadrons' C.O.s: 1st Squadron: Cpt Antoni Wczelik, 2nd Squadron: Cpt Juliusz Frey;
- Lts: Tadeusz Czerwiński, Robert Janota, Julian Kowalski, Jan Obuchowski, Wacław Wilczewski, Zdzisław Zadroziński;
- S-Lts: Jerzy Czerniak, Eugeniusz Fiedorczuk, Bolesław Gładych, Czesław Główczyński, Jerzy Godlewski, Leon Jaugsch, Lech Lachowicki-Czechowicz, Witold Łanowski, Marian Łukaszewicz, Bolesław Skibiński, Marian Szalewicz, Aleksiej Żukowski;
- NCO: Antoni Markiewicz, Andrzej Niewiara, Jan Palak, Mieczysław Parafiński, Edward Paterek, Lucjan Szempliński, Edward Uchto, Ernest Watolski, Piotr Zaniewski, Jerzy Zieliński.

The Wing had at its disposal 20 MS 406s and completed training on these aircraft, but was not yet declared operational pending the promised re-equipment with MS 152 fighters. During the German bombardment of Lyon-Bron at dawn on 10 May 12 Polish airmen were killed, 16 wounded and extensive damage caused, and the Wing immediately established three- aircraft stand-by flights on its MS 406 fighters for protection of the town and its airfields. Three take-offs were made during the day in response to air raid warnings and nine aircraft sorties were flown, but no Germans were encountered. In the evening the Wing moved to the Mions airfield near Lyon.

During the period 11 to 18 May the Wing operated from Mions on its own initiative and continued to provide air cover for Lyon and its airfields, taking-off to the sound of the Lyon raid-warning sirens. Mobile "Sadir" radio stations were put at its disposal, but unreliable R/T sets on Morane fighters made communications very erratic and radio messages were backed up by the traditional Polish method of ground signaling sheets, which, because of lack of other suitable fabric, had to be sewn up from spare white items in airmen's kits.

Cpl Edward Uchto, I/145 Warsaw Fighter Wing, depicted by MS 406-C1 '4', c/n 901, which was usually flown by S-Lt Czesław Główczyński. Uchto was killed in action on 9 June 1940 (C. Główczyński)

All aircraft were well dispersed and carefully camouflaged with tree branches. Renewed French attempts to split the Wing were successfully resisted, but due to a shortage of Bloch fighters the French decided to equip it temporarily with the Caudrons. On 14 May a few selected Polish pilots were ordered to go to Villacoublay to collect the first five CR 714s, which they flew to Mions on 15 and 16 May. By 18 May the Wing had carried out six operations from Mions, a total of 16 sorties in 19 hours flying time – all on Moranes – but with no result.

On 18 May in the evening the Wing left Mions for Villacoublay airfield to be fully armed with the CR 714 fighters. The move was made by rail and road transport, with four pilots flying over on the Caudrons.[28] Lt Janota was left behind and became involved in a formation of one of the new Chimney Flights (No.7 Team, the IX Chimney Flight-Angers). On 20 May the Wing took on charge a further 16 Caudrons and became fully operational. Its Daily Order No.1 of the same date allotted nine of these to 1st Squadron and seven to 2nd Squadron.[29]

Between 22 May and 2 June the last 18 Caudrons were received and distributed between the Wing's Command and its two squadrons,[30] bringing the total delivered to 39. Four CR 714s were written off in landing accidents at Villacoublay, including I-230 and I-238 which collided on the ground on 25

[27] Main sources: Wing's official records, Daily Orders Nos.1 to 27 (20 May-17 June 1940) and Cpt Łaguna's reports (IPMS, Lot.A IV, 2/2f and g files).

[28] The fifth CR 714 (then presumably unserviceable) was to follow, but it crashed at Bron on 19 May, killing its pilot, S-Lt Witold Dobrzyński from the Fighter Training Squadron.

[29] The following examples were received by 1st Squadron – serial number, c/n (and, in brackets, CR 714 consecutive airframe number): I-224,8566(34), -226,8568(36), -230,8572(40), -233,8575(43), -234,8576(44), -235,8577(45), -236,8578(46), -238,8580(48), -240,8582(50); and by 2nd Squadron: I-201,8543(11), -209,8551(19), -213,8555(23), -214,8556(24), -222,8564(32), -229,8571(39), -241,8583(51). The first five included I-198,8540(8), -202,8544(12), -204,8546(14), -217,8559(27) -the identity of the remaining one being obscure.

[30] The 18 examples were allocated as follows – the Wing's Command: I-223,8565(33) to the C.O., -212,8554(22) to the Deputy C.O.; -1st Squadron: I-191,8533(1), -207,8549(17), -208,8550(18), -227,8569(37), -237,8579(47) -246,8588(56); -2nd Squadron: I-200,8542(10), -210,8552(20), -215,8557(25), -220,8562(30), -225,8567(35), -228,8570(38), -231,8573(41), -242,8584(52), -243,8585(53), -244,8586,(54) (as per Daily Orders Nos.3, 6, 8, 10, and 13).

The CR 714 Cyclone I-222 '13', c/n 8564, of the 2nd Squadron, I/145 Warsaw Fighter Wing, at Villacoublay.

The CR 714 Cyclone I-204 '2', c/n 8546, of the 2nd Squadron, I/145 Warsaw Fighter Wing, wrecked on landing at Villacoublay by S-Lt Fiedorczuk.

May. On 23 May Col Stefan Pawlikowski joined the Wing to assist in its operations as adviser.

At Villacoublay the Wing was put at the disposal of the 23rd Fighter Grouping, its operational tasks being defence of the airfield and the southern approaches to Paris. Having no radio station, the Wing scrambled its aircraft in response to telephone messages from the French GC I/4 Wing stationed on the same airfield.

However, the Caudrons proved extremely unreliable and were plagued by numerous defects. The airscrew's variable-pitch gear suffered frequent failures, the flimsy engine covers buckled in a dive, flight instruments and R/T radios went dead after violent manoeuvers, oil and petrol leakages were often experienced, the undercarriages were fragile and their release mechanisms jammed. Although the manufacturer sent its own engineers to iron out these problems, a large proportion of the fighters remained unserviceable and on 25 May the French Air Minister Guy le Chambre inspected the Wing and its Caudrons and in view of the faults suspended further use of the aircraft on the spot. Since no replacement machines were promised and the ministerial decision would have meant grounding of the Wing, the Poles protested and resolved to continue to fly the Caudrons at their own risk.

The French Air Staff tacitly accepted the situation and two days later agreed to make the Wing officially operational on completion of its equipment with the CR 714s, when it was to be transferred to the 21st Fighter Grouping, charged with the defence of Paris. On 30 May a decision was taken to move I/145 to Dreux, where it was to be installed and officially declared operational on 2 June. By then the Wing had already flown 23 operational missions from Villacoublay, logging 86 aircraft sorties in 92 flying hours, but without contact with the enemy.

The order to move to the Maison la Blanche airfield at Dreux, west of Paris, was received by the Wing on 2 June at 13:20 hours. It read "1. -If Paris is attacked, I/145 shall provide cover to Villacoublay for half-an-hour. 2. -After that, it will land at Dreux. 3. -During take-off of the first squadron, the second shall be put on alert. 4. -The second squadron shall then provide the cover, carrying on with the mission and proceeding as at One. 5. -In any event, the flying detachment shall depart for Dreux by 18:00 hours."[31]

At 16:00 hours the Wing took off for Dreux and upon landing was welcomed by a departing RAF contingent of No.59 (Blenheim) Squadron. The Poles found the airfield well defended by anti-aircraft artillery and machine-gun emplacements, but generally in a terrible mess and with no means of liaison with the air warning network. The civilian engineers moved with the Wing to keep the Caudrons flying, but at the time of the move only 18 fighters were combat-worthy out of 34 on the unit's inventory. The Polish mechanics worked on the aircraft throughout the night and by the morning of 3 June 28 were ready for combat. Orders for that day requested the Wing to keep a duty flight on stand-by from 09:00 hours onwards, and then one squadron from 13:00 hours onwards. All machines were well dispersed and camouflaged on the airfield's perimeter and the duty flight stood in readiness with engines running.

At 13:10 hours the alert flight, led by Mjr de Marmier, with Czerwiński and Żukowski as his wingmen, was ordered to take off to patrol the Dreux area and look out for enemy bombers operating near Paris. During the sortie Czerwiński spotted three He 111s below him, waved his wings and dived on the bombers. Żukowski followed suit, but de Marmier did not notice their manoeuver and flew on. The Poles claimed to have shot down two Heinkels in the Etampes area, but this was never confirmed.

4 June was a quiet day. The Wing was provided with a mobile radio station, which enabled it to receive warnings from the air defence center at the Eiffel tower in Paris. With the German armor launching a massive assault on the Somme and the Aisne rivers, 5 June became a critical day. At 05:25 hours the Wing was ordered to send one squadron to patrol at 06:00 hours over Rouen at 4,000 m, to provide help for the exhausted GC II/10 – this task being carried out by 2nd Squadron. At 06:30 I/145 was requested to dispatch its other squadron – 1st Squadron, with 11 serviceable aircraft – to the Bretigny airfield, a few miles south of Paris, to protect the city. Similar, inconclusive missions were flown by both Squadrons on the next day and in the evening 1st Squadron returned to Dreux. During reorganization of the fighter force on that day I/145

[31] Edouard Mihaly, *Journal de Marche du GC I/145, L'Album du fanatique de l'Aviation*, no.34, p.7.

Wing was assigned at 23:30 hours to the newly established 42nd Fighter Subgrouping, forming part of 110th Air Force of the 10th Land Army, and entrusted with the defence of the lower Seine between Vernon and Meulan.

On 7 June I/145 was asked to perform a rather unusual task: a double-strength reconnaissance patrol (two times three aircraft) at 6,500 m over the Meulan-Magny-Fleury-Vernon area between 15:40 and 16:35 hours, and two double-strength reconnaissance patrols in succession over Dreux and along the Seine river between 19:00 and 21:15 hous.

The French expected heavy enemy bomber activity on 8 June and requested the whole I/145 Wing to stand by and be ready to join forces with GC I/4, II/10 and III/10 for a major fighter operation, when called upon. At 10:00 hours 1st Squadron was ordered to move to Bernay, to be at the disposal of GC II/10. The anticipated German air attack did not take place, but the Poles flew two five-aircraft patrols. The first of these, carried out between 13:30 and 14:30, was uneventful. The second, between 15:54 and 17:10 hours, was led by Cpt Wczelik, with Lt Czerwiński, S-Lt Godlewski, S-Lt Żukowski and Cpl Zaniewski as his wingman, and covered the Vernon-Meulan section of the Seine river. The pilots noticed to the North-East, in the direction of Rouen, anti-aircraft artillery bursts and flying towards them at 4,000 m sighted about 15-20 Messerschmitt Bf 110s below. The Poles dived on the Germans and after a brief fight Wczelik and Czerwiński claimed one Bf 110 each, and these two claims were put forward to the PAF High Command for approval.[32] Subsequent reports of French officers on the ground indicated that in fact five Bf 110s crashed at that time in woods north of Rouen and eventually, after a further scrutiny of pilots' combat reports, Czerwiński had two Bf 110s confirmed, Wczelik and Żukowski one each, and Godlewski and Zaniewski, 1/2 each.[33] This was the first victorious fight of I/145 Wing. One Caudron was damaged.

The Wing's Daily Order No.20 from 8 June makes a mysterious reference to "the aircraft type 39, no.57," received on that day and put at the disposal of Mjr de Marmier. There is no further mention of this machine in any of the unit's subsequent records. Was this the Arsenal VG 39 prototype (perhaps temporarily allocated to I/145 for operational trials)?

9 June saw the Wing's most intensive combat activity and heaviest fighting. At about 05:00 hours most aircraft from 2nd Squadron, led by Mjr Kępiński, took off for Bernay to join 1st Squadron. Frey's fighter, grounded by engine difficulties and another unserviceable had to remain at Dreux. Three defensive patrols were flown over the same section of the Seine as before between 06:30 and 07:30 hours, and one patrol was dispatched to protect the area North of Dreux between 06:40 and 07:45 hours, but no enemy was sighted.

A Polish pilot resting beside his CR 714 Cyclone fighter at one of the operational airfields.

At 14:30 hours Wing's 17 aircraft were scrambled from Bernay and formed over Evreux for a fighter sweep of the Seine valley between Meulan and Rouen, with Kępiński, de Marmier, Łaguna and Wczelik participating in the sortie. A large formation of about 50 Dorniers escorted by about 20 Messerschmitt Bf 109 fighters was encountered over Vernon. Because of malfunctioning radios, the Polish attack was badly coordinated, with every pilot for himself. Whilst Wczelik's flight attacked the Dorniers, a number of individual duels were fought with the Bf 109s. Three Bf 109s and one Dornier were seen to crash, two south of Andelys and two near Louviers, and one Do 17 and one Bf 109 escaped trailing smoke. The victories were credited as follows: Główczyński-one Bf 109 destroyed and one Bf 109 and one Do 17 as probables; Czerniak – one Bf 109 destroyed; Parafiński – Bf 109 destroyed; Wczelik, Kowalski and Markiewicz 1/3 Do 17 each.[34] The Wing suffered heavy losses: Lachowicki-Czechowicz, Obuchowski and Uchto failed to return and neither they nor their aircraft were ever found. Three more Caudrons were lost in forced landings, one of them at Evreux, but their pilots, Czerniak, Godlewski and Skibiński, survived, although Godlewski failed to rejoin the Wing. Kowalski, wounded in the right arm, managed to successfully

S-Lt Czerniak's CR 714 Cyclone I-217 '6', c/n 8559, after a forced landing at Dreux on 9 June 1940.

[32] The claim report from the Wing's Command, L.dz.47, dated 13 June 1940 (copy: J.B. Cynk Polish Aircraft Archives, the French File). Since none of the pilots saw the Bf 110s crashing to the ground, the Kępiński's preliminary report (IPMS, Lot.A IV, 2/2c-10) referred to only one Bf 110 as "probably shot down."

[33] As listed by Col Bajan's Fighter Claim Commission, L.dz. FC/S.5/1/Air/CPLO, dated 11.07.45.

[34] Ibid.

land at Bernay in a damaged Caudron. When the exhausted Wing returned to Dreux, the day's last sortie was flown by five aircraft between 20:10 and 20:55 hours in response to an air raid warning. German bombers destroyed the Dreux railway station and departed before they could be intercepted.

10 June was also a day of heavy engagements and the last one of I/145's combat on Caudrons. In the morning the Wing had 12 serviceable CR 714s out of 13 still at its disposal. By then, of 34 fighters of this type in its strength when it arrived at Dreux, 21 had been lost: six were shot down by the enemy (three missing and three destroyed in crash landings), four were severely damaged in the fighting, two were extensively damaged on landing and nine were unserviceable because of mechanical failures. The Armée de l'Air accordingly decided to withdraw the Caudrons from operational use and rearm I/145 Wing with Bloch MB 152s.

At 08:00 hours the Wing's 12 CR 714s took off after receiving warning of a raid from the Eiffel Tower. The signal was transmitted too late and the fighters met the enemy, about 15 Dorniers protected by some 12 Bf 109s, South of Dreux at about 4,000 m, while still climbing and before they could get into position for an attack. Inoperative radios compounded the difficulties. The Poles went for the Dorniers head-on and were pounced upon by the Bf 109s from above. Heavy fighting developed and scattered duels lasted for nearly half-an-hour. Two Do 17s were destroyed North of Dreux, one each by Czerwiński and Żukowski, and one Bf 109 near Henonville, North-West of Paris, after a persistent chase by Łaguna.[35] Mjr Kępiński was attacked by Bf 109s and was wounded, but despite great loss of blood succeeded in landing his damaged Caudron wheels-up in a field. Łaguna took over the command of the Wing, nominating Cpt Wczelik as Deputy Commander and Lt Czerwiński as the C.O. of 1st Squadron in his place. The French radio broadcast an announcement that the Polish Wing deserved a special commendation for the heroic defence of Dreux against an overwhelming enemy force.

The whole Wing was scrambled for the second time at 10:40 hours, to intercept a reported Luftwaffe raid from the direction of Chartres and patrolled the Dreux area until noon. One suspect aircraft was sighted, but this turned out to be a RAF machine. The mission was the last one to be flown by I/145 Wing from the Maison la Blanche airfield.

During its eight-day operations from Dreux the Wing carried out 19 missions – 14 to intercept enemy raids and five to provide air cover for the airfield – logged 161 aircraft sorties in 188 flying hours and shot down 12 German aircraft, with two more "probables," this being the most intensive and successful period of its work.

On 11 June in the morning the Wing was ordered to move to the Surmaize airfield, 15 km South-East of Etampes, to begin conversion to the MB 152 fighters. On leaving Dreux the Wing formally handed over its 11 defective CR 714s in need of major repairs to the French Compagnie de l'Air 40/104, which promptly burned the aircraft. Its 12 remaining Caudrons flew over to Surmaize, one crashing on landing. There the Wing returned to the control of the 23rd Fighter Grouping and was entrusted with the defence of the airfield. Its three-aircraft stand-by flight took off twice on that day, each time patrolling for an hour with no contact with the enemy.

[35] Ibid.

An unserviceable Polish CR 714 Cyclone captured by the Germans.

The next day, on order from the Air Commandant of the Northern Sphere, the Wing took on charge three MB 152s[36] and began familiarization training. The order also requested a temporary posting of eight pilots from 1st Squadron to the Bloch-equipped GC I/1 and eight pilots from 2nd Squadron to GC I/8, to give them combat experience with the type. The first group, under the command of Cpt Wczelik, departed on 14 June, but the second, under Lt Wilczewski, which was to leave on 16 June, stayed with the Wing because of increasing chaos as French military resistance collapsed.

During the night of 13/14 June the Wing was withdrawn to Chateauroux and ordered to protect the airfield. The first mission flown from the new airfield was flown on 14 June between 19:30 and 20:30 hours in response to a raid warning, but no enemy was encountered.

Three take-offs were made the next day and three more on 15 June, all without result. Altogether seven missions, in as many flying hours, comprising 34 aircraft sorties, were made from Chateauroux on Blochs and Caudrons. One of the MB 152 fighters was lost in an emergency landing at Meinvillers.

On 17 June I/145 Wing was ordered to transfer to Rochefort and guard the local airfield. The move of the Wing's transport detachment over congested French roads proved to be slow and difficult. At Rochefort the Wing began to receive contradictory orders; nevertheless, by nightfall of 18 June it had still managed to carry out three fighter and one reconnaissance missions – its last ones – involving ten aircraft, the Blochs and Caudrons, in ten flying hours. The French abandoned the airfield during the night of 18/19 June and the next morning the Wing formally surrendered its remaining 11 CR 714 – according to the official records[37] seven of them still combat-worthy – and two MB 152 fighters, plus various spares, firearms and other equipment, and all the Polish aviation personnel gathered at the Rochefort airfield left for the nearby port of La Rochelle.

[36] Received from a French unit, with 25-30 flying hours each.

[37] "Index of equipment and technical materials left behind in France" (IPMS, Lot.A IV, 2/2d-12) and also "Report on the Wing's combat activity 10.V.40 to 19.VI.40" (IPMS, Lot.A IV, 2/2g-1).

On 18 June at 17:40 hours GC I/1, the French Wing to which the eight Polish pilots from I/145's 1st Squadron were posted four days earlier, flew its last combat patrol from the Rochefort airfield. The flight, led by a French pilot, Sgt Delagay, with Wczelik and Markiewicz as his wingmen, met and attacked a solitary He 111 bomber, which crashed into the sea. However, this last victorious flight of the Polish pilots in the skies of France was not rewarded: the contributions of the two pilots to the destruction of the Heinkel were never officially credited.

On 19 June at noon the Poles embarked in a French ship to be taken to England, only to disembark two hours later when its Captain received different orders. Thanks to the personal intervention of Col Pawlikowski, a British ship docked at La Rochelle at 18:00 hours to pick up the Poles and sailed the next morning for England.

Between 10 May and 18 June I/145 Wing carried out 64 operational missions, flying 322 aircraft sorties in 358 flying hours. Although its combat activities were hampered by the severe shortcomings of its Caudron fighters and a generally inefficient early-warning system, the Wing achieved 12 confirmed victories with two further "probables." Its own losses amounted to at least 11 aircraft, all CR 714s, totally destroyed or severely damaged in air battles, and three pilots reported missing and later counted as dead, one missing and two wounded.

(2.) The Fighter Exercise Squadron (GMCP)

The Fighter Exercise Squadron, under the command of Cpt Wiórkiewicz, remained in a state of self-imposed operational readiness on the Bron airfield from 3 March, to afford protection to the Lyon base and the town itself.

After the bombardment of the airfield on 10 May at dawn, of which no warning was received, the Squadron placed its 11 MS 406 fighters on stand-by. Its equipment was supplemented by a solitary Dewoitine D.501 fighter and the only combat-worthy Caudron CR 714, out of eight given to the Lyon CWL Centre for familiarization training. A few days later four more CWL Caudrons were allotted to the Squadron.

On 22 May the Squadron, with 13 pilots on its strength and 24 (from Pamuła's Wing, in the final stages of training) in reserve, became known as le Groupe de Marche de Chasse Polonaise de Lyon (GMCP or the Mobile Polish Fighter Wing of Lyon) and was attached to GC III/9, which, operating within the 24th Fighter Grouping, was charged with the defence of Lyon).[38] The GMCP was asked to protect the Lyon area, the Lyon-St. Etienne railway link and the Rhone Valley, and during May flew a number of defensive patrols without making contact with the enemy.

On 1 June at about 13:00 hours the GMCP scrambled eight MS 406s to intercept an approaching raid. Lt Gabszewicz led the main group of six fighters, with Radomski and Poniatowski following about a minute later.[39] The Moranes, still climbing, encountered nine He 111s north of Lyon and had to attack without height advantage. Gabszewicz was the first to come close enough to fire and selected one of the bombers of the last trio. The Heinkel dived to the ground, its crash being witnessed by other pilots. Moments later Barański and Radomski accounted for another He 111 and were subsequently credited with 1/2 each.

In the late afternoon the unit's five MS 406s, with Gabszewicz in the lead and Dudwał, Kalpas, Poniatowski and Rytka as wingmen, flew the day's second mission. Again the alert came too late; nine Heinkels dropped a few bombs on the airfield and the Poles found it difficult to catch up with them. Kalpas dived in desperation into the middle of the enemy formation and shot down one He 111, but was himself mortally wounded by gunners of the other bombers and his fighter spun to the ground. He was posthumously awarded the Croix de Guerre avec Palme.

In the evening Gabszewicz was given command of the newly forming front-line No.10 "Ga" Flight and three other pilots of the Wing departed with him two days later to GC III/10, to which the Flight was to be attached. At the same time Wyrwicki began to organize No.7 "Wy" Flight, which also included pilots drawn from the GMPC and was destined to reinforce GC II/10.

The Lyon-Bron base was again heavily attacked on 2 June, when three airmen were killed in the bombardment. However, because of a further drain of personnel to other units, the GMCP ceased to exist around 6 June, its remaining pilots flying to Perpignan on 19 June, to be evacuated to England. The Squadron's operational effort, including the GMCP, amounted to 50 aircraft sorties – 36 involving defence patrols and 14 interception sorties and combat with the enemy, during which three confirmed victories were achieved. One of its own pilots was killed in action.

B. The Polish Front-line Flights with the French Fighter Wings

(1.) No.1 "Ła" Flight (GC III/2)[40]

The three-aircraft "Ła" Flight, commanded by Cpt Łaszkiewicz, was one of the six flights of the "Montpellier Squadron" (1st Squadron of the 2nd Poznań Fighter Wing). The Flight began its combat activities on joining the 1st Squadron of GC III/2 at Cambrai airfield, 25 miles from the Belgian border, on 29 March, where its Polish ground personnel of 11 men under the command of Cpt Franciszek Gomuliński were already awaiting its arrival.

The Wing was given responsibility for patrolling the Valenciennes-Maubeuge section of the border, but poor weather in April severely limited air activity on both sides. During good weather the Germans flew reconnaissance sorties at extreme

[38] French sources regard 22 May as the day when the Squadron became operational.

[39] Radomski's report (IPMS, Lot.A IV, 2/1d-4)

[40] Main source: Flight's combat diary (IPMS, Lot.A IV, 2/1a-9).

altitudes of 10,000 m or more and the intercepting fighters, climbing to about 9,000 m, were unable to make contact.

On 10 May at dawn the GC III/2 Wing was surprised by a German raid, which inflicted substantial damage to the Cambrai airfield and set a number of Morane fighters on fire, including Łaszkiewicz's machine. The other two MS 406s of the Flight were damaged. A few French fighters rose to intercept the second German attack at noon, but the "Ła" Flight, with no serviceable aircraft, could not participate. Flanek's fighter was repaired by the next morning and Łaszkiewicz and Zantara went to Chateaudun to receive new Moranes.

In the first days of the German offensive the Flight operated with GC III/2 over the Namur-Liege area, giving cover and ground support to French and Belgian troops and protecting bombers over the battlefield, but the very heavy and accurate German anti-aircraft fire claimed three French fighters and some bombers. On 14 May the Wing's nine MS 406s, which included the "Ła" Flight, came to grips with 20 Bf 110s during a fighter sweep over Belgium. Flanek, after a long and persistent chase, which started at 4,000 m and ended at 1,000 m, brought down one of the Messerschmitts over Maubeuge. Two days later the Wing, operating over the Laon-Vervins area, engaged a formation of Dornier bombers flying with Bf 110 escort. In the ferocious battle Zantara, flying in a French flight, attacked and shot down a Bf 110 over Laon jointly with a Frenchman, Lt Lansoy; each pilot was credited with 1/2 of the aircraft. One of the French pilots reported the destruction of a Do 17 by Flanek, who failed to return from the sortie. A few hours later the "Ła" Flight received a message that Flanek had lost his bearings but had landed safely in Belgium. However, he never reached his unit and after the war it was found that he was killed when his aircraft crashed in the Ardennes.

On 17 May GC III/2 hastily evacuated the relentlessly bombarded Cambrai airfield because of a sudden threat from German armour spearheads and moved to Beauvais. On 20 May Beauvais was heavily attacked by 60 Luftwaffe bombers flying with strong fighter cover. Over 400 bombs fell on the airfield, causing havoc and destruction. Cpls Feliks Kolasiński and Leon Lasek were severely wounded. Zantara, strapped in the cockpit of his Morane in preparation for take-off,[41] survived the raid unscathed, but his aircraft suffered extensive damage. The next day the Wing moved to Persanne-Beaumont airfield and its Polish pilots went to Romorantin to collect two replacement MS 406 fighters. The GC III/2 task was to protect the Somme Estuary and Abbeville. With the cannon and guns of their Moranes loaded with armour-piercing ammunition, the Flight flew a ground attack sortie from Beaumont against advancing German tanks. On 31 May the Wing surrendered its Moranes and moved to Bourges, to begin conversion training to the Curtiss Hawk 75A fighters. This lasted from 1 to 6 June.

French and Polish pilots exchange their "wings" at Lyon-Bron on 27 March 1940. Cpt Rougevin-Beauville, of GC III/2, pins his badge on Cpt Łaszkiewicz, C.O. of No.1 "Ła" Flight (with S-Lt Gnyś, facing the camera, behind Łaszkiewicz). (IPMS)

On 6 June the Curtiss-equipped III/2 Wing was ordered to move to the La Ferte-s/s-Jouarre airfield, to provide fighter cover and ground support in the critical Battle of Somme and to protect Reims. On 8 June, Łaszkiewicz, patrolling his sector of the front, shot down a Hs 126 over territory already held by the Germans, at La Ferte Milon. During the next day the Flight recorded heavy Luftwaffe activity. Łaszkiewicz, flying in a French six-aircraft patrol, counted about 150 enemy aircraft and made eight inconclusive attacks on the German bombers within half-an-hour. He was eventually chased off by Bf 110s, after expending all his ammunition. Later the same day he attacked, together with two Frenchmen, a He 111 which, presumably damaged, left the formation and turned for home. Some time early in June S-Lt Jan Bury-Burzymski joined the "Ła" Flight.

Between 10 and 14 June the Poles flew over the front twice daily – each sortie averaging about two hours – and encountered large enemy formations and very strong anti-aircraft defences. Normally six to nine GC III/2 fighters were involved in these missions and a number of attacks were carried out, but these were often badly coordinated with little cooperation between the pilots and remained largely inconclusive. On 13 June 18 Curtiss Hawks of III/2, led by the Wing's C.O., Mjr Geille, made a fighter sweep over the front and came under very heavy and accurate fire from the ground. Ju 87s were sighted, but at that moment the Hawks were surprised by Bf 109s diving from the sun and had to scatter. Geille was shot down and running fights developed. Zantara, himself a target, managed to shake off his pursuers and gave chase to one of the Messerschmitts, firing long rounds until the German burst into flames and crashed behind enemy lines. Its destruction was confirmed by witnesses' accounts and eventually credited to Zantara.

German advances forced the Wing to move to Romilly-sur-Seine and then, on 14 June, to Auxerre. On that day Łaszkiewicz, his aircraft damaged by ground fire, made a forced landing at Romilly. On 15 June GC III/2 withdrew to Issoudun and the Polish pilots of the "Ła" Flight were all awarded the Croix de Guerre. The Flight patrolled the Loire River until the

[41] In his reminiscences, written down some 40 years after the event, Zantara confused the date and place of this bombardment, incorrectly recalling that it happened on 5 June at Beaumont.

Wing moved to Perpignan on 18 June, prior to an intended transfer to the Italian front.

With the capitulation of France, a large group of Curtiss fighters, the "Ła" Flight included, were ordered to fly from Perpignan to Oran in Algeria. The flight, made on 21 June, was at the limit of the Hawk's range, and Bury-Burzymski came down in the sea close to the African coast, Zantara landed with a dry tank on the beach and Łaszkiewicz turned back to Perpignan and crossed to Africa the next day, making for the nearer Maison Blanche base near Algiers. The Poles traveled to Casablanca to join a Polish transport which sailed for Gibraltar on 2 July.

Between 29 March and 17 June (both dates inclusive) No.1 "Ła" Flight carried out about 60 operational missions, flying about 110 aircraft sorties in some 140 flying hours. Its pilots were credited with 3 1/2 confirmed air victories and the Flight lost four of its MS 406 fighters, destroyed or damaged in the air or on the ground. One of its pilots, initially listed as missing, was killed.

(2.) No.2 "Pe" Flight and the "Ce" ("Kościuszko's") Flight (GC II/6)

The three-aircraft "Pe" Flight, commanded by Cpt Pentz and originally of the "Montpellier Squadron," commenced operational duties on its arrival at the GC II/6 Wing's Anglure-Vornarce airfield in the afternoon of 27 March. The Flight was incorporated in the Wing's 4th "Jeanne d'Arc" Squadron and had a Polish ground crew of one technical officer, Cpt Antoni Gärtner, and 11 men. The Poles were initially involved in a few patrols and airfield protection sorties. Between 27 April and 5 May GC II/6 temporarily operated from Dunkerque.

Heavy bombardment of the Vornarce airfield on 10 May at dawn caused a lot of disruption but little damage. The next morning the Wing was moved to a combat airfield at Maubeuge, close to the Belgian border and on that day at about 18:00 hours two French, one Czech and the Polish "Pe" flights took off to provide escort for LeO 45 bombers, whose task was to destroy a bridge at Maastricht on the Belgian-German border, 50 miles behind the front. Nine Bf 109s attacked the formation and were engaged by the French fighters. Individual duels followed and one Czech and two French pilots were shot down. Pentz evaded the Messerschmitts' fire by going into a cloud and when he came out, he saw one Bf 109 below him. His well-aimed bursts found their mark and the enemy fell to the ground in a spin. This uncorroborated victory was never credited to him.

On 12 May the Flight gave upper cover to Potez aircraft on photographic reconnaissance of the front. The following day a fighter patrol of six MS 406s came across a small formation of Heinkel bombers and Karwowski, attacking jointly with Riess, a French pilot, shot down one He 111 over the Namur area, the pilots sharing the credit for its destruction. On 14 May, as the Germans advanced, the Wing had to leave Maubeuge and moved to Vertaine airfield near Beauvais. On the 16th a massive Luftwaffe raid, arriving over Vertaine without any warning, caused tremendous destruction and demolished or set on fire some 40 Moranes. Only five machines escaped with minor damage. Sgt Mjr Józef Żak, the highly regarded chief of the Polish mechanics, was killed by a bomb. With GC II/6's equipment wiped out, the French decided to withdraw the Wing to the Bloch Deols factory airfield at Chateauroux for rearmament with MB 152 fighters.

A Bloch MB 152-C1, c/n 577, damaged in a forced landing on a training flight near Chateauroux on 18 May 1940. The aircraft was flown by S-Lt Władysław Radwański, who did not complete his training in time to join operations. (K. Chołoniewski)

Meanwhile, the PAF High Command was requested by the French to organize new Polish front-line-flights at the Lyon-Bron base. These were to be allocated separately to various French fighter wings. Around 15 May Lt Arsen Cebrzyński was instructed to form and command one of these flights – each consisting of three pilots – which became known as the "Ce" Flight. At the last moment he persuaded his superiors to allot Cprs Eugeniusz Szaposznikow and Michał Brzezowski to him, in preference to others. To continue the traditions of his former 111th Kościuszko Fighter squadron, he unofficially adopted the name of "Kościuszko's" Flight for his unit.[42]

On 17 May the "Ce" ("Kościuszko's") Flight, still without aircraft, received orders to report immediately to GC II/6, which it was to find in the Beauvais area. The three pilots went to Beauvais the next day in the early afternoon, but the exact location of the Wing proved difficult to establish. When Cebrzyński eventually tracked it down (in Vertaine), he learned of the destruction of almost all of the Wing's fighters in the recent bombardment of the airfield and was ordered to follow the unit by rail to Chateauroux. At the Chateauroux-Deols airfield, where the Flight arrived on 20 May, Cebrzyński met Cpt. Pentz and reported to him, as to a higher ranking PAF officer.

Both Polish Flights began familiarization training on MB 152 fighters and within a few days the "Ce" ("Kościuszko's") flight was attached to the GC II/6's 3rd "Cigogne" Squadron, but the French attempt to have its Polish pilots separately attached one each to Squadron flights was firmly rejected by Cebrzyński, who insisted on the unity of his flight. On 29 May Cpt Pentz fell sick and had to go to the hospital. Karwowski

[42] Cebrzyński's "Account on conduct of works of the Polish Patrol at a French fighter wing" (IPMS, Lot.A IV, 2/1a/4-3).

took over the command of the "Pe" Flight and Cebrzyński assumed overall responsibility for the Poles in the GC II/6. His efforts to obtain a replacement pilot for Pentz were unsuccessful.

On 5 June, when GC II/6 had at last received a full complement of its MB 152 fighters, the Wing was ordered to the front, back to the Anglure-Vornarce airfield. On that day two of its flights, including the "Ce" Flight, were on stand-by at Chateauroux from 13:00 hours. They were scrambled half-an-hour later and guided by ground control to intercept a Luftwaffe formation reported in the Bourges area. Cebrzyński, having a better machine, lost his wingmen during the climb and encountered about 20 Heinkels at 5,000 m South-East of Bourges. In a solitary attack he closed on the rear right-hand bomber and opened fire. The He 111 immediately released its bombs, and, after Cebrzyński's second burst, one of its engines became engulfed in flames and the crew started to bail out. The Bloch was by now the target of a concentrated fire from enemy gunners and Cebrzyński had to break off the fight. A few moments later he came across a lone Heinkel which, as it subsequently turned out, was being pursued by Lt Henneberg from the IV "He" Chimney Flight-Chateauroux, this Flight having also been despatched to deal with the raid. Both fighters kept firing at the bomber until it crashed near Vatan.

By this time the other two pilots of the "Ce" Flight caught up with the main enemy formation. Brzezowski approached first and attacked its middle aircraft. His Bloch was badly damaged while going through a wall of defensive fire and he was forced to land in a field near Issoudun. Szaposznikow repeatedly fired at the rear aircraft and after exhausting all his ammunition returned to the airfield, picked up another Bloch and pursued the Heinkels, but without success. Cebrzyński returned safely in his extensively damaged MB 152, which had to be replaced. He had the destruction of 1 1/2 He 111 eventually confirmed.[43] After the battle the indifferent attitude of the French pilots towards the Poles underwent a radical change. The heroic action of the Poles in defence of the Bloch factory was also greatly appreciated by its management, which promised to provide them with the best MB 155 as a matter of priority.

In the afternoon GC II/6, including the "Pe" Flight, left for Anglura, except for the "Ce" Flight, which had to wait until the next day for the repair or exchange of its fighters. Since it was not possible to have the PAF chessboard insignia painted on their MB 152s the pilots inscribed Polish names on the fuselages: Cebrzyński's '6' becoming "Kuma," Szaposznikow's '15'-"Iwaś," and Brzezowski's '16'-"Miś."

On 6 June in the morning Cebrzyński was scrambled from Deols to intercept a single reconnaissance Do 215, flying at 8,000 m. He fired at the Dornier with inconclusive results. In the afternoon the "Ce" Flight rejoined its Wing at Anglure. Cebrzyński flew one sortie with a French flight which provided cover for bombers operating over the Amiens area. On 7 June he flew his second mission with a French flight, which this time was participating in an 80-aircraft fighter sweep aimed at giving air cover to ground forces in the Compiegne-Soissons sector. Cebrzyński and the flight were involved in a brief fight with Dornier bombers, which immediately jettisoned their bombs and escaped. During the next day Cebrzyński led his "Ce" Flight on a ground-protection sortie over the Noyon-Compiegne area.

The two Polish Flights participated in the operations of their French squadrons independently of each other. On 9 June Karwowski, of the "Pe" Flight, and a French pilot patrolling the Compiegne front, attacked a small formation of Do 17s and jointly destroyed one of the bombers. The "Ce" Flight was engaged in a "free hunt" sortie over the Rethel area.

On 10 June the "Ce" Flight was ordered to fly an air-cover mission over the Rethel-Neufchatel front. Brzezowski's "Miś" was undergoing servicing, so the pilot took off in another Bloch, although it was spraying oil and was deemed unserviceable by the French, so that his C.O. "would not feel lonely in the air" and "the Polish Flight would be complete." On the way back the Poles met 15 Bf 109s over Reims, which, however, did not look for a fight. On 12 June both Polish Flights were to take part in escorting bombers in an attack on a German fighter airfield, but the bombers failed to appear. The Wing landed on the Pont-s-Yonne combat airfield, an ill-prepared and muddy field of clay. The following day the fighters had to fly to the Wing's next airfield at Auxerre with wheels down, because their clay-covered landing gear could not be safely retracted. A great concentration of aircraft at Auxerre made the airfield unsafe and on 14 June, after a low-level "free hunt" fighter sweep over the Troyes area, the Wing landed at Les Aix-d'Angillon.

On 15 June both Polish Flights took part in the GC II/6's last combat operation, a "free hunt" sweep behind the front line over the Troyes-Estissac region. During the sortie the French fighters failed to intercept a Hs 126 army reconnaissance machine, supporting the advancing German armour. The "Ce" Flight went after it and after a chase of some 30 miles over ground heavily defended by German anti-aircraft defences shot the Henschel down, each pilot being credited with 1/3 of the aircraft. The next day the Wing was ordered to move to Avignon, to the Italian front, to provide air cover for Marseille and Toulon. A few uneventful patrols followed.

On 20 June two French pilots and the five Poles were sent to Toulouse to take on charge seven Dewoitine D.520 fighters and begin training on the aircraft. These were provided with additional fuel tanks for the anticipated transfer of the Wing to North Africa. One operational take-off was made. By this time the Polish Flights had earned the admiration of their French comrades and the Wing's C.O., Mjr Foutanet, assured the Poles that whatever happened, he would help them to leave and continue the fight. Nevertheless, the pilots were prepared to risk escaping to England on the D.520s, but on the night of 23/24

[43] Henneberg protested against the allotment of 1/2 He 111 to Cebrzyński and initially one He 111 was awarded to him, but Bajan's Fighter Claim Commission reversed this decision and restored 1/2 He 111 to Cebrzyński.

June they received orders to depart immediately to Port-Vendres to be evacuated by sea to Oran. On the way to the port they were joined by the Flights' ground personnel.

No.2 "Pe" Flight, operating on the front since 27 March, made about 20 missions, flying about 40 aircraft sorties in 45 hours' time. One of its pilots, Lt Karwowski, was credited with the destruction of two 1/2s of enemy aircraft.

The "Ce" ("Kościuszko's") Flight, commencing combat activities on 5 June, carried out about 15 missions, comprising some 35 aircraft sorties, of which 27 involved encounters with the enemy, in 40 flying hours. Its pilots had 2 1/2 enemy aircraft confirmed as destroyed.

(3.) No.3 "Su" Flight and the "Ba" Flight (GC III/6)

The three-aircraft "Su" Flight, commanded by Cpt Sulerzycki and formerly of the "Montpellier Squadron," joined 2nd Squadron of the GC III/6 Wing at its Vez-Thuizy airfield, 20 miles East of Reims, in the afternoon of 27 March. A Polish ground crew of 11 men under Sgt Edward Kulikowski, was attached to the Flight. The Wing's main task was to provide defence of the airfield on which a RAF bomber squadron was also stationed. Very limited initial activities consisted of a few operational take-offs against high-flying German reconnaissance machines.

At the end of April GC III/6 was transferred South, to the Arc-et-Senans, about 50 miles from the Swiss border. On 14 May, soon after take-off at about 11:00 hours the "Su" Flight was directed by radio to intercept enemy bombers raiding Dijon. Kawnik had to turn back because of an undercarriage malfunction and Rychlicki got lost in the clouds. Sulerzycki found a lone He 111 flying at about 7,500 m and attacked it head-on. However, his machine-guns froze and only the cannon fired. The bomber was hit and one of its engines began to smoke. With his ammunition exhausted, the pilot had to abandon the fight, but reported the exact position of the bomber by radio and a French flight finished it off. Meanwhile, Kawnik, who upon landing changed aircraft and took off again, encountered the returning enemy formation over the Vesoul area and in a lone attack shot down one Ju 88 bomber.

On 20 May in the morning, before the Wing left for Coulommiers, 30 miles east of Paris, the "Su" Flight flew a defensive patrol. Kawnik became detached from the unit in misty weather and on the way back sighted a single Do 17, which he attacked and forced to the ground in the vicinity of Dole. This was confirmed by the French and the victory was credited to him.

During operations from Coulommiers, GC III/6 was engaged in a particularly difficult combat on 21 May. At 16:00 hours the Wing's 18 MS 406 fighters took off for a sweep of the Cambrai-Le Catelet area, with GC II/3's Dewoitines as the top cover. Heavy clouds and intensive ground fire dispersed the formation into isolated flights, which were attacked by Bf 109s and Bf 110s. In the savage fighting some French machines were brought down and others began to run short of fuel. Sulerzycki, flying alone, noticed a Bf 110 between the clouds and opened fire. His enemy was soon engulfed in smoke and dived to earth. At that moment Sulerzycki himself became the target of two Bf 109s, but managed to break away and nurse his bullet-riddled Morane down to a field. French pilots reported that the Bf 110 fired at by Sulerzycki crashed on the German side. Sulerzycki claimed it as "probable," but this was not recognized by the PAF Command. A few days later

No.4 "Bu" Flight pilots, S-Lt Gnyś (standing second from the left), S-Lt Chciuk (standing fourth from the left) and Lt Bursztyn, C.O. of the Flight (standing sixth from the left), and mechanics, by Bursztyn's MS 406 'I', c/n 1031, on the GC III/1 Wing's airfield in May 1940. (J.F. Lansdale)

The MS 406 'I', with the badge of the 5e Escadrille "Le Renard Clignotant" ("The Winking Fox") of GC III/1 on the fuselage below the cockpit, with Lt Bursztyn and his mechanics in readiness by the aircraft. (J.F. Lansdale)

S-Lt Gnyś' MS 406 "III," with the pilot and ground crew in readiness in May 1940. The 5e Escadrille's "Fox" badge is well visible on the Morane's fuselage. (J.F. Lansdale)

Rychlicki engaged a Do 17 bomber which, according to unconfirmed reports, came down in the German-held territory. Coulommiers airfield was heavily bombed and by the end of May GC III/6 had only a few serviceable Moranes left. The "Su" Flight was awarded the Croix de Guerre avec Palme.

Around 1 June the Wing was transferred to the Italian front, to Le Luc, some 30 miles North-East of Toulon, and after a few days began to re-equip with Dewoitine D.520 fighters. A number of defensive patrols were flown over the Toulon-Nice area, initially on Moranes and later on D.520s. On one occasion Italian Fiat CR 42 fighters, sighted at a distance, hastily turned back.

On 9 June the "Ba" Flight arrived at Le Luc to reinforce the GC III/6 Wing. The Flight was formed at Lyon in response to a French Air Ministry order of 6 June and comprised Lt Wieńczysław Barański, the Flight's C.O., and Lt Jan Borowski and NCO Michał Cwynar. The pilots departed from Lyon by rail on 8 June in the afternoon. At Le Luc they started familiarization training on D.520s, but the "Ba" Flight only received its three Devoitine fighters on 15 June and flew no operational sorties.

On 16 June both Polish Flights were ordered to move with the Wing to Perpignan. Four days later the Poles flew to Algeria, where they surrendered their D.520s, and subsequently sailed to Gibraltar.

During the period of 26 March-19 June No.3 "Su" Flight carried out about 25 missions involving about 60 aircraft sorties in some 75 flying hours. Its pilots were officially credited with 2 confirmed victories and 1 enemy aircraft damaged.[44]

(4.) No.4 "Bu" Flight (GC III/1)

The three – aircraft "Bu" Flight, under the command of Lt Bursztyn and initially forming part of the "Montpellier Squadron," reached its operational destination, the GC III/1 Wing at the Toul-Croix de Metz airfield on 29 March. An 11-man strong ground crew, under Sgt Stanisław Basiński, was assigned to it. Apart from an occasional patrol of the Luxembourg border and one sortie providing escort for Potez 63s, the Flight had little to do. On 16 April the Wing moved to the Boos-Rouen airfield, to defend the Rouen-Le Havre area, and on 5 May flew to the Auchy-au-Bois airfield near Norrent Fontes, close to the Belgian border.

On 10 May the airfield was bombed and a few Moranes destroyed. The Wing was transferred to Belgium, to Moerbecke, east of Brugge, and it was while traveling there that the Flight's transport column was attacked by Heinkels. Cpl Stefan Lisiak instantly pulled a heavy machine-gun from a lorry, set it up by the roadside and shot down one of the bombers. He was decorated with the Croix de Guerre for his resolute action. On 12 May the "Bu" Flight, operating from Moerbecke, provided top cover in defensive patrol. During the mission its pilots engaged and shot down a Heinkel bomber, which landed on water near Antwerp with a dead gunner. The rest of the crew swam ashore and were taken prisoner. However, Bursztyn, with his engine on fire, came down in difficult terrain, overturned and was taken to a hospital in Ghent with minor injuries. His Morane, white 'I' no.1031, was destroyed. Gnyś, flying the white 'III', was also hit and crash-landed. The three pilots were awarded 1/3 of the He 111 each.

Five days later GC III/1 returned to Auchy-au-Bois and the next morning, on 16 May, carried out a sweep over Brussels. The "Bu" Flight, with Lt Marche, a Frenchman, in the lead and Chciuk and Gnyś on his flanks came to grips with a formation of Do 17s at 3,500 m. The pilots attacked the rear bomber, set one of its engines alight and followed it down until it crashed. They were credited with the destruction of 1/3 of the aircraft each. In the final exchange of fire, Chciuk's Morane was hit in a fuel tank and burst into flames. Soon afterwards Gnyś fought a lonely duel with a He 111. He saw his bullets exploding into the German, but his machine was also holed. Seeing German fighters in the distance, he broke off the attack. He was credited with one Heinkel "damaged."

Chciuk force-landed near Gembloux, but his eventful return journey during which he was taken for a German spy, took three days to complete and when he arrived at Auchy-au-Bois, he found it abandoned. Only a few derelict aircraft and a French guard about to set them on fire remained on the field. To the guard's astonishment, Chciuk commandeered one of the Moranes which had no armament, a faulty airscrew-pitch

[44]As listed by Col. Bajan's Fighter Claim Commission, L.dz. FC/S.5/1/Air/CPLO.

mechanism and inoperative landing gear and flaps, and took off without a parachute. Flying at tree-top level with wheels down, he attracted considerable attention from the German anti-aircraft defences when passing over German lines, but managed to reach the Paris Le Bourget airfield. He found out where GC III/1 was stationed and arrived at the Plessis Belleville airfield on 19 May.

The same afternoon he took off with two Frenchmen to patrol the Beauvais area. Near Montdidier the patrol encountered six Do 17s. The Frenchmen turned back and Chciuk attacked alone. Concentrated fire from the Dorniers set his MS 406 alight and Chciuk, forced to land near Clermont, went missing again. The following day he turned up at the airfield on a bike!

On 25 May Lt Bursztyn returned from the hospital and the "Bu" Flight was complete once again, but only very briefly. Although Bursztyn's injury was still giving him trouble, he insisted on leading his Flight later the same afternoon on a mission in which the Wing's 12 Moranes flew as escort to two reconnaissance Potez 63s. At about 15:00 hours 14 Bf 109 fighters approached the French formation over the German lines near Cambrai and the "Bu" Flight, providing the top screen, accepted the challenge, whilst the rest of the French aircraft dispersed and ran for home. Having a height disadvantage, Bursztyn made a climbing turn to meet the Messerschmitts and began to fire, but he was instantly hit. His Morane's landing gear came out and the fighter went into a spin from which it never recovered. In a brief but vicious battle one Bf 109 was seen to crash to the ground and the other two Polish pilots, their aircraft riddled with bullets, succeeded in breaking away and returning to base. Since it was not clear who shot down the Messerschmitt, each of the three Poles were awarded 1/3 of the aircraft. A few days later the airfield was heavily bombed. No warning had been received and several Moranes were destroyed. After Bursztyn's death, S-Lt Gnyś assumed the command of the Flight.

On 5 June GC III/1 was withdrawn to Valence, south of Lyon, for rest, but its Commandant protested and requested a return to the front. This was granted the next day and the Wing was sent to the Rozay-en-Brie airfield, south-east of Paris. On 8 June the "Bu" Flight, the two Polish pilots and Sgt Gaguaire, operating east of Paris, engaged a Ju 87 and shot it down, but the aircraft had been earlier attacked by a French Lieutenant, who received full credit for its destruction.

On 11 June GC III/1 was ordered back to Valence and later went to Orange. The Poles were once or twice scrambled on a false alarm and flew a few patrols and reconnaissance missions without meeting the enemy. On 16 June the Polish pilots learned about the award of the Croix de Guerre avec Palme. The Flight ended its operational activities on 20 June and the following day S-Lt Gnyś and the ground crew departed to Port-Vendres, from where they were evacuated to Algeria. Later they sailed for Gibraltar and eventually reached England. Chciuk found his way to England separately.

During its front-line service between 29 March and 20 June No.4 "Bu" Flight flew about 35 missions, comprising about 70 aircraft sorties in some 85 flying hours. Its pilots were credited with 2 2/3 enemy aircraft destroyed and 1 damaged. The Flight lost five of its own MS 406s and one pilot was killed in action.

(5.) No.5 "Br" Flight (GC I/2)[45]

The three-aircraft "Br" Flight, commanded by Lt Brzeziński and originally part of the "Montpellier Squadron" was assigned to the GC I/2 Wing stationed at Xaffevilley airfield, about 30 miles from the German border. The Flight arrived on 29 March and joined the Wing's 1st Squadron. A Polish ground crew of 11 men, under W/O Władysław Kubecki, was attached to it. Before the German offensive the Flight flew two escort missions from Xaffevilley over the border (Wissembourg) area and, after transfer to the Ochey airfield near Toul in April, eight missions, mainly over the Saarbrücken-Zweibrücken area.

Between 10 and 14 May the Poles were scrambled five times to protect the airfield and on the 10th engaged three Ju 88 bombers east of Nancy and five Ju 88s the next day, but without result. On 12 May the Ochey airfield was bombed without warning and all three Polish Moranes were damaged, two of them only slightly. On 13 May the Flight provided escort for a reconnaissance over Zweibrücken.

From 15 May onwards the Wing's major task was the air cover of the Montmedy sector of the front. On 17, Chałupa's MS 406 suffered damage to the cooling system and pressure installation on a combat sortie. During the wheels-up emergency landing the pilot was heavily bruised and spent over a week in a hospital. When he returned, Brzeziński became ill and on 1 June was admitted to hospital. Chałupa took over the command of the Flight. S-Col Leopold Pamuła, who about this time joined the staff of GC I/2 to widen his frontline command experience, was not directly involved with the "Br" Flight.

---

[45]Main source: Chałupa's combat report (IPMS, Lot.A IV,2/1a/1-4).

One of the damaged Polish MS 406-C1s, c/n 944, believed to be that of No.5 "Br" Flight of GC I/2, captured by the Germans. (T.E. Willis)

Earlier, on 27 May at about 13:00 hours, the Damblain airfield near Neufchateau, to which the Wing had moved a day or two earlier, was strafed by Bf 110s. Eight fighters were set on fire and others were damaged. The gunners of the anti-aircraft batteries were having lunch and opened fire only when the Messerschmitts were leaving.

Replacement Moranes were received and on 1 June the Wing encountered a large enemy formation in the Colmar area. Chałupa and Beda fought with no apparent success, but the next day both pilots, flying as wingmen to the Squadron C.O., Cpt Yvernand, attacked returning bombers over Selestat and shot down jointly with him a He 111. They were credited with 1/3 of an aircraft each. On 5 July at about 13:00 hours, Beda destroyed a Ju 88 bomber near Chaumont after a bitter fight.

On 8 June the Wing was temporarily transferred to Coulommiers, operating from this airfield for two days. Its first mission, at about 15:00 hours, was a fighter sweep east of Beauvais, during which its 1st Squadron, led by Cpt Yvernand with Chałupa and Beda as wingmen, met nine Bf 109s flying in the opposite direction some 200 m below. The Germans did not notice the Moranes against the sun. The Squadron's C.O. let two flights through and dived on the last three Messerschmitts, shooting down two in a lightning attack. Chałupa opened fire on the remaining aircraft at very close range. It burst into flames, pieces of metal flew from its tail, and spun into the ground. The fight was over before the rest of the Squadron could join it.

The day's second mission, at 19:45 hours, carried out by two flights of 1st Squadron over the Soissons area, led to a battle with a formation of more than 20 Ju 87 dive-bombers. Chałupa, attacking head on from a height advantage, destroyed one Stuka and then fired on another, which trailing smoke tried to escape at low level. Two French pilots joined in the chase and the enemy crashed into trees while attempting to land. Chałupa was awarded 1 1/3 of the two Ju 87s shot down during the encounter. On 9 June the Flight had nothing to report and on the next day GC I/2 returned to Damblain.

On 13 June the Wing moved south via Dijon to the Chalous Champforguil airfield. A few patrols were flown and on 15 June at 05:30 hours the Poles were scrambled to intercept the enemy approaching the airfield. Beda's Morane was attacked by a Bf 110 soon after take-off and received several hits which forced it to land. Chałupa came to his rescue and fended off the Germans, firing blindly because his aiming gear had failed. On 17 June the Wing landed at the St. Symphorien-d'Ozon. Three days later, with the transfer to Nimes, its operations came to an end. After a warm farewell ceremony, the Poles departed by road to begin their evacuation journey.

No.5 "Br" Flight, operating with GC I/2 from 29 March until 20 June, flew about 40 missions involving some 80 sorties in about 100 flying hours. The Polish pilots destroyed four enemy aircraft and the Flight lost four Morane fighters. One of the pilots was wounded.

Gen Sikorski, in company of Gen Zając (on his right), Gen Denain and Col Pawlikowski (on his left), saluted by three pilots of No.6 "Go" Flight, left to right: Lt Goettel, Flight's C.O., S-Lt Król and Cpl Nowakiewicz, during the opening of the CWL Training Centre at Lyon-Bron on 27 March 1940. (IPMS)

(6.) No.6 "Go" Flight and the "Jan" Flight (GC II/7)

The three-aircraft "Go" Flight, under the command of Lt Goettel, was the last of the six flights of the "Montpellier Squadron" (1st Squadron of II/8 Kraków-Poznań Fighter Wing) which were separately distributed among the French wings at the front. The Flight arrived at the GC II/7's Luxeuil airfield on 29 March, where a Polish ground crew of one technical officer, S-Lt Henryk Busen-Schmidt, and 11 men under Sgt Mjr Bolesław Psujek, was already installed, and joined the Wing's 3rd Squadron. Its three MS 406 fighters were allocated as follows: No.959, to Goettel, No.956, to Król and No.953, to Nowakiewicz. Appalling weather prevented flying and the Poles spent their first week with the Wing acclimating to new surroundings and making friends with their French colleagues.

The Flight carried out its first operational task on 7 April and by the end of month the Poles, always flying together, logged 15 missions, comprising 44 sorties in 50 hours 55 min. flying time.[46] Most of these involved defensive patrols over the Luxeuil area, occasional sweeps over the Belfort-Besançon region and one flight to escort Potez 63 bombers attacking Lörrach.

During a sortie on 19 April Heinkel bombers were engaged over Mulhouse and one appeared to be hit, but in the exchange of fire Goettel's Morane was damaged. On 28 April a second engagement took place, when the Poles and a French flight together attacked Do 17s over Besançon. Destruction of one Dornier was credited to the French flight. Towards the middle of April Mjr Mieczysław Mümler joined the staff of GC II/7 to gain command experience with a French wing. He insisted upon flying with the "Go" Flight, but the French were reluctant to provide him with a fighter. Eventually he was allotted an MS 406, but not until mid-May.

A similar pattern of activities continued at the beginning of May. On the 4th day of that month the whole Wing provided cover for 15 Potez 63 bombers raiding a Freiburg factory and fought a battle with Bf 109 fighters which attacked the returning French formation. Four days later, during a rou-

[46] As per Flight's logbook (IPMS, Lot.A IV, 2/1a/5-2).

tine patrol, the Poles made an inconclusive attack on a He 111 during which Goettle's Morane was hit and had to make a forced landing at Belfort.

On 10 May soon after sunrise Luxeuil was the target of a heavy attack by German bombers and fighters, which bombed and strafed the airfield at will, totally wrecking seven Moranes and damaging several others. The defenders were taken completely by surprise: the pilots were just arriving on the airfield in their cars and the anti-aircraft batteries, initially unmanned, began to fire only when the Germans had dropped their bombs. Well dispersed MS 406s of the "Go" Flight survived intact. Later during the day a defensive patrol incorporating the Polish pilots fought six Bf 110s over the airfield.

On 11 May at about 08:00 hours Nowakiewicz was scrambled with French pilots to intercept an enemy raid from the direction of Vesoul. The take-off was chaotic, with every pilot for himself. Flying alone, he encountered a formation of 28 Heinkels, but managed to engage it only past Dijon. He attacked one of the rear bombers and shot it down north-west of Dijon. Running out of fuel, Nowakiewicz landed at Cuisery. There he abandoned his aircraft and returned to the Wing.[47] He was credited with the destruction of one He 111 and awarded the Croix de Guerre. In the afternoon the Wing's airfield was again bombed without warning. Król's Morane no.956 was set on fire and completely destroyed. A Polish mechanic, Wacław Kępa, died on the airfield from fatal injuries. After the bombardment GC II/7 was ordered to a landing field at Saint Sauver, near Luxeuil. Here it was brought back to its full operational strength with replacement MS 406 fighters.

12 May was a day of very intensive activity from the new airfield, with the Polish pilots flying several missions, Król alone recording four sorties in 4 hours 35 min flying time. On 19 May Mjr Mümler logged his first operational flight with the Wing, flying MS 406 no.220 on a patrol sortie lasting 1 hour 40 min.[48] Later on the same day GC II/7 moved to Marey-s-Tille airfield (north of Dijon). Goettel, who became progressively ill, stopped flying and eventually went to hospital for treatment on 2 June. His Morane was subsequently used by Król and Mümler. On 24 May Król took off in no.959 with four French pilots to intercept a reported enemy aircraft. Król, flying at the rear of the patrol, which was led by Cpt. Papin, sighted a He 111 below, but his radio was out of order. He attacked alone, unnoticed by the rest of the flight and brought down the bomber near Istre-s-Tille. The German crew parachuted to safety and were captured by the French. Mümler flew no.959 on operational sorties on 25 and 26 May.[49] On 30 May the Wing was rearmed with the Devoitine fighters; Mümler received D.520 no.119, Król, no.181 and Nowakiewicz, no.62.

On 21 May three Polish pilots of the "Jan" Flight reinforced GC II/7, with Mjr Mümler taking over command of the Polish pilots within the Wing. The Flight, which was formed at Lyon a day earlier, comprised Lt Wojciech Januszewicz, its C.O., S-Lt Wiktor Strzembosz and Cpl Henryk Szope, and was attached to the Wing's 4th Squadron. On 30 May the pilots, who arrived without aircraft, received their D.520 fighters (nos.196, 216 and 238) and the Flight became operational.

1 June was another day of hard combat. During several encounters with the enemy Nowakiewicz, flying with two French pilots in a three-aircraft flight, engaged a He 111 at close range. After his second burst the Heinkel dived almost vertically to the ground in the Lousle-Saulieu area. He was credited with the destruction of 1/3 of the bomber. In the afternoon Mümler, promoted by then to the rank of S-Col, fought with a formation of Heinkels over Epinal and was awarded one He 111 confirmed as destroyed. Król, taking part in a ten-aircraft fighter sweep by the 1st Squadron led by Cpt Papin, came to grips with a large formation of Dorniers and Bf 110s east of Epinal, close to the Swiss border. The right engine of the Dornier he attacked had begun to smoke when his own fighter was hit by a Bf 110. Król had to break off the attack and eventually force-landed with a dry tank. He reported one Do 17 as probably destroyed.

On 3 June in the afternoon Król and four French pilots carried out an offensive patrol over the Strasbourg section of the Siegfried Line. A Dornier was shot down by W/O Walentin, leading the flight, but on the way back Król spotted a single Do 17 near Besançon, some 2,000 m below, and went after it alone. The enemy crashed in flames and Król was credited with its destruction. The next day Nowakiewicz shot down one He 111 in the same area. The bomber made a belly landing and two of its crew members tried to escape to a forest nearby, but Nowakiewicz shot them dead from the air.

During the first four days of June the "Jan" Flight was also involved in heavy air fighting, but details of its activities are incomplete. Strzembosz mentions in passing[50] his participation in a battle with a large formation of Ju 88s over the Swiss border region, during which two of the bombers came down in Switzerland, close to the Lake of Geneva, and another occasion (possibly on 3 June) when the Flight, part of a five-aircraft patrol under the command of W/O Pontens, had a second encounter with Ju 88s and Bf 110s over the same area. This time Januszewicz led the attack, with Strzembosz following him closely. One Junkers was shot down but its destruction was never officially confirmed.

On 4 June in the afternoon the Wing moved to Meaux, east of Paris. During the next day several sorties were flown over the collapsing front, two of them by Mümler in D.520 no.119, and notable for the first confirmed victory of Strzembosz of the "Jan" Flight.

The 4th Squadron, led by its C.O. Cpt Hugo in whose flight Strzembosz was to occupy no.2 position, took off from the bomb-riddled Meaux airfield at about 20:00 hours on the day's last mission: a patrol over the Amiens area. Strzembosz, avoid-

[47] Nowakiewicz's report (IPMS, Lot.A IV,2/1a/5-7).
[48] Mümler's Combat Flights Record (IPMS, Lot.A IV,2/1a/5-4).
[49] Ibid.

[50] Only Strzembosz's account of the Flight's operations (Lot.A IV, 2/1a/5-6) is available in the IPMS Aviation Archives.

ing the bomb craters on take-off run, became separated from the rest. Continuing towards Amiens at 5,000 m, he noticed five D.520s involved in a fight with several Bf 109s. One of the French fighters, chased by a Messerschmitt, flew towards him some 500 m below. Strzembosz dived on the enemy's tail and fired his first burst but his guns jammed and he had to continue the attack with cannon fire. The Bf 109 began to lose parts of its tail and wing and went to earth out of control. Strzembosz climbed to 7,000 m to reload his guns, when he was suddenly attacked from the rear and his machine burst into flames. He jumped, and seeing Messerschmitts circling his falling plane, he delayed opening his parachute until he was close to the ground. He landed safely, but could not return to the Flight and was posted missing (he later found his way to England).

On 9th June GC II/7 was hurriedly evacuated to Ouanne. By this time the Meaux airfield had been bombed twice and the Wing lost four D.520s. The "Go" Flight received D.520 no.242 as a replacement. A number of missions were flown from Ouanne and some inconclusive encounters fought, but on 15 June, the last day of operations from the airfield, the Polish pilots were credited with two Germans shot down and one damaged. During the afternoon sweep of a 40-mile section of the front by four flights of GC II/7, Mümler and Nowakiewicz[51] spotted a few Dorniers in the Neufchateau area, which attempted to seek refuge in the clouds. Both pilots managed to attack the last machine, Nowakiewicz firing from a 90 degree angle. The bomber re-emerged from the clouds and dived towards the ground with one crew member entangled with his parachute in its tail. Two more airmen attempted to jump, but they were both killed.

Each pilot was awarded 1/2 of the Dornier. Mümler and Nowakiewicz also claimed a second Do 17 as damaged, but Mümler's D.520 no.119 was hit in an exchange of fire and the pilot was forced to land close to the German lines. Later the same afternoon, during a "free hunt" in the Nancy area, Nowakiewicz in a lightning attack shot down a Hs 126 before any of the other pilots had a chance to engage. The rapidly advancing German armour forced the Wing to withdraw further south before nightfall.

After spending the night at Feurs (north of St. Etienne) GC II/7 moved to Carcassonne, where S-Col Mümler rejoined the Poles on 18 June, arriving with the Wing's transport column. The pilots were ordered to proceed to Perpignan, from where they were to fly across the Mediterranean to North Africa. In the early hours of 21 June a formation of about 20 Dewoitines, including Mümler, Januszewicz, Król, Nowakiewicz and Szope, flew to Algeria, most aircraft landing at Bone (Annaba). Mümler, who had to turn back short of fuel, arrived two days later at the Oued airfield near Tunis, to which other fighters of GC II/7 had by then been transferred. On 26 June the Polish pilots departed to Casablanca, to join the Polish evacuation transport sailing for Gibraltar.

No.6 "Go" Flight, operating on the front from 29 March to 20 June, carried out about 55 missions, comprising some 110 sorties in about 145 flying hours, these figures including S-Col Mümler's operational flights which amounted to 13 sorties in 25 hours 05 min flying time. The Flight's combat record included 7 1/3 enemy aircraft confirmed as destroyed, 1 probable and 1 damaged.[52] The Flight lost four of its own machines.

The "Jan" Flight, which also participated in the combat activities of GC II/7 during the period 30 May-20 June, logged about 25 missions, involving some 45 sorties in approximately 60 flying hours, and was credited with 1 confirmed air victory. One of its D.520 fighters was lost in combat and one pilot went missing.

(7.) No.7 "Wy" Double Flight (GC II/10)

No.7 "Wy" Double Flight, which began to organize at Lyon at the very end of May, comprised six pilots, drawn in part from the Fighter Exercise Squadron: Mjr. Eugeniusz Wyrwicki, the Flight's C.O., S-Lts Hieronim Dudwał, Jerzy Poniatowski and Jerzy Radomski, and NCOs Kazimierz Sztramko and Mieczysław Adamek. The pilots, ordered to join GC II/10, departed for the Wing's Bernay airfield on 2 June at 21:00 hours. There is no mention in the available Polish records of any Polish ground personnel being allocated to the unit.

After a brief conversion training on the GC II/10's Bloch fighters on 4 and 5 June, the Poles commenced operational duties on 6 June. Wyrwicki and Poniatowski were sent that day with a French squadron to set up a stand by patrol unit at the Boos airfield near Rouen. The rest of the "Wy" Flight stood in readiness at Bernay and took part in emergency scrambles.

7 June was a black day for the "Wy" Flight. Soon after 04:00 hours seven French pilots and four Poles: Wyrwicki, Dudwał, Poniatowski and Radomski, attached one each to the four French flights of the 3rd Squadron, were scrambled to carry out a fighter sweep over the Amiens area. Soon after crossing the Seine the fighters spotted six Bf 109s, which flew parallel to them some distance away. Approaching Amiens the formation came under extremely heavy German anti-aircraft fire and dispersed. The pilots, whose attention was totally absorbed by the ground fire and the shadowing Bf 109s, were suddenly attacked by Bf 110s, which dived on them from the sun. The French pilots lost one of their comrades and immediately headed for home. Only the Poles accepted the fight. Radomski, flying the slower MB 151, fell behind the others. He saw one of the MB 152s going down, but lost sight of the remaining two and turned back short of fuel. He landed on one of the airfields occupied by the RAF and after refueling returned to Bernay. Wyrwicki, Dudwał and Poniatowski did not

[51] Nowakiewicz in his account of the events (IPMS, Lot.A IV,2/1a/5-7) quotes the date of 14 June, but Mümler's combat report (IPMS, Lot.A IV,2/1a 5-4), which contains entries from his logbook, states that the flight took place on 15 June.

[52] Pierre Riviere, who researched PAF operations in France, credited the pilots of the "Go" Flight with certain destruction of 13 enemy aircraft, with four further as "probable."

return and were initially posted missing, presumed dead. The shot-down French pilot, who managed to reach the Wing three days later, reported seeing a wrecked MB 152 with the body of a Polish pilot still inside; he read the name Poniatowski on the pilot's helmet.[53]

On the same day in the afternoon Sztramko and Adamek flew a patrol sortie with the 4th Squadron and on the way back fought a victorious battle with Bf 109s South-east of Amiens, Sztramko being credited with the destruction of one of the enemy fighters. However, running out of fuel, he was forced to make an emergency landing and damaged his machine. In the evening, after the tragic experiences of the day's fighting, the GC II/10's pilots were ordered to engage only unescorted German bombers and avoid all combat with German fighters.

On 8 June a number of patrol flights were carried out along the Seine river south of Rouen. This time Adamek ran short of fuel, but he managed to bring his machine down in a field without damage. On the next day GC III/10, incorporating the Polish "Ga" Flight, arrived at Bernay. On 11 June GC II/10 moved to Caen where it operated for four days, flying several missions over the collapsing front. On the 15th the Wing withdrew to Dinard, where it stood idle for three days. On the 18th the Wing was transferred to Rennes and the following day it proceeded via la Roche-s-Yon to la Rochelle. Its combat activities were over.

On 20 June the fighters flew over to Arcachon. Two days later the three Polish pilots were transferred by air to Perpignan, from where Radomski was taken to Oran and Adamek and Sztramko to Alger.

The "Wy" Double Flight, engaged over the front from 6 to 18 June, flew about 15 missions, comprising some 40 sorties in about 50 flying hours. It was credited with 1 confirmed victory, but lost three of its pilots and three MB 152 fighters in action at the very beginning of its combat activities.

(8.) No.8 "Ja" Double Flight (GC II/1)[54]

No.8 "Ja" Double Flight, comprising six pilots: Cpt Franciszek Jastrzębski, the Flight's C.O., Lt Marian Trzebiński, S-Lt Jan Jakubowski, and NCOs Stanisław Brzeski, Tadeusz Krieger and Bolesław Turzański, was formed at Lyon on 18 May.

It was allocated to GC II/1, which was stationed at the Buc airfield in the Paris area. The pilots traveled there by rail the following night and on 20 May began training on the Wing's MB 152 fighters, which lasted a few days and totaled 26 flights in 29 hours 35 min. There is no record of any Polish ground crews being attached to the Flight. Details of the Flight's operations are sketchy, although the Poles flew numerous sorties. Highlights of their activities are summarized below.

On 3 June Trzebiński managed to take off as German bombs began to fall on the Wing's airfield. He chased the raiders and in a lone attack destroyed one He 111 south-east of Paris. He was then engaged by the Bf 109 fighters and his Bloch sustained moderate damage, but he succeeded in nursing it back to base. The next day Krieger was hit by enemy ground fire and had to make an emergency landing, but later returned home. On 5 June the "Ja" Flight took off together with a French squadron, except for Jastrzębski. The engine of his fighter was started with some delay and he eventually managed to get airborne about 15 minutes late, hoping to catch up with the formation. He was suddenly pounced upon by several Bf 109s and his aircraft was set on fire. Jastrzębski, burnt on the face and wounded in the legs, jumped to safety and was taken to hospital. He never rejoined the Flight.

On 10 June GC II/1 moved to Anglure and on the 12th Trzebiński, flying with a French flight on a fighter sweep in the Reims area, shot down a Hs 126 north of the town. He shared the victory with a Frenchman, each pilot being awarded 1/2 of the aircraft. On the next day the Wing was ordered to move to Nimes in Southern France, to the Italian front. Flying in bad weather, the fighters scattered and became lost, some turning back, others making emergency landings en route. Only Trzebiński reached Nimes. The Wing's aircraft were not reassembled until 17 June and made only a few more operational flights before its combat activities came to an end three days later. Trzebiński was decorated with the Croix de Guerre. On 23 June the Poles arrived in Port-Vendres and sailed for Oran.

The "Ja" Double Flight, operating on the front during the period 20 May-20 June, took part in about 40 missions, logging 121 sorties in 137 flying hours and was credited with 1 1/2 enemy aircraft destroyed. The Flight lost two of its own fighters and one of its pilots was wounded in action.

(9.) No.10 "Ga" Double Flight (GC III/10)

No.10 "Ga" Double Flight started to form at Lyon on 1 June with six pilots, including three officers recalled from the Fighter Exercise Squadron: Lt Aleksander Gabszewicz, the Flight's C.O., Lt Tadeusz Sawicz, S-Lt Bogusław Mierzwa, S-Lt Włodzimierz Miksa, C-Off Marian Rytka and NCO Jan Musiał. A 15-man strong Polish ground crew was attached to it.

The Flight was ordered to join GC III/10 at the Deauville airfield and departed for its destination on 3 June in the evening. After a brief conversion training within the Wing to the Bloch fighters, the Poles were apparently allocated four MB 152s and became operational two days after their arrival.

The Flight moved with GC III/10 to Anglure, Bernay, Les Gr.- des-Chapelles, Montbard, Clermont Ferrand, Ussel and Bergerac, but saw only limited combat because of the short stay at the front and frequent airfield changes. Details of its operations are lacking. According to available information the "Ga" Double Flight participated in seven missions,[55] flying about 20 sorties in some 30 flying hours and fought two inconclusive encounters.

[53] The above description of the fight is based on Radomski's letter of 10 June 1940 and his other contemporary accounts (IPMS, Lot.A IV, 2/1a/7-1 & -2).

[54] Main source: Trzebiński's account (IPMS, Lot.A IV,2/1a/3-3).

[55] Gabszewicz's account (IPMS, Lot. A IV,2/1a/12-3).

(10.) The "Pa" Flight (GC II/8)[56]

The three-pilot "Pa" Flight, with Lt Ludwik Paszkiewicz its C.O., and NCOs Leon Nowak and Kazimierz Wünsche, began to organize simultaneously with the "Ce" Flight in the middle of May. The Flight, without Polish ground personnel, was allocated to GC II/8 and left Lyon on 18 May, but bombed railway lines and difficulties in finding the Wing delayed its arrival at the GC II/8's Villacoublay airfield by a day.

The Poles were incorporated in the 4th Squadron and after brief retraining on MB 152 fighters started to participate in combat activities. These at first consisted mainly of emergency take-offs to intercept incoming enemy aircraft in the areas of Paris, Le Havre and Cherbourg. Towards the end of May the Wing moved to Deauville and on the 28th of that of that month Nowak was killed in MB 152 no.298 in an accident on the airfield. A few days later GC II/8 flew to Falaise and then to Cherbourg.

On 9 June, operating from Falaise, the 4th Squadron, including the "Pa" Flight, provided escort for French bombers which were to destroy bridges over the Seine at Rouen. During the repetition of the task the next day, German motorized columns were found on a pontoon bridge crossing the river, but the bombers gave up the attack because of heavy fire from German ground defences and bombed instead only the secondary target, an undefended railway bridge. The Flight's later missions included air cover while the Le Havre airfield was evacuated. On 18 June, while in transit from Cherbourg to Nantes, the fighters were instructed to make a low-level strike on German armoured and motorized spearheads, but only Paszkiewicz carried out the order (Wünsche had no serviceable aircraft on that day and traveled to the new airfield on a transport plane), the French pilots asserting that it was not worth the risk to attack armoured vehicles. This was the Wing's last combat mission. Two days later GC II/8 proceeded to Rochefort, from where the Poles were evacuated to England via La Rochelle.

Early in June an attempt was made to expand the "Pa" Flight into a Double Flight. Three more pilots: Lt Franciszek Skiba, the C.O. of the second Flight, Lt Eugeniusz Antolak and NCO Marcin Machowiak, were ordered to join GC II/8 and left Lyon on the 7th of that month. However, because of the French retreat the pilots were unable to reach the designated airfield nor to find the Wing. They returned to Lyon on 10 June and the next day were redirected to join Krasnodębski's "Kr" Chimney Flight with ECD I/55 at Etampes.

The "Pa" Flight, operating as part of GC II/8 from 20 May to 18 June, flew about 20 missions involving some 35 sorties in about 45 hours of flying time. No victories were reported by the Flight, which lost one pilot and one aircraft in an accident.

(11.) The "Wi" Double Flight (GC III/9)[57]

The six-pilot "Wi" Double-Flight, comprising Mjr Edward Więckowski, the Flight's C.O., Lt Kazimierz Rębalski, S-Lt Wiktor Szulc, and NCOs Marek Słoński-Ostoja, Henryk Pietrzak and Stanisław Piątkowski, was formed on 8 June. Two days later the Flight was ordered to join GC III/9 and the pilots departed by road from La Foire to Venisieux (some 20 miles north-east of Lyon) on 11 June at 08:00 hours. No Polish ground crew was attached to the Flight.

GC III/9, which moved from Columbies to Venisieux to give fighter support to the Dijon front, possessed two MB 152 squadrons of four flights each. The Poles, after a few training flights on the Wing's Blochs, were attached separately to various French flights. The fighters, taking-off in response to messages from the local early-warning posts, were usually scrambled too late and operated without guidance from the ground control. When German planes were sighted, the French pilots avoided contact with stronger enemy formations. On 16 June Więckowski spotted 18 Heinkels flying without escort between Dijon and Lyon. His French colleagues turned back on seeing the raiders and Więckowski attacked alone with the height advantage. He shot down one He 111 and saw it crash, but because of the German advance this could not be confirmed and the destruction of the Heinkel was never credited to him.

On 19 June GC III/9 was transferred to the Le Luc airfield, to provide air cover for the Alps Army, but its operational activities were very limited and soon ceased altogether. In view of the breakdown in communications with the PAF Command, Więckowski, acting on his own initiative, insisted on going with his pilots to Port-Vendres. The Poles reached the port on 23 June and joined the Polish evacuation transport.

The "Wi" Double Flight, flying with GC III/9 from 11 to 20 June, took part in about 30 missions amounting to some 40 sorties in 50 hours' flying time, and reported one victory which was not confirmed.

C. The Polish Home-Defence Flights with the French Defence Squadrons (ECD & ELD)

(1.) No.9 "Ko" Local-Defence Flight (No.1 Team)-Chateaudun (ELD/Chateaudun)[58]

With the beginning of the German offensive, the special auxiliary Parisian Fighter Wing was established under the command of Col Chatelin to strengthen the air defences of the capital. The Wing included Light Defence Squadrons for the protection of Chateaudun, Chartres and Orleans. On 15 May two Polish pilots who were completing their fighter training at Chateaudun, NCOs Henryk Kowalski and Stefan Tomicki, were ordered by the PAF Command to form No.9 "Ko" Flight (the unit being conceived as No.1 Team with Kowalski in command) assigned to ELD/Chateaudun. The two Poles were incorporated in the MB 152 Flight of W/O Herve and lost all further contact with the PAF.

[56] Main source: Paszkiewicz's report (IPMS, Lot.A IV,2/1a/6-3).
[57] Main source: Więckowski's report (IPMS, Lot.A IV, 2/1a/11-2).
[58] Main source: Kowalski's report (IPMS, Lot.A IV,2/1b/3-2).

A Bloch MB 152-C1, piloted by a Polish pilot of an unidentified unit, comes to land over MS 406 fighters of a French squadron.

Herve was a very keen fighter pilot and his Flight was involved in very intensive operational activities, being kept on stand-by from dawn to dusk and making between two and five take-offs each day. On 17 May, as the Squadron patrolled in pairs on various levels between layers of heavy clouds, Kowalski and Tomicki spotted a Do 17. Kowalski fired first and when the Dornier dived into clouds and re-emerged below them, Tomicki was waiting for it to continue the attack. The enemy machine was damaged, with explosions in one of its engines clearly visible, but in an exchange of fire Tomicki's Bloch was hit and the pilot had to break off the engagement. He was forced to land at Orleans short of fuel.[59] The damaged Dornier, chased by Kowalski past Orleans, was then attacked head-on by a flight of French Curtiss fighters and eventually crashed. After a long argument its destruction was credited jointly to both Polish pilots – 1/4 of the aircraft each – and to two Frenchmen.

At the end of May, with the collapse of the Somme front, 28 RAF Hurricanes landed at the airfield and for a brief period operated with the ELD/Chateaudun. Kowalski was then given the command of a flight which included two Hurricanes and on 28 May one of the RAF pilots of his flight shot down a He 111.

On 4 June at 06:00 hours the Poles were scrambled with Herve in the lead and encountered Heinkel bombers flying at very high altitude (some 8,000 m) from the direction of Tours towards Chartres. Herve attacked first with the Poles following suit and one He 111 was shot down, but in the event the Polish share in this victory was never officially recognized. Immediately after landing the flight was ordered to go up again to intercept a reported unidentified aircraft which was approaching the airfield from the direction of Tours. The machine, first fired at by Tomicki and then by Kowalski, turned out to be a French LeO 45 bomber. The pilots realized their mistake, but the bomber was damaged and its pilot wounded.

On 12 June the Poles of the "Ko" Flight were ordered to leave the ELD/Chateaudun and join the X. "Fa" Chimney Flight at Cognac.

From 15 May to 12 June, during its operational activities with the ELD/Chateaudun, No.9 "Ko" Flight participated in about 50 missions and flew some 80 sorties in about 85 flying hours. Its pilots were officially credited with the destruction of 1/2 enemy aircraft.

(2.) The I. "Kr" Chimney Flight – Etampes (ECD I/55)

The Armee de l'Air Directive no. DM 1525 I/IS EMAA from 11 May 1940 to the Commander of the Lyon Mixed Aviation Base and the DIAP (CWL) Training Centre requested the immediate formation of six Polish Chimney Flights. The basic establishment of such flights, which were regarded by the French as auxiliary combat elements rather than proficient first-line units, was to comprise a C.O. and three pilots, plus a 13-strong ground crew. The equipment was to consist of three fighters, usually drawn from training centres or reserves. In the next stage the formation of double-strength Chimney Flights was envisaged, these including a C.O. and two three-pilot flights.

[59] Two Polish pilots, NCOs Mizgojan and Kasprzyk, undergoing fighter training at Cheteaudun, were killed on the same day when a trio of the Spad 510 fighters, lost in fog, ran out of fuel and crashed during emergency landing in open country. Pierre Riviere gives the name of the third pilot, who was apparently wounded in the catastrophe, as Tomicki, but this is obviously not possible in view of Kowalski's account above. It is likely that Tomicki, who also had to land short of fuel, was confused with the third pilot involved in the training accident.

In response to the Directive DM 1525 I/IS the PAF Command commenced formation of the requested Chimney Flights, initially allocating six pilots (including a C.O.) to each, but in most cases the surplus personnel had to be recalled to Lyon within a day or two of the unit's arrival at its operational airfield. Later some of the Flights were enlarged by the addition of a second three-pilot team and a few others were organized as double units from the start.

S-Col Leopold Pamuła, the C.O. of the emergent 3rd Fighter Wing, ordered his Deputy, Mjr Zdzisław Krasnodębski, to organize and command the first of these units, which became known as the I."Kr" Chimney Flight.

The Flight as initially established comprised Mjr Krasnodębski, S-Lts Mirosław Ferić and Jan Zumbach, and NCOs Marian Bełc, Stanisław Karubin and Stefan Wojtowicz. A ground crew of 14 men, under the command of W/O Kazimierz Mozoł, was attached to the Flight, which was directed to Chateaudun and left Lyon by rail on 16 May, arriving at its destination the following day. At Chateaudun the Flight received four MS 406 fighters and on 17 and 18 May flew its first two operational missions patrolling the local area. On 19 May Ferić, Wojtowicz and two mechanics were recalled back to the Lyon base, leaving the Flight with four pilots and a 12 man ground crew. No further sorties were carried out from the airfield and on 21 May the Flight received new orders posting it to the ECD I/55 Squadron which was forming at Etampes.

While the rest of ECD I/55 moved to Tours, the I."Kr" Chimney Flight remained at Etampes and on 22 May began intensive operations in defence of the town, which frequently involved two to three protective patrol missions in a day. With the Flight still subordinated to ECD I/55 in Tours, lack of adequate French command in the area made the conditions of work chaotic. The early-warning system was totally ineffective and the fighters were always "scrambled" too late to be able to intercept. The French flights with which the Poles were to cooperate were unwilling to engage the Germans and this led Krasnodębski to request that his Flight be sent up only with the Czechs and to ask for more Polish pilots to double the strength of his unit. Aircraft of the French units were poorly utilized and the French mechanics, inadequately trained, left most of the major or difficult repairs to their Polish counterparts. It also appears that the "Kr" Flight had no aircraft for its own exclusive use, but most of the time flew different machines, usually the worst available. The first C.O. of ECD I/55, Mjr Merlin, was fair and helpful to the Poles, but both his successors, Dhomme and Rabatel, were quite the opposite.[60]

The "Kr" Flight's first encounter with the enemy came on 3 June, during the bombing of the airfield. The Dorniers were already overhead when the take-off order was given. Sgt Mieczysław Żołnowski started the engine of Zumbach's MS 406 and ran to the next machine when an exploding bomb tore his arm, which had to be amputated. A few fighters managed to get airborne one by one, one Frenchman among them, but Krasnodębski's landing gear failed to retract and he was unable to take part in the fight. Zumbach found his guns to be jammed and made three inconclusive passes at the bombers firing only his cannon. Each time he was chased away by the escorting Bf 110s and after running out of ammunition he had to withdraw. Karubin also attacked a Dornier and probably damaged it.

On 10 June the Poles made a few flights on the Bloch MB 152s and on the following day the "Kr" Flight converted to Curtiss Hawk 75A fighters and went to the Villacoublay airfield to strengthen the defences of the Paris area. On 11 and 12 June the Flight carried out three missions from Villacoublay and later on the second day returned with its Curtiss Hawks to Etampes.

Krasnodębski's efforts to obtain more Polish pilots eventually bore fruit. The three-pilot flight, comprising Lts Skiba and Antolak and NCO Machowiak, which was initially posted to GC II/8, was reassigned to the "Kr" Chimney Flight and joined it on 12 June at Etampes. The new arrivals received three MS 406s and Krasnodębski's unit was reformed, with Zumbach taking command of the second flight incorporating Antolak and Machowiak. On 13 and 14 June the double-strength "Kr" Flight patrolled the Etampes area and the roads leading to Paris. The airfield was bombed again on the 14th. On the same day Karubin and Machowiak took off on their own initiative and spotted a few flights of Bf 109s. Machowiak, flying higher, attacked one of the flights from the rear and Karubin engaged another, but with no visible results. After a brief dogfight with the scattered Messerschmitts, contact with the enemy was lost.[61]

On 15 June ECD I/55 was instructed to move all its flights to Orleans, where the Poles were withdrawn from the Squadron. The "Kr" Flight was split. Krasnodębski with Skiba, Bełc and Karubin, provided with three Hawks, were ordered to Limoges, and Zumbach's flight, reinforced with (three?) Czech pilots, to Clermont-Ferrand. Both flights made an interim landing at Chateauroux and made two patrols over the local area in the direction of Orleans, Zumbach flying on this occasion an Arsenal VG 33 fighter.

Two Hawks of Krasnodębski's flight arrived at Limoges on 16 June, with Skiba having to leave his defective fighter at Chateauroux. The French transport column serving the flight never reached the Limoges airfield, which was in chaos and short of fuel and ammunition. No further operations were possible. Two days later Krasnodębski received French permission to fly to Rochefort. Skiba, with the Polish mechanics, went by road to Bordeaux, and then directly to the port of La Rochelle, to join Krasnodębski and the evacuation transport to England.

[60] Main source of the above: Zumbach's report (IPMS, Lot.A IV, 2/1a/2-4).

[61] Based on Skiba's accounts (IPMS, Lot.A IV,2/1a/2-2 & -5), Machowiak's account (A IV,2/1a/2-3) and Antolak's account (A IV, 2/1a/2-6).

Zumbach's flight was redirected from Clermont-Ferrand to Bordeaux, from where Zumbach flew his last operational sortie (on the Hawk) soon after his arrival on 17 June. There he received orders to proceed to North Africa, but the next day the French C.O. commandeered all the aircraft flown by the Poles and left them to travel to evacuation ports as best they could.

During the period of its operational activities, from 17 May to 17 June, the I. "Kr" Chimney Flight carried out about 35 missions, which comprised some 120 sorties in about 130 flying hours, these totals including 15 sorties in 17 flying hours logged by the three additional pilots who joined the Flight on 12 June.

(3.) The II."Op" Chimeny Flight (No.2 Team) – Romorantin (ELD/Romorantin)

The II."Op" Chimney Flight began to form – initially as No.2 Team – under the command of Cpt Tadeusz Opulski in response to the Armee de l'Air Directive no. DM 1525 I/IS EMAA of 11 May 1940. Apart from Opulski, the Flight originally comprised five pilots, S-Lts Jan Daszewski and Witold Łokuciewski and NCOs Ryszard Górecki, Karol Krawczyński and Leon Nowak, as well as a 15-strong ground crew under Sgt Mjr Julian Nowicki. The unit was entrusted with the defence of the EAA 301 Air Army Supply Store and aviation industrial establishments at Romorantin, and left Lyon on 16 May.

The II."Op" Chimney Flight arrived at Romorantin on 17 May and initially received three MS 406 fighters. Two days later Daszewski and Nowak were recalled to Lyon (Order no. R.135/40). However, as the Flight was intended to provide the main force of the emergent ELD/Romorantin, the French – bearing in mind the importance of this area – requested that its strength be doubled. Consequently, three pilots from the organizing V."Ła" Chimney Flight intended to be posted to Bourges, Lt Wacław Łapkowski (C.O.), S-Lt Jan Daszewski (two days earlier recalled from Romorantin) and NCO Zygmunt Rozworski, were re-assigned to the II."Op" Flight on 21 May. The fourth pilot originally allocated to the planned "Ła" Flight stayed behind and eventually went to Bourges. The enlarged "Op" Flight comprised: Cpt Opulski (C.O.), Lt Łapkowski (deputy C.O.), S-Lts Daszewski and Łokuciewski, and NCOs Górecki, Krawczyński and Rozworski. Three more Moranes were added to the Flight.

The Flight patrolled the Romorantin-Vierzon-Salbris area, the intensity of its operations increasing towards the end of May. On 1 June the Polish pilots engaged 14 Heinkels which attacked the airfield and Opulski and Daszewski achieved one confirmed victory each. On 4 June an enemy aircraft was reported in the vicinity of Romorantin. Opulski, who led the patrol, sighted a single-engined low-wing machine which he took for a Ju 87 and shot it down, killing one of its crew. This, however, proved to be a tragic mistake, since the aircraft, a Vought V-156F, belonged to the French Naval Aviation.

On 10 June Romorantin and its airfield were bombed for the second time. Four Polish fighters, flown by Opulski, Łokuciewski, Krawczyński and Rozworski, were scrambled around 14:00 hours and came into contact with 18 Heinkels escorted by Bf 110s. In the outcome of a bitter air battle Łockuciewski, Krawczyński and Rozworski were credited with the destruction of one He 111 each. A fourth Heinkel was also claimed by the Poles, but it was officially awarded to the French C.O. of the ELD/Romorantin who joined the flight. During one of the take-offs Opulski lost a wheel off his Morane, but made a successful landing on one wheel.

The II."Op" Chimney Flight ceased to operate on 18 June having flown about 20 combat missions, which involved some 75 sorties in 80 flying hours. With five confirmed air victories, the unit was the most successful Polish Chimney Flight in France. All five victorious pilots were nominated to receive the Croix de Guerre.

(4.) The III. "Ku" Chimney Flight (No.3 Team) – Nantes (ECD/Nantes)[62]

The III."Ku" Chimney Flight, known in its preliminary stages as No.3 Team, was organized under Directive DM1525 I/IS EMAA. The Flight consisted of four pilots, Cpt Kazimierz Kuzian as its C.O., S-Lt Mirosław Ferić, NCO Ludwik Lech and NCO Stefan Wojtowicz (Ferić and Wojtowicz being recalled from the I."Kr" Chimney Flight), and a ground crew of 13 men, with Sgt Michał Jarmoc in charge.

The III."Ku" Chimney Flight was assigned to the ECD/Nantes Squadron with the task of defending the SNCAO LeO 45 assembly plant and left Lyon on 20 May arriving at the factory's Nantes-Bouguenais airfield the following day. The Poles were to receive their aircraft and full ancillary equipment at the factory, but nothing was ready.

They were sent instead to the Tours air base to collect their Moranes and waited three days there before the fighters were made available. The three MS 406s were decrepit, unarmed, ex-training machines, which still had to be fitted with machine guns and had no provision for cannon armament. The pilots returned to Nantes only on 30 May, but had no liaison of any sort (not even a telephone link) with early-warning observation posts. Because of the lack of suitable operating facili-

[62]Main sources: Kuzian's accounts (IPMS, Lot.A IV, 2/1b/6-7 & -8) and Ferić's report (IPMS, Lot.A IV, 2/1b/6-9).

A Polish mechanic standing by MS 406-C1, c/n 1038, of the III. "Ku" Chimney Flight at Nantes in early June 1940. The aircraft features the Bronzavia flame-damper shrouds.

ties and the passive attitude of the French ECD/Nantes pilots, on 1 June the Flight began to work in cooperation with the locally based RAF Fighter Squadron. A further two days were needed to bring the Polish Moranes to acceptable operational standards and eventually the Flight achieved its full combat capability on 3 June, standing in readiness from dawn to dusk.

On 5 June, at the request of Kuzian, two more Polish pilots, Cpt Władysław Szczęśniewski and S-Lt Sergiusz Szpakowicz, from the Tours air base, reinforced the III."Ku" Chimney Flight. From 6 June onwards the Polish mechanics also began to service the fighters of a newly arrived French flight. By 10 June activity considerably increased, amounting to two or three patrols over the Nantes-Angers-St. Nazaire area a day, but most take-offs proved to be either false alarms or were ordered too late to permit interception and the Flight made no contact with the enemy.

On 14 June the French flights suddenly left Nantes and Kuzian ordered Ferić and the two youngest pilots to fly to Bordeaux. The rest of the "Ku" Flight followed by road using transports from the RAF base which was also closing down. All the personnel eventually went via Port-Vendres to Casablanca.

The III."Ku" Chimney Flight carried out about 12 missions, amounting to some 37 sorties in 40 flying hours.

(5.) The IV. "He" Chimney Flight (No.4 Team) – Cheteauroux (ELD/Cheteauroux)[63]

Conceived in response to the Directive DM1525 I/IS EMAA as No.4 Team, the IV."He" Chimney Flight, as initially organized in the middle of May, comprised four pilots, Lt Zdzisław Henneberg, the Flight's C.O., S-Lt Stanisław Wielgus, NCO Paweł Gallus and NCO Ryszard Lewczyński, and 15-strong ground crew with W/O Gustaw Pokrzywka in charge of the mechanics and W/O Wacław Cesarz in charge of the armourers. With its assignment to the ELD/Chateauroux for the defence of the strategically vital SNCASO aircraft factory at Chateauroux-Deols (producing the Bloch fighters) and No.103 Aviation Base, the strength of the Flight was more than doubled by the inclusion of five more pilots, Lt Marian Imiela, and S-Lts Brunon Kudrewicz, Witold Retinger, Mieczysław Waszkiewicz and Benedykt Zieliński, but its ground personnel was reduced to 13 men.

The whole "He" Flight arrived at Cheteauroux-La Martinerie airfield on 20 May and received from Deols six unarmed MB 152 and 151 fighters (including MB 152 c/n 315). At about 20:30 hours La Martinerie was bombed, but since none of the Blochs were combat ready the Flight was unable to intervene. A similar situation prevented the Poles from intercepting the next enemy raid, which took place on 25 May before noon. The Flight achieved combat readiness only on the next day with all its Blochs fully armed and equipped, and a number of abortive scrambles followed.

The Flight's major test came on 5 June. After a take-off at 13:25 hours, the pilots were directed by R/T to fly first a southerly course and then back towards the airfield. While still climbing, they sighted about 17 Heinkels approaching La Martinerie, but before they reached sufficient height to intercept, the Germans had dropped about 150 bombs. Lewczyński had to abandon the sortie because of oxygen failure, but Henneberg, with Gallus as his wingman, attacked the rear of the formation, Henneberg firing at the middle Heinkel and closing to less than 100 m. The German, with one engine on fire, went into a shallow dive, banking from side to side. At the same time he was also engaged by Cebrzyński of the "Ce" Flight attached to the GC II/6 and based at the Deols factory airfield. He had already set the rear right-hand bomber on fire and then followed Henneberg's victim. Both pilots repeatedly attacked the Heinkel until it crash-landed 3 miles southwest of Vatan. Its five-man crew was captured on the ground. Both pilots were credited with 1/2 the aircraft each. Meanwhile, Gallus continued to fire on the rear left-hand bomber and damaged it, but when his cannon and one of his two guns jammed and the ammunition from his other gun was expended he was forced to break-off the fight.

Intensive patrol activities continued during the next 11 days, the last operational mission from La Martinerie being on 16 June at 19:00 hours in very poor weather conditions. The following day the French left the airfield, leaving the Poles without orders, supplies or any transport. At 19:00 hours four Polish pilots led by Lt Henneberg took-off in one MB 152, two MB 151s and one Caudron Simoun from Bordeaux, and the rest of the Flight's personnel followed by road in a requisitioned lorry. This group, under the command of S-Lt Waszkiewicz, reached the port the next morning, boarded a ship and sailed to England.

The four-aircraft "He" Flight moved from Bordeaux to Nantes on 18t June in the morning and at 14:00 hours took off for England, landing at Tangmere airfield two and a half hours later. At about 19:30 hours the Flight, the only PAF unit to fly over with its aircraft from France, was ordered to proceed to Andover. The next afternoon the pilots surrendered their machines to the British and were posted to West Kirby.

During its operations between 26 May and 17 June the IV."He" Chimney Flight flew some 25 missions entailing about 90 sorties in 100 flying hours, and was credited with one shared victory.

(6.) The V. "Kos" Chimney Flight (No.5 Team) – Bourges (ELD/Bourges)[64]

The defence of Bourges, as originally proposed in fulfillment of Directive DM 1525 I/IS EMAA, was to be entrusted to Łapkowski's four-pilot flight, provisionally referred to as the V."Ła"Chimney Flight. However, as a number of Polish pilots had just completed their operational training at the Avord

[63]Main sources: Henneberg's account (IPMS, Lot.A IV, 2/1b/4-2) and Imiela's account (A IV, 2/1b/4-3).

[64]Main source: Kosiński's account (IPMS, Lot.A IV, 2/1b/2-2).

Fighter School near Bourges, the regional I° Subdivision Aerienne asked for a local defence squadron to be formed from these graduates, and Łapkowski with two of his pilots were incorporated instead into the II."Op" Chimney Flight, which required reinforcement.

The Avord pilots, selected on 16 May for the forming ELD/Bourges Squadron – No.5 Team, which became the V."Kos" Chimney Flight – comprised Cpt Bronisław Kosiński, appointed to command the Flights, Lts Zbigniew Moszyński and Marian Wesołowski, and NCOs Wacław Giermer, Jan Kremski and Władysław Majchrzyk. They were joined by the seventh pilot, NCO Adolf Pietrasiak, and a ground crew of 13 men under W/O Aleksander Gajewski transferred from the proposed Łapkowski's Flight, who arrived from Lyon on 21 May to complete the establishment of the "Kos" Chimney Flight. The unit's main task was to protect the SNCAC factory assembling the Curtiss Hawk 75A fighters.

The "Kos" Chimney Flight was initially provided with three Curtiss Hawks, and, after brief familiarization training, Giermer became its first pilot to fly an operational sortie – with a French flight. On 24 May Wesołowski, Giermer and Kremski were scrambled and intercepted a formation of 14 Heinkels. After a determined chase, the Poles brought down one He 111 and claimed two more damaged, although Wesołowski had to make an emergency landing with his badly shot up machine on the airfield. The other two pilots who pursued the raiders for as long as their ammunition lasted, ran out of petrol and also had to make emergency landings – Giermer in a damaged fighter. They returned to base after refueling. The destroyed bomber was at first credited to Kremski, but eventually the victory was officially shared between the three pilots (1/3 each) and so were the two damaged Heinkels (2/3 each). Following this successful first encounter, the "Kos" Flight received "privately" four more Hawks from the grateful management of the SNCAC factory, with no.188 allocated to Kosiński.

Two or three take-offs a day became a normal routine, but most of these were ordered too late because of belated warnings from the observation network. On 5 June four Polish pilots operating in pairs engaged head-on an approaching raid by 18 Heinkels and then continued the attack from the rear. After the bombers by-passed Bourges, the Poles noticed a second formation of nine He 111, which began to bomb the airfield. Kosiński ordered his fighters to abandon the chase after the first wave and concentrate its attack on the second, which was followed and fired on for 50 minutes, until all ammunition had been expended. A French flight, composed of test pilots and commanded by S-Col Haegelen, also participated in the battle. Three Heinkels were confirmed as destroyed and two as damaged; 2 2/5 of the destroyed He 111s and 1 3/5 of the damaged ones were credited to the Poles: 3/5 destroyed and 2/5 damaged each to Kosiński, Wesołowski, Kremski and Pietrasiak. The remaining 3/5 destroyed and 2/5 damaged went to the French Commander.

All subsequent patrols proved abortive. The last fighter sortie from Bourges was carried out on 16 June against a solitary Do 17 which escaped. Wesołowski's reconnaissance revealed German troops about 30 miles from the airfield. Pietrasiak, who became ill, was sent to hospital. On the morning of 17 June the "Kos" Flight was ordered to proceed to Perpignan. During a landing at Limoges to refuel the Poles learned about the capitulation of France. A few days later the Flight crossed to Algeria where it surrendered its aircraft. Its personnel were eventually shipped to England.

Operating from Bourges from 20 May to 17 June, the V."Kos" Chimney Flight carried out about 30 missions, involving approximately 120 sorties in 130 flying hours, and achieved 3 2/5 confirmed air victories, with a further 3 3/5 counted as damaged. Kremski, its first victor, was awarded the Croix de Guerre avac Palme and Kosiński, Wesołowski and Giermer were nominated to receive the decoration when the end of the French campaign intervened.

(7.) The VI. "Sa" Chimney Flight (No.6 Team) – Toulouse (ELD/Toulouse)[65]

The last Polish flight to be organized to comply with Directive DM 1525 I/IS EMAA was the VI."Sa" Chimney Flight – initially No.6 Team – which included Lt Czesław Sałkiewicz,[66] the unit's C.O., S-Lt Bronisław Kłosin, NCO Tadeusz Nastorowicz and NCO Władysław Wieraszka, and a ground crew of 21, with Sgt Mjr Franciszek Lipski in charge of the mechanics. The Flight was to provide air defence for the SNCAM plants at Toulouse-Blagnac and Toulouse-Francazal, which produced the Dewoitine D.520 fighters. The four pilots and servicing personnel left Lyon on 20 May in the evening and after arriving at Francazal airfield the next day, were reinforced by three test pilots from the Dewoitine factory and received three D.520s.

The Poles immediately set up a stand-by flight, but had no links of any description with early-warning posts and encountered great functional and administrative difficulties, which took almost two weeks to be fully resolved. On 30 May at 20:10 hours, Lt Sałkiewicz, returning from a patrol, was killed while coming in to land in his D.520 no.99. On 1 June S-Lt Stanisław Szmejl was ordered to take over the command of the Flight and reached Toulouse a few days later. On 9 June S-Lt Stanisław Czarnecki, C-Off Adam Damm and C-Off Longin Majewski, who two days earlier were withdrawn from the VIII. Chimney Flight (1° Escadrille Koolhoven Salon-Istres) and posted to Toulouse to gain operational experience with the VI."Sa" Chimney Flight, came to Francazal. The next day Szmejl obtained three Morane MS 406 fighters to train the new arrivals to operational standards. The three D.520s, with experienced Pol-

[65]Main source: Szmejl's account (IPMS, Lot.A IV, 2/1b/10-2).

[66]Most French sources misspell this name as Talkiewicz and some Polish publications repeat the same mistake.

ish pilots on stand-by, were kept in constant readiness from dawn to dusk, while the Moranes were used for combat proficiency training, with Czarnecki joining the operational rotation within a few days. The SNCAM Toulouse factories were never bombed and the Flight did not report any contact with the enemy.

On 21 June the three D.520s, piloted by Szmejl, Kłosin and Czarnecki, proceeded to Perpignan and then flew to Algeria. The rest of the Flight's personnel traveled by road to Port-Vendres and were evacuated by sea.

Between 21 May and 21 June, the VI."Sa" Chimney Flight made about 15 missions, amounting to some 35 aircraft sorties in 40 flying hours. The Flight lost its first C.O. in a landing accident after an operational sortie.

(8.) Jasionowski's VII. and VIII. Chimney Flight (No.8 and No.9 Teams) – 1° and 2° Escadrille Koolhoven/Clermont and Salon

The Armee de l'Air Directive DM 1825 I/IS EMAA of 22 May 1940[67] called for the formation of five more "Polish fighter patrols for the defence of sensitive points of the Interior" at the Lyon Air Base by 25 May, with assignments as follows:

> I – One patrol for the defence of Angers, to be composed of six pilots including two (or, if possible, three) officers and four (one reserve) Bloch 151 fighters, with the ground personnel as specified in the Wartime Tables annexed to the Note No.1578 I/IS EMAA from 14 May 1940, to be based at Angers under the Commanding General of the 3rd Aviation Region.
>
> II – Two patrols for the defence of Clermont-Aulnat, to be composed of one officer in overall command of the two patrols and six pilots including two (or, if possible, three) officers, with the ground personnel as under Paragraph I; to be provided with six Koolhoven fighters and based at the Aulnat Base under the Commanding General of the 4th Aviation Region (these patrols are to replace the patrol provisionally set up from the pilots of the base).
>
> III – Two patrols for the defence of Salon-Istres, to be staffed and equipped in the same way as under Paragraph II for the defence of Clarmont Aulnat.
>
> IV – The patrol for the defence of Angers was to have top priority. As far as the four other patrols were concerned, their personnel and equipment (the Koolhoven fighters) were to be directed to the Salon Evaluation School for appraisal, and the first two patrols approved were to go to Aulnat and the remaining two were to stay at Salon. The patrols were to be operative by 30 May, if at all possible.

[67] Copy of the document: The French File, J.B. Cynk, Polish Aircraft Archives.

S-Lt Kłosin in the cockpit of a Dewoitine D.501 of the Toulouse Fighter School before the formation of the VI. "Sa" Chimney Flight.

For practical reasons, the execution of this Directive proved rather prolonged, varied in various details and did not follow the desired order of priorities. Although the selection of personnel for the Angers Patrol (No.7 Team) seemed to have been given first priority by the PAF Command, of the two other double-patrols (No.8 & No.9 Teams), under the overall command of Cpt Walerian Jasionowski, were organized more quickly.

On 22 May Cpt Jasionowski was called upon to form immediately two seven-pilot teams for familiarization training on Koolhoven fighters at the Salon School. He selected eight Officers, five Cadet-Officers and one NCO, and a 34-strong ground crew of 16 NCOs and 18 airmen of various trades.[68] The teams were as follows:

> - The VII. Chimney Flight (No.8 Team) – 1° Escadrille Koolhoven, intended for Clermont-Aulnat: Lt Bogdan Grzeszczak, the Squadron's C.O., S-Lts Wiesław Czapliński, Stanisław Łukaszewicz and Stanisław Wandzilak, C-Offs Jerzy Cechrzycki and Bogdan Muth, and NCO Jan Lipiński; ground crew of 17 men, with Sgt Kazimierz Moczulski in charge.
>
> - The VIII. Chimney Flight (No.9 Team) – 2° Escadrille Koolhoven, intended for Salon-Istres: Lt Walerian Żak, the Squadron's C.O., S-Lts Stanisław Czarnecki, Andrzej Malarowski and Jerzy Wolski, and C-Offs Adam Damm, Tadeusz Hegenbarth and Longin Majewski; ground crew of 17 men, with Sgt Błażej Żulik in charge.

Jasionowski's group left Lyon on 25 May in the evening. On 28 May the first three Koolhoven FK 58 fighters were taken on charge (including nos. 6 and 14) and intensive training began the same day. On 30 May Jasionowski received the first operational order, to patrol the Marseille area, although the Koolhovens lacked oxygen masks and their armament was not yet fully aligned and tested. The fourth FK 58 was delivered in the first days of June. During the nine-day period ending on 7

[68] Jasionowski's account (IPMS, Lot.A IV, 2/1c-6).

A Hispano-Suiza-powered FK 58 fighter of the 1e Escadrille Koolhoven at Clermont-Aulnat in June 1940.

One of the FK 58A fighters with the Gnome-Rhone radial engine of Cpt Jasionowski's Chimney Flights. The aircraft is fully armed and bears the individual number 11.

June Jasionowski's Chimney Flights, flying from Salon, carried out 27 operational sorties, patrolling the Salon-Avignon-Istres-Marseille area.[69] Some of these were flown in MS 406 fighters belonging to 130th Battalion de l'Air.

However, the demands of the Directive DM 1825 were highly unrealistic. The Koolhoven fighters, 18 of which were delivered to France from the Netherlands in June-September 1939 (seven early production Hispano-Suiza-powered FK 58s and 11 Gnome-Rhone-powered FK 58As, all armed with four Belgian 7.5-mm FN-Browning guns in underwing housings) were originally intended for use in French colonial territories. The aircraft were at first distributed unarmed among various training establishments in Metropolitan France, and later, in the middle of May, some were allocated to Chimney Flights. Pulling in the 12 Koolhovens needed at Salon turned out to be a lengthy and difficult process and the aircraft as they arrived still had to be armed and fully equipped to become combat-worthy. The situation gave rise to acidic Polish comments on the "marvels" of Allied cooperation: "Dutch airframes, French engines, Belgian guns and Polish pilots under French skies," and no effective unit to fight the Germans!

Because of a shortfall in the required number of fighters, the Commanding General of the 4th Aviation Region on 7 June ordered three Polish pilots – Czarnecki, Damm and Majewski – and six mechanics to Toulouse to strengthen the VI."Sa" Chimney Flight of the ELD/Toulouse. At the same time Jasionowski's VII. and VIII. Chimeny Flights, reduced in strength to 12 pilots and 28 ground crew and equipped with four Koolhovens, were ordered to move to Clermont-Aulnat as one squadron and provide air cover for the base and the Morane-Saulnier factory.

Immediately as he arrived at Aulnat on 8 June, Jasionowski set up a combat flight of three aircraft with a rota of pilots, although the practical ceiling of the FK 56s was still limited to some 5,000 m because of a continued lack of oxygen masks. In fact the masks and spares for the fighters, which were ordered when the Poles were still at Salon, never reached the unit. One further FK 58A was received at Aulnat (no. 8, flown mainly by Wandzilak), but urgent requests for more machines remained unanswered. Jasionowski learned by accident that a few Koolhovens were undergoing a refit at Nevers and tried in vain to obtain their release. Only after a sharply worded complaint to the Commandant of the VIIe Subdivision Aerienne was he allowed to collect three aircraft, none of which was armed. Polish mechanics were to bring them to operational standards.

On 16 June three Polish pilots were flown to Nevers to ferry the fighters to Aulnat, but Czapliński, the last to leave Nevers, at 14:45 hours, never arrived and was at first posted "missing." It was later established that he was shot down and killed by German fighters. On the same day S-Lt Kolubiński and S-Lt Hoyden from the dissolved Rennes Chimney Flight joined Jasionowski's Squadron.

On 17 June Jasionowski obtained two more Koolhovens, which, brought to the base's Maintenance Park in a damaged condition, were repaired by his mechanics. Learning that Browning machine-guns suitable for Koolhovens were available at the Curtiss assembly plant at Bourges (Hawk 75A fighters used the same type of guns), he sent Wolski to Bourges in a borrowed transport plane and got 24 guns. At the same time he despatched Grzeszczak in another aircraft to Nevers for spare parts for the fighters. The plane was attacked upon landing by 15 Messerschmitts, but Grzeszczak survived and brought the spares in a lorry. By nightfall of 17 June his Koolhoven Squadron had nine FK 58 and FK 58A fighters but only six of them were armed. The guns obtained from Bourges needed modifications before they could be installed in Koolhovens.[70]

Jasionowski's resourcefulness earned him the admiration of the French Commanders and on 17 June he was ordered to take all available combat-worthy aircraft from the base, regardless of type, and to organize them into a special assault squadron to attack German tanks. Alarming stories, which began to reach Aulnat on 18 June, prompted Jasionowski to act on his own initiative and he sent Grzeszczak on a low-level reconnaissance of the Vichy-Moulins area. The pilot, who was fired at from the ground, returned at 20:45 hours, reporting many fires and destroyed bridges but no immediate danger from enemy troops. From 8 to 18 June, flying from Aulnat, the Poles logged 20 sorties.

---

[69] Ibid.
[70] Ibid.

On 19 June the Poles were ordered to move to Rodez. In addition to the nine Koolhovens, Jasionowski took with him four light planes which carried the remaining pilots and five mechanics, the rest following by road. In Rodez he learned about Gen Sikorski's order for the evacuation of the Polish troops to England. On 20 June he was instructed to fly to Montpellier, and the following day to Perpignan where the Poles were requested to surrender all their machines and advised to proceed by road to Port-Vendres. They found the port already closed and had to travel by rail across France, to Bayonne. There they managed to board a ship bound for England on 24 June.

Between 30 May and 13 June the VII. and VIII Chimney Flights, also known as 1° and 2° Escadrille Koolhoven, put up some 20 operational missions, accounting for 47 sorties in about 60 hours flying time. Although no encounters with the Luftwaffe were formally recorded, one fighter was lost and one pilot killed due to enemy action.

(9.) The IX. "Jan" Chimney Flight (No.7 Team) – Angers (ELD/Angers)[71]

Organization of the IX. "Jan" Chimney Flight for the defence of Angers, initially referred to in some of the Polish documents as No.7 Team, was prompted by the Directive DM 1825 I/IS EMAA of 22 May 1940. The unit comprised: Lt Robert Janota, recalled from the I/145 Warsaw Fighter Wing to command it, Lt Władysław Walendowski, S-Lt Edward Kowalski and NCOs Antoni Siudak, Marian Wędzik and Stanisław Zięba, as well as a ground crew of 15 men, with Sgt Bronisław Mitek in charge of the mechanics. Angers was the seat of the Polish Government in Exile and HQ of Gen Sikorski, and it was only appropriate that a Polish Flight should be involved in its protection.

The IX."Jan" Chimney Flight left Lyon on 25 May in the evening and arrived in Angers late the next afternoon. Janota had to await an order detailing allocation and collection of aircraft. This was not issued until 31 May and required the Polish pilots and ground personnel to proceed to Tours on 1 June, to accept four MB 151 fighters (three to provide combat equipment and one as reserve) and undergo a familiarization and combat-proficiency training at the Tours air base. The Blochs were received the following day and training began in earnest on 3 June, with Janota making the first operational sortie with the French ECD I/55 Squadron, which had in its strength a solitary Polish pilot, C-Off Ryszard Budrewicz.

On 4 June the Poles made three emergency take-offs. The following day at about 13:30 hours, before the "Jan" Flight commenced its daily stand-by duty, Ju 88s bombed the airfield without warning. The French fighters gave chase immediately afterwards and claimed two bombers shot down. The Polish pilots were scrambled at 18:00 hours but made no contact with the enemy. During the period up to 14 June the "Jan" Chimney Flight operated from Tours in conjunction with the ECD I/55. On 14 June the Poles finished their last "readiness" rota at Tours at 14:00 hours. An hour later the airfield was the target of another heavy raid which caught the defences unawares.

On 15 June at 7:00 hours the Flight was ordered to return to Angers with the task of defending the city, its bridges and its airfield. Three Blochs took off immediately, with the fourth, temporarily unserviceable, following later. Three missions were flown on that day from Angers, one of them in response to a mistaken early warning of the RAF Blenheims, which, however, were correctly identified by the fighters and not attacked. On 16 June four take-offs were made, two of them on the Flight's initiative, on warnings from its own listening post. Three more missions followed on the 17th before noon, none of them leading to contact with the enemy. At 12:00 hours the Flight was ordered to withdraw to Nantes, where it landed at 14:30 hours. That same evening three pilots, Janota, Walendowski and Zięba, proceeded to La Rochelle. Kowalski in the fourth aircraft was unable to start its engine. On 18 June in the afternoon the three fighters moved to Bordeaux, with the rest of the personnel traveling by road. On 19 June almost the whole Flight, with the notable exception of Janota who went missing, gathered in the port and embarked for England. Kowalski managed to fly from Nantes directly to England.

During its operational period 3 to 17 June, the IX."Jan" Chimney Flight carried out about 25 missions, involving some 55 sorties in 60 flying hours, but reported no engagements.

(10.) The X."Fa" Chimney Flight – Cognac (ELD/Cognac)[72]

The X."Fa" Chimney Flight began to form in response to the Directive DM 1952 I/IS EMAA of 28 May 1940. This asked for a fighter patrol for the defence of the EAA 302 Air Army Supply store at Cognac. The Flight was composed of Lt Jan Falkowski, the unit's C.O., S-Lt Tadeusz Kratke, S-Lt Ryszard Malczewski and five NCOs, Tadeusz Andruszkow, Zygmunt Gruszczyński, Stefan Krzyżagórski, Stanisław Łoś and Mieczysław Popek; its ground personnel was 23 strong, with Sgt Józef Popkowski in charge of the mechanics.

Arrival of the Flight at Cognac came as a complete surprise to the Base Commander, who had no idea what to do with the Poles and sought help from the Tours air base to find out. In the meantime the Flight personnel was used for digging anti-aircraft trenches on the airfield perimeter. On 7 June the pilots were ordered to go to Bordeaux "for training," although they all produced their log-books, which proved their capabilities. At Bordeaux no one knew why the Poles had been sent there. After due deliberations, each pilot was instructed to make one control flight on a Dewoitine D.501 fighter-trainer at Bordeaux. The group was then posted back to Cognac to await delivery of MS 406 fighters. Digging of trenches was resumed.

On 15 June the "Fa" Chimney Flight at last received its aircraft: five weary MB 152s, only three of which were com**D.**

[71]Main source: Walendowski's account (IPMS, Lot.A IV, 2/ab/1-2).

[72]Main source: Falkowski's account (IPMS, Lot.A IV, 2/1b/5-3).

bat-worthy. Falkowski immediately set up a stand-by flight with rota of pilots. Up to and including 18 June about ten patrols were carried out, but the Germans flew only reconnaissance sorties by single aircraft at very high altitudes and no contacts with the enemy were reported.

On 19 June the Flight learned about the Polish evacuation order and traveled at night by road to Bordeaux and then continued via Biarritz to the port of St. Jean-de-Luz. There its personnel boarded the Polish ship "S/S Batory" and sailed for England on 21 June at dawn.

The X."Fa" Chimney Flight, operating for only four days – 15 to 18 June – flew about ten missions, entailing 25 sorties in 30 flying hours. No contact with the enemy was made.

(11.) The XI. "Kow" Chimney Flight – La Rochelle (ELD/La Rochelle)[73]

The XI. "Kow" Chimney Flight was established to fulfill requirements of the Directive DM 2022 I/IS EMAA of 31 May 1940. This called for a fighter patrol for the protection of "the La Rochelle Grouping of the Fighter Schools from enemy reconnaissance and bombardment." The Flight embraced Cpt Adam Kowalczyk, its C.O., S-Lts Włodzimierz Klawe, Jan Marciniak and Tadeusz Szumowski, and NCOs Władysław Kiedrzyński and Stanisław Widlarz, as well as 23 ground personnel with the Technical Officer, Lt Wacław Wiórkiewicz, in command.

The Flight left Lyon on 6 June, arriving at La Jarne airfield near La Rochelle the following day. After a few days of waiting, the Poles received four unarmed Bloch fighters of the MB 151 and 152 type (one intended as the reserve machine) in very poor condition. Kowalczyk's aircraft caught fire on its first flight from La Jarne due to a perforated carburetor and leaking fuel pipes. The mechanics encountered great difficulties in bringing the fighters to combat-worthy standards because of a lack of spare parts. A shortage of ammunition was another factor hampering the work of the Flight. Three machines were kept on stand-by most of the time, with pilots on duty from 4:15 until 21:00 hours, changing every four hours. Enemy aircraft were sighted over the airfield four times, but without early warning the fighters could not take off in time to make contact. On 19 June the Poles left the airfield on their own initiative in response to the general Polish evacuation order. They were joined by Col Pawlikowski in the port of La Rochelle and reached England on 23 June.

The XI. "Kow" Chimney Flight, operating for about a week, carried out about eight missions, involving some 20 sorties in 22 flying hours.

S-Lt Hoyden standing by a Bloch MB 152-C1 fighter during training at Rennes.

(12.) The "Kol" Chimney Flight – Rennes[74]

The Flight commenced its existence on 6 March 1940 as a fighter training unit, when S-Lt Antoni Kolubiński arrived at the Rennes Training Center with three pilots, S-Lt Tadeusz Andersz, S-Lt Tadeusz Hoyden and NCO Józef Kędziora, under his command. On 14 March S-Lt Eugeniusz Maliszewski joined the Flight.

After control flights on Morane MS 230 trainers, the Poles received three Dewoitine D.501 fighters and began intensive fighter training which also included a few flights on the Bloch fighters. Early in May Maliszewski was posted to Dinard, taking with him one of the D.501s. Another D.501 needing a major engine overhaul was temporarily withdrawn from service, and the only remaining Dewoitine was armed when the German offensive began and put at the disposal of the four Polish pilots, who from 15 May onwards became known as the "Kol" Chimney Flight and stood in readiness to defend the airfield. On 30 May NCO Kędziora, sleeping on the grass of the airfield, suffered a tragic death when a French bomber rolled over him.

The second D.501 fighter returned to the Flight and the pilots operated a duty rota from dawn to dusk until 13 June. During this time the Flight was scrambled three times, the pilots involved being Kolubiński and Andersz. On 13 June the Flight was dissolved and Kolubiński and Hoyden were ordered to fly its two D.501s to the Clermont-Aulnat Base. Three days later both these pilots were assigned to Jasonowski's 2^{e} Escadrille Koolhoven.

[73]Main source: Kowalczyk's account (IPMS, Lot.A IV, 2/1b/8-2).
[74]Main source: Kolubiński's account (IPMS, Lot.A IV, 2/1b/7-1).

Single Polish Pilots in the French Fighter Units

Available Polish and French documents indicate that at least two Polish pilots, Budrewicz and Larczycki, served individually in French operational fighter units. How and why this came about is not clear from records. The most likely explanation appears to be that these Poles were encouraged to join French units while in training in French establishments.

C-Off Ryszard Budrewicz, who was posted from Lyon to the Tours Training Centre on 1 March 1940, flew operationally with the ECD I/55 Squadron – Tours, from late May until 18 June, logging about 30 sorties on Blochs. According to Lt Walendowski's report, the pilot received no service pay and his behavior and abilities earned him the respect of his French colleagues.

NCO Józef Larczycki flew operationally with the ELD/ Chateauroux from its formation on 11 May until 5 June. Piloting the Bloch MB 151 no.93, he took part in a fight with a large enemy formation over the airfield on 5 June and was shot down by Bf 110s and killed.

5. Attempts at Air Liaison with Occupied Poland

The Polish Government in Exile was very anxious to establish air liaison with occupied Poland, where a military underground movement had already begun to take shape when Warsaw capitulated at the end of September 1939. By Gen Sikorski's decree of 13 November 1939, Gen Kazimierz Sosnkowski was appointed to chair a special ministerial Home Affairs Committee and to command a new central Polish underground organization, the ZWZ (Związek Walki Zbrojnej, or Union for Armed Struggle), the forerunner of the AK (Amia Krajowa, or Home Army), and the need for courier and currency transfers to Poland became an issue of great urgency.

Air liaison was the obvious solution and Gen Sikorski's order of 28 November 1939[75] instructed Gen Zając to:

> "- organize permanent secret air communication links with major centers of the occupied country: Lwów, Warsaw, Kraków and, if possible, Pozań;
> - found such an organization on own equipment (light aircraft adapted for long-distance flying and airliners of the PLL "LOT") and on aircraft obtained from the Allies, who should be at once approached in connection with this matter;
> - establish, in consultation with Gen Sosnkowski, a list of personnel to operate such air communication links with the homeland (with the selected personnel to become members of a secret military organization and specially sworn in)."

YR-TIT, the Rumanian-registered long-range RWD 15, c/n 335, formerly SP-KAT (depicted at Saltz near Odessa in September 1941 with Bf 109s behind), which the Polish authorities wanted desperately to buy back to establish clandestine links with Poland. (A. Glass)

Two days earlier Gen Sikorski broached the subject of clandestine flights to Poland and special equipment required with the British Chief of the Air Staff, Air Chief Marshal Sir Cyril Newall, and asked him for a suitable aircraft to be instantly put at the disposal of Polish Government, so that the first flight could be made by 2 December at the latest. A similar approach was also made to Gen Romatet, the French Deputy Chief of the Air Staff. The proposed timescale was, however, quite unrealistic and responses from both Allies were discouraging. They pointed out that no aircraft with sufficient range to land in Poland and return home (parachute-drop techniques were at that time not yet developed by the Allies) were available and tremendous risks would be involved. Both Air Chiefs suggested that the Lockheed L14H airliners of the PLL "LOT" offered the best and quickest solution, but would have to be specially adapted. Air Chief Marshal Newall added that if the mission was that urgent, the aircraft could be destroyed after landing in Poland or fly on to the nearest neutral country.

As early as the middle of November Gen Zając had begun to explore the possibility of obtaining two Polish RWD 15 light tourers, which had been evacuated to Rumania, for flights to Poland. The SP-KAT, the long-range RWD 15 variant, specially developed for Stanisław Karpiński's attempt at the international distance record and a tour of Australia, Tasmania and New Zealand, was particularly suitable for the purpose. Unfortunately, the machine was sold by the DWL directors at a give-away price of £350 to a Rumanian owner in order to provide desperately needed funds for evacuation of DWL personnel to France and the new owner refused all Polish offers to buy back the tourer. The same applied to the second RWD 15, the SP-ALA.

The "LOT"'s Lockheeds remained the only option open and two of them, the SP-BNF and -BPM, were requisitioned in Great Britain by the Polish Aviation Command, the former to be speedily adapted for liaison flights to Poland and the latter to be overhauled and transferred to Paris for possible conversion later. The third L 14H, the SP-LMK, was to be used for training selected flight crews in Greet Britain.

Instructions for the modification of the -BNF were issued on 20 December 1939 and entailed installing additional fuel and oil tanks to increase its range to about 3,000 km (1,860 miles), alterations to the fuselage doors and fitting two fixed

[75] Order L.81/tjn (IPMS, Lot.A IV,1/3c file).

Two PLL "LOT"'s Lockheed L 14H airliners, the SP-BPM and -BNF, which the Poles intended to adapt for clandestine liaison flights to Poland.

pilot's guns in the forward fuselage and two flexible lateral guns in the rear side windows. The work, to be carried out by Airwork at Heston, was to be completed by 10 February 1940. At the same time Gen Zając sought to acquire in England six fast warplanes with 3,000 km (1,860-mile) range and three or four three-seat touring planes with 2,000-km (1,240-mile) range, for one-way flights to Poland. The Air Ministry was asked to help but declined on the grounds that no suitable aircraft were readily available and the aircraft industry was overloaded with defence orders.

The -BNF was not delivered to Heston until 17 January 1940 and preparation of detailed specification and workshop drawings took several days. By then the British civil aviation authorities, anxious about the safety of their existing transport routes in Europe, began to protest about the proposed use of civil airliners for military purposes. Despite Gen Zając's assurances that all civil markings would be removed from the -BNF and that the aircraft would be camouflaged in RAF style and would carry the Polish military white-and-red chessboard insignia, Imperial Airways continued to oppose its use and raised the question of legal ownership of the three "LOT" L 14s. The airliners were formally deposited with them as surety against loans granted by the Corporation to the PLL "LOT" for its transport operations outside Poland.

As the wrangle continued, the British authorities began to use delaying tactics. Deliveries of the parts and armament needed for modifying the -BNF were persistently delayed and the Poles, alarmed by the situation, removed the second L 14H, the -BPM, which required a major overhaul, to France in the middle of March 1940 and asked for British mechanics to be sent to Paris to carry out the work. The British refused, but Lockheed's European representative came to the rescue and organized maintenance and spares service for the airliner.

Early in May the Air Ministry began fresh negotiations over the Lockheeds, and expressed the desire to purchase the L 14s in Heston and Paris for the newly established BOAC. In return, the Ministry would exercise maximum diplomatic pressure on the Rumanians to persuade them to release all the "LOT"'s transport aircraft held in Rumania. They also undertook to provide Polish squadrons with Wellingtons which, according to the Ministry, were better suited for flights to Poland.[76] The Poles eventually agreed on the condition that two long-range Wellington bombers with British and Polish crews would be put at the disposal of the PAF Command for such operations. The sale of the Lockheeds was completed towards the end of May. Airwork was instructed by the BOAC to bring the -BNF at Heston back to its original airliner standards and a British crew was to fly the -BPM from the Paris Le Bourget airfield back to England, but the collapse of France prevented this.

In parallel with the negotiations over the Lockheeds, the Poles continued to search for other aircraft which could fulfill their air-liaison requirement. Approaches made to the French early in 1940 drew a similar response to that of the British: France had no bombers or light aircraft suitable for the task. However, in March 1940 Gen Picard, Chief of the Air Staff, offered the PAF Command a Percival Vega Gull four-seat tourer, which was available in Paris. Wyrwicki was to bring the machine to Lyon for modifications, but technical investigations of its capabilities were discouraging and the project was not pursued.

In the winter and spring months of 1940 a number of schemes for air support and liaison with Poland were put forward. These included the use of seaplanes (such as the Catalina flying boat) which could alight on large lakes in central Poland, possibly in the Province of Lublin, and a very ambitious long-term scheme of March 1940 for a Polish Airborne Corps and Aviation for the Insurgence Support composed of six trans-

[76] This was a blatant ploy: Wellingtons were not suited for the task and first flights to Poland eventually had to be made in Whitleys, and the forming Polish squadrons were at first provided with Battles, not with Wellingtons.

port wings of 50 aircraft each and two command and supply wings of similar strength, or 400 aircraft in all.

In the months preceding the German offensive a few German transport and liaison planes had been forced to land in the Allied territories and in the middle of May Gen Gustaw Paszkiewicz proposed the formation of a special Polish courier squadron on the captured aircraft to be provided by the French and the British. Paszkiewicz also produced a study for air support for a Polish uprising.

All these proposals led to the decision to establish a Polish parachute task force and the first steps towards its formation were taken with the organization of the Polish parachutists school at St. Etienne in June. The collapse of France dictated a readjustment of Polish military planning to the new strategic situation.

The substantial efforts of the PAF in France to achieve air liaison with Poland produced no tangible results. However, while it would be true to say that the Allies declined to do anything to assist the Poles in establishing such a link, the PAF Command did not seem to have fully appreciated the tremendous technical difficulties and the real lack of readily available machines capable of the task. This was reflected in totally unrealistic time-scales and the superficial approach to the problems of modifying aircraft.

6. The Polish Technical Contribution to the French Aviation Industry[77]

About a thousand members of Polish design teams, skilled production personnel of the Polish aviation establishments and staff of the ITL arrived in France from Rumania in the closing months of 1939. Only a small number stayed behind in Rumania, including a group of 26 former PZL engineers, headed by Wilhelm Gibałka and Jerzy Hoffman, who were engaged by the IAR Brasow factory to initiate quantity production of the IAR 80 fighter. The Polish Department of War Industry (at first at the Polish Treasury in Exile, later transferred to the Ministry of Industry and Commerce) was responsible for preparing a register of the technical and industrial personnel who came to France. Early in November 1939 the Association of Polish Engineers in Exile was established, with Zygmunt Cyma as the first Chairman of its Aviation Branch. At about the same time the Independent Section of Aviation Industry, briefly led by S-Col Bronisław Wojtarowicz and then, after his transfer to the Polish Air Base at Lyon, by Mjr Roman Suryn, was instituted at the Polish Aviation Command.

These three bodies worked closely together to make the best possible use of the available specialists in the war effort, with the Independent Section of the Aviation Industry having the ultimate responsibility for appointments of technical officers as well as the civilian staff who were all subordinated to the Polish Aviation Command.

[77]Based on Jerzy Płoszajski's scripts (in Polish): "Polish engineers in the PAF in France" and "France's industry."

The SE 100 three-seat, twin-engined multi-purpose fighter prototype, to the development of which the Polish design team of the SNCASE at Argenteuil made a major contribution.

Most of the officers who before the outbreak of war held technical posts within the LW organization, in research institutes or in industry, were directed to the CWL Lyon or to various French centres where Polish airmen were trained as instructors and technical translators. Others went to work in aviation parks to gain experience with French equipment and maintenance practices. However, because of a high ratio of technical officers in relation to the relatively small Polish aviation in France, not all of them could be found suitable postings and overdue promotions of technical Cadet-Officers to officer rank were deferred even further.

About 750 Polish design, production and workshop specialists were placed in employment with the French aviation industry by March 1940. These included about 150 experienced designers and engineers holding Master Diplomas of technical universities, 70 engineers and technicians with half-diplomas and the rest, highly skilled craftsmen. The largest number was engaged by the SNCASE grouping of aviation establishments in the factories at Argenteuil near Paris, at Amberieu-en-Bugey in the Lyon area and at Marignane near Marseille.

At the Argenteuil plant (formerly Liore-et-Oliveir) 18 Poles – including among others Wsiewołod Jakimiuk, who headed the Polish design team, and such familiar names as Aleksander Grzędzielski, Kazimierz Korsak, Jerzy Płoszajski, Józef Medwecki, Szczepan Grzeszczyk and Jan Luboiński – worked from early November 1939 on redesign and preparation for mass production of the unconventional SE 100 twin-engined multi-purpose fighter armed with nine to ten 20-mm cannon, which was developed from the LeO 50 project to replace the Potez 63. The Polish design team was involved in the development of an extensively re-worked fuselage with a revised crew, fuel tank and undercarriage layout and provision for the increased cannon armament and internal bomb stowage. The team's other work included a re-design of the tail surfaces of the LeO 451 bomber, which was aimed at replacement of the Duralumin alloy in the structure with non-strategic metals, and a general improvement of the production model in the light of reported shortcomings.

The Amberieu plant employed over 100 Poles on production of the LeO 451 bomber, with Zygmunt Fabierkiewicz in overall charge of the group, but relations with the French management were rather strained. Only a small Polish team was given an opportunity to participate in the design and development activities. The rest of the highly qualified engineers and technicians had to work as semi-skilled labour on the assembly lines; understandably, this caused a lot of resentment. Another group of 25 Poles was engaged on assembly of the LeO 451 bombers at the Marignane plant.

The Aviation Division of the Somua concern at Venicieux in the Lyon area, which was involved in the shadow production of the LeO 451 wings, engaged 74 Polish engineers and craftsmen, led by Eryk Kosko.

Numerous Polish specialists worked on the development and production of the Dewoitine D.520 fighter at the SNCA du Midi establishment in Toulouse where a Polish engineer living in France, Stanisław Makowiecki, was the Deputy to the Chief Designer, Emile Dewoitine. The Polish group, under overall supervision of Zygmunt Cyma, included several well-known figures: amongst the design team, led by Wacław Czerwiński, were Andrzej Anczutin, Leszek Dulęba, Władysław Fiszdon and Kazimierz Nowicki; Jerzy Teisseyre was in the production team led by Zbysław Ciołkosz. The main task of the design team was the substitution of the Duralumin-alloy tail assembly of the D.520 with a wooden structure. The production team was engaged on preparing the improved D.521 variant for mass production. Several Poles, including Lucjan Bartel, also helped in other departments of the plant. Over 40 Polish craftsmen worked on the factory's shop floor and its subsidiary in Tarbes. Two Polish engineers worked at the SNCA du Midi's Latecoere plant in Toulouse.

Potez, incorporated under the Law for the Nationalization of Military Industries in the SNCAN concern, enrolled 74 Polish aviation engineers and specialists, including a group of the former employees of PLL "LOT." This team, under the leadership of Jerzy Rudlicki, was directed to Casablanca in French Morocco to assemble and bring to delivery standard the American bombers Douglas DB-7 (Boston) and Martin 167F (Maryland). Between April and July 1940 the team completed about 50 bombers, mostly of the DB-7 type.

Another factory of the SNCAN concern, the Amiot plant at Caudebec-Caux near Le Havre, employed over 40 Poles: a few engineers led by Zbigniew Krzywobłocki-Leliwa and the rest, mostly former PZL craftsmen.

The CAPRA office at the Avions Bernard plant in Le Courneuve near Paris employed a small Polish team, comprising Zygmunt Jabłoński, Tadeusz Tarczyński and Stanisław Sznajder, who were later joined by Andrzej Anczutin, Franciszek Janik and Władysław Wielkoszewski. The team worked on a design of the R.40 twin-engined long-range fighter, which incorporated welded-steel box wings forming integral fuel tanks.

Individual Polish engineers were engaged by Bronzavia in Paris, including Wiesław Stępniewski; by Kellner-Bechereau in Billancourt-Wiaczesław Gołubowski; and by SNCAO in Bordeaux-Mieczysław Kurman.

Substantial numbers of Polish engineers and specialists worked in the French aeroengine, aviation armament and ancillary industries. Gnome-Rhone employed 38 Poles in its Paris establishment, these including ten under the leadership of Stanisław Olszewski, in the design and development offices and 28 on the production lines, and over 30 in its Le Mans factory, in production departments and on the shop floor.

At the Hispano-Suiza factory at Bois-Colomber near Paris Poles were on the staff of the design offices, with Adam Tupalski in charge of the Polish team, and in the test and production departments. About 20 more Polish employees were engaged in the firm's production plant in Tarbes. The later famous Turbomeca establishment, formed in Billancourt by a Polish-born engineer Józef Szydłowski and making turbo-compressors for Hispano-Suiza engines intended for the D.520 fighters, employed four Polish engineers.

The SIGMA Company, specializing in preparing the Bristol radial engines for production under licence in France, employed several Polish specialists in its Paris office and its factory in Venissieux. These included the well-know engine designers Jerzy Bełkowski and Aleksander Seńkowski among others. A few smaller firms working for the aeroengine industry employed Polish engineers and the Chauviere airscrew factory had three on its staff.

About ten Polish aircraft armament specialists were engaged in design and development work in various research and production centers, the most prominent among them being Władysław Świątecki, the inventor of the patented Świątecki-system bomb racks and release gear, who was employed by the SADEA Company.

A strong Polish team worked in the OLAER Company's Paris office and its La Couronne factory. The firm specialized in landing gears and hydraulic installations for aircraft and was engaged mainly in the manufacture of undercarriages for the D.520 fighters. Its Polish design team in Paris, led by Jerzy Chrzanowski, included among others Wiktor Narkiewicz, Jerzy Szabłowski and Zbigniew Żabski. Poles participated in the development and production of aircraft component parts, accessories and equipment in various other factories in France.

Although the Polish contribution to the French aircraft design and production effort was of a very short duration and therefore difficult to measure in terms of specific achievements, the French companies quickly learned to appreciate the capabilities and ingenuity of their Polish employees and entrusted them with demanding and responsible tasks. However, in several instances working relations between the French and their Polish colleagues left much to be desired. The prevailing French easy-going attitudes were resented by the Poles, who were conscientious and fast workers, fully committed to their jobs, and this caused feelings of deep frustration among some of them.

7. Capitulation of France and Evacuation of the PAF to Great Britain

Within five days of the beginning of the German offensive on 10 May 1940, which brutally violated the neutrality of

Holland, Belgium and Luxembourg, the Dutch were defeated and the fate of Belgium, France and the British Expeditionary Force was sealed. Seven divisions of German tanks, concentrated at the weakest point of Western defence, the Ardennes Forest, and supported by the Ju 87 Stuka dive bombers, achieved a big break-through of the French front. On 15 May early in the morning the French Premier Paul Reynaud in Paris telephoned Winston Churchill and told him: "We have been defeated! We are beaten!" Churchill was astounded. The great French Arny crushed and the impregnable Maginot Line, the biggest fortification in the world, rendered useless in a week? That was beyond comprehension for the Allied leaders. As the German Panzer avalanche headed swiftly for the Channel behind the Allied forces in Belgium, reaching Abbeville on 20 May, the stubborn resistance of the Polish troops, fighting alone in September 1939, was suddenly looked at from a different perspective and recognized as an outstanding feat.

The trapped Belgians battled very well and lasted until 28 May, when King Leopold III surrendered unconditionally to the Germans. Notwithstanding the "Miracle at Dunkirk," the astonishingly successful Operation "Dynamo," which resulted in the evacuation of 338,226 men, the defeat of France was a foregone conclusion. On 5 June the Germans launched a lethal strike on the Somme, which developed into a massive attack all along a 400-mile front stretching across France. Five days later the French government fled from Paris and on 14 June the undefended capital was occupied by the German XVIIIth Army. On 10 June Italy entered the war, but, in contrast to the Germans, the Italian troops were unable to move a step forward, being halted on the Alpine front by a small French force of only six divisions.

On 16 June Premier Reynaud resigned and was replaced by Patain, who the next day asked the Germans for an armistice. Following this, the rapidly withdrawing French divisions began to disintegrate in chaos and confusion. On 21 June 1940 Hitler presented the French emissaries with harsh surrender terms on exactly the same spot in the Forest of Compiegne and in the same railway carriage where the German Empire had capitulated to France on 11 November, 1918. The French objected to some of the terms, but the following day were forced to accept them. On 24 June the Franco-Italian armistice was signed in Rome and the cease-fire came into force on 25 June at 01:35 hours.

German Army losses for the whole of the offensive amounted to 27,077 killed, 18,384 missing and 111,034 wounded.[78] No comparable French figures are available, except for the number of Frenchmen taken prisoner, which was around 1,900,000.

French air resistance varied sharply from very determined or even reckless opposition from certain individual French and Allied pilots and crews, including Poles and Czechs who fled to France to continue the fight, to near-apathy among many French wings and squadrons, whose main concern was not to take risks and to avoid any casualties. Those who did not hesitate to challenge the Luftwaffe's air supremacy took a relatively heavy toll of German aircraft, with the RAF squadrons leading the field. Luftwaffe losses amounted to a formidable total of 1,279 machines destroyed or critically damaged, including 184 transport planes. Of these 304, including 157 Ju 52 transports, were lost on 10 May, the first day of the offensive, largely to anti-aircraft fire, during the simultaneous air strikes on the Low Countries and France which involved extensive use of paratroop units.

The figure of 304 aircraft, the highest German daily loss of the entire war, also included a few machines lost on the Norwegian front. By the end of the French campaign at least 1,000 further aircraft were either in need of extensive repair or replacement in order to be fit for operational use. The Luftwaffe's actual combat strength in the middle of July 1940 was about 1,200 aircraft below the corresponding figure for 10 May. Flying personnel losses during the period 10 May-30 June 1940 amounted to 1,272 killed, 580 missing and 1,156 wounded (these totals also include the small losses sustained in Norway).

The Germans claimed 1,841 enemy aircraft shot down in the fighting, but this figure seems to be considerably exaggerated. As far as can be ascertained the French Air Force actually lost 757 aircraft destroyed from all causes. The cost to the RAF amounted to 959 aircraft, including 477 fighters destroyed in air combat, on the ground, or abandoned during evacuation. RAF personnel losses totaled 1,382 killed, missing and wounded in action, of whom 915 were aircrew, including 534 pilots. The fighter losses, which, after adding 32 fighters lost during the same period in the Norwegian Campaign, represented the equivalent of nearly 40 squadrons, were the ones the RAF could least afford.

Although the French Air Force had suffered heavy aircraft losses, surprisingly, they had been more than made good by increased production and American deliveries, with more machines actually available for combat in the middle of June than on 10 May. The deepening shortage of pilots proved to be the most serious problem. The French tried in vain to overcome this by demanding a greater RAF commitment to the battle. Yet they made no real attempt to utilize the considerable personnel resources of the PAF close at hand.

When the armistice talks began the PAF operational strength comprised:

- I/145 Warsaw Fighter Wing,
- 14 Front-line Flights,
- 10 Local Defense and Chimney Flights.[79]

In all 174 PAF fighter pilots were engaged in the fighting out of a total of 1,600 Polish pilots and trainee pilots available.

[78] H.A. Jacobsen: Dokumente zum Westfeldzug 1940, Berlin-Frankfurt, 1960.

[79] For full details refer to the Appendix 7.

One of the MB 131-RB4 bombers used by Polish aircrews for training at St. Synphorien in June 1940. (P. Riviere/K. Chołoniewski)

The flying personnel of other specialties (about 600) were not involved in operations apart from a few officers on operational training with French unit commands and a handful of Polish pilots and observers who were posted individually to French bomber and reconnaissance units in the last days of the campaign.

The anticipated establishment of the remaining three Fighter Wings, of two squadrons each, which were in various stages of formation at the beginning of the German offensive, made little progress. If anything, it became further delayed by the dispersal of some of their pilots to separate Polish flights intended to strengthen the depleted French combat squadrons, which made the operational deployment of the Wings as complete all-Polish entities even more remote. The nomination of six officers: Brzeziński, Goettel, Jastrzębski, Łaszkiewicz, Pentz and Sulerzycki, by Gen Zając early in June to command the proposed squadrons of the Wings was more of a token gesture than an indication of real progress.

The formation of Polish reconnaissance and bomber units proved equally difficult, although the PAF had at its disposal 120 multi-engined aircraft pilots (of whom 43 in Toulouse and 24 in other centers were engaged as ferry pilots), 54 gunners (30 in Toulouse, 11 in Clermond-Ferrand and 13 in Rennes) and 68 observers (34 in Clermond-Ferrand and 34 in Rennes) already fully trained on French equipment.[80] More personnel was under training in various French centers, including the Observers' School at St Synphorien.

The 1st Reconnaissance Wing, to be commanded by Mjr Edward Młynarski (who was temporarily posted to a French wing for operational command experience), became in effect an operational training unit. The Wing was to be equipped with the new Hanriot NC-530 three-seat tactical reconnaissance aircraft (which had yet to be built)[81] after completion of its operational training on Bloch MB 131s at St. Etienne by the end of June and the French designation GAO 523 (Polish) was reserved for it. In the meantime it had to make do with the only two Potez 63.11s (nos. 132 and 138), which, adorned with the PAF chessboard markings, were available at the Lyon CWL Training Centre. The organization of the 2nd Reconnaissance Wing, to be commanded by Mjr Bogdan Kleczyński, was even further away. Early in June Mjr Stanisław Olszanik and Mjr Julian Wojda were appointed to command the first two proposed tactical reconnaissance squadrons which were to provide the nuclei of the planned wings.

Formation of bomber units looked somewhat more promising. Towards the end of April, in view of the expected declaration of war by Italy, the French command of the Lyon base asked the PAF to select four crews for immediate, intensive

[80] As per Gen Zając's secret report to the C.-in-C. French Air Force from 7 June 1940, L.dz.1943/Tj. (IPMS, Lot.A IV,1/12a-42).

[81] The NC-530 existed only in prototype form and its quantity production was by no means certain.

A group of Polish and French airmen under training at St. Synphorien on top of a MB 131 bomber in June 1940. (P. Riviere/K. Chołoniewski)

training on Bloch MB 131 bombers. Cpt Kazimierz Szyszkowski was given command of the flight, which was composed of experienced bomber personnel from the former 1st Air Regiment in Warsaw. The crews were to be made ready at Lyon's satellite airfield at Corbas for night bombing raids across the Alps against Italian targets, because the obsolete MB 131s were restricted to nocturnal sorties only. However, the Polish crews were not trained for night operations, especially over the Alps, and the Corbas airfield was not equipped for night-flying training.

These circumstances and availability of the superior American Martin 167F bombers resulted in a decision to switch the Polish crews to training on the Martins and to increase their number. In the middle of May, in response to an order from S-Col Stanisław Luziński, Deputy Commander of the CWL, 60 officers and airmen: pilots, observers, radio operators and gunners, were selected to what became known as the OWP unit (Oddział Wyszkolenia Powietrznego, or Aerial Training Unit). On 24 May they were formed into 15 crews to be trained on Martin bombers, with Cpt Szyszkowski in overall command, Cpt Eugeniusz Prusiecki in command of the 1st eight-crew training squadron and Cpt Stefan Alberti in command of the 2nd seven-crew squadron.

After a brief training on MB 131s and Caudron Goelands, the OWP unit left Lyon for the Toulouse-Francazal Bomber Training Centre on 29 May and commenced training on the Centre's Martins on 1 June. With the beginning of armistice talks, there could be no question of the Poles receiving any of the bombers.[82] A few Polish crews participated in the evacuation of the Centre's Martins, first to Perpignan and then to Oran in Northern Algeria. Prusiecki's crew, flying Martin no.146, was forced to land on the Spanish coast because of engine malfunction and were interned (but eventually managed to escape to England in 1942). The rest of the OWP unit traveled by road to Port-Vendres on 19 June and evacuated via North Africa.

Meanwhile, Gen Zając, outlining revised plans for composition of the PAF early in June, nominated Cpts Włdysław Dukszto, Fidelis Łukasik, Marian Dychut and Jan Gaździk (all of whom were temporarily attached to French bomber wings for experience in operational command) to command the four proposed bomber squadrons, which were later to be formed into two Bomber Wings, the 1st, under S-Col Jan Jankowski, and the 2nd, under S-Col Jan Biały.[83]

Twelve new squadrons – six fighter, two reconnaissance and four bomber – were intended to be formed at the PAF Lyon establishment. The Lyon CWL Training Centre was expanded on 1 June when two forward training bases at Roanne (intermediate training) and St. Etienne (basic and parachute training) were subordinated to it, and became the Polish Aviation Training Grouping with increased autonomy. Its previous, difficult French Commander, Col Hug, was replaced on that day by Gen Baltus, who established good working relations with Pawlikowski's successor, Col Wacław Iwaszkiewicz. Deliveries of the American Curtiss fighters and Martin bombers to Lyon were promised, but no tangible results of these changes and promises had emerged before the capitulation of France, with the actual formation of the new Polish squadrons still in the distance.

Establishment of the Polish anti-aircraft artillery units, which, in line with French military structure, were incorporated within the PAF organization (under the Command of Anti-Aircraft Defences), suffered a serious setback in May 1940, when the French requested the formation of anti-tank batteries from anti-aircraft artillery personnel. Gen Zając resisted this move and managed to retain a sufficient cadre to man four heavy (75-mm) anti-aircraft batteries (for defence of fixed targets) and four light (25-mm) batteries (for defence of Polish Army units in the field), but the AA artillery, concentrated at St. Nazaire, was removed from PAF control in the middle of June.

The PAF signals and early warning/observation personnel were undergoing training in four groups, the largest of them – comprising early in June over 360 officers and men – at the PAF Communications Training Centre in Matha, at the French St. Jean d'Angely air base.

In view of the alarmingly deteriorating situation the RAF pulled back all its remaining squadrons from France in the middle of June, completing the move by the 18th. At the same time the French began to evacuate their combat wings to North Africa, expecting that the war would be continued from there. The defeatist attitudes of Marshal Petain and Gen Weygand prevented this and brought about the swift surrender of France. The news astounded Gen Sikorski and came as a bitter blow to all Polish and some French troops eager to fight the Germans to the end.

The PAF Command left Paris on 12 June and withdrew to Chateau-Londigny. Gen Sikorski, whose HQ moved from Angers to Libourne, met with Zając in Libourne on 18 June and gave him directives for the evacuation of all PAF units to Great Britain. That same evening Sikorski flew to England and on 19 June, after a momentous conference with Winston Churchill, made a broadcast from London ordering all Polish forces to proceed forthwith to the nearest ports and to evacuate with the British. Since the PAF Command lost all contact with most of its units after withdrawal from Paris and Zając failed to communicate the directives to them, the Polish airmen learned about Sikorski's broadcast and evacuation instructions mainly from their French commands. Although the French prevented their Polish comrades from flying in most instances directly to England, they proved very willing to give them all possible help to reach the Atlantic and Mediterranean ports and assisted with their evacuation.

British and Polish ships were dispatched to French ports to take as many troops as possible and some French vessels,

[82] Contrary to assertions in certain Polish publications, the unit did not "receive" any bombers; it only used the Centre's machines for training.

[83] Gen Zając's report from 7 June (IPMS, Lot.A IV,1/12a-42).

Evacuation of Polish airmen by the "Arandora Star" from St. Jean-de-Luz on 24 June 1940. (T. Kopański)

Polish airmen on board the "Arandora Star" on the way to England.

including "Ille de France," also participated in the evacuation. President Raczkiewicz and the Polish Government boarded the British cruiser HMS "Arethus" at St. Nazaire on the night of 17 June and sailed for England on 20 June. Polish soldiers, including small groups of airmen, were picked up from St. Nazaire, La Pallice (La Rochelle) and Le Verdon (Royan) ports, but the main Atlantic boarding point for the PAF personnel was St. Jean-de-Luz in the Biarritz area, from where two Polish ocean liners, M/S "Sobieski" and M/S "Batory" and the British ship "Arandora Star" lifted well over a thousand Polish airmen (besides other troops) and brought them to Plymouth and Liverpool. Gen Zając and most of the PAF Command sailed on the "Batory."

The largest number of Polish airmen, comprising 906 officers, 197 Cadet-Officers, 3,104 other ranks and 98 aviation industry workers, or 4,305 in all, left France through the Mediterranean ports of Port-Vendres and Argeles (in the Perpignan area) between 21 June and the early hours of the 25th, most of them sailing to Oran in North Africa and the rest to Great Britain via Gibraltar. They included all the PAF personnel from the Lyon base and its Polish Aviation Training Grouping, a number of front-line and Chimney Flights, small groups of Polish observers (under Cpt Lucjan Fijuth) and bomber pilots (under Cpt Kazimierz Śniegula) from French combat squadrons and various training units, as well as 80 French cadets and several Armeé de l'Air officers, all dressed in Polish uniforms, who wanted to carry on the fight.

The Polish airmen who arrived in Oran and other North African locations were sent to Casablanca and joined the PAF personnel from African training establishments who were gathering there. They sailed from Casablanca to Gibraltar, where they were transshipped on 5 July to a small convoy bound for Glasgow.

According to Returns of the Chief of Armed Forces Staff's Personnel Branch from 1 June 1940, the PAF had in its strength in France 8,351 men, as follows:[84]

[84]Col O. Tuskiewicz, Polskie Siły Powietrzne na Obczyźnie; London, 1947. With 2,164 PAF personnel (165 Officers and 1,999 Other Ranks) in Great Britain, the PAF overall strength then totaled 10,515 men.

- Aviation units: 1,449 Officers and 5,414 Other Ranks, total 6,863;
- AA artillery: 181 Officers and 922 Other Ranks, total 1,103;
- Signals: 33 Officers and 352 Other Ranks, total 385.

With the withdrawal of the AA artillery from the PAF organization just before the armistice, PAF personnel strength in France in the middle of June remained at about 7,250 officers and men (including the Signals units). Of these, 6,220, or almost 86 percent, were evacuated to England.

The Polish Armed Forces as a whole managed to evacuate from France about 27,600 men, or some 33 percent of their overall strength of almost 84,500 men. After separating the PAF personnel from these totals, it appears that the Polish Army – in sharp contrast to the PAF – succeeded in rescuing only 28 percent (21,400 men) of its strength. This was partly because Polish Army units were tied down in fighting on a wide front right up to the end of hostilities and partly because of an error of judgment by Gen Sikorski; he could not believe that France was losing the war and failed to issue timely evacuation orders.

8. Assessments and Conclusions of PAF Operations in France

The great fighting ability and overwhelming desire to get even with the Germans, which characterized the Polish airmen flocking in their thousands to the reborn Polish forces in France, were largely ignored by the French. Months elapsed before training of the Poles began and the training provided was generally inappropriate, slow and haphazard, and conducted on obsolete or inadequate equipment. Polish personnel trained in the French centres were mainly used to fill vacancies at these establishments rather than to form new PAF combat units. The formation of PAF squadrons and wings, painfully slow and difficult, was resisted by the Armee de l'Air commands, on the whole unwilling to apply the provisions of the "Procedural Accord" of 17 February 1940.

The appointment of Gen Zając to the post of C.-in-C. PAF

by Gen Sikorski, dictated by the force of circumstance rather than by choice, did not help. Zając, lacking sound knowledge of air force procedures and requirements, seemed unable to formulate a consistent development strategy in France for the PAF and to follow it through. Studies of his memoranda to the Armée de l'Air Command seem to indicate a superficial and indecisive approach which made it all the easier for the French to manipulate the "Accord" to suit their own convenience.

In the event, no Polish pilots participated in any air operations in France until the end of March 1940, when the first 18 fighter pilots were attached to the French front-line wings in groups of three, and no more Poles joined combat units until the beginning of the German offensive. The first and the only Polish wing to be fully organized in France – I/145 Warsaw Fighter Wing – was equipped with inferior aircraft, rejected by the French as unsuitable for combat.

By the beginning of June the number of Polish pilots engaged in the battle had increased to about 130, but more than half of this total was deployed in the improvised, auxiliary local defence flights and flew dilapidated or obsolete fighters. By the end of hostilities a total of 174 Polish fighter pilots had figured on the rosters of operational fighter units at one time or another during the duration of the French Campaign (see Appendix 8), although some of the pilots who joined these units in the middle of June might not have flown any operational sorties.

The maximum number of Polish fighter pilots participating in operations at any one time hardly exceeded 140. Only a handful of Polish bomber and reconnaissance aircrews saw any action and even this was limited to the last days of the campaign.

Less than 10 percent of the available Polish flying personnel were utilized in the fighting.

Despite these limitations, the Polish airmen in France gave an excellent account of themselves and achieved a number of outstanding successes in air battles and in individual encounters. They claimed no enemy aircraft during their very restricted involvement in operations before 10 May. However, during the period of the German offensive (10 May-24 June) Polish fighter pilots were officially credited with 51 9/10 aircraft confirmed as shot down, 3 "probables" and 6 3/5 "damaged" (for details see Appendix 7 and 8).

These scores were finalized and endorsed by Col Bajan's Fighter Claim Commission in Great Britain,[85] which – after careful scrutiny of all relevant pilots' reports and independent operational documentation (French and Polish) – brought all the Polish fighter claims to the same pattern as that of the RAF (i.e.: destroyed for certain-probable-damaged). The French scoring systems recognized only aircraft shot down for certain and "probables," so the approved PAF score in France, appropriately adjusted and rounded up to the nearest whole number should be 52 destroyed and 10 probably destroyed. Of these, 10 victories and 4 "probables" were achieved between 10 and 31 May and the rest – 42 victories and 6 "probables" – between 1 and 24 June.

These figures are of great significance. The Armée de l'Air fighters – exclusive of the multi-engined and the Naval Aviation units – were officially credited with 696 confirmed air victories and 253 "probables" for the entire duration of the war (3 September, 1939-24 June, 1940). During the first phase of the German offensive, 10-31 May, the French fighters scored 350 confirmed victories and 145 "probables," and during the closing phase, 1-24 June, about 260 victories and 90 "probables." For the first period, when the Polish pilots were only gradually entering combat and engaged in the fighting in greater numbers only towards the end of May, their contribution to the confirmed air victories represented only about 3 percent of the total. In the final period, their share shot up to a staggering 16 percent.

The above percentages highlight what the Poles could have achieved, had they been fully utilized in the battle from the beginning. Furthermore, at least five more enemy aircraft claimed by the Poles as shot down and seven more claimed as "probables," were never credited to them because of a lack of independent, officially acceptable corroboration, which proved particularly difficult to obtain in the chaotic conditions of the closing stages of the campaign.

The PAF fighters' share of the French operational effort was also formidable. During the 1939-40 war French planes flew 12,443 aircraft sorties (about a quarter of the Luftwaffe's total), of which 9,987 were by fighters. Approximately 9,000 of the fighter sorties were carried out during the German offensive. The available PAF documentation indicate a total number of aircraft sorties by the Polish fighters between 27 March and 24 June, of about 1,916 (see Appendix 7). Between 80 and 90 of these were recorded by the 18 Polish pilots with the French front-line wings prior to 10 May, leaving about 1,830 for the period of the German offensive.

However, the Polish records also include the missions flown during the self-imposed operational readiness of the PAF units in the opening stages of the German attack, before they were officially declared operational by the French Aviation Command. These comprised about 150 sorties by the I/145 Warsaw Fighter Wing and the Fighter Exercise Squadron before the beginning of June and perhaps as much as half – about 350 – of all sorties by the Polish local-defence and Chimney Flights. Such missions would not be included in the official statistics of the Armée de l'Air. When account is taken of these, which would align the PAF fighters' war effort with the French criteria for operational sorties, the Polish contribution would amount to about 1,300 aircraft sorties, or to about 14 percent of the French total of approximately 9,000.

The PAF pilots' losses sustained in action, which comprised 11 killed, 2 missing and 5 wounded, were very light.

[85] The full Bajan Commission's listings also record one He 111 shot down by a Polish mechanic, Cpr Stefan Lisiak, firing a machine gun from the ground on 10 May. With this addition, the total PAF score in France rises to 52 9/10-3-6 3/5.

This could be explained by the superb tactics and flying skills of the Poles and by the fact that in many instances, particularly towards the end of the French campaign, the Luftwaffe bombers operated without fighter escort. The estimated loss of 44 fighters of the Polish units from all causes, including wear and tear and abandonment of equipment during evacuation, compared very well with the 52 air victories officially credited to them in the battle.

The final conclusions to be drawn from the air operations in France are as follows:

- The Allies had learned nothing from the German onslaught of Poland in September 1939. The French concept of air warfare – outdated and perhaps even more naive in some respects than that of the Poles – remained unchanged, with no attention paid even to the most elementary air defence precautions. Only after France's total inability to withstand the German armour and air forces and her rapid collapse, resulting in the radically altered strategic situation in Europe, was the true magnitude of Poland's lonely resistance in 1939 recognized in the Free World.
- Gen Sikorski, true to his long-held convictions of the decisive role of air power in modern war – which he had begun to preach as early as the 1920s – aimed to establish the PAF as an independent element of the reborn Polish Armed Forces, with its numerical strength equal to that of the pre-war aviation even though personnel resources in the West were less than 10 percent of the Polish Army's pre-war mobilization strength. Despite the disappointing progress with the formation of Polish aviation units on French soil, the steps taken in France laid foundations for the successful re-establishment of the PAF in Great Britain.
- Considering the severe constraints put upon the PAF's ability to fight by the French Aviation Command's attitudes – which resulted in the appallingly poor utilization of the Polish airmen, in reluctance to provide the Poles with acceptable, up-to-date combat aircraft and in unwillingness to assist in the formation of the agreed all-Polish units – the Poles gave an excellent account of themselves in the difficult conditions of the French campaign and made a significant contribution to the air victories and operational effort of the Armee de l'Air.
- The attitude of French commanders and airmen towards their Polish colleagues, at first patronizing and mistrustful, underwent a dramatic change when the Poles began to take part in operational flying. The relations became respectful and affectionate, with many real friendships being struck. In several instances the Poles were entrusted to lead into action flights composed of French pilots. Specific exploits of Polish pilots were praised in official French communiqués and received widespread publicity in the French press, which hailed them as heroes. Evacuation of the Polish pilots from France in such large numbers would not have been possible without the whole-hearted help of their French comrades.

PART II

The PAF's Part in the Struggle for Air Supremacy

(1940-1943)

Chapter I

ESTABLISHMENT AND DEVELOPMENT OF THE PAF IN GREAT BRITAIN

1. Arrival and Reception

First suggestions for reviving the Polish air arm on British soil were put forward to the British authorities by Col Bogdan Kwieciński, the Polish Air Attaché in London, on his own initiative at the end of September 1939. In October, as the Polish Government's plans for the Polish Armed Forces in Exile were taking shape, arguments were put forward for the formation of PAF in Great Britain rather than in France. These pointed to a general superiority of British aviation equipment and to the fact that Polish airmen and mechanics were largely familiar with it. British aero-engines were licence-manufactured in Poland and powered most of the Polish aircraft, and British armament and other aviation gear were widely used in Polish designs. However, the language barrier weighed against this solution, because while French was fairly widely known in Poland, Poles who had a satisfactory command of English were very few and far between.

Whatever the arguments, it became abundantly clear that the British were prepared to take only a limited number of Polish aviation personnel at that time and on terms that were very restrictive. At the Anglo-Franco-Polish conference at the HQ of the Armée de l'Air in Paris on 25 October, 1939, AVM Evill, representing the British side, handed to Gen Zając a Memorandum "Outline of the arrangements proposed for the reception and organization of Polish Air Force personnel in England," which confirmed the British readiness to accept immediately only 300 pilots and 2,000 of other aircrew and ground personnel and offered to form "Two Medium Bomber Squadrons (Battle aircraft); (and) Two reserve Bomber Squadrons (Battle aircraft)." The RAF Station Eastchurch was designated as the receiving base for Polish personnel. However, the PAF units, which were to organize within the RAF structure and to remain under the RAF command were denied any independence (For the full text of the Memorandum see Appendix 9).

The first Polish group of 50 airmen departed from France for England on 6 December, 1939, and arrived at Eastchurch two days later. Other, larger transports followed at ten- to 14-day intervals and by 1 June, 1940, 2,164 reached Great Britain, with a further 65 awaiting shipment, but coming with the first evacuation parties. G/Cpt A.P. Davidson became the C.O. of the Station with S-Col Bolesław Stachoń in command of the Polish airmen, but largely in name only. The new arrivals, who also included a few fighter pilots, prepared to join the bombers for the sake of coming to England, were registered, divided into units and commenced intensive language courses with special emphasis on aviation and technical terminology, and familiarization with the rules and regulations governing the Royal Air Force. In February pilots began to be sent in small groups to refresher flying courses at Redhill and navigators and radio operators/gunners started ground training at Eastchurch. To ease congestion at Eastchurch some newcomers were sent to Manston.

Order WAR/BC 127, dated 14 March, 1940, announced the establishment of the Polish Training Unit as part of No.18 Operational Training Unit (OTU) within No.6 Bomber Group at RAF Hucknall. The first Polish party, commanded by Col Wacław Makowski and comprising 23 flying officers and 14 NCOs, one technical officer and one clerical officer, arrived at Hucknall on 1 April. The Polish flying personnel were divided between Squadrons 'A' and 'B' and the maintenance personnel formed Squadron 'D'. Flying began a fortnight later. Eventually 31 Fairey Battles, crewed by pilot, navigator and radio operator/gunner were put at the disposal of the Poles.

In view of excellent progress with training, the Polish unit was moved to RAF Bramcote towards the end of June 1940, where the crews selected to provide personnel for the first two Polish operational Bomber Squadrons, No.300 ("Mazovian") and No.301 ("Pomeranian") underwent final training. The surplus airmen still remaining at Eastchurch were moved to Blackpool at the end of May.

The British originally intended to use the Poles exclusively for bomber duties, because, as they later frankly admitted, they had various misgivings regarding the ability of the Polish fighter pilots to cope with British operational methods and the highly sophisticated centralized British defence system. They also believed that fighter pilots who had once been defeated would lose their fighting zest and aggressive spirit. However, the developing war situation compelled the RAF to put increased emphasis on strengthening the fighter force and in April 1940 pilots were invited to apply for transfers to Fighter Command. This provided a welcomed opportunity for a number of Polish pilots to join the fighters.

The collapse of France and massive evacuation of the Polish aviation personnel to England altered the circumstances drastically. By early July main camps for the Polish airmen evacuated from France comprised: Blackpool – 1,188 men, including 534 officers; Gloucester – 1,215 men, including 242 officers; Kirkham – 970 men, including 13 officers; and Weeton – 773 men, including 24 officers and about 50 civil technical personnel. By the end of July 1940 6,220 Polish airmen had reached the United Kingdom, with Blackpool becoming the largest Polish centre. The newcomers increased PAF personnel strength in Great Britain to a total of 8,384 men.

The British, continuing the fight alone in a dramatically worsened strategic situation, found themselves in a perilous position. Poles were among their staunchest Allies, and the RAF's whole attitude towards them became more accommodating. This was reflected in a new Anglo-Polish Agreement of 5 August, 1940, which gave the PAF more autonomy and envisaged the early formation of Polish Fighter and Army Co-operation Squadrons in addition to an increased number of Bomber Squadrons.

Gen Zając, in company of S-Col Stachoń and G/Cpt Davidson, takes a parade of the early Polish arrivals at Eastchurch on 19 February 1940. (IPMS)

Inspecting newcomers at Manston in April 1940 left to right, S-Col Stachoń, G/Cpt Davidson, Gen Kalkus and, last on the right, Mjr Floryanowicz. (IPMS)

2. Legal Basis and Status of the PAF (Agreements of 11 June and 5 August, 1940)

In contrast to the PAF's situation in France, personnel training in Great Britain was well organized, purposeful and progressed according to plan, but the conditions initially offered were not acceptable to the Polish Government.

The Memorandum of 25 October, 1939, was to provide the basis for a more comprehensive Anglo-Polish Air Force agreement, but since British constitutional provisions prohibited the stationing of foreign troops in the United Kingdom without the prior consent of Parliament, the prospective PAF units were denied the requisite autonomy.

That apart, the Polish airmen were to be initially enrolled in the RAF Volunteer Reserve: all officers (regardless of their Polish rank) in the lowest Pilot Officer (P/O) rank and all NCOs in the lowest Aircraftman 2nd Class (AC2) rank and in Group V (Aircrafthands, or Unskilled). The right to promote Polish officers and NCOs was to rest with the RAF. Despite these restrictions, all costs relating to the establishment and equipment of the PAF unit in Great Britain were to be born by the Polish Treasury.

Gen Sikorski refused to agree to these terms and protracted negotiations with the British followed. Compromise proved difficult and took almost six months to achieve. The proposed agreement in its final form was submitted to him in May 1940. Sikorski still hesitated, but eventually, for expediency's sake, decided to accept.

The Agreement, which was not signed until 11 June, 1940, reaffirmed broadly the principles set out in the early Memorandum regarding the size, composition and main conditions of the PAF establishment in the United Kingdom, but included a number of concessions in the Appendix I, which are summarized below:

- It formalized the setting up of the PAF Inspectorate in Great Britain to serve as a liaison organ between the Polish and the British authorities. Formation of such an Inspectorate had already been ordered by Gen Sikorski on 1 May (with an establishment of five officers) as a result of the Anglo-Polish talks preceding the signing of the Agreement, and Gen Władysław Kalkus arrived from France in London to head it.
- It gave the C.-in-C. Polish Armed Forces and the C.-in-

C. Polish Air Force (both of whom then had their HQs in France) the right to inspect the PAF units in the United Kingdom.
- It agreed that the PAF units would not be used outside Europe without prior permission of the Polish Government.
- It further agreed that the enrolled airmen would take an oath of allegiance to the King and to the Polish Republic.
- It confirmed that: the RAF uniforms would be worn with the PAF Eagle on the caps and the inscription "Poland" on the shoulders, but wearing of distinctive Polish badges by the flying crews, denoting their specialties, would not be allowed; aircraft used by the PAF would carry the RAF markings with the PAF chessboard markings on the fuselage; and the PAF ensign would be flown below the RAF ensign at the stations at which PAF units were based.

This Agreement remained in force until 6 August, 1940, when it was replaced by a new Agreement between the Governments of the United Kingdom and the Polish Republic signed the previous day.

Discussions, conducted by Gen Sikorski with Winston Churchill at the time of the collapse of France, opened a new chapter in Anglo-Polish relations, which found their expression in Churchill's famous words spoken to Sikorski on 19 June, "We are now united for better or for worse." The political and military implications were far reaching. For the PAF, they opened the way to achieve a degree of independence which nourished the essentially Polish character of the force. Its HQ was set up in London, and by the end of July all its personnel were united in Polish units in one country for the first time since leaving Poland. The first enemy aircraft had already been shot down by a Polish pilot fighting over England.

By 3 August the new Anglo-Polish Agreement concerning the organization and employment of the Polish Land and Air Forces was fully determined and ready for submission to Parliament for legislation. It was signed by the two Governments in London on 5 August, 1940. On 22 August the Allied Forces Act was passed by Parliament, legalizing the accomplished fact of the PAF establishment in Great Britain. The additional Protocol relating to jurisdiction over the PAF was concluded between both parties on 22 November, 1940.

By the Agreement of 5 August the PAF became an integral part of the Polish Armed Forces and all the Polish airmen previously enlisted in the RAF Volunteer Reserve were released to rejoin the PAF and were given the same rights as RAF personnel. The PAF Inspectorate in Great Britain was replaced by the Inspectorate General of the PAF with greatly upgraded status (the formation of which had already been ordered on 18 July). Its duties included inspection of the PAF units and communication with the Air Ministry on all aspects of their activities. PAF stations commanded by Polish officers were envisaged and all promotions were to be made on the authority of the C.-in-C. of the Polish Armed Forces.

The Polish airmen were to take an oath of allegiance to the Polish Republic only. RAF uniforms with dual rank insignia, Polish and British, were to be introduced and all restrictions on wearing Polish badges were lifted. The PAF ensign was to be hoisted alongside the RAF ensign at the stations where Polish units were based (For the full text of the Agreement see Appendix 10).

Gen Sikorski met with HM King George VI soon after his arrival from France. (IPMS)

However, for practical and operational reasons the PAF was still to be incorporated in the RAF in organization, supply, discipline and service matters and the RAF still reserved the right to have a say in selection of personnel for service. This was resented by a number of older PAF officers, some of whom had to be transferred to the army and others who lost their seniority because of dualism of ranking. As many of the younger men were promoted quickly, their British ranks were soon higher than the British ranks of some of their senior Polish officers. The positive effect of this was to bring new blood to the PAF chain of command through keen young officers, maturing in combat and with progressive ideas. Difficulties arising from these anomalies were largely overcome by tact and a good military upbringing of both sides.

Various amendments to the Agreement resulted in fairer regulations governing terms of service and pay of the PAF personnel. New directions of the Air Ministry coming into force on 1 March, 1941, introduced so-called permanent war ranks for Polish airmen. All lower-ranking officers and NCOs were given British ranks equal to their Polish ones and officers from Captain onwards, one rank lower than their Polish ones. War-rank promotions could be earned after six months service in a post carrying a higher acting rank on the recommendation of the Inspector General of the PAF. The Allied Forces Order of 26 March, 1941, stipulated that the Air Force Act should also apply to Polish personnel, this being of major importance for matters of rank, the administration of justice and the application of King's Regulations.

Apart from these changes, the Agreement of 5 August, 1940, remained in force in the main until 6 April, 1944, when the coming invasion of Europe and forward planning for the anticipated return of the PAF to Poland led to its fundamental revision.

3. The PAF Organization, Strength, Training and Equipment, 1940-43

The Inspectorate General of the PAF, the successor to the former French-based PAF Command, was established in London in place of the earlier embryonic, regional PAF Inspectorate in Great Britain by order of Gen Sikorski (L.dz.116/tjn/40/NW) from 18 July, 1940. The new Inspectorate was to be directly subordinated to him and was to cooperate closely with his HQ Staff as "the specialist adviser in all air force matters." It. was to provide a liaison with the British Air Ministry and the RAF, and to exercise powers arising from the Aviation Agreements.

At the same time Gen Stanisław Ujejski replaced Gen Józef Zając[1] as C.-in-C. PAF and became Inspector General of the PAF, and Gen Kalkus, the former head of the PAF Inspectorate in Great Britain was moved to Bramcote as the joint Polish C.O. of the Polish bomber training unit.[2]

The Inspectorate, as initially set up, comprised 12 officers and its structure was as follows: Inspector General of the PAF (Brig Gen Stanisław Ujejski), First Deputy (Col Wacław Iwaszkiewicz), Chief of the Bureau (Col Stefan Sznuk), the Organization and Training Department (under S-Col Zygmunt Wojciechowski) with three officers; the Technical Department (under S-Col Władysław Bohuszewicz), with two officers; the Personnel Department (under Col Franciszek Wiedeń), with two officers; and two Orderly Officers.

The new Anglo-Polish Agreement of 5 August, 1940, referring to the formation of "an Inspectorate of the Polish Air Force," legitimized its existence and defined its function, but its statute was not officially approved by the British Air Ministry Until 1 October, 1940. The Inspectorate was to communicate with the Air Ministry via the Directorate of Allied Cooperation (later the Directorate of Allied Air Cooperation and Foreign Liaison, or DAAFL, of the Air Ministry), where a Polish Department, incorporating a permanent Polish Liaison Officer, was organized. On 23 October, 1940, a post for a PAF Liaison Officer at Fighter Command HQ in Stanmore was set up and the following January similar posts were introduced at Bomber Command and Technical Training Command HQs.

As the Inspectorate's activities grew with the formation of new PAF units and expansion of training, its staff had to be enlarged rapidly. By the early months of 1941 it numbered 44 officers, ten NCOs and seven clerks, excluding 23 liaison officers attached to various British departments and RAF Commands. Its new establishment was authorized by the Air Ministry on 1 March, 1941 (Air Ministry/Establishment No.WAR/TT/159).

During the following months the Inspectorate's responsibilities also extended to civil aviation. To deal with the recruitment of Polish volunteers in America and Polish airmen evacuated from Russia, PAF Liaison Officers were assigned to North American (six officers) and the Soviet Union (one officer) Military Missions, and one was posted out to Canada. A minuscule Air Staff began work on cooperation with the Polish underground army and commenced studies in the fields of aviation development and technical progress, intelligence, communications and armament. The Air Ministry, which under the terms of Agreement of 5 August was entitled to the decisive voice in all PAF affairs and initially looked upon the Inspectorate as little more than an advisory body, quickly learned to trust its judgment and allowed it increasing freedom of action. Questions of personnel management and transfer up to the rank of F/Lt for aircrews and S/Ldr. for other specialties were left to the Inspectorate.

A reorganization of the Inspectorate from 1 July 1941 made it more independent and reflected some of these widening responsibilities. Its new structure comprised[3]: Headquarters, Orderly Room (later Unit Staff), the Organization and Training Department, the Technical Department, the Personnel (Records) Staff (later the Personnel Department.), the Supply and Accounts Department, the Experimental Studies Department (newly formed, for technical and operational analysis, and to collaborate with the Polish Armed Forces HQ in planning the future independent use of the PAF), the General Branch, the Medical Branch, the Department of Law, Liaison Staff Officers, Officers at the Polish Armed Forces HQ, Army Corps Staff, and North American Mission. Its staff was increased to comprise 80 officers and 23 airmen and civilian clerks. At the same time S-Col Wojciechowski took over from Col Sznuk as the Chief of the Bureau (Chief of Staff) and S-Col Kazimierz G. Winicki became Head of the Organization and Training Department in his place (See Appendix 12 "Command structure of the Polish Air Force in Great Britain, August 1941").

Further changes were introduced at frequent intervals and these were incorporated in the Air Ministry's revised Establishment (No. WAR/TT/169A) of 1 October, 1941, which authorized enlargement of the Inspectorate's staff to 92 officers and 24 airmen and civilian clerks and reaffirmed its revised structure of 1 July with the addition of the Propaganda and Welfare Section, the USSR Mission and the Duty Officer in Canada.

In the first half of 1942 two top-level personnel changes took place. Col Stanisław Karpiński was appointed to the post

[1] Because of various criticisms and personal frictions, Gen Zając was about to tender his resignation when he was relieved of his duties.

[2] In 1941 Gen Kalkus was appointed to the post of the Inspectorate's Delegate at the Polish Depot in Blackpool.

[3] IPMS, Lot.A IV, 1/3, XVIIb.

of Inspector's Deputy, replacing Col Iwaszkiewicz, and S-Col Władysław Madejski succeeded S-Col Wojciechowski as the Chief of Staff. A Provisional Organization Directive of 7 June, 1942 (L.dz.1369/Tjn/42/Org), amended the Inspectorate's structure and progressive expansion of its responsibilities continued, especially in the fields of technical studies, air force organization and structures, and supply, with the aim of preparing a proficient command cadre to run the independent PAF in readiness for its return to Poland. The Inspectorate's Legation for the Middle East was established in Cairo (in 1942), and in connection with a partial transfer of PAF training to Canada, the PAF mission there received Legation status (in 1943).

Two supplementary Directives (L. dz. 1200/ and L.dz. 1350/Tjn/43/Org), both issued in May 1943, were to prepare the Inspectorate for its anticipated transformation into the Polish Air Force Command, the work on the shape and establishment of which commenced in the summer of 1943. The Inspectorate then comprised:[4]

- Headquarters, under the Chief of Staff S-Col Władysław Madejski and incorporating six Departments: Organization (S-Col Kazimierz G. Winicki), Training (Mjr Marceli Pustelniak), Personnel (S-Col Jerzy Bajan), Operational Studies (S-Col Władysław Bohuszewicz), Equipment and Accounts (S-Col Augustyn Gruszka), and Counter-Espionage (Cpt Roman Niewiarowski); Unit Staff (Cpt Kazimierz Stoy), Officer of Law (Cpt Jerzy Słoński), and Liaison Officer at the DAAFL, Air Ministry (Mjr Czesław Korbut);
- three further Departments: Technical (S-Col Jan Szczerski), Experimental Studies (S-Col Mieczysław Lewandowski) and Signals (S-Col Stanisław Pociask); the Independent Civil Aviation Branch (Mjr Zygfryd Piątkowski); the Independent Armament Branch (?); the Chief of Medical Service (S-Col Dr Antoni Fiumel); the Chief of Meteorological Service (S-Col Kazimierz Zacharewicz); and Technical Teams at the Air Ministry and the Ministry of Aircraft Productions, all subordinate to the Chief of Staff, but carrying out work for the Deputy Inspector General, Col Stanisław Karpiński (who, after the acceptance of Gen Ujejski's resignation by Gen Sosnkowski, the new C.-in-C. Polish Armed Forces, on 17 August, 1943, became for a short time the Acting Inspector General of the PAF).[5]

The Inspectorate's personnel more than doubled as compared with the autumn of 1941 levels, reflecting the importance attached to it by the Air Ministry.

[4] Inspectorate's Establishment Tables and the Organization Chart of 24 August, 1943, issued by the Acting Inspector General, Col (W/Cdr) Stanisław Karpiński (Ibid.).

[5] For personnel changes at the Inspectorate's HQ after Gen Sikorski's death in an air disaster on 4 July, 1943, refer to Part III, Chapter I, "2. The Polish Air Force's new organization and strength."

The Polish camp in Blackpool, which became the largest and most important assembly and clearing centre for the PAF rank and file in Great Britain, received much attention from the Inspectorate. At the end of May 1940 groups of Polish airmen who were not required by the Polish squadrons then forming, arrived from Eastchurch and Manston at the RAF Station in Blackpool, providing the nucleus of the camp. Since the end of June, with the influx of evacuees from France, their numbers grew rapidly and the expanded Polish Camp in Blackpool, commanded by Col (W/Cdr) Stefan Pawlikowski, was transformed into the RAF Polish Depot, Blackpool, with the Station's C.O. G/Cpt Jarvis in overall command.

In August 1940 the Centrum Wyszkolenia Ziemnego (CWZ, or Ground Training Centre) was formed in Blackpool under the command of S-Col (W/Cdr) Władysław Bokalski, to provide ground training for engineering and servicing personnel as well as for flying crews in its Technical Training School. The Centre comprised the Technical Training Wing, Administration Wing and Translation Offices. In September the Komenda Uzupełnień Nr.2 (KU 2, or the Clearing Military Office No.2) began to form in Blackpool and, under the command of S-Col (W/Cdr) Wiktor Zawadzki, was fully established on 6 October, 1940 as an organ of the Inspectorate General, with the purpose of collecting and maintaining personal records of all rank and file enlisted in the PAF (The British kept separate but similar records of all Polish personnel at the RAF Polish Records Office).

S-Col (W/Cdr) Stanisław Luziński was appointed as the Inspector's Delegate to Blackpool, to work independently of the Inspectorate, but according to its directives and in liaison with the Polish Camp and the KU 2. Polish airmen from Gloucester and Kirkham were transferred to the Polish Camp in Blackpool, which, after Col Pawlikowski's departure to HQ Fighter Command as the PAF Liaison Officer in the second half of October 1940, was commanded by S-Col (W/Cdr) Wiktor Pniewski. The Polish airmen in the Weeton Camp were incorporated in the establishment of the Polish Camp in Blackpool, which in December 1940 was reorganized into the Supply Wing and included the non-effective pool.

On 1 July, 1941, the revised Establishment of the Polish Depot, Blackpool, comprised: the Ground Training Centre (CWZ), the Clearing Military Office No.2 (KU 2), the Supply Wing-Reception and Dispatch Centre (replacing the non-effective pool for rank and file), and several other smaller elements such as Medical Pool, Interpreter Pool, PAF Police Pool and Chaplains Pool. At the same time the HQ of the Inspectorate's Delegate at the Polish Depot was expanded and included: Welfare Section, Intelligence and Counter-Espionage Section, Historical Section (newly formed, to gather, preserve and work on documentary material relating to all activities and achievements of the PAF), enlarged Translation and Printing Centre, and Medical Services.[6]

[6] IPMS, Lot.A V, 1/3, XVIIb.

Gen Kalkus succeeded Luziński as the Inspectorate's Delegate in Blackpool and his HQ was further expanded during 1942 by the addition of Postal Section, Selection Commission for the Flying Personnel, and Meteorological Pool, although due to changes in the training requirements, the numbers of pupils at the Ground Training Centre decreased steadily and the Centre was eventually moved to Halton on 31 July, 1942. In the autumn of 1942 formation of the Inspectorate's Legation, Blackpool, with an establishment of 100 officers, 245 airmen and six civilian clerks, was approved by the Air Ministry (on 26 October) and the Polish Depot was formally renamed the PAF Inspectorate's Legation, Blackpool, on 25 November. The Legation was renamed the Polish Air Force Base in April, 1944.

One of the Inspectorate's most important tasks was the management of available PAF personnel, which after arrival of the evacuees from France rose in July 1940 to 8,384 men, this total including the Signal units. On 1 October, 1940, without the Signals men whom were detached from the PAF, PAF personnel strength totaled 1,526 officers and 6,628 other ranks, or 9,154 in all, but on 1 December it dropped to 7,832, mainly because of the transfer of vehicle drivers and mechanics not required by the PAF to the Army and the departure of aircraft construction and engineering specialists to British industry. Operational and training losses accounted for 79 airmen of whom 52 were killed in action. These outgoing personnel amounted to 712, and, after adding 230 men of the Signals units, totaled 942, but the arrival of 390 airmen from the Balkan countries, who were originally expected to go to France from Palestine in the autumn of 1940, partly redressed the balance, reducing the net decrease in PAF personnel strength to 552 men.

During 1941 (December 1940 – November 1941 period) the PAF received 945 newcomers: 400 from Palestine, remnants of the Balkan evacuation; 199 from the USSR, the first group of airmen to be released from the Russian camps after the signing of the Soviet-Polish Agreement in July 1941; about 225 volunteers from both Americas; and the rest, volunteers from the Army sent to primary flying training schools. During the same period the PAF lost 699 men, including 341 killed in action. On 1 December, 1941, the PAF strength totaled 8,078 men.

Extremely heavy operational casualties in 1942 (December 1941-November 1942 period), the highest of the war, included 503 killed in action and accounted for most of an overall loss of 655 men. Only hard to come by personnel reinforcements, totaling 2,315, averted wholesale reduction of the Polish bomber force, in which losses of combat crews reached critical levels. The new arrivals included about 1,500 former, trained or partly trained, LW airmen from the Soviet Union. The rest were volunteers from America and from the Polish Army. On 1 December, 1942, the PAF strength totaled 9,738. The next twelve-month period (December 1942-November 1943) saw a fall in the numbers killed in action to 347, but the overall wastage amounted to 699 men. However, in connection with an anticipated restructuring of the PAF and the proposed formation of an independent PAF Group, the Inspector General requested reinforcements of about 10,000 men. He was promised 3,200 men, but eventually only 2,456 men were actually received by the PAF, and on 1 December 1943, its personnel strength stood at 11,495 men. In April 1943 the first 36 Polish women were selected to join the British Women's Auxiliary Air Force (WAAF) and were enrolled as members of the WAAF on 17 June, their number rising to 143 by 1 December 1943. The PAF overall personnel strength, plus the Polish WAAF contingent then increased to 11,638[7] (For personnel strength changes see Appendix 13 "Polish Air Force personnel strength in Great Britain, 01.06.1940-01.05.1945.").

The first three PAF squadrons – two bomber and one fighter – were formed in July 1940 and by the end of the year their numbers had risen to ten, with seven reaching operational status. The operational squadrons were:

- No.300 "Mazovian" Bomber Squadron,
- No.301 "Pomeranian" Bomber Squadron,
- No.302 "Poznań" Fighter Squadron,
- No.303 "Warsaw Kościuszko's" Fighter Squadron,
- No.306 "Toruń" Fighter Squadron,
- No.307 "Lwów Eagle-Owls" Night Fighter Squadron,
- No.308 "Kraków" Fighter Squadron.

The three remaining squadrons, formed but still under training, were:

- No.304 "Silesian" Bomber Squadron,
- No.305 "Wielkopolan" Bomber Squadron,
- No.309 "Czerwieńska Land" Army Cooperation Squadron.

A squadron had been the basic unit of the Royal Air Force since its establishment on 1 April 1918, and of the Royal Flying Corps before it. A squadron was normally subdivided into three flights of three aircraft each. Later, during the expansion in the 1930s, squadrons were reorganized into two flights with six aircraft each for operations plus a few in reserve. Heavy bomber squadrons usually had eight aircraft each. When aircraft were in short supply, squadrons did operate at less than their nominal strength. In 1940 the establishment of most squadrons comprised 16 aircraft as Initial Equipment (IE) – 12 for immediate combat use, four as operational back up – and two as Immediate Reserve. Each squadron usually comprised 22 pilots or aircrews, including six spare ones. The establishment of some heavy bomber squadrons still comprised eight aircraft as IE.

The early RAF "War Establishment of a Fighter Squadron" was adjusted at the end of 1940 and early in 1941 to suit

[7] For the PAF personnel strength in the years 1944-1945 refer to Part III, Chapter I "Regrouping for the invasion of France."

the specific needs of Allied squadrons. Such squadrons required more personnel because of duplication of command and certain other posts and the need for a greater number of spare pilots (14 instead of six, to allow for conversion training). The resulting Establishments – (Polish) (Hurricane) WAR/FC/231 and (Polish) (Spitfire) WAR/FC/222 – set the personnel strength of a Polish single-seat fighter squadron at 23 Polish officers and 186 other ranks, and six RAF officers and 36 other ranks.

RAF structures above the squadron level were generally less formal and rigid than in most other air forces. For operational purposes squadrons were grouped into wings, which varied in size, and were interchangeable between wings.

The basic unit of the Polish LW was historically eskadra, which, as initially envisaged (in December 1918) was to have six operational aircraft and two in reserve. In 1919 the establishment of eskadra was increased to ten operational aircraft and five in reserve and ten flying crews. In practice, because of acute equipment shortages, only a few eskadras achieved their nominal strength. The most notable exception was the 7th Kościuszko Eskadra manned by American volunteers, which from August 1919 operated 12 fighters, subdivided into two flights of six.

The next LW structure was grupa, which in April 1920 was renamed dywizjon, or dyon for short. This was composed of two to five eskadras, to suit the operational needs of the Russo-Polish war and eskadras were interchangeable between dyons. For practical purposes the Polish eskadra was directly comparable with the British squadron – or with the French escadrille, from which it derived its name – and the dyon was equivalent to the wing – or the French groupe.

In the interwar era various adjustments were made to the structures of the LW elements for different tactical purposes and, mainly due to the chronic shortage of equipment during the 1930s, no reserve aircraft were included in the established strength of units. Immediately before the war, the strength of the fighter (including the planned heavy fighter) and army cooperation eskadras was ten aircraft each, with the actual or proposed addition of a machine for the C.O. of each eskadra, of the bomber eskadra, nine aircraft each, and of the observation eskadra, still seven aircraft each but, with the pending re-equipment with new types, this was also to be increased to ten.

Klucz[8] – or, in the case of observation ('companion') units, pluton – usually of three aircraft, was the nearest equivalent to the British flight, but no eskadras were formally subdivided into kluczes, except for the bomber ones. These were initially formed with two three-aircraft kluczes. Observation eskadras were earlier composed of a C.O.'s aircraft and two, or occasionally three or even four, three-aircraft plutons, with the plutons usually operating independently. Notwithstanding this, the fighter eskadras sometimes formed ad hoc kluczes for specific operational tasks.

In 1939 the eskadra still remained the basic tactical unit of the army cooperation and observation elements of the LW, but the dyon, composed normally of two eskadras, became the basic fighting unit of the fighter and bomber elements, similar to that of the French groupe, or the British wing. When the Polish Air Force was reborn in France, the groupe was officially translated to dywizjon (i.e. I/145 Groupe de Chasse Polonaise de Varsovie was Dywizjon I/145 Warazawski).

During the organization of the PAF in Great Britain discussions took place on how to best translate the term squadron: to eskadra, or to dywizjon? After much deliberation, the PAF Command elected to call a squadron a dywizjon. Prestige considerations, particularly important to project a powerful image of PAF strength in occupied Poland, undoubtedly influenced this decision. This led, however, to difficulties in comparisons with the structures of other European air forces or the prewar LW.[9] The Polish Fighter Dywizjon (Groupe de Chasse) I/145 in France in 1940 had on strength 38 fighters and consisted of two Eskadras (Escadrilles) of 16 aircraft (12 for immediate combat use, and four in reserve) and six machines of the command. The strength of the Polish Fighter Dywizjons (Squadrons) in Great Britain – with Initial Equipment of 16 aircraft each – was the same as that of the Escadrille. By adopting the term dywizjon (and not eskadra) for a squadron, the PAF also lost a name for the next level of its structure, there being no equivalent to the RAF Wing in the Polish aviation vocabulary. To resolve this problem the rather odd new Polish term, 'skrzydło,' was concocted, arising from a direct translation of the word 'wing.'

The four PAF Fighter Squadrons – Nos. 302, 303, 306 and 308 – formed in 1940 were each initially equipped with 16 Hawker Hurricanes Mk I. The Hurricane Mk I (3,790 built) was provided with the eight-gun wing and powered by one 1,030 hp Rolls-Royce Merlin II or Merlin III twelve-cylinder Vee liquid-cooled engine. Its maximum speed was 324 mph at 16,250 ft (521 km/h at 4,950 m), maximum climb rate 2,420 ft/min (12.3 m/sec) and service ceiling 34,200 ft (10,425 m). Each Squadron usually had 26 to 30 pilots and about 180 maintenance and ground personnel, although these figures fluctuated with the effect of intense operations on casualty rates and replacements. The expected levels of operational effort for a fighter squadron, calculated on the assumption of 20 days of operations each month, were: average, 240 sorties (12 sorties per flying day); high 360 sorties (18 sorties per flying day); and maximum, 480 sorties (24 sorties per flying day).

[8] The term klucz was also used to describe the V flying formation of any number of aircraft.

[9] The RAF was a totally independent part of the British forces, whereas the pre-war LW was an integral part of the Polish Army, with the structures – generally smaller in size – administratively and tactically closely linked. RAF squadrons, with reserve aircraft on strength, were larger than eskadras, but if one discounted the reserves, the difference in numbers (usually 12 in the RAF, as compared with 10 in the LW) was minor, except for the RAF heavy bomber squadrons the size of which progressively increased. However, an eskadra was distinctly larger than a RAF flight and tactically more akin to a squadron than to a flight. For the generally recognized counterparts of various European air force structures and their LW equivalents refer to Glossary of terms – Basic units, p. 22 of this work.

The one PAF Night fighter Squadron, No.307, was initially provided with 16 Boulton Paul Defiant Mk I two-seat fighters (712 built) which had no fixed front guns and were armed only with a four-gun power-operated turret. They were powered by a 1,030 hp Rolls-Royce Merlin III twelve-cylinder Vee liquid-cooled engine. The Defiant's maximum speed was 304 mph at 17,000 ft (490 km/h at 5,180 m), maximum climb rate 1,880 ft/min (9.5 m/sec), and service ceiling 30,350 ft (9,250 m). The Squadron had on average 20 flying crews and about 300 maintenance and ground personnel. For night-fighter squadrons, bearing in mind more demanding night sorties of longer duration, monthly operational norms were lower: average, 180 sorties; high, 210 sorties; and maximum 240 sorties.

The two PAF Bomber Squadrons, Nos.300 and 301, at first received 18 Fairey Battle Mk I three-seat medium bombers each. The Battle carried a bomb-load of 1,000 lb (454 kg) and was armed with one fixed pilot's gun and one flexible gun in the rear cockpit. It was powered by one 1,030 hp Rolls-Royce Merlin II or Merlin III twelve-cylinder Vee liquid-cooled engine. Its maximum speed was 241 mph at 13,000 ft (388 km/h at 3,950 m), service ceiling 23,500 ft (7,160 m) and normal tactical range 900 miles (1,448 km). In October 1940 both squadrons were rearmed with eight – later increased to 16 – Vickers Wellington Mk IC six-seat heavy bombers which carried a bomb-load of up to 4,500 lb (2,041 kg) and were armed with two-gun front and rear turrets and two flexible beam guns. The Wellington Mk.IC (over 2,500 built) was powered by two 1,000 hp Bristol Pegasus XVIII nine-cylinder radial air-cooled engines and had maximum speed of 250 mph at 15,500 ft (402 km/h at 4,725 m), service ceiling 22,000 ft (6,700m) and tactical range with a maximum bomb-load 1,540 miles (2,480 km). The squadrons had on average 20 bomber crews and about 300 maintenance and ground personnel each. Since an average bomber sortie was expected to last seven to nine hours and most of this time was spent over enemy territory, monthly operational norms for a bomber squadron were considerably lower and amounted to: average, 60 sorties; high, 90 sorties; and maximum, 120 sorties.

Hurricane Is of the newly formed No.303 "Warsaw Kościuszko's" Fighter Squadron featured at first the PAF chessboards under the cockpits. By September 1940 these were replaced by the Kościuszko squadron badges. A ground crew by the aircraft are, left to right standing on the ground, Cpl Leśniewicz, LAC Oręziak, LAC Kowalski, Sgt Busko, and on the aircraft, LAC Skiba and LAC Gulski.

The sole PAF Army Cooperation Squadron, No.309, was equipped with 12 Westland Lysander Mk III two-seat army cooperation high-wing monoplanes, which were armed with two fixed guns in wheel fairings and a single flexible gun in the rear cockpit, and could carry a bombload of up to 500 lb (227 kg). The power was provided by one 870 hp Bristol Mercury XX or Mercury XXX nine-cylinder radial air-cooled engine. The Lysander Mk III had maximum speed of 229 mph at 10,000 ft (368 km/h at 3,050 m), service ceiling 26,000 ft (7,925 m) and normal range 600 miles (965 km). The squadron had on average about 20 flying crews and 300 ground personnel.

Hurricane Is received by No.306 "Toruń" Fighter Squadron in the middle of September carried the PAF chessboards on the rear fuselage and the squadron badges on the forward part, as shown in this photo from November 1940, with V7118 UZ•V nearest to the camera.

By the end of 1940 the PAF had on its operational strength: 64 single-seat fighters, 16 two-seat night fighters and 32 heavy bombers, or 112 combat aircraft in all.

During 1941 six more PAF squadrons achieved operational status. These were:

- No.304 "Silesian" Bomber Squadron,
- No.305 "Wielkopolan" Bomber Squadron,
- No.309 "Czerwieńska Land" Army Cooperation Sqdn,
- No.315 "Dęblin" Fighter Squadron,
- No.316 "Warsaw" Fighter Squadron,
- No.317 "Wilno" Fighter Squadron.

The three new Fighter Squadrons were all formed on the Hurricane Mk I. In the middle of the year No.315 changed over to the Spitfire Mk IIA and Nos.316 and 317, to the Hurricanes Mk IIA and IIB. All three squadrons received the Spitfires Mk VB or Mk VB and VC later in the year.

The original four Fighter Squadrons, Nos.302, 303, 306 and 308, frequently re-equipped, operated in turn the Hurricane Mk IIA and IIB (nos.302 and 306), Spitfire Mk I (Nos.303 and 308) and Spitfire Mk IIA and IIB (Nos.303, 306 and 308). Later all four Squadrons converted to the Spitfire Mk VB and VC. By the end of 1941 the entire PAF day fighter force flew Spitfires of these two marks and all its squadrons were at their full strength of 18 aircraft each – 16 Initial Equipment and two Immediate Reserve.[10]

The Hurricanes Mk IIA and IIB (almost 5,500 of both Marks built) were usually fitted with the more powerful, supercharged 1,280 hp Rolls-Royce Merlin XX engine, but while the Mk IIA retained the eight-gun wing of the earlier Mk I, the Mk IIB featured an improved and strengthened twelve-gun wing, which could carry two 250-lb (113-kg) or two 500-lb (227-kg) bombs. The Hurricane Mk IIB (without bombs) had maximum speed of 342 mph at 22,000 ft (550 km/h at 6,700 m), maximum climb rate 2,950 ft/min (15.0 m/sec) and service ceiling 40,000 ft (12,190 m).

The Supermarine Spitfire Mk I (1,566 built) was provided with the eight-gun "A" wing[11] and one 1,030 hp Rolls-Royce Merlin II or Merlin III twelve-cylinder Vee liquid-cooled engine. Its performance included a maximum speed of 362 mph at 19,000 ft (582 km/h at 5,790 m), maximum rate of climb 2,500 ft/min (12.7 m/sec) and service ceiling 34,000 ft (10,360 m.). The Spitfire Mk II (920 built) was powered by 1,175 hp Rolls-Royce Merlin XII engine and its Mk IIA variant embodied the eight-gun "A" wing and the Mk IIB had the "B" wing with two 20-mm cannon and four guns. The Spitfire Mk IIA had a maximum speed of 357 mph at 17,000 ft (575 km/h at 5,180 m) and a service ceiling of 37,200 ft (11,340 m). The Spitfire Mk V (6,479 built) was normally fitted with the 1,470 hp Rolls-Royce Merlin 45 engine. The Mk VB featured the "B" wing with two cannon and four guns, while the Mk VC incorporated the universal "C" wing, which could take either two cannon and four guns or four cannon and two 250-lb (113-kg) bombs. The Mk VC was also provided with a centerline rack for a 500-lb (227-kg) bomb or fuel tank. Its maximum speed was 374 mph at 13,000 ft (602 km/h at 3,960 m), maximum climb rate 3,300 ft/min (16.8 m/sec) and service ceiling 37,000 ft (11,270 m).

One of the PAF Fairey Battle bombers comes in to land.

Two Polish Fighter Wings, No.1 at Northolt, comprising No.303, 306 and 308 Squadrons, and No.2 at Exeter, comprising No.302, 316 and 317 Squadrons, were organized in April and August 1941 respectively. Their size and composition varied in the months ahead depending on operational needs.

The No.307 Night Fighter Squadron converted in August 1941 to the two-seat Bristol Beaufighter Mk IIF. The Beaufighter Mk IIF (450 built) had fixed armament of four 20-mm cannon and six guns and was powered by two 1,280 hp Rolls-Royce Merlin XX twelve-cylinder Vee liquid-cooled engines. It had a maximum speed of 330 mph at 20,000 ft (531 km/h at 6,090 m), maximum climb rate 1,900 ft/min (9.6 m/sec) and service ceiling 30,000 ft (9,140 m).

The two Bomber Squadrons, Nos.304 and 305, became operational on the Vickers Wellington Mk IC bombers, which, originally counted as "heavies," were then officially re-classified as medium bombers. By the summer of 1941 No.305 Squadron had exchanged its Mk ICs for the Wellington Mk II (585 built), which were powered by two 1,145 hp Rolls-Royce Merlin X twelve-cylinder Vee liquid-cooled engines.

The two earlier Bomber Squadrons, Nos.300 and 301, converted in the autumn to the Wellington Mk IV (220 built and most of them delivered to the PAF squadrons), which were fitted with two 1,200 hp Pratt & Whitney R-1830-S3C4-G Twin Wasp fourteen-cylinder two-row air-cooled radial engines. Apart from engines, the Wellingtons Mk II and Mk IV were basically similar to the Mk IC. The strength of Wellington squadrons increased to 18 aircraft – 16 Initial Equipment and two Immediate Reserve.

By the end of 1941 the PAF had on its strength: 126 single-seat fighters, 18 two-seat night fighters, 72 medium bombers and 12 army cooperation aircraft or 228 in all.

[10] As per War Establishment of a Fighter Squadron – (Spitfire) (Polish) No.War/FC/222 and /222A.

[11] A small number of Mk Is were either produced as, or converted to the Mk IB variant, with the "B" wing accommodating two 20-mm cannon and four guns, but none of these were delivered to the Polish squadrons.

1942 saw comparatively little change in PAF equipment. The fighter squadrons retained their Spitfires Mk VB and VC, except for Nos.306 and 315, which in the autumn converted to the superior Spitfire Mk IXC, No. 306 being among the first squadrons in Great Britain to receive this new model. The Spitfire Mk IX (5,665 built), based on the 1,600-1,700 hp Rolls-Royce Merlin 60 series of engines with a two-speed two-stage supercharger, was developed to counter the German Focke-Wulf Fw 190 fighters. The early Mk IX variants retained the universal "C" Wing. The Mk IXC had a maximum speed of 408 mph at 12,250 ft (657 km/h at 3,740 m), maximum climb rate 4,100 ft/min (20.8 m/sec) and a service ceiling of 42,000 ft (12,800 m). However, squadrons posted for recuperation and rest periods, during which they were expected to perform less demanding operational duties and train replacement pilots, usually reverted temporarily to older fighter models.

No.307 Night Fighter Squadron changed over to the later Beaufighter Mk VIF (1,830 Mk VIF and VIC built), which was powered by two 1,670 hp Bristol Hercules VI or Hercules XVI fourteen-cylinder two-row air-cooled radial engines. The Mk VIF had a maximum speed of 333 mph at 15,600 ft (536 km/h at 4,755 m) and a service ceiling of 26,500 ft (8,080 m). At the end of December the Squadron began to convert to the two-seat De-Havilland Mosquito NF Mk II which was armed with four 20-mm cannon and four guns in the front fuselage, and was powered by two 1,460 hp Rolls-Royce Merlin 21 or Merlin 23 twelve-cylinder Vee liquid-cooled engines. Its performance included a maximum speed of 370 mph at 14,000 (595 km/h at 4,260 m), a maximum climb rate 2,550 ft/min (12.9 m/sec) and service ceiling 34,500 ft (10,515 m).

In August 1942 No.305 became the third PAF Squadron within Bomber Command to receive the Pratt & Whitney-powered Wellingtons Mk IV. Three months earlier No.304, which had suffered very heavy casualties, was transferred from Bomber Command to Coastal Command, but retained its Wellingtons Mk.IC – the only Polish unit to do so – until the middle of 1943.

Early in 1942 the operational strength of No.309 Army Cooperation Squadron was increased by the new (third) 'C' Flight. However, what was to be of far greater consequence, was the conversion in the middle of 1942 of its 'B' Flight to the North American P-51 mustang Mk I single-seat fighter-reconnaissance aircraft. The Mustang Mk I, armed with four heavy (12.7-mm) guns (two of them mounted below the engine) and four 7.69-mm guns, was powered by one 1,150 hp Allison V-1710-F3R twelve-cylinder Vee liquid-cooled engine. Its maximum speed was 387 mph at 8,000 ft (623 km/h at 2,440 m) and service ceiling 30,000 ft (9,140 m).

By the end of 1942, with the enlargement of No.309 Squadron to 18 aircraft, the overall strength of the PAF rose to 234 aircraft.

Major changes in the composition of the PAF, largely associated with the regrouping of the RAF for the anticipated invasion of France and the progress of war in the Mediterranean, took place in the middle quarters of 1943. Between March and July 1943 No.303 and 316 Fighter Squadrons converted to the Spitfires Mk IXC and Nos.306 and 315 exchanged their Mk IXCs for new ground-attack variants of the Spitfire Mk V, the LF Mk VB and VC, armed with two 20-mm cannon and four guns and a 500 lb (227-kg) bombload. The LF Mk VB and VC had their wingtips removed and were fitted with the low-altitude 1,585 hp Rolls-Royce Merlin 45M, 50M or 55M twelve-cylinder Vee liquid-cooled engine. In the autumn Nos.303 and 316 were also re-equipped with the LF Mk VB and VC, so that by the end of the year four PAF Fighter Squadrons operated this low-altitude fighter. Later in the year the remaining three fighter Squadrons, Nos.302, 308 and 317, still flying the earlier Spitfires Mk VB and VC, received the Spitfire Mk IXC.

No.3 Polish Fighter Wing, initially comprising No.302 and 317 Squadrons, was organized in June 1943.

In the summer of 1943 No.307 Night Fighter Squadron had its equipment supplemented by the Mosquito FB Mk VI fighter-bombers (nearly 2,600 built), which in addition to four-cannon and four-gun armament could carry a bombload of up to 2,000 lb (907 kg). The FB Mk IV, usually fitted with two 1,635 hp Rolls-Royce Merlin 25 twelve-cylinder Vee liquid-cooled engines, had a maximum speed of 378 mph at 13,200 ft (608 km/h at 4,025 m) and provided the Squadron with intruder capability. At the beginning of August the squadron was assigned to Coastal Command for a period of three months.

In March 1943 a new Fighter-Reconnaissance Squadron, No.318 "Gdańsk," was formed to cooperate with the 2nd Polish Corps in the Middle East, its initial equipment comprising 16 Hawker Hurricane Mk IC. In the autumn the Squadron, still in training, received the Hurricane Mk IIB, but did not become operational until the following spring.

Early in 1943 No.309 completed conversion to the North American Mustang Mk I, when its 'A' Flight received a full complement of these aircraft and its 'C' Flight, still flying Lysanders, was disbanded. With the disbandment of Army Cooperation Command on 1 June 1943, the Squadron, which had in its strength 14 Mustangs, was transferred to Fighter Command.

No.300 became the only PAF Squadron to remain in Bomber Command. At the beginning of 1943 the Squadron converted to the Wellington Mk III (1,519 built), which was fitted either with two 1,375 hp Bristol Hercules III or with two 1,500 hp Hercules XI fourteen-cylinder two-row air-cooled radial engines, but two months later it began to receive the improved Wellington Mk X (3,804 built), powered by two 1,675 hp Hercules VI or Hercules XVI radials. The Wellington Mk X carried a bombload of up to 6,000 lb (2,720 kg), had a maximum speed of 255 mph at 14,500 ft (410 km/h at 4,420 m), service ceiling 24,000 ft (7,310 m) and a maximum range (with reduced bombload) of 1,885 miles (3,030 km).

In April 1943 No.301 had to be withdrawn from bomber operations because of extremely high crew losses and was disbanded. Its most experienced surviving crews were posted to the 'C' Flight of No.138 Special Duties Squadron, to expand air support operations for the Polish resistance movement. The 'C' (Polish) Flight was provided with six specially adapted Halifax B Mk II Series IA bombers, which were powered by four 1,480-1,640 hp Rolls-Royce Merlin XX series twelve-cylinder Vee liquid-cooled engines and had a maximum speed of 270 mph at 9,500 ft (435 km/h at 2,900 m) and a maximum range of 2,100 miles (3,380 km). In the autumn the Flight received its first three American Consolidated Vultee B-24 Liberator Mk IIIs, and became independent as No.1586 Special Duties Flight. The Liberators were powered by four 1,200 hp Pratt & Whitney R-1830 Twin Wasp series fourteen-cylinder two-row air-cooled radial engines, and had a maximum speed of 290 mph at 25,100 ft (467 km/h at 7,650 m) and a maximum range 3,200 miles (5,150 km).

In May 1943 No.304 Squadron was transferred to Coastal Command. Its aging Wellington Mk ICs were supplemented by a few Wellington Mk Xs and Mk XIs adapted for maritime operations, but within two months the squadron was re-equipped with the specialized Wellington GR Mk XIII torpedo-bombers (843 built) which were provided with ASV Mk II ("Stickleback") radar with mast aerials. In the autumn it received the Wellington GR Mk XIVs (841 built) fitted with the more refined ASV Mk III radar under the nose and a retractable Leigh Light (for illuminating targets by night) in the rear bomb-bay. The GR Mk XIII and Mk XIV were both powered by two 1,650 hp Bristol Hercules XVII fourteen-cylinder two-row, air-cooled radial engines, and their performance were similar to those of the Wellington Mk X.

In September 1943 No.305 Squadron left Bomber Command to join the 2nd Tactical Air Force. With its transfer to day bombing the Squadron converted to the North American B-25 Mitchell Mk II bombers, which could carry a bombload of up to 3,200 lb (1,450 kg) in the bomb-bay, or up to 5,200 lb (2,360 kg) with additional bombs on external racks.

The Mitchell Mk II, powered by two 1,700 hp Wright Cyclone R-2600-13 fourteen-cylinder two-row air-cooled radial engines with two-speed superchargers, had a maximum speed of 305 mph at 16,400 ft (490 km/h at 5,000 m), service ceiling of 24,500 ft (7,470 m) and a maximum range of 1,350 miles (2,170 km). At the turn of the year the Squadron began conversion to the Mosquito FB Mk VI fighter-bomber and received a few Mosquito Mk III trainers to facilitate the transition to this new type.

By the end of 1943 the PAF's operational strength consisted of:

> 126 single-seat fighters, 18 two-seat night fighters, 16 single-seat fighter-reconnaissance aircraft, 54 medium bombers (including 18 in Coastal Command) and nine Special Duties heavy bombers, or 223 in all.

In the years 1940-43 a number of balloon, training and service units came into being, supplementing the PAF combat elements.

On 14 September 1940, the Air Ministry asked the Inspectorate General of the PAF to form a trans-African transport unit to ferry aircraft from Takoradi (Ghana) on the Gold Coast to Cairo. The first group of 22 Polish aircrews arrived in Ghana on 21 November, 1940, to establish the Polish Detachment RAF Takoradi under the command of W/Cdr (S-Col) Mateusz Iżycki. The following year its establishment rose to 60 pilots. At the end of August 1941, after numerous previous applications, 12 of the Takoradi pilots were allowed to join No.112 "Shark" Fighter Squadron, which flew the Curtiss P-40 Kittyhawk single-seat fighter. After conversion training they flew operationally with the Squadron on the Libyan front from February 1942, but a few months later had to return to transport duties. At the end of 1941 the unit was renamed the Polish Air Force Detachment RAF Middle East and moved to Cairo to be incorporated in No.216 Transport Group, with which it joined Transport Command in 1943.

In October 1940 organization of the Polish Balloon Flight began in Scotland. From the beginning of 1942 the Flight manned between 11 and 14 balloon sites, providing protection for Glasgow, and later in the year, for the Forth Bridge, until it moved to the London area in July 1944.

Organization of training, which began in 1940 with formation of a Polish Training Unit at No.18 OTU in Hucknall (later moved to Bramcote and becoming self-contained by the beginning of 1941) and the establishment of the CWZ in Blackpool, was greatly expanded in the following years.

In January 1941 No.1 Polish Flying Training School (FTS) was activated in Hucknall. The School incorporated the Elementary Flying Training Squadron and the Intermediate Training Squadron. In the middle of 1941 the school was completely reorganized. A part of it which provided the basic training was transformed in June into No.16 (Polish) Service Flying Training School (SFTS), commanded by G/Cpt (Col) Wacław Iwaszkiawicz. In the middle of the following month No.16 (Polish) SFTS, composed of two squadrons, was moved to Newton. At the same time No.25 (Polish) Elementary Flying Training School (EFTS), formed under the command of W/Cdr (S-Col) Jerzy Garbiński at Peterborough and equipped with the De Havilland Tiger Moth biplane trainer, came to Hucknall. At the end of the year the Air Crew Training Center (ACTW), giving introductory, pre-elementary training to Polish air crews was also established in Hucknall.

Towards the end of 1941 a Polish training squadron was established at No.12 Initial Training Wing (ITW) at St. Andrews in Scotland. In the middle of 1942 the Polish squadron left Scotland for Brighton to become the Polish Initial Training Squadron.

In 1942, when Polish staff became fully proficient in RAF training methods, No.16 (Polish) SFTS at Newton became entirely Polish. Early the following year the School had begun to

A Polish servicing crew of No.18 OTU prepares a Battle for a practice bomber sortie at Hucknall under the watchful eye of a RAF instructor. (IPMS)

receive the Pratt & Whitney-powered Miles Master Mk III basic trainer. In the summer of 1942 the Polish Technical Training School was organized at No.1 RAF School of Technical Training, Halton, as an independent unit.

By July 1942 No.18 OTU evolved into a wholly Polish unit with a Polish Station Commander, W/Cdr (S-Col) Piotr Dudziński, at Bramcote. Towards the end of the year a Polish Squadron was formed at No.6 OTU in Silloth, to provide operational training for Polish Coastal Command crews.

In 1943 new training facilities were created to meet increasing needs. In February No.1 (Polish) Air Apprentice Technical School – for boys who had survived the ordeal of the Soviet camps and been evacuated to Persia – was opened in Heliopolis in Egypt. Five months later No.2 (Polish) Air Apprentice School was established at Halton. In April the Wyższa Szkoła Lotnicza (Air Force Staff College) was formed, initially as a section of the Polish Army Staff College in Eddleston near Peebles in Scotland. In September the Polish Instructor's School was opened at Newton. In the middle of the year the first Polish contingent of the Women's Auxiliary Air Force (WAAF) commenced training at the RAF Station at Wilmslow.

The first PAF service units emerged in the years 1942-43 with the formation of No.1 Polish Repair Depot at No.30 Maintenance Unit (MU), Sealand, at the beginning of 1942, and No.1 Mobile Repair and Salvage Unit (RSU) at No.71 MU in Slough – for fighter aircraft – and No.2 Mobile Repair and Salvage Unit at No.58 MU in Newark – for bomber aircraft – in the middle of 1943. A further expansion of the servicing and maintenance units took place in the second half of 1943 in connection with the regrouping of the PAF for the invasion of France. Details of this reorganization are given in Part III, Chapter I "Regrouping for the invasion of France."

For further details of organization and activities of the balloon, training and technical formations refer to the appropriate chapters in Part IV "Supplementary and non-operational units and activities of the PAF."

4. The Polish Air Force Colours

The LW as a whole had no standard of its own before the war. The only formation to possess its own colours was the Aviation Cadet-Officers' School at Dęblin, which received its standard from Marshal Śmigły-Rydz in 1937. It perished in September 1939.

At the end of 1939, during the rebirth of the Polish forces in France, Cpt Jan Hryniewicz, a native of Wilno, conceived the idea of a standard for Military Aviation, to be made in Poland and smuggled out to the Polish airmen fighting for the freedom of their country on foreign soil. Such a standard would be an endearing symbol of unity between the airmen and their occupied homeland. Design sketches for the standard were pre-

AVM Ujejski, C.-in-C. PAF, receives the Polish Air Force standard from Gen Sikorski at Swinderby on 16 July 1941. (IPMS)

Gen Sikorski and ACM Sir Charles Portal salute the Polish Air Force standard at Swinderby on 16 July 1941. (IPMS)

pared by two Cadet-officers, Kazimierz Karaszewski and Zbigniew Wojda. In January 1940 Hryniewicz sent letters with copies of the drawings via Switzerland to Soviet-occupied Wilno.

One of the letters reached Zofia Wasilewska-Świdow, who, helped by Rev Father Kazimierz Kucharski, arranged for the standard to be made. The idea captivated the imagination of Polish women in Wilno and seemingly insurmountable obstacles were successfully overcome. The damask and gold and silver embroidery thread were not available locally; they were obtained with the help of neutral diplomats from (of all places) Berlin. The work was divided among many families and religious communities and after months of painstaking labour the standard was ready in the middle of 1940, by which time France had capitulated and the Polish airmen had come to Great Britain.

Delivery of the standard to the Polish Air Force posed even greater problems than its production and took longer. After some abortive Polish attempts, Japanese diplomats, who had earlier assisted the Wilno Poles in obtaining some of the materials for the standard, came to the rescue again. The Japanese Consul in Kowno (Kaunas), Lithuania, Mr. Sugihara, arranged to smuggle it out via Berlin to Stockholm in the diplomatic luggage of one of his couriers – almost certainly Col Bin Yamamoto – earning the deeply felt gratitude of the Poles. The standard eventually reached London in March 1941.

Following the traditional pattern of other colours of the Polish Armed Forces, the standard's design was in the form of the red Polish Knights' cross on a white background, with Our Lady of Ostra Brama,[12] the national White Eagle and the words: "God Honor Fatherland" superimposed on one side; on the other side was St. Theresa of Lisieux[13] in the center, the Polish Air Force white and red chessboard, flying badges and the crest of the Cadet-officer School on the white fields, with the inscriptions "Wilno 1940" and "Love demands sacrifice" on the red fields of the cross.

The standard was ceremonially presented by the. C.-in-C. Polish Armed Forces, Gen Sikorski, to the Air Officer C.-in-C. Polish Air Force, AVM Ujejski, in the presence of the President of the Republic of Poland, Władysław Raczkiewicz, ACM Sir Charles Portal and AMs Peirse, Douglas, Barratt and Walsh – all of whom hammered their names in silver nails into the flag-mast – at RAF Swinderby on 16 July, 1941. After a drumhead service, the standard was handed on to No.300 Bomber Squadron as the senior PAF squadron. Each PAF squadron was to hold the standard for three months in rotation.

After the war, under a Decree of President Raczkiewicz of 27 June, 1945, all ensigns of the Polish Armed Force were transferred to the ownership of Gen Sikorski's Institute in London "for eternal safe-keeping." There the PAF Standard remained until 22 August, 1992. On that day it was handed back to the Polish Air Force Association in Great Britain at its request, for return to Poland at the beginning of September.

[12] Patron of Wilno; the image is credited with miraculous powers.
[13] Patron of France; the standard was originally intended to be used in France.

Chapter II

POLISH FIGHTER OPERATIONS

1940-1943

1. Formation and Deployment of the First Polish Fighter Squadrons

After disengaging in France, the exhausted RAF fighter force had only a couple of weeks to recover. By early July 1940, when increased Luftwaffe pressure against Britain became apparent, Fighter Command was still rearranging its Order of Battle and desperately trying to rebuild its strength to meet the mounting German challenge which developed into what became known in history as the Battle of Britain.

(1.) No.302 "Poznań" Fighter Squadron

The Air Ministry's decision to form the first two Allied fighter squadrons – No.302 (Polish) (Fighter) Squadron at Leconfield and No.310 (Czech) (Fighter) Squadron at Duxford – was communicated by Fighter Command HQ to No.13 and No.12 Groups on 10 July 1940. These units were to form to War Establishment No.WAR/FC/158, which, however, had to be modified to allow for certain command posts to be doubled up by British and Allied personnel. The aircraft establishment of each Squadron was to be 16 Hurricanes as Initial Equipment and two as Immediate Reserve with the addition of one Miles Master for training.[1]

No.302 Squadron was formed at Leconfield on 13 July 1940 (within No.13 Group) and was allocated code letters WX. The RAF personnel charged with its initial command and tactical and technical training included: S/Ldr William A.J. Satchell, C.O. of the Squadron; F/Lt James N.W. Farmer, 'A' Flight C.O.; F/Lt William Riley, 'B' Flight C.O., and F/O Peter E.G. Carter, Flying Instructor. The British contingent assigned to the Squadron totaled 11 officers, 16 NCOs and 17 men.

Gen Sikorski's order authorizing organization of the Squadron referred to it as "Ist Polish Fighter Squadron." S/Ldr (Mjr) Mieczysław Mümler was given command of the Squadron's Polish personnel, who came from evacuees from France, former members of the "Montpelier Squadron" of the 2nd Poznań Fighter Wing, I/145 Warsaw Fighter Wing and various Chimney Flights. The Squadron was named No.302 Dywizjon Myśliwski Poznański ("Poznań" Fighter Squadron) and adopted the badge of the pre-war III/3 Fighter Dyon of the 3rd (Poznań) Air Regiment, from which a large number of its personnel originated. The badge (not officially authorized by the RAF) incorporated the unit numbers I/145 at the top and 302 at the bottom.

The initial Polish personnel, arriving at Leconfield on 23 July 1940, comprised 16 officers, 12 cadet-officers, 55 NCOs and 80 men. The officers included 12 pilots: S/Ldr (Mjr) Mümler, doubling Polish C.O. of the Squadron, F/Lt (Cpt) Piotr Łaguna, 'A' Flight Commander, F/Lt (Cpt) Franciszek Jastrzębski 'B' Flight Commander, F/Lt (Cpt) Tadeusz Chłopik, P/Os (Lts) Tadeusz Czerwiński, Julian Kowalski, P/Os (S-Lts) Stanisław Chałupa, Czesław Główczyński, Włodzimierz Karwowski, Stanisław Łapka, Edward Pilch and Stefan Wapniarek. Officers in charge of the engineering and ground crews were F/Lt(Cpt) Adam Jaworski and F/O (Lt) Bartłomiej Popławski. NCOs included five pilots: Sgts (Sgts) Antoni Markiewicz, Jan Palak, Edward Paterek, Antoni Siudak and Marian Wędzik. With the RAF personnel, the total strength of the Squadron amounted to 207.

The Squadron received its first Hurricanes on 26 July and its first Daily Order was issued on 28 July. Intensive training began on the ground and in the air, with a Kestrel-powered Miles Master I advanced trainer facilitating transition to the Hurricane. A Gipsy-powered Miles Magister light trainer was also allocated to the Squadron. The in-flight and technical training progressed very well, but learning English presented a problem, which proved difficult to overcome. On 7 August Gen Sikorski accompanied by Gen Ujejski arrived in Leconfield in a De Havilland Flamingo to inspect the Squadron.

On 15 August, the day the Leconfield base was raided by Ju 88 bombers, the Air Officer Commanding No.13 Group declared No.302 Squadron operational. By then a few person-

[1] RAF Form 1924; Ref.No.FC/S.2 0437/AOA (IPMS, Lot.A V.48/1a-1).

LOCATIONS OF THE SQUADRONS

ACKLINGTON	QU	309, 315, 316, 317
ANDREWS FIELD	WM	303, 306, 309, 315, 316
APPLEDRAM	WQ	302, 308, 317
BAGINTON	VK	308
BALLYHALBERT	QX	303, 315
BRADWELL BAY	WM	309*)
BRENZETT	WR	306, 315
CASTLE CAMPS	WM	307
CATTERICK	QZ	306
CHAILEY	WQ	302, 308, 317
CHARTERHALL	QU	303*)
CHILBOLTON	VU	308
CHURCH FENTON	WA	306, 307, 308
CHURCH STANTON	VT	302, 306, 316
COLEBY GRANGE	WF	307
COLERNE	VU	307, 316, 317
COLTISHALL	WG	303, 306*), 307, 309*), 315*), 316, 318*)
COOLHAM	WQ	306, 315
CROYDON	WQ	302, 306, 316, 317
DALCROSS	QJ	309
DEANLAND	WQ	302, 308, 317
DEBDEN	WM	303
DERBY	VK	302
DETLING	WR	318
DREM	QU	307, 309
DUNINO	QP	309
DUXFORD	WL	302
EXETER	VT	307, 308, 317
FAIRLOP	WL	302, 317
FAIRWOOD COMMON	VO	302, 306*), 307, 308, 315*), 316*), 317
FINDO GASK	QO	309
FORD	WQ	302, 306, 308, 315, 317
FRISTON	WQ	306, 308, 316
GATWICK	WQ	309
GRAVESEND	WR	306
HARROWBEER	VY	302
HESTON	WL	302, 303, 306, 308, 315, 316, 317
HETHEL	WG	303*), 308*), 316*), 317*)
HOMSLEY SOUTH	VU	306, 315
HORNE	WQ	303
HORSHAM ST. FAITH	WG	307
HUTTON CRANSWICK	WA	302, 306, 308, 315, 316
IBSLEY	VU	302
IPSWICH	WM	308
JURBY	QX	302, 307
KENLEY	WQ	302
KIRKNEWTON	QT	309
KIRTON-IN-LINDSEY	WA	302, 303, 306, 307, 317
LECONFIELD	WA	302, 303
LLANBEDR	VJ	306, 308, 315
MARTLESHAM HEATH	WM	303, 317
MERSTON	WQ	303
NORTHOLT	WL	302, 303, 306, 308, 315, 316, 317
OUSTON	QZ	317
PEMBREY	VN	307, 316
PERRANPORTH	VX	302, 317
PETERHEAD	QK	309, 315
PREDANNACK	VX	307
REDHILL	WQ	303, 308
RENFREW	QS	309
SNAILWELL	WM	309
SOUTHEND	WM	302, 317
SPEKE	VD	303, 306, 308, 315
SQUIRES GATE	VD	307
TANGMERE	WQ	302
TERNHILL	VJ	306
TURNHOUSE	QT	303*)
VALLEY	VH	302
WARMWELL	VU	302
WELLINGORE	WF	309
WEST MALLING	WR	316
WESTHAMPNETT	WQ	302, 303
WICK	QD	303*), 316*)
WOODVALE	VD	308, 315, 316, 317

*) only after the end of war

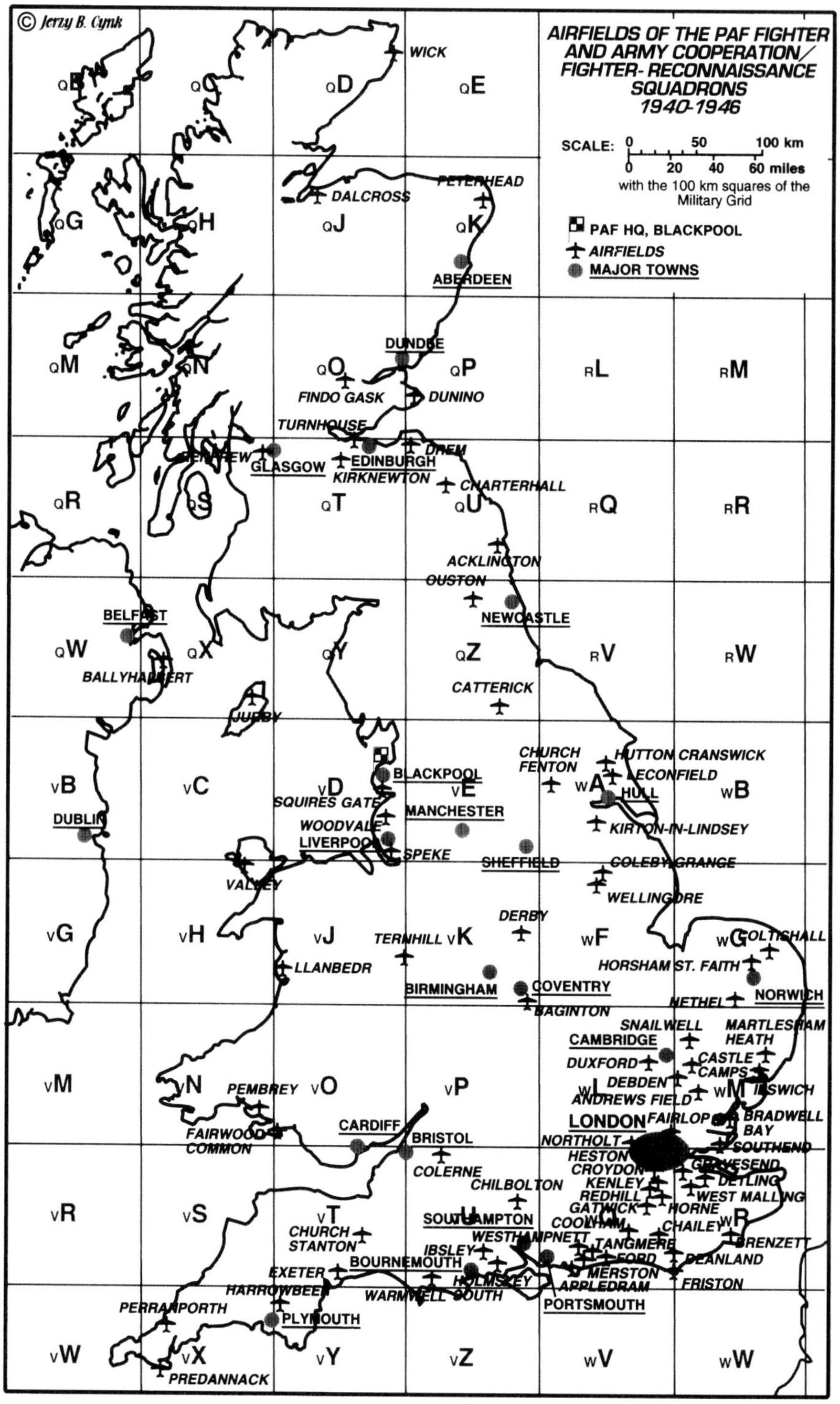

nel changes had taken place and some pilots were still completing their training. On 17 August Hurricane P3927 flown by P/O (S-Lt) Główczyński burst into flames on a training flight at a low altitude. The pilot, seriously burned, was thrown out of the aircraft on crash landing, but managed to shout a warning to approaching help about a danger of explosion which happened seconds later. He spent almost three months in hospital. Between 20 and 28 August 12 more pilots were posted to the Squadron to provide it with adequate reserves. These were: F/Lt (Cpt) Jan Czerny, F/O (S-Lt) Wacław Król, P/O (Cpt) Antoni Wczelik, P/Os (S-Lts) Jerzy Czerniak, Władysław Gnyś, Jan Maliński, Zbigniew Wróblewski, Aleksiej Żukowski and Sgts (Sgts) Antoni Beda, Wilhelm Korsarz, Antoni Łysek and Eugeniusz Nowakiewicz.

(2.) No.303 "Warsaw Kościuszko's" Fighter Squadron

An order of Inspector General of the PAF of 22 July 1940 authorized the formation of "II Warsaw Fighter Squadron" with F/Lt (Mjr) Zdzisław Krasnodębski appointed to command it. Its personnel were selected largely from former members of the 111th and 112th Eskadras of the pre-war III/1 Fighter Dyon of the 1st (Warsaw) Air Regiment in the Polish Depot in Blackpool. They initially comprised 178 officers and other ranks, including 24 pilots.

The Squadron was formed within No.11 Group at Northolt on 2 August 1940 as No.303 Dywizjon Myśliwski Warszawski im. Tadeusza Kościuszki ("Warsaw" Thaddeus Kościuszko's Fighter Squadron) and was allocated code letters RF. It adopted the 111th Edkadra's "Kościuszko" badge (not officially authorized by the RAF), with the number 303 added at the bottom. The RAF command staff included: S/Ldr Ronald G. Kellett, C.O. of the Squadron, F/Lt John A. Kent, C.O. of the 'A' Flight, and F/Lt Athol Forbes, C.O. of the 'B' Flight. Its British component was similar in size and composition to that of No.302 Squadron.

The Polish personnel, of 13 officer and 143 other ranks, including eight NCO pilots, arrived at Norholt from Blackpool on 2 August in the afternoon. The pilots comprised:[2] Acting S/Ldr (Mjr) Krasnodębski, doubling Polish C.O., F/Lt (Cpt) Tadeusz Opulski, 'B' Flight Commander, F/Os (Lts) Zdzisław Henneberg, Wojciech Januszewicz, Ludwik Paszkiewicz, P/O (Lt) Wacław Łapkowski, P/Os (S-Lts) Jan Daszewski, Mirosław Ferić, Witold Łokuciewski, Jan Zumbach and Sgts (Sgts) Marian Bełc, Josef František – a Czech who joined the LW in July 1939, evacuated with Polish airmen to Rumania and went with them to France – Paweł Gallus, Stanisław Karubin, Karol Krawczyński, Eugeniusz Szaposznikow, Stefan Wojtowicz and Kazimierz Wünsche. F/Lt(Cpt) Witold Żyborski became Adjutant, F/Lt (Cpt) Jarosław Giejsztowt, Intelligence Officer and F/O(Lt) Wacław Wiórkiewicz, Engineering Officer. F/O (Lt) Witold Urbanowicz, destined to command 'A' Flight was temporarily posted to No.145 Squadron, and some 30 officer pilots and other ranks assigned to the Squadron remained for the time being in Blackpool. The personnel at Northolt were allocated to Flights 'A' and 'B' for operational flying, and to Flight 'C' for maintenance duties.

Training began on 3 August, each pilot being tested on the Link Trainer and on a Master I (N7888). Later a Battle (R7399) was also used for training. English lessons were immediately arranged and a number of pilots were sent on the R/T course at Uxbridge. Sir Archibald Sinclair, Secretary of State for Air, visited the Squadron and wished it success in combat. First solo flights on lent Hurricanes were made the following day.

During the next few days in-flight training intensified. All pilots completed dual-control training on 7 August. On the 8th the first four Hurricanes were received by the Squadron and on the same day two landing accidents occurred, but both pilots, Sgt (Sgt) Bełc and Sgt (Sgt) František, escaped unhurt. The following day F/O (Lt) Paszkiewicz damaged his aircraft while taxiing. On 10 August 'A' Flight practiced R/T procedures in the air and two days later formation flying began. On 12 August Sgt (Sgt) Mirosław Wojciechowski arrived and went to

Pilots of No.302 "Poznań" Fighter Squadron by a Hurricane I WC•V, bearing the PAF chessboard insignia below the cockpit. The photograph, taken early in 1941, shows, left to right, P/O Neyder, Sgt Beda, F/O Czerwiński, P/O Gnyś, P/O Bernaś and P/O Chałupa.

'B' Flight, and four days later Sgt (Sgt) Leon Świtoń joined the same Flight. On the 18th Sgt (Sgt) Jan Rogowski reinforced 'A' Flight. In the middle of August 17 airmen came from Blackpool to supplement the Squadron's ground personnel.

The final contingent of Polish pilots reached Northolt on 21 August and included:[3] F/O (Lt) Urbanowicz, 'A' Flight Commander, F/Lt (Cpt) Stanisław Pietraszkiewicz, who briefly succeeded Opulski as 'B' Flight Commander,[4] F/Os (Lts) Arsen Cebrzyński, Bodgan Grzeszczak, Marian Pisarek, Walerian Żak, P/Os (S-Lts) Bogusław Mierzwa, Jerzy Radomski, and Sgts (Sgts) Tadeusz Andruszkow, Michał Brzezowski, Józef Kania and Jan Kowalski. P/O (Lt) Zygmunt Wodecki, Medical Officer, and two Medical Orderlies arrived the following day. In the meantime a few pilots had to leave the Squadron for various reasons. As training progressed pilots grew accustomed to the retractable undercarriages and to poor visibility from the Hurricane's cockpit on taxiing but one more landing accident occurred in the middle of the month.

In view of greatly increased Luftwaffe activity over South-East England an order to maintain one section at "Available"[5] during the "Day State" for Station defence purposes was received by No.303 Sqdn from the Senior Controller of No.11 Group on 16 August. At 12:30 hours on that day three aircraft were put on stand by for the first time as the rest of the Squadron proceeded to complete a shortened period of training to achieve full operational capability as soon as possible. On the following day, however, S/Ldr Kellett, the British C.O. of the Squadron, warned No.11 Group Controller about the practicability of the "Available" status because of language difficulties on R/T. His view was strongly endorsed by the Station Commander, G/Cpt Stanley Vincent, who believed – for the same reason – that the risk to Polish lives would be too great.

[2] Sources: RAF Form 540 (Public Records Office, AIR 27-1663), IPMS, Lot.A V, 49/31.

[3] RAF Form 540 (Public Records Office, File AIR 27 – 1663).

[4] Opulski was sick at the time and Pietraszkiewicz was soon recalled to Blackpool to take command of No.307 Squadron.

[5] States of operational readiness included: "Available" – ready to "scramble" (take-off) within 15 minutes; "Readiness" – ready to "scramble" within 30 minutes; "Released" – squadron or section standing down.

The British officers were openly dubious about the wisdom of assigning the Poles to No. 11 Group responsible as it was for the defence of some of the most vital Sectors, including London. In addition, the strain and exertion of intensive training had been such that three NCO pilots were declared "medically unfit" and temporarily suspended from flying duties.

These factors were not appreciated by the Poles, obliged to watch helplessly the furious combat in the skies above. Their spirits sunk lower and their resentment of the British Command grew with every day of waiting. On 24 August at 15:40 hours the whole Squadron – 12 aircraft – was "scrambled" for the first time to patrol the Northolt base and although no contact was made with the enemy, their hopes began to revive. Two days later the squadron took-off for the second time with a similar result. At the same time night air raid warnings became longer and more frequent.

On 27 August Gen Ujejski visited the Squadron. During the next two days the pilots exercised at the Sutton Bridge shooting range and on 30 August in the afternoon, in the course of an interception exercise with Blenheims, the Squadron sighted German aircraft and fought its first victorious encounter (described in the following section). The next day No.303 was declared operational.

Two more Polish fighter squadrons – Nos. 306 and 308 – and one night fighter squadron – No. 307 – were formed during the Battle of Britain, but none became operational before its end.

(3.) No. 306 "Toruń" Fighter Squadron[6]

On 21 August 1940 Col Iwaszkiewicz, Deputy Inspector General of the PAF, issued an order to F/Lt(Cpt) Tadeusz Rolski to begin organizing a fighter squadron – the third one in Britain – at the RAF Polish Depot in Blackpool. The squadron became No. 306 (Polish) Fighter Squadron and under the relevant RAF order of 28 August was formed at Church Fenton, within No.13 Group. The unit was assigned the code letters UZ. Its personnel were mainly drawn from officers and men of the former 4th (Toruń) and 3rd (Poznań) Air Regiments and the Squadron adopted the name No. 306 Dywizjon Myśliwski Toruński ("Toruń" Fighter Squadron). The "Wild Duck" emblem of the Toruń III/4 Fighter Dyon on the diamond background of the 3rd Regiment's insignia formed its badge (not officially authorized by the RAF) which also carried at the top the Bear of No.605 Squadron, whose former C.O., S/Ldr Scott, became the British commander of No.306. S/Ldr (Mjr) Jerzy Orzechowski was earmarked to take over the Polish command from Rolski at a later date.

On 2 September Rolski, who became the Acting S/Ldr, was instructed to proceed with the selected personnel to the RAF Station Church Fenton near Tadcaster (Yorks), but because of the dispersal of Polish airmen in various camps, he was able to muster only 12 pilots – out of the 25 assigned him – and 109 ground personnel. They moved to Church Fenton on

[6]Main source: IPMS, Lot.A V, 50/3 and 50/50-51

P/O Jan Zumbach by Hurricane I P3975 RF•U of No.303 "Warsaw Kościuszko's Fighter Squadron" at Northolt at the beginning of September 1940. The Kościuszko squadron badge under the cockpit, which replaced the earlier PAF chessboards, is clearly visible on the aircraft. (IPMS)

4 September. Six more Polish pilots arrived on the 19th and the remaining seven and the rest of the ground crew on the 24th.

The pilots commenced flights on the dual-control Fairey Battle and Miles Master I (serial N8007) trainers on 10 September and during the following ten days received eight Hurricanes and one Hurricane airframe for familiarization training. On 19 September the Polish flag was officially hoisted and the Squadron's Hurricanes went on parade, proudly displaying the PAF chessboard markings and "Wild Duck" badges on the fuselage. Despite language difficulties and the small number of aircraft available, 15 pilots converted to Hurricanes by the end of September.

The Squadron was due to move to another airfield and one of its Flights to become operational early in October, but this order was revoked, presumably because of decreased German activity in the final phase of the Battle of Britain. Towards the end of October the Polish command of the Squadron was thrown into some confusion. S/Ldr (Mjr) Orzechowski arrived from Blackpool to take over the command on 23 October from Rolski, who reverted to F/Lt. After three days at Church Fenton, Orzechowski returned to Blackpool and Rolski remained the senior Polish officer on the station. Orzechowski came back to Church Fenton on 1 November but was finally recalled on the 14th of that month and Rolski became the Polish Commander of No.306 again.[7]

Beginning with 28 October 1940, a section of Hurricanes was put at "readiness" every day before sunset. At the end of the month the organization of the Squadron comprised:

- Headquarters, under the overall command of S/Ldr Douglas R. Scott;
- 'A' Flight: F/Lt Hugh C. Kennard, Flight Commander, F/Lt (Cpt) Tadeusz Rolski, doubling Commander – temporarily replacing F/O (Lt) Franciszek Skiba, who went to hospital for a week – F/O Peter L. Dawbarn, Deputy at

[7]Overall command of the Squadron passed to Rolski on 3 March 1941.

Commander, F/Os (Lts) Władysław Nowak, Stanisław Zieliński, P/Os (S-Lts) Bohdan Bielkiewicz, Edward Jankowski, Kazimierz Rutkowski, Edward Suszyński, Sgts. (Cpls) Władysław Gmitrowicz, Otton Pudrzycki, Bolesław Turzański, Jan Śmigielski, Stanisław Wieprzkowicz and Sgt (LAC) Leon Kosmowski.

- 'B' Flight: F/Lt Tennant, Flight Commander, F/Lt (Cpt) Jerzy Zaremba, doubling Commander, F/Os (Lts) Jerzy Słoński-Ostoja, Kazimierz Woliński, P/Os (S-Lts) Jan Czapiewski, Zdzisław Langhamer, Józef Żulikowski, Sgt (LAC C-Off) Marek Słoński-Ostoja, Sgts (Cpls) Wawrzyniec Jasiński, Brunon Kroczyński, Henryk Pietrzak, Kazimierz Waśkewicz and Sgt (LAC) Marcin Machowiak.

The Polish maintenance and ground personnel numbered four officers – including F/O (Lt) Rafał Dabrowski, Engineering Officer – and 140 other ranks. With the RAF personnel of 36, the overall strength of the Squadron amounted to 205 officers and men.

By 1 November 22 of the 26 pilots had completed training on the Hurricanes – these figures excluding S/Ldr (Mjr) Orzechowski – and six pilots began night-flying training. The Squadron had on strength 11 Hurricanes, including one damaged in a wheels-up landing accident by Sgt (Cpr) Waśkiewicz and one without an engine. Most of the Hurricanes were of the early production series (L and P serials) and required frequent repairs. On average only three to four aircraft were serviceable. On 2 November the squadron was ordered to move to Ternhill and its forward party left for the new airfield the following day. Bad weather delayed the aircraft and its eight Hurricanes,[8] one Battle and one Master did not reach Ternhill until the 7th. The next day No. 306 Squadron was declared operational.

(4.) No.308 "Kraków" Fighter Squadron[9]

No.308 (Polish) Fighter Squadron began to organize at the RAF Polish Depot in Blackpool on 9 September 1940 under the command of F/Lt (Cpt) Stefan Łaszkiewicz, who became Acting S/Ldr, and was ordered to form at the RAF Station Speke in Liverpool. The Squadron was given the code letters ZF. Its

A Hurricane of No.306 "Toruń" Fighter Squadron, showing to advantage the unit's "Wild Duck" emblem on the forward fuselage, Ternhill, November 1940. Note the black (port) and white (starboard) underside of the fighter, with the colors split along the fuselage center line. (IPMS)

personnel came from the former III/2 Fighter Dyon of the 2nd (Kraków) Air Regiment and its full name, No.308 Dywizjon Myśliwski Krakówski ("Kraków" Fighter Squadron), and its "Winged Arrow" badge[10] reflected a continuation of this Dyon's traditions.

The Squadron's Polish personnel, comprising 11 officers – seven pilots, Adjutant, and Engineering, Intelligence and Medical Officers – and 118 other ranks (including six NCO pilots) proceeded to Speke on 12 September. The pilots were: F/Lt (Cpt) Stefan Łaszkiewicz, F/Lt (Cpt) Mieczysław Wiórkiewicz, P/Os (S-Lts) Władysław Bożek, Władysław Chciuk, Ryszard Koczor, Feliks Szyszka and Stanisław Wielgus, and Sgts (Cpls) Tadeusz Hegenbarth, Paweł Kowala, Jan Kremski, Władysław Majchrzyk, Mieczysław Parafiński and Ernest Watolski. The rest, including 13 pilots, three reserve pilots and 39 other ranks (17 from Kirkham and 22 from Innsworth Lane camps) were to follow later. The squadron's Daily Order No.1 was issued on 13 September.

[8] These included L1870, L1956, P3069, P3089 and P3408. Two unserviceable Hurricanes remained at Church Fenton and one, flown by a RAF pilot, crashed en route.

[9] Main source: IPMS, Lot.A V, 52/1 File and RAF Form 540.

[10] For some unexplained reason the "Winged Arrow" emblem as used by No.308 Squadron was reversed, with the Arrow pointing towards the rear and not to the front of aircraft as on III/2 Dyon fighters.

Hurricane Is of No.306 "Toruń" Fighter Squadron at Ternhill in November 1940. V7743 UZ•D and R4101 UZ•F bring to light a different positioning of the PAF chessboards on the rear fuselages, which at that time varied a little on almost every aircraft of the squadron.

No.306's R4101 UZ•F, displaying a different underside colour scheme, with only the wing undersurfaces black (port) and white (starboard), Ternhill, November 1940. (IPMS)

Speke airfield was bombed during the night of 14/15 September, but no casualties were sustained. F/Lt Younghusband was charged with the initial organization of the squadron at Speke and S/Ldr John A. Davis assumed overall command on 18 September. The following day Polish pilots began training flights with British instructors. On 21 September seven officer and five NCO pilots were posted to the squadron: F/O (Lt) Stefan Janus, F/O (Lt) Zbigniew Moszyński, P/Os (S-Lts) Tadeusz Hoyden, Erwin Kawnik, Bronisław Skibiński, Stanisław Wandzilak, Jerzy Wolski, Sgt (LAC C-Off) Bohdan Muth, Sgts (Cpls) Tadeusz Krieger, Stanisław Piątkowski, Stanisław Widlarz and Piotr Zaniewski. More British personnel arrived on the same day.

In view of frequent air attacks on Liverpool, the squadron was ordered to move to the RAF Station Baginton, near Coventry, on 25 September. Its strength then comprised two British and 18 Polish officers and 21 British and 123 Polish airmen. Two days later intensive flight training began at Baginton on the Fairey Battle and Miles Master I trainers, with a few pilots making their first solos. F/Lt Łaszkiewicz crashed in the Master and was admitted to hospital suffering from a concussion. F/Lt Wiórkiewicz assumed command of the Polish personnel until Łaszkiewicz's return to duty on 10 October.

On 16 October S/Ldr Davis, flying in clouds, struck a barrage balloon cable above Coventry. He was killed instantly and his Hurricane (P3399) was destroyed. The overall command of the squadron passed to S/Ldr Morris. F/Lt (Cpt) Walerian Jasionowski, posted to No.308 as Flight Commander, arrived at Baginton on 17 October and while in training, crashed his Hurricane (L2092) in a landing accident on the 30th of that month; he was unhurt.[11] With Łaszkiewicz's departure to No. 245 Squadron on 10 November 1940, Jasionowski took over as doubling C.O. of No.308.

The fearfully damaging terror raid on Coventry on the night of 14/15 November claimed the lives of three Polish mechanics: Edward Cebula, Józef Jurkowski and Franciszek Krzemiński, who – together with other Polish airmen – went to help fighting fires and rescuing civilians in the burning city. Two other mechanics were wounded.

During November Polish pilots fully completed conversion to Hurricanes and the squadron finalized its organization, with 'A' Flight commanded by F/Lt Younghusband and F/Lt (Cpt) Wiórkiewicz, and 'B' Flight commanded by F/Lt John R. Young and F/O (Lt) Stefan Janus. On the 24th of that month, while practicing formation flying, the squadron encountered a Ju 88 over Coventry and shot it down, thus achieving its first victory (described in the following section). A week later, on 1 December, it was declared operational.

(5.) No-307 "Lwów Eagle-Owls" Night-Fighter Squadron[12]

HQ Fighter Command Signal of 24 August 1940 ordering the formation of No.307 (Polish) Fighter Squadron was received by the RAF Station Kirton-in-Lindsey (in No.12 Group) and the RAF Polish Depot in Blackpool. The Squadron was to form at Kirton-in-Lindsey near Lincoln on 5 September 1940,

[11] The accident became the subject of a very critical report by S/Ldr Morris.

[12] Main source: RAF Form 540, IPMS, Lot.A V,51/6, and 51/1-4 files.

equipped with Hurricanes. HQ Fighter Command letter of 30 August advised that the Squadron was to be a Night Fighter Squadron and was to be provided with the Defiant two-seat night fighter aircraft and not with the Hurricanes as previously notified. The code letters EW were allocated to it.

F/Lt (Cpt) Stanisław Pietraszkiewicz in Blackpool was appointed as the Polish C.O. of the Squadron and immediately began selecting personnel, mainly from former members of the 5th (Lida) and 6th (Lwów) Air Regiments. The Polish name given to the Squadron was No.307 Dywizjon Myśliwski Nocny "Lwowskich Puhaczy" (No.307 "Lwów Eagle-Owls" Night Fighter Squadron). Initially, the Squadron had no badge. A competition for a badge resulted in the winning design – an "Eagle-Owl" in front of a crescent, holding an aircraft in its claws – produced by F/O (Lt) Antoni Waltoś in July 1941 (not authorized by the RAF).

Most of the RAF personnel, including S/Ldr George C. Tomlinson, posted from No.6 OTU as Squadron Commander, reported for duty at Kirton-in-Lindsey between 30 August and 6 September and by 9 September the Battle (R7411) and Master (N8009) trainers were ferried to the Squadron. In the middle of September F/Lt Cole joined No.307 as 'A' Flight Commander and F/Lt Hubert P.F. Patten as 'B' Flight Commander.

The initial Polish personnel, comprising 12 pilots and 122 ground crew, arrived at the Station on 10 September. The pilots included Acting S/Ldr (Cpt) Stanisław Pietraszkiewicz, as doubling C.O., F/Lts (Cpts) Kazimierz Kosiński and Władysław Szczęśniewski, as Flight Commanders, five other officers and four NCOs; the ground personnel, six officers, including F/O (Lt) Wilhelm Gierasimowicz, Engineering Officer, 14 NCOs and 102 men. Seven more NCOs and men reported for duty the following day and intensive flying and ground training began the same afternoon.

During the next few days another Battle (L5126) was lent to the Squadron to facilitate training, and additional Polish personnel, including six officer and seven NCO pilots, arrived from Blackpool. Eighteen Defiant I night fighters were delivered to the Squadron between 14 and 19 September, six being flown to Caistor for dispersal.

When the Poles began to examine their Defiants at Kirston-in-Lindsey, they found to their amazement that the aircraft carried no forward-firing armament. The pilots, full-blooded fighter types, would be nothing more than "drivers," to carry gunners, on the whole strangers, but including some who were experienced observer-navigators and qualified as aircraft captains under the former Polish regulations. All this was a matter of grave concern to the pilots who were selected in the belief that the Defiant was a front-gun fighter and when the squadron was inspected by the Air Officer Commanding No.9 Group (which was beginning to take over certain Sectors of No.12 Group) on 14 September, nearly all of them asked for transfers to Hurricane or Spitfire fighter squadrons. Representations were made to the Air Ministry by the Polish Inspector General seeking official consent to replace the pilots. This was granted, and in the first ten days of October most of the pilots originally assigned to No.307 – including both Flight Commanders – were posted to Old Sarum for further training.[13]

On 14 October S/Ldr (Mjr) Kazimierz Benz took over the command from Pietraszkiewicz. At about the same time new pilots who volunteered to join the Squadron – among them former test pilots of the ITL, older reserve officers and university graduates driven by an overwhelming desire to fight – began to arrive at Kirton-in-Lindsey. These included F/Lts (Cpts) Stanisław Grodzicki and Józef Tański, who became the new Flight Commanders. Training had to start all over again, and all this considerably delayed combat readiness. However, the replacement pilots soon created a good team spirit and the friendly atmosphere, which helped the Squadron to overcome serious difficulties facing it at the beginning of its operations.

Training intensified again in the middle of October, with night flying commencing on the 15th. A Link Trainer was made ready for use at the same time. By the turn of the month total flying hours started to approach 30 per day when weather conditions were favorable and firing practice began at the Sutton Bridge range. A few accidents occurred, mainly due to bad atmospheric conditions: Sgt Grondowski crashed a Defiant (N1560) during a night landing, P/O Kazimierczuk and Sgt Adamiecki, unable to land their Battle (R7411) in fog, had to parachute to safety, and Sgt Wójcik, landing a Defiant (N1696) in gale conditions struck a stationary Defiant. On 26 October the Secretary of State for Air, Sir Archibald Sinclair, accompanied by the AOC No.12 Group, AVM Leigh-Mallory, visited the Station.

On 7 November the squadron flew its 16 Defiants, one Battle (L5126) and one Master (N8009) to Jurby on the Isle of Man. Two Harrow transports ferried essential ground crews, and the main party proceeded by road and sea. Training recommenced at Jurby the following day, but subsequent heavy rain and mud on the airfield grounded the squadron for several days. On 14 November the Polish command of No.307 passed to F/Lt Grodzicki. Miles Master trainer (T9843) was allotted to the squadron on 21 November. On the 26th of that month high-ranking PAF officers – G/Cpt Iwaszkiewicz and S/Ldr Wiedeń from the Polish Inspectorate General, accompanied by G/Cpt Pawlikowski from HQ Fighter Command and F/Lt Brzezina – visited the squadron. By the end of November some confusion regarding staffing levels was finally cleared up with the issue of a specially amended Establishment (Polish) (Defiant) WAR/FC/230.

Serious maintenance problems at Jurby – lack of Defiant spares and tools – severely affected aircraft serviceability. Official misgivings regarding No.307's night-fighting capability came to the surface on 30 November. A message from No. 9 Group advised that the Squadron was to train "for day opera-

[13] This was also unfortunate, since at Old Sarum only initial training could be offered and after a brief period of complete inactivity the pilots were eventually posted to various fighter squadrons.

tional state and not for night operations." The Defiant aircraft were to be retained.

No.307 Squadron became operational on 4 December 1940 and was temporarily assigned to convoy protection duties. Its flying personnel, under command of S/Ldr Tomlinson and Acting S/Ldr (Cpt) Grodzicki, then comprised:

- 'A' Flight: F/Lt Cole, Flight Commander, F/O (Lt) Roman Smok, doubling Commander; pilots.: P/Os (S-Lts) Antoni Alexandrowicz, Jan Buczyński, Kazimierz Kazimierczuk, Jan Krzyżanowski, Sgts (Sgts & Cpls) Mieczysław Adamiecki, Kazimierz Bocheński, Witold Dukszto, Antoni Joda, Ryszard Lewczyński, Jerzy Malinowski, Ludwik Mirończuk and Stefan Wójcik; air gunners: F/O(Lt) Lech Karwowski, Sgts (Cpls & LACs) Jan Banyś, Zygfryd Blachowski, Kazimierz Frąckiewicz, Wiktor Gandurski, Stanisław Jarzembowski, Stanisław Kaliszewski, Marian Krzystyniak, Mieczysław Łydka, Kazimierz Osieleniec, Stefan Ostolski, Edward Podgajny and Józef Starosta.
- 'B' Flight: F/Lt Patten, Flight Commandeer, F/Lt (Cpt) Józef Tański, doubling Commander; pilots: F/Os (Lts) Marian Koterla, Maksymilian Lewandowski, P/O (S-Lt) Stanisław Szabłowski, Sgts (Sgts & Cpls) Wacław Bernatowicz, Juliusz Bilau, Maksymilian Frychel, Stanisław Grondowski, Jan Mikszo, Stanisław Pietniunas, Mieczysław Piwko, Lucjan Szempliński and Jan Wisthal; air gunners: Sgts (Cpls & LACs) Jan Broda, Jerzy Karais, Stanisław Kondras, Jan Laskowski, Józef Lipiński, Marian Niewolski, Henryk Ostrowski, Stanisław Sadawa, Jan Stęgierski, Maksymilian Trawicki, Jan Woźny and Mieczysław Zakrocki.

2. The Polish Contribution to Victory in the Battle of Britain

After the spectacular success of the German offensive in the West, Hitler seemed to have had no doubt that, with France defeated, Britain would be anxious to make peace. The German terms, from his point of view, were to be most "generous." They were to "guarantee" the existence of the British Empire. All Hitler wanted in return was a free hand on the Continent. He was so confident of British compliance that he made no plans for continuing the war with Britain and ordered partial demobilization of the Army (from 160 to 120 divisions). He considered that the war had been won and became more concerned about Russian threats in the Balkans and the Baltic than about the British. However, the peace feelers, involving Sweden, the United States, and the Vatican, were firmly rebuffed by Churchill who was not in the mood to compromise on the enslavement of Europe.

On 2 July 1940 Hitler issued his first – and as yet hesitant – directive on the war against Britain. He had decided that "a landing in England is possible, providing that air superiority can be attained ... The date of commencement is still undecided." He added that "the invasion is still only a plan ... " A fortnight later – as Britain showed no willingness to come to terms – his resolve grew stronger. He decided "to prepare a landing operation against England and if necessary to carry it out" (Directive No.16). The code name for the operation was to be "Sealion." Preparations for it were to be completed by mid-August. However, both Hitler and the Army – which was very skeptical about the prospects of a successful landing operation – still believed that intensification of the German sea and, in particular, air war would paralyze and finally break the will of the British people to resist and force the British government to capitulate.[1]

The Battle of Britain was set to begin.

Because of Hitler's initial reluctance and lack of any preparatory work on an invasion of Britain, the Luftwaffe assault was a tale of divided counsels, conflicting aims and divergent plans. Methods of attack were changed and various phases of German operations overlapped, making their precise dates difficult to define. The first official Air Ministry publication in 1941, "The Battle of Britain, August-October 1940," distinguished four phases of the Battle:

1st phase- 8 to 18 August,
2nd phase-19 August to 5 September,
3rd phase- 6 September to 5 October,
4th phase- 6 to 31 October.

A subsequent Air Ministry Pamphlet 156 of 1943 "The Battle of Britain," identified three phases:

1st phase (subdivided into two parts) -10 July to 18 August;
2nd phase (subdivided into two parts)- 24 August to 27 September. (after a lull between 19 and 23 August);
3rd phase- 28 September to 31 October.

Further studies of German war directives and changes in Luftwaffe tactics led to a division of the Battle into five phases which is now accepted by most historians as definitive:[2]

1st phase – 10 July to 7 August: attack on Channel shipping and Naval installations, testing of British air defences;
2nd phase – 8 to 23 August: intensified attacks on shipping and the initial attempt at elimination of Fighter Command:
3rd phase – 24 August to 6 September: the major assault on the RAF and aviation industry in an all-out bid to destroy the basic organization of British air power;

[1] Gen Jodl's paper "The Continuation of the War against England."

[2] With the notable exception of Francis K. Mason who argued in his book "Battle over Britain" that the Battle opened on 1 July with the occupation of the Channel Islands by the Germans. However, the occupation might be arguably regarded as the epilogue of the French campaign, and Hitler's first directive on the war against Britain was not issued until the following day.

4th phase – 7 to 30 September: Blitzkrieg on London as the supreme target;
5th phase – 1 to 31 October: retreat from daylight raids on London by large bomber formations and recourse to more dispersed, fast fighter-bomber attacks.

The PAF General Headquarters war records assessed Polish participation in the Battle of Britain from the perspective of the early (1941) division of the battle into four phases. Consequently, various PAF inspired publications[3] presented Polish fighter score figures and other statistical data which are not compatible with those applicable to the definitive five phases of the Battle. This present account, based on close analysis of PAF and RAF records, sets out to correct the figures accordingly.

The Luftwaffe strength deployed against Britain at the beginning of the Battle (2nd week of July) amounted to about 1,200 twin-engined bombers, 280 dive bombers, 760 single-seat fighters, 220 twin-engined (heavy) fighters, 50 long-range reconnaissance and 90 short-range reconnaissance aircraft (some 2,600 in all) of two Luftflotten, 2 and 3, whose boundaries extended from Germany across Holland, Belgium, and Northern France, beyond the mouth of the Loire. Luftflotte 2 established HQ in Brussels and Luftflotte 3, in Paris. Luftflotte 5 in Norway – with HQ at Stavanger – provided additional strength of 115 twin-engined bombers, 75 single- and twin-engined fighters and 30 reconnaissance aircraft (some 220 in all). Actual serviceable combat strength of the three Luftflotten was about 70 percent of the above figures.

At that time Fighter Command aircraft states showed a total of about 870 fighters in 55 operational squadrons (19 Spitfire, 26 Hurricane, eight Blenheim and two Defiant), of which some 660 were combat-ready, with a pilot availability of about 1,300. Fighter Command was organized geographically in three Groups:

- No.11 Group in South-East and Southern England, with HQ at Uxbridge, Middlesex, under AOC AVM Keith Park;
- No.12 Group in Central and Northern England, with HQ at Watnall, Nottingham, under AOC AVM Trafford Leigh-Mallory;
- No.13 Group in Northern England and Southern Scotland, with HQ at Newcastle, Northumberland, under AOC AVM Richard E. Saul.

A fourth Group, No.10, in the South-West, with HQ at Box, Wiltshire, to be commanded by AVM Sir Quintin Brand, was already authorized and became operational on 18 July with the transfer of the South-Western Sectors of No.11 Group to its control.

[3] E.g. "Polski Siły Powietrzne 1940-1945" ("Polish Air Force 1940-1945"), London, 1947, and "Poland's part in the air (1939-1945)", "The Aeroplane" 12 and 19 September 1947. The original PAF figures are continually repeated in most current publications.

(1.) 1st Phase - 10 July to 7 August

Polish pilots who arrived in England under the original Anglo-Polish arrangements and in April 1940 had volunteered for transfer to Fighter Command, began to go into action with the RAF during the opening phases of the Battle. The first four joined British squadrons on 16 July: F/O (Lt) Antoni Ostowicz and F/Lt (Cpt) Wilhelm Pankratz went to No.145 Squadron, and P/O (S-Lt) Tadeusz Nowak and P/O (S-Lt) Włodzimierz Samoliński, to No.253 Squadron. Ostowicz became the first Polish pilot to fight a victorious combat over England and the first to die (less than four weeks later).

On 19 July Ostowicz shared a He 111. His combat report read:[4]

> "I was No.3 Red Section, No.145 Sqdn; We took off from Tangmere at 17:50 hours with orders to patrol Brighton at 10,000 ft. ... Shortly afterwards we turned slightly South towards the sea and ... sighted an aircraft flying South Southeast at approx. 3,000 ft. I followed Red 1[5]. Red 2[6] apparently was unable to keep up with us, so I took his place as he fell back behind me. The aircraft was identified as a He 111. We closed to within 300 yards and after Red 1 had fired at it, the enemy turned slightly right and I was able to fire a burst. Smoke then issued from his starboard engine. ... A few moments later I fired another burst and broke away. ... At the same time I looked to see what Red 2 was doing and saw him turning to the North with black smoke coming from his plane. I then saw the enemy aircraft turning further to the right. It went into a glide and subsequently landed on the sea. It sank in about 3 minutes and I saw four of the crew in the water.[7] I ... was then ordered to "pancake."[8] I landed at Tangmere at 18:40."

Ostowicz was officially credited with half of the He 111. He claimed a Do 215 as "damaged" on 31 July and further increased his score in the 2nd phase of the Battle before being killed on 11 August.

Several more Polish pilots were posted to British squadrons at the end of July and early in August and by the end of the 1st phase of the Battle, on 7 August, their number had risen to 24.

(2.) 2nd Phase – 8 to 23 August

The 2nd phase of the Battle (8 to 23 August) was the prelude to the all-out assault on Britain. With no signs that the British would be willing "to listen to reason," Hitler resolved to destroy England. His top secret Directive No.17, issued on 1 August 1940, included orders that the Luftwaffe "must, with

[4] IPMS, Lot. A V,11/15B.
[5] F/Lt Roy Dutton.
[6] P/O Michael Newling.
[7] This was a He 111P of III/KG 55. The crew, commanded by Oblt Westhaus, all perished.
[8] Code name for "return to base" and land.

all means at its disposal and as quickly as possible, destroy the English air force" (1.); that the increased air war is to be carried out in preparation for Operation Sealion for which the Luftwaffe is to stand by in force (4.); and that the strengthening of the air war may start on or after 5 August (6.). Simultaneously, another directive, signed by Keitel on behalf of Hitler, ordered that preparations for the Sealion invasion were to be continued and completed by 15 September.

Consequently, the three Luftflotten (2, 3 and 5) received a directive for "Adlerangriff" ("Attack of the Eagles"), to start on a day to be nominated as "Adler Tag" ("Eagle Day"). Arguments between the Luftwaffe chiefs on how to best execute the raids and bad weather delayed "Adler Tag" until 10 August. This was further postponed until finally fixed for the morning of 13 August.

The Luftwaffe overall strength was then about 4,630 aircraft of which 3,300 were serviceable and the strength of the three Luftflotten deployed against Britain was increased to 3,358 first-line aircraft of which 2,550 were serviceable (as on the morning of 10 August).[9] This force was opposed by about 710 serviceable fighters of Fighter Command (out of some 1,100 available; one third of them Spitfires), with a pilot availability of 1,420. Göring believed that four days of good weather would be sufficient for the Luftwaffe to establish air superiority.

The weight of the Luftwaffe attacks markedly increased on 8 August. In fierce fighting in the first days of the 2nd phase Polish pilots in British squadrons achieved their first spectacular successes and also suffered great losses. On the 8th at midday, in a ferocious dogfight involving about 150 aircraft over the Convoy PEEWIT in the Channel, F/O Michał Stęborowski, flying with No.238 Sqdn, claimed one Bf 110 and Sgt Marian Domagała of the same Squadron one Bf 110 and one Bf 109. In the late afternoon, as the battle over the convoy continued, F/O Antoni Ostowicz shot down a Bf 109 – in addition to one claimed as "damaged" in the morning; Witold Urbanowicz, flying briefly with No.601 Sqdn, who the following day moved to No.145 Sqdn and later became the highest-scoring Polish fighter ace of the Battle, achieved his first kill, a Bf 110, and Sgt Józef Kwieciński, of No.145 Sqdn, accounted for a further Bf 109. The RAF claimed 60 enemy aircraft destroyed on that day of which six were credited to the Poles.

Three days later, on the 11th, the Luftwaffe again returned in force and in vicious fighting in the morning over the Portland area the first Polish pilots fell victim to the Battle. At about 10:30 hours F/O Ostowicz crashed into the sea off Swanage and shortly afterwards F/O Stęborowski shared a similar fate. At the same time Sgt Domagała brought down one Bf 109.

The next day saw heavy attacks on the coastal radar stations and Fighter Command airfields, and a large fighter force was sent into battle to intercept the raids. Fierce combat claimed the lives of three Poles: P/O Kazimierz Łukaszewicz of No.501 Sqdn over Dover, and F/Lt Wilhelm Pankratz and Sgt Józef Kwieciński both of No.145 Sqdn, off the Isle of Wight. Polish pilots claimed four of the 62 Germans deemed by the RAF as destroyed: F/O Witold Urbanowicz – a Ju 88, P/O Paweł Zenker – a Ju 87 and P/O Karol Pniak and Sgt Wojciech Kłoziński – one Bf 109 each. Sgt Kłoziński also claimed one more Bf 109 as "probable" (The actual German losses amounted to 31 aircraft and the British lost 22).

The launching of "Alderangriff" on 13 August was badly coordinated and largely misfired due to confusing signals and serious mistakes by German Intelligence. An early morning raid by Dorniers of KG 2 on targets in the Thames Estuary, whose escort was recalled to its bases, was intercepted by Spitfires of No.74 Sqdn Its two Polish pilots, F/Lt Stanisław Brzezina and P/O Henryk Szczęsny, reported one Do 17 shot down plus one damaged, and one Do 17 shot down respectively. In the afternoon F/O Tadeusz Nowierski, flying with No.609 Sqdn, claimed one Bf 109 destroyed.

15 August, with well over 2,000 sorties flown, became the day of Luftwaffe's greatest effort and witnessed perhaps the bitterest fighting of the whole Battle. RAF bases were targeted, but several of those attacked were of little or no importance to Fighter Command and the Luftwaffe lost 76 aircraft – this causing the day to be known within the Luftwaffe as "Black Thursday." RAF fighters claimed 183 enemy aircraft destroyed, six and half of which were credited to Polish pilots in British squadrons. Fighter Command lost 34 aircraft and 17 pilots killed, including P/O Mieczysław Rozwadowski, and 16 wounded.

A hard-fought combat between two Spitfires, flown by F/O (Lt) Piotr Ostaszewski-Ostoja of No.609 Sqdn and P/O (S-Lt) Janusz Żurakowski of No.234 Sqdn, and Bf 110C (M8+BP) of 6/ZG 76 piloted by Fw Josef Birndorfer escorting Ju 88 bombers, was among the most notable Polish victories of 15 August. The German pilot used all his skill to shake off his attackers, who forced him down to "naught feet" and stuck to his tail, closing on him to less than 100 yards, until he crashed at North Ashley Down, on the Isle of Wight, ripping the Bf 110 apart. The pilots were awarded one Bf 110 each.[10] The other Polish scores comprised: P/O Stefan Witorzeńc, credited with two Ju 87s, Sgt Antoni Głowacki, with one Ju 87 destroyed one Do 215 damaged, P/O Bolesław Własnowolski, with one Bf 109, and Mieczysław Rozwadowski, with 1/2 Bf 109. In addition, P/O Karol Pniak claimed one Bf 109 "probable" and one "damaged."

The following day, as the Luftwaffe attacks continued with unabated ferocity, Sgt Zygmunt Klein added to the Polish score, claiming destruction of a Bf 109 off the Isle of Wight.

After a brief respite on the 17th, the Germans returned in great strength on 18 August in a bid to deliver the decisive blow to Fighter Command. Instead they met with unintimidated

[9] Luftwaffe Quartermaster General 6th Abteilung returns.

[10] It appears that each pilot reported the destruction of one Bf 110 to his Squadron's Intelligence Officer and at the time no connection was established between these two claims.

resistance. According to the contemporary figures of the Air Ministry, the air defences accounted for 155 enemy aircraft destroyed on that day, of which seven were awarded to the Polish pilots – two each to P/O Pniak and P/O Własnowolski, and one each to P/Os Franciszek Czajkowski, Stefan Witorzeńc and Paweł Zenker. The actual German losses of 71 aircraft destroyed in combat[11] were nevertheless high enough to alarm the Luftwaffe and dispel any impression that Fighter Command was beaten. Exceptional vulnerability of the Ju 87 dive-bombers forced the Luftwaffe to pull them out of the battle. The RAF lost 27 of its fighters, with ten pilots killed including F/O Franciszek Gruszka. Another Polish pilot, P/O Franciszek Kozłowski, was severely wounded and did not return to the Battle.

Victories reported by individual RAF pilots greatly exceeded the numbers of enemy aircraft actually shot down. Despite stringent sifting of all claims by the RAF, the official score tables of confirmed victories still exaggerated the true German losses.[12] On the other hand, the destruction of some enemy aircraft was never reported and went unnoticed. Such was the case of F/O Franciszek Gruszka of No.65 "East India" Sqdn, reported "missing" on 18 August 1940.

Gruszka, flying Spitfire R6713, vanished without a trace on the 18th at about 13:30 hours over Kent, as did one of the Messerschmitts of 8/JG 3. Only in the 1970s was it established for certain that both these aircraft buried themselves in the soft earth on impact. Kent aviation archaeologists, puzzled by the unsolved mystery and inspired by local stories, began searching ground near Preston (east of Canterbury) and Stodmarch and in 1971 located the site of the Spitfire crash near Preston. During complete excavation in the spring of 1975 Gruszka's remains and several parts of the R6713 were recovered. Two years later the wreck of JG 3's Bf 109, with unidentified human remains inside, was discovered in a field – formerly a peatbog – near Stodmarsh. According to television interviews with local inhabitants, a solitary Spitfire fought a battle with four Messerschmitts at a high altitude on Sunday, 18 August. One German was seen to come down in the Stodmarsh area and a few minutes later the Spitfire shared the same fate crashing not far from Preston. Gruszka's lonely fight and his victory was not listed in any of the official RAF war records.

A deterioration in the weather severely restricted German air activities from 19 August onwards until dawn on the 24th.

The first Polish fighter squadron to enter the Battle, No.302, became operationally fit at Leconfield on 15 August. Its task was to guard the East Coast and relieve No.11 Group as necessary, but, like most other squadrons of No.12 Group, its work was initially confined to routine patrols. In the evening of 20 August the squadron encountered the enemy for the first time. Six Hurricanes took off at 18:50 hours to look for a reported raider off Flamborough Head. A section led by S/Ldr Satchell, with P/O Stefan Wapniarek as No.2 and F/Lt Franciszek Jastrzębski as No.3, sighted and chased a Ju 88 through clouds.

Sgt Głowacki and P/O Witorzeńc, the victorious Polish pilots of No.501 Fighter Squadron at Gravesend in August 1940.

Satchell reported:

> "Just as he came out of the cloud and I was going into another, I gave him a burst. The next thing I knew I was so close I had to break away to avoid crashing into him. Then a Polish pilot blew in and gave him a burst from astern ..." Wapniarek's report read:[13] "The Commander attacked first from the rear and above and I followed immediately from the same position, closing to 90 yards. I saw flashes of my hits dancing on the enemy's wings. My target banked into a cloud and I had to break off the attack. I spent 300 rounds. In the circumstances, No.3 had no chance to fire. Climbing above the cloud I noticed the Commander's aircraft some 300 ft below me. He dived on the German again as he emerged from the cloud and I followed suit firing this time about 450 rounds. I observed some movement on the fuselage of the enemy aircraft as the German crew began to bale out. At this point deteriorating visibility obscured the target ..."

Soon after returning to base the pilots received a message confirming that a Ju 88 of 8/KG 30 had been shot down into the sea off Withernsea and its crew picked up and made prisoner. The destruction of the Junkers was credited wholly to S/Ldr Satchell.

The following afternoon the patrolling Blue Section of No.302 Sqdn, led by F/Lt Riley, reported two encounters with single Ju 88 bombers over the sea. F/Lt Riley and P/O Stanisław Chałupa each claimed one Ju 88 as "probably destroyed." As Chałupa's Hurricane, hit during an exchange of fire, began to develop engine trouble, the pilot was forced to make a belly landing before reaching the base.

---

[11] With aircraft lost or damaged in accidents, the Luftwaffe's overall losses amounted to 89 machines, the highest daily total of the Battle.

[12] The Luftwaffe's exaggeration of the actual enemy combat losses was almost three times as high as the RAF.

[13] IPMS, Lot.A V,48/8 -abridged and translated from Polish by the author.

During the 2nd phase a further 46 Polish fighter pilots entered the Battle – 18 in British squadrons and 28 in No.302 Sqdn. Nine were lost to the squadrons – seven killed in action, one severely wounded and one posted away (see Appendix 14). By the end of this phase 61 Polish pilots were participating in the battle, a net increase of 37. Out of 755 enemy aircraft deemed by the RAF as "destroyed for certain" (as against the actual German losses of 403 destroyed and 127 damaged),[14] 29 1/2 were credited to the Polish pilots (see Appendix 15).

(3.) 3rd Phase – 24 August to 6 September

Flustered by the continued strength of the British air defences, the Luftwaffe High Command resolved to strengthen its offensive operations to the scale necessary to force the RAF to abandon its airfields in the South and South-East and commit all its reserves to the battle. Göring did not doubt the Luftwaffe's victory claims, but thought that the British resilience might be due rather to an underestimation by German intelligence of the output from British factories. He ordered vigorous action against the British aviation industry as well as Fighter Command airfields. The period 24 August to 6 September, the 3rd phase of the Battle, devoted to achieving these objectives, indeed proved to be the most critical for Fighter Command.

Polish pilots in British squadrons continued to make substantial contributions to the successes of their units. On 24 August, during a day-long battle with large enemy formations, Sgt Antoni Głowacki, flying with No.501 Sqdn, accomplished a remarkable feat, claiming five Germans in three combat sorties in one day. A passage from Głowacki's memoirs, describing the destruction of one Bf 109 and one Ju 88 during his second sortie of the day, reflects the drama and fervour of the combat:

> "Blast! We won't make it before the bombardment. Junkers bombers drop their loads and climb away. We close up to the firing distance. In my earphones rings the battle cry "Tally Ho." First three fighters are already firing. The gunners respond. I can see gun flashes and swarms of glow-worms flying towards us. Under us flares of explosions, smoke and flying debris. I catch up with the nearest Junkers, sit on his tail and pump a long burst into his fuselage. Pieces of metal fly away and his left engine begins to smoke. He slows down. I open up my throttle, jump to a side and repeat the attack. I spray his whole length with bullets from above. He is on fire; both engines and the fuselage. Suddenly a Defiant[15] with a Messerschmitt 109 on its tail flash across my path between me and the Junkers. I am now firing at the Messerschmitt and see my bursts sink into its fuselage and wings. He is hit and goes down closely behind the Defiant, which trails black smoke. Both aircraft crash into the sea below. I watch my Junkers. Three parallel wakes of smoke mark his passage. I want to finish him off. No need! A sudden flash, curls of smoke and some litter descending into the sea. Nothing else. Explosion and the end. Shivers run down my spine thinking about such a death, but such is war. I return towards the coast. Far to my side I see three Junkers bombers running for their lives to France."

Out of 50 enemy aircraft confirmed by the RAF as destroyed on that day, five were credited to Głowacki. P/O Paweł Zenker, another Polish pilot serving with No.501 Sqdn, was killed in the morning combat over the Dover area. Three other Polish pilots claimed three enemies as "probable" (The actual German losses amounted to 38 aircraft.). Sgt Głowacki scored another victory, his seventh, on 28 August.

The seven weeks of almost uninterrupted fighting took a terrible toll of Fighter Command's most experienced men – 181 pilots had been killed and 145 wounded by the end of August. The shortage of trained fighter pilots became the critical problem for Dowding. The Poles were needed now as never before.

The Polish pilots of No.303 Sqdn, most of them veterans of the Polish and French campaigns, were awaiting action with desperate impatience. They were held back for the second week from becoming fully operational because of ill-founded misgivings of their British superior officers.

On 30 August the Squadron's most cherished desire, to come to grips with the enemy, was realized by a blind chance. Six fighters led by S/Ldr Kellet, took off from Northolt at 16:15 hours on a routine interception exercise. They were to rendezvous with six Blenheims north of St. Albans in Hertfordshire and carry out mock attacks on them. Twenty minutes later F/O Ludwik Paszkiewicz (of 'B' Flight, flying the RF•V) spotted a large formation of aircraft in the distance. His report read:

> "After climbing to 10,000 ft we flew northward. After a while we noticed ahead a number of aircraft carrying out various evaluations. The center of the commotion seemed to be about 1,000 ft below us to starboard. I reported it to the Officer Commanding, S/Ldr Kellett, by R/T and, as he did not seem to reply, I opened up the throttle and went in the direction of the enemy. I saw the rest of the flight some 300 yards behind me; below me were the burning suburbs of a town and a Hurricane diving with smoke trailing behind it. Then I noticed, at my own altitude, a bomber with twin rudders – probably a Dornier – turning in my direction. When he noticed me he dived sharply down. I turned over and dived after him. When turning over I noticed the black crosses on the wings. Then I aimed at the fuselage and opened fire from about 200 yards, later transferring it to the port engine, which I set on fire. When I drew very close I pressed down under for a new attack and then I saw another Hurricane attacking

[14] Luftwaffe Quartermaster General 6th Abteilung returns.
[15] From No.264 Sqdn.

and a German baling out by parachute. The Dornier went into a steep turn, and I gave him another burst. He dived and then hit the ground without pulling out of the dive and burst into flames. I then approached the other Hurricane and saw his markings: UC J.[16]

I have been firing at an enemy aircraft for the first time in my life."

F/O Ludwik W. Paszkiewicz, of No.303 "Warsaw Kościuszko's" Fighter Squadron, who achieved the unit's first air victory on 30 August 1940.

In all contemporary records – British, including the RAF 540 Form[17] and Polish – which confirm Paszkiewicz's victory, the aircraft shot down by him is identified as a Dornier Do 17, although the pilot himself was clearly not quite certain of the type. However, recent investigations of the combat by the author permitted to establish beyond any reasonable doubt that the machine in question was in fact the Messerschmitt Bf 110C M8+MM (c/n 3615) of II/ZG 76 – not a Do 17. This Bf 110C formed a part of the escort (involving ZG 76 and ZG 26) of He 111 bombers of KG 1 and KG 53, which raided targets in Luton and Radlett between 16:30 and 17:00 hours. It crashed in flames on Barley Beans Farm, Kimpton, at about 16:30 hours,[18] killing its pilot Oberfw Anthony; its gunner, Uffz Nordmeier, who baled out, suffered a broken spine and was paralyzed. The other attacking Hurricane mentioned by Paszkiewicz was flown by P/O Wicks of No.56 Sqdn (code letters US, apparently misread by Paszkiewicz as UC).

The rest of the No.303 pilots did not follow Paszkiewicz. S/Ldr Kellett heard his warning, but replied too late to restrain Paszkiewicz. He did not allow the other pilots to go with the Pole, having in mind the Blenheims which – in view of the presence of enemy fighters – had to be safely escorted back to base. Paszkiewicz was reprimanded upon landing for not observing discipline and at the same time congratulated on his – and the Squadron's – first victory. Kellet considered the matter carefully and, recognizing the fever-pitch eagerness of his Polish pilots, immediately recommended that the Squadron be declared operationally fit forthwith. Dowding agreed without hesitation, and on the next day No.303 formally supplemented the combat strength of No.11 Group.

P/O Włodzimierz Samoliński and P/O Stanisław Skalski, flying with British squadrons, succeeded in destroying two more Germans on 30 August. Samoliński's aircraft sustained damage but the pilot managed to nurse it back to base. Two further enemy aircraft were claimed as "probably destroyed" by P/O Franciszek Surma of No.151 Sqdn (He 111) and P/O Tadeusz Nowak of No.253 Sqdn (Do 215), but Sgt Feliks Gmur of No.151 Sqdn was shot down and killed in the afternoon fighting.

Pressure on Fighter Command became even heavier on 31 August with the increased use of large formations of Bf 109 fighters, which were regrouped and concentrated in the Pas de Calais for sweeps in strength over the South-East. No.303 Sqdn was thrown into the Battle on the first day of its operational readiness to fight in demanding fighter-versus-fighter combat and emerged from it covered in glory.

Several airfields and radar stations were hit by bombers and strafed by fighters. Biggin Hill, a key Sector Station south of the Thames, was attacked twice on that day. No.303 was among the squadrons ordered off during the second of these raids and was vectored to the area. 'A' Flight was scrambled at about 18:00 hours and 'B' Flight, five minutes later. After about 20 minutes flying, 'A' Flight, coming from the sun, spotted a large enemy formation and went after its fighter screen. Red Section, with S/Ldr Kellett in the lead and Stanisław Karubin and Eugeniusz Szaposznikow as his wing mates, surprised three Bf 109s. Each pilot selected his victim and all three Germans were shot down in flames.

Karubin related:

" ... It did not take long to find the enemy. There was a moment of joy as we saw them. I went ahead, pointing their direction to the Flight. We increased speed and chased them. Revenge was coming. The three Huns ahead of us split up their formation and dived. We dived after them and I gave one of them a burst in the dive. He pulled up – that was what I wanted. A new burst. Fire and smoke broke out in the white belly of the Me 109. It went down like a flaming torch. I fired another burst just in case. I was very excited by the victory and after a few circles round the spot where he fell I turned back. ... "

Szaposznikow reported:

" ... A flight of Me 109s popped up on the port side. There was little time to pick one for each of us and I got the left-hand fellow. I sent him a burst just to scare him, just as we did it at home. He climbed steeply and in doing

[16] In all published versions of Paszkiewicz's report the code letters VC I (or J) appear repeatedly, but the hand-written original (IPMS, Lot.A V, 49/3g) shows quite clearly the letters UC J.

[17] Public Records Office, AIR 27-1663.

[18] During excavation of the site in 1982 several parts with aircraft identification stencils were recovered.

so came within my sights, as I was above him. That was the end of him; all I had to do was to press the trigger. He turned over and went down followed by a trail of smoke. It was all over very quickly."

Yellow Section, led by F/O Zdzisław Henneberg, saw other Messerschmitts higher up and attacked them as they were coming to the rescue of their comrades. After individual chase three more Bf 109s were dealt a fatal blow – one each by F/O Henneberg, P/O Mirosław Ferić and F/Sgt Kazimierz Wünsche.

Henneberg recorded:

"... after a few minutes' chase I caught up with four Me 109s going towards the Channel. They broke formation and one of them swung away to one side. I went after him, keeping an eye on the others. I had to hurry, as the three Jerries were already behind and above me to the port side. I opened fire at 300 yards. After the first burst he started to smoke. I gave him another two bursts and the Jerry went down with a big trail of gray smoke. It was high time to scram, because of the other three Me 109s behind my back. I noticed a white smudge on the water about seven miles south of Newhaven. ... "

Ferić wrote in his journals:

"... They (Messerschmitts) dived down and were passing in front of us, as we were some 300 yards behind our first Section. That was lucky, especially as one of the Huns was already firing at Sgt Karubin, who was busy with the Me 109 ahead of him. Wünsche took him on. I chased after another, putting on full throttle. I caught up with him easily. He grew in my sights until his fuselage filled the whole luminous circle. It was certainly time to fire. I did this quite calmly – I was not even excited, rather puzzled and surprised at the ease of it, quite different from Poland when you had to scrape and try until you were in a sweat, and then, instead of getting the b ... ... , he got you."

'B' Flight did not encounter the enemy and returned disappointed about an hour after take-off. 'A' Flight Hurricanes began to land singly at Northolt shortly afterwards. None of them suffered any damage.

Congratulations started to roll in. The first was from Air Officer Commanding No.11 Group, AVM Park, congratulating the Squadron "on their excellent fighting." The Chief of the Air Staff, Sir Cyril Newell, signaled: "Magnificent fighting 303 Squadron. I am delighted. The enemy is shown that Polish pilots definitely on top." The message from the Station Commander, G/Cpt Vincent read: "Congratulation on magnificent first operational day."

No.303 Sqdn was eventually credited with the destruction of six Bf 109s,[19] one each to every pilot of 'A' Flight – one British and five Polish. Their victims were presumably all from JG 77. The Poles in British squadrons accounted for three Germans confirmed as shot down – one each by P/O Nowak, P/O Skalski and Sgt Głowacki – and P/O Czajkowski claimed one Bf 109 as "probable." The RAF assessed German losses on that day at 94 aircraft destroyed for certain, of which eight were attributed to the Poles, but in fact the enemy lost 39 aircraft, with a further 14 damaged. Fighter Command suffered its heaviest daily loss of the Battle, 39 aircraft, reaching parity with the Luftwaffe; 14 of its pilots were killed.

After a day's respite, No.303 Sqdn was in action again on 2nd September. With the Luftwaffe aiming to saturate the RAF fighter defences with large successive and diversion raids,

[19] The Squadron's score as initially recorded comprised four Bf 109s shot down for certain and two others "probably destroyed." As a result of a later scrutiny of the fighting and pilots' reports, this was changed to six confirmed victories.

Hurricane I P3700 RF•E takes off from Northolt in the early part of September 1940. The aircraft carries the Kościuszko unit badge under the cockpit, but no PAF chessboards. Contrary to the "Battle of Britain" film, the PAF chessboards did not appear on the noses of Polish fighters until the middle of 1941, and on No.303's aircraft, until early 1942.

Squadron's both Flights, 12 Hurricanes in all, took-off from Northolt at 17:30 hours to intercept two enemy formations over Kent and then to patrol the Ashford area, where one of the biggest dogfights, involving about 90 British fighters from ten squadrons and 160 Bf 109s, developed.

No.303 Sqdn, waiting directly in the path of the enemy, attacked a flank German fighter Staffel head-on. The Messerschmitts turned back towards the Channel and climbed, leading the Polish Hurricanes which followed them into a trap. Nine Bf 109s dived out of the setting sun on the Poles. Sgt Jan Rogowski, Green Section No.3 ("Rear Guard"), spotted them at the last moment and flew straight at them, splitting their formation and largely ruining the effect of their surprise attack. In the ensuing dogfight, he destroyed one German which crashed into the sea 10 miles from the French coast. In the melee P/O Ferić's machine was hit and the pilot, who claimed one Messerschmitt as "probably destroyed," turned for home. Sgt Rogowski, who noticed that Ferić's engine was smoking, escorted him back towards Dover and circled above until he made a successful forced landing at Eythorne. F/O Henneberg and Sgt František pursued the Germans over the French coast and saw one of the Bf 109s come down in flames and the other to lose height smoking heavily. Then the Poles had to break off the chase due to heavy anti-aircraft gun fire. F/O Henneberg was credited with one Messerschmitt damaged and Sgt František with one destroyed. The new victories added to the reputation of No.303, but AVM Park signaled: "The Group Commander appreciates the offensive spirit that carried two Polish pilots (Henneberg and Ferić) over the French coast in pursuit of the enemy to-day. This practice is not economical or sound now that there is such a good shooting within sight of London." Polish pilots in British squadrons shot down three enemies on 2 September, thus increasing the Polish score to five Germans "destroyed for certain": P/O Skalski, who had to force land in a damaged aircraft at Horton Priory, Sellindge, slightly wounded in a battle with several Bf 109s, claimed two, and P/O Witorzeńc, one.

As the Luftwaffe continued unrelenting pressure on Fighter Command on 3 September, No.303 Sqdn reported two encounters with the enemy. Its 12 Hurricanes were scrambled at 10:28 hours and patrolled various areas, as ordered. During a brief contact with Bf 109s off Dungeness Sgt Wojtowicz was attacked and force-landed near Tenderden slightly wounded, and F/O Henneberg's aircraft was also hit, but the pilot safely returned to base. At 14:15 hours nine Hurricanes took off from Northolt on the day's second sortie. Sgt František, Green Section No.2 rearguard, departed from the Squadron to investigate a formation of aircraft below, which turned out to be Spitfires. Then below clouds he sighted a solitary enemy, which he identified as a He 113 and shot it down in mid-Channel near Dover.

The following day No.303 flew two patrols, but made no contact with the enemy. Polish pilots in British squadrons were luckier. Three pilots flying with No.234 Sqdn scored four victories – Sgt Józef Szlagowski two (one Bf 110 and one Do 17), F/O Zbigniew Oleński and Sgt Zygmunt Klein, one Bf 110 each; Oleński's Spitfire was damaged in the fight. P/O Tadeusz Nowak and P/O Włodzimierz Samoliński, of No.253 Sqdn, also claimed one Bf 110 each. The relevant extract from RAF Form 540 described No.253's action: " ... Plt. Officer M.W. Samolinski (Polish) attacked a Me 110 and saw a fire in the cockpit, after which the enemy turned port and went into a dive. Plt. Off. T. Nowak (Polish) attacked a Me 110, saw it catch fire and crash." P/O Jerzy Jankiewicz, of No.601 Sqdn, reported one Bf 110 as "probable"; he was shot down and wounded, and force-landed near Worthing. P/O Janusz Maciński, of No.111 Sqdn, was shot down in a dogfight with Bf 109s near Folkestone and baled out but was killed by enemy fighters while descending.

A ground crew servicing RF•C in a blast pen at Northolt in September, 1940: left to right, LAC Bieniek, Cpl Nabakoft, AC1 Rossochacki, LAC Noculak and Sgt Ruczka. (IPMS)

On 5 September, when the exhausted squadrons of Fighter Command were reaching the limit of their endurance, the Germans adopted new tactics, launching 22 separate raids in an almost continuous onslaught over a period of eight hours. The enemy repeated the same pattern of activity during the night, dropping bombs on almost 50 towns and cities, including London, Liverpool and Manchester.

No.303 Sqdn was ordered off at 09:45 hours, but met no enemy. At 14:53 its nine Hurricanes were scrambled again and after various vectors, S/Ldr Kellett, Red 1, patrolling near Gillingham, spotted anti-aircraft fire across the Thames Estuary and led the squadron to attack. Red Section engaged Bf 109s swarming around Ju 88 bombers and with about a dozen more Messerschmitts which descended on them from above.

In a lightning attack S/Ldr Kellett downed one Bf 109 and claimed a second as "probable," and Sgt Kazimierz Wünsche destroyed another. Sgt Stanisław Karubin dealt with another and watched him crash, when a stream of tracer bullets suddenly flashed by his cockpit. He went after his attacker, which was already being chased by two Hurricanes. He forced the Bf 109 lower and lower and pursued it at "naught" feet, firing short bursts from 70 yards and closing. The other Hurricanes, unable to keep up, fell behind. His ammunition ran out, but the

Grooming of P3700 RF•E at Northolt before one of its last flights. The aircraft, participating in a number of air battles, was shot down on 9 September, flown by F/Sgt Wünshe. Three days earlier P/O Ferić destroyed in it a Bf 109.

A damaged tailwheel of R4175 RF•R receives urgent attention of ground crew at Northolt at the height of the Battle.

German was still going. In a frenzy of desperation Karubin overtook the enemy and flew straight at him, passing above the Messerschmitt with a yard or two to spare. The German's nerve gave way, he hit the ground and shattered to fragments in a cloud of dust and smoke.[20]

Meanwhile, Yellow Section, confused by circling Messerschmitts, mistook the enemy's direction and after taking evasive action, lost contact. Blue Section, led by F/Lt Forbes, went in under Red to get at the Ju 88s but encountered Bf 109s. Sgt Jan Rogowski, the Section's rear guard, rescued F/Lt Forbes from his attackers, and F/Lt Forbes and F/O Wacław Łapkowski accounted for one bomber each and Sgt Josef František for one Ju 88 and one Bf 109. F/O Łapkowski, shot down near Rochford, baled out with a broken arm. The squadron's aircraft became scattered and returned independently.

G/Cpt Vincent, the C.O. of Northolt, became skeptical of No.303's soaring victories, especially as its entire score consisted so far only of Bf 109s. Hurricanes, generally inferior to the Messerschmitts, were expected to deal successfully mainly with bombers. He asked his Intelligence Officer to treat the Polish claims with reserve and decided to check the squadron out for himself. On the afternoon of 5 September he followed the Poles with the Station Defence Flight.

What he saw at 21,000 ft over the Thames Estuary, trailing behind No.303 Sqdn some 1,000 ft below, he would remember forever: two Hurricanes diving on the leading Vic of Ju 88s almost to the point of collision and forcing them to break and turn back; other Poles swooping down on the dispersed bombers and riddling them with bullets from only a couple of dozen yards; Hurricanes and Messerschmitts zooming and diving disconcertingly close together. Sgt Wünsche rushing to save Kellett with a 109 on his tail; disabled machines and parachutes criss-crossing the sky. Vincent, pushed aside by Polish fighters, was not given a chance to fire even a single burst. He was now convinced. On returning to Northolt, he told his Intelligence Officer: "My God, they are doing it; it isn't just imagination."[21]

Out of 39 enemy aircraft reckoned by the RAF as destroyed on 5 September, eight were credited to No.303 Sqdn: six to its Polish pilots and two – plus one "probable" – to Kellett and Forbes. P/O Janusz Żurakowski, flying with No.234 Sqdn, claimed a Bf 109. The seven Polish victories represented 18 percent of the total. In reality the Germans lost 23 aircraft. The defence was not able to prevent the Germans from setting five of the Thames Haven's oil tanks on fire.

On 6 September the Luftwaffe returned to the phased massive raids. These included attacks on aircraft factories, which, fortunately, did not cause any serious damage. For No.303 Sqdn this was the day of the heaviest and most difficult fighting so far. The squadron's nine Hurricanes took off at 08:40 hours to patrol the Kenley area. No.1 Sqdn, which had been ordered to join up with them, met the enemy on the way and failed to rendezvous.

Over the Kent coast the Poles sighted above them a colossal German formation moving westward – a mass of fighters and bombers emerging from a hazy sun, estimated at between 300 and 400 aircraft and covering an area of 20 miles by five.[22] The squadron encountered the enemy too late to gain height and had to attack on the climb, at a mere 140 mph. Bf 109s pounced upon the fighters and S/Ldr Zdzisław Krasnodębski's machine, leading the second (Yellow) Section was immediately set alight. Its pilot baled out over Dartford with his face horribly burned and with his overall cloth on fire.[23] He hit the ground severely injured and barely conscious and it took him a year to recover fully. F/O Witold Urbanowicz, Yellow 2, was attacked from starboard, but managed to draw the German into a short dogfight and to send him down to earth on fire in a

[20] The official Air Ministry Pamphlet 156 "The Battle of Britain" specifically referred to this incident (p.72) to illustrate the aggressiveness of the Polish pilots, although it did not mention Karubin by name.

[21] According to the story by Richard Collier in "Eagle Day," p.177. In his book Collier incorrectly referred to the Ju 88s as "Dorniers."

[22] As reported by S/Ldr Kellett – the Combat Diary of Royal Air Force Station Northolt; also RAF form 540 (IPMS, Lot.A V,49/30).

[23] The wrecked Merlin III engine of his Hurricane P3974, recovered from Langley Park, was presented to the Sikorski Museum in London in 1990.

vertical dive. He lost touch with his section, attacked a bomber and fought a inconclusive duel with three more Messerschmitts, before turning for home.

S/Ldr Kellett, the leader of Red Section, destroyed a German bomber,[24] but was himself slightly wounded when his aircraft was hit in the tail and wings. With ammunition boxes exploding under the enemy fire, he had to force-land on the cratered Biggin Hill airfield. F/Lt Forbes at the head of Blue Section, destroyed one Bf 109 and claimed another as "probable," before being slightly injured by splinters and forced down by petrol pouring into his cockpit. He overshot the field at Northolt and completely wrecked his Hurricane. In the confused fighting P/O Mirosław Ferić got one Bf 109, and Sgt Wünche, one destroyed and one "probable." Sgt Josef František also brought down a Bf 109. Although the tail of his Hurricane was damaged by a cannon shell, he managed to land at the Northolt base. Sgt Stanisław Karubin, attacking a He 111 and claiming it as "probably destroyed," fell victim to an exchange of fire and crashed near Pembury wounded in the thigh.

A peculiar incident was reported by No.303 pilots. They heard a strange voice on their headphones urging "all Apany aircraft to pancake." "Apany" was the squadron's radio call-sign, and "to pancake" meant to return to base. On landing the Poles found that no such order had been issued by any controller. Apparently German radio monitors listening to the British R/T chat learned to recognize the Polish accents, associated them with the "Apany" call-sign and tried to mislead the pilots by broadcasting false messages.

After the morning sortie on 6 September the squadron was credited with six confirmed victories and three further "probables" – including one "destroyed" and one "probable" by a British pilot (F/Lt Forbes) – but lost four of its Hurricanes and a fifth machine returned damaged. During a later defence patrol by three aircraft at 13:10 hours, F/Lt Kent's engine caught fire, but the pilot landed successfully at Northolt.

At 17:45 hours F/O Wojciech Januszewicz flew the last patrol of the day with No.1 Canadian Squadron, which encountered a small enemy formation. Januszewicz attacked three unidentified twin-engined warplanes over Lenham (Kent), but he was shot down – reportedly by a He 113 fighter – and crashed. He returned to the squadron on 8 September.

Two Polish pilots in British squadrons added 1 1/2 aircraft to the overall Polish score: P/O Janusz Żurakowski, one Bf 109 and F/O Juliusz Topolnicki, half Bf 109, the latter pilot also claiming one Bf 109 as "probable." The RAF claimed 46 Germans destroyed – the six and half Polish victories making up 14 percent of this total – as compared with the actual German loss of 35.

During the 3rd phase of the Battle 36 more Polish pilots joined operational squadrons and six – including three killed and three seriously wounded – were lost (see Appendix 14), this increasing the total number of Poles engaged in the fighting to 91. The Air Ministry put the number of enemy aircraft destroyed during this period at 643 and the Polish score amounted to 41 1/2, or 6.45 percent of this total (see Appendix 15). In reality the Luftwaffe losses totaled 378 aircraft destroyed and 115 damaged.

By the end of the 3rd phase combat fatigue and heavy losses among the squadrons bearing the brunt of the Battle – such as those suffered by No.303 Sqdn on 6 September – could no longer be offset by replacements. With 103 pilots killed and 128 wounded in two weeks, wastage reached almost 25 percent of all available Fighter Command fighting strength. The Command's ground organization had also suffered severely. Bomber attacks severely damaged six of the seven No.11 Group Sector Stations, with Biggin Hill being almost knocked out of operations, and five forward airfields. Unbeknownst to its leaders, the Luftwaffe was virtually within sight of achieving its operational objective – air superiority over Southern England. Had this pressure against Fighter Command targets been maintained for another week or two, the German air victory would have been very likely, and with it the launching of an invasion almost certain.

(4.) 4th Phase – 7 to 30 September

An accident resulted in the issue of new tactical directives by Göring, which demanded an unexpected switch in targets and allowed the RAF to recover. Hitler, certain that Britain would sue for peace, consistently forbade attacks on London, but on the night of 25 August a few German bombers strayed over London due to a navigational error and dropped a number of bombs on the city. The RAF retaliated immediately and carried out a few night raids on Berlin. These caused little damage, but infuriated Hitler and Göring. A decision to make a massive reprisal strike on London and to continue with day and night attacks on the capital as the key objective, and also on other large centres of population, was taken by them at the beginning of September. The tactical justification for this was a belief that such action would force the RAF to throw in its last reserves and bring the battle of attrition with British fighters to its final conclusion.

The effects of the change were the opposite. As the heavy daylight raids on London developed into a pattern, Fighter Command could rebuild its shattered defences almost unhindered and make an efficient use of its available strength on a single and relatively simple task – the straightforward defence of one anticipated target.

The 4th phase, frequently called the Battle of London, which quickly developed into a round-the-clock air offensive with night bombers taking over from the daytime raiders and stoking the fires they had started, began on 7 September. After the customary morning reconnaissance sorties and small raids, the skies over England remained strangely empty despite the fine weather. This sinister quiet, combined with the issue of the Air Ministry warning – Invasion Alert No.1, meaning "At-

[24] Reported by S/Ldr Kellett as "Do 215" and recorded as such on RAF Form 540, but almost certainly a Ju 88 of KG 76.

tack Imminent" – increased the foreboding of the RAF defenders. Just before 16:00 hours first glimpses of the enemy began to appear on the radar screens and built up rapidly to an enormous mass of aircraft: 348 bombers and 617 Bf 110 and Bf 109 fighters – the biggest Luftwaffe raid of World War 2. Göring, who arrived at the Pas de Calais in his lavish personal train and assumed himself command of operations – for the first time since the outbreak of war – watched through binoculars as his aerial armada passed over Cap Gris Nez. Its target: Central and East London, and banks of the Thames, with their docks and warehouses.

The raid adopted new techniques. It advanced across the Channel on a 25-mile front in two tidal waves, stepped up in solid layers at a higher than usual altitude of 15,000 to over 25,000 ft, and covered about 800 square miles. Its huge escort was divided into two parts, some of the fighters staying close to the bombers, above, behind and on both sides of their formations, and others flying at a distance of a few hundred yards and at a greater height. The bombers made a determined effort to maintain their formation in the face of fighter attacks and heavy anti-aircraft barrage, and the skillfully evolved new escort tactics made interception extremely difficult.

By 16:30 hours all available 21 squadrons of No.11 Group were urgently ordered to the Thames Haven-Tilbury area, and Polish No.303 – Kościuszko's Squadron – was the most successful among them on that day.[25] Its 11 Hurricanes, led by F/Lt Forbes, were "scrambled" from Northolt at 16:20 hours to join up with No.1 Sqdn and vectored to North-East London. F/O Urbanowicz, succeeding S/Ldr Krasnodębski as the Polish Commander of the squadron, led the rear (fourth) section. Over half an hour later they met about 40 Dornier bombers flying over Essex at 20,000 ft, with a fighter escort above. Whilst Spitfires engaged the top Bf 109 cover and Hurricanes of No.1 Sqdn drew away Bf 110s and attacked the rear of the bomber formation forcing it to turn back, "the Poles, who had been flying in sections of three one behind the other, swung around, so that the whole Squadron was almost broadside to the bombers. They dived 4,000 ft out of the sun, each pilot selecting his own victim."[26]

The Polish pilots' reports from this action indicate that F/Lt Forbes was slow to react as the opportunity to attack developed and F/O Paszkiewicz, who led the second section, decided to take the initiative. He dipped his wings to signal his intention to the aircraft behind and broke formation, turning to starboard and diving at the bombers. F/O Henneberg's third and F/O Urbanowicz's fourth Sections followed suit, and F/Lt Forbes, noting the manoeuver, turned and dived last. The fighters went in practically in one straight line, and, with all their guns blazing away, finished the attack at point-blank range, tearing nearly a quarter of the Dorniers to pieces. The engagement was a first-class example of the kind of teamwork on the part of various squadrons that AVM Park wanted.

The first Dornier burst into flames under F/O Ludwik Paszkiewicz's fire, and his wingmates, Sgt Eugeniusz Szaposznikow and Sgt Stefan Wojtowicz, got one each almost at the same time. Two more Dorniers fell to the guns of P/O Witold Łokuciewski and P/O Jan Zumbach. Then two pilots of the first section, F/Lt Forbes and P/O Jan Daszewski, shot down two further bombers. Forbes, hit by return fire from a Do 17 and slightly wounded, brought his damaged Hurricane back to Northolt. F/O Witold Urbanowicz set yet another Dornier alight. Paszkiewicz succeeded in destroying his second victim, and Wojtowicz and Zumbach also added one more bomber each to their scores. Zumbach's second prey, escaping in a blind panic from another Hurricane, ran right across his line of fire at a distance of a few yards and under his burst lit like a torch. The pilot, trying not to be engulfed in the bomber's fanning flame, banked so violently that he lost consciousness. He recovered only a few hundred feet above the burning roofs of a London suburb. Daszewski and Łokuciewski claimed additionally one bomber each as "probable."

The Dorniers were attacked so rapidly that the Bf 109s, locked in combat with other British fighters, intervened too late to be of help to them. Henneberg's section had time to turn to face the Messerschmitts, and F/O Zdzisław Henneberg – flying in Hurricane V6605 YO•N, borrowed from No.1 (RCAF) Sqdn – destroyed one and reported another as "probable," while F/O Marian Pisarek accounted for one more Bf 109. However, Henneberg's machine was slightly damaged and Pisarek was shot down in flames over Loughton. The pilot jumped, but as he did so, his left foot lodged itself on the edge of the cockpit frame and he kept falling with his aircraft hanging head down. He eventually managed to pull his foot out of his shoe and made a safe landing.[27] Daszewski became another casualty of the Messerschmitts and baled out with a severe thigh wound. His aircraft crashed in Selsted. Meanwhile, Urbanowicz and Szaposznikow joined the dogfight, the former claiming one Bf 109 as "probable" and the latter, one Bf 109 as destroyed for certain. Urbanowicz had to land at Detling to refuel. The aircraft returned to Northolt individually having expended all their ammunition. Two fighters of 'B' Flight took off again at 17:50 hours, immediately after being re-armed and five of 'A' Flight, a few minutes later, to patrol over the base, but no contact with the enemy was made.

No.303's score – the highest among the squadrons participating in combat on that day – amounted to 14 enemy aircraft destroyed,[28] including one accounted for by a British pilot, and four "probably destroyed." Polish pilots in British squadrons reported two more Germans as "probably destroyed":

[25] According to the Air Ministry Pamphlet 156 "The Battle of Britain" p.71.

[26] Ibid, p.72.

[27] The shoe, left behind in the cockpit, was found in May 1976, when aviation archaeologists excavated fragments of Pisarek's Hurricane R4173 in the back garden of 40 Roding Road, Loughton.

[28] These included 11 Dorniers, but it appears that some of these, reported as Do 215s, might have been in fact the escort Bf 110s of ZG 2, flying at the rear of the Dornier formation.

F/O Tadeusz Nowierski, a Do 17, and Sgt Zygmunt Klein, a Bf 110. P/O Walenty Krępski, flying with No.54 Sqdn, was killed in action, crashing into the sea off Flamborough. The Air Ministry put the number of the German aircraft destroyed at 100, of which 22 were attributed to the A.A. batteries and 78 to the fighters – the Polish score of 13 aircraft destroyed amounting to 17 percent of the fighters' total. The true German losses included 40 destroyed and 13 damaged. Fighter Command lost 28 fighters and 19 pilots.

Despite the valiant efforts of the defence London suffered terribly. The East End had its little streets demolished and engulfed in flames, Silvertown became a raging inferno and the fires spread rapidly along miles of the waterfront, burning out of control. Göring, delighted, broadcast that this was "a historic hour" and German radio stations kept up an exuberant commentary on the attack, but added that it demanded "heavy sacrifices." The fires were fueled by a night attack by 318 bombers, this resulting in the fiercest blaze ever recorded in Britain – the first fire-storm in history from aerial bombardment.

On 8 September, after the exertions of the previous 24 hours, the Luftwaffe mounted only a few small attacks. Enemy activities during the following day remained fairly limited. The Luftwaffe received directives to disengage in the case of substantial opposition – an admission on its part that Fighter Command was far from being overwhelmed, despite Göring's repeated assurances that the RAF was left with only a handful of fighters – and to carry on with a systematic "round-the-clock" destruction of London. Such orders were hardly appropriate to the needs of an imminent invasion, the timetable for which, set by OKW, was as follows: orders for the launching of the invasion to be issued on 11 September (these including the laying of a minefield carpet across the Channel); final operational orders to be given a week later, with embarkation beginning on 19 September, sailing of the invasion fleet on the 20th, and landing assault on the 21st.

The biggest single raid on 9 September comprised 26 Heinkel bombers flying with an escort of 80 Bf 110 and Bf 109 fighters, and was closely followed by another by Ju 88 bombers. British fighters intercepted the enemy in force. Both Flights of No.303 Sqdn, 12 Hurricanes in all, took off from Northolt at 17:25 hours and met the He 111 and Ju 88 bombers, already escaping southwards at great speed, near Beachy Head. F/Lt Kent – flying V6665 – opened the attack by damaging a Ju 88, which escaped into a cloud, and fought a duel with a Bf 110, which he shot down into the Channel. P/O Jan Zumbach received the attention of several Bf 109s, but managed to get on top of them, probably destroying one, which disappeared into a cloud, and caught another banking in front of him. He fired into its cockpit and engine from above at 30 yards, seeing tracers hit the target. The Messerschmitt turned on its back and fell down. Sgt Josef František destroyed one Bf 109 and a He 111 of KG 1, but his radiator, fuselage and wings were riddled with bullets and he force-landed outside the Downs Hotel in Woodingdean near Brighton. Sgt Kazimierz Wünsche was shot down and baled out wounded.[29] His Hurricane RF•E (P3700) crashed at Saddlescombe Farm, Poynings, near Hove.[30]

In view of the spirited defence put up by Fighter Command, Hitler deferred the invasion timetable by three days, reserving his decision on its launch until 14 September.

With poor weather conditions on 10 September big raids developed again only the following afternoon. No.303's 12

[29] As described in RAF Form 540 (IPMS, Lot.A V,49/30).
[30] Several fragments were recovered from the site by the Wealden Aviation Archaeological Group in September 1979.

V6665 RF•J was initially flown by S/Ldr Kellet, the British commander of No.303 Squadron and later by F/Lt Kent, C.O. of 'A' Flight. The Hurricane sports a diagonal strip forward of the tail unit presumably denoting the aircraft of a squadron commander. (R. Gretzyngier)

Hurricanes left Northolt at 15:30 hours on 11 September and half an hour later encountered a large enemy formation of 150 Dornier and Heinkel bombers with a strong fighter escort at 25,000 ft over Horsham and forging towards London. The squadron, operating with No.229 Sqdn, wheeled in to attack the rear of the formation and managed to penetrate the fighter screen, inflicting heavy casualties on the Germans, but lost two of its own pilots and one other wounded.

The scattered bombers jettisoned their bombs on a wooded country well to the south of London and ran for home chased by the fighters right over the coast. F/Lt Forbes, leading the Squadron, destroyed two Dorniers. Himself wounded, he force-landed his damaged machine at Heston. F/O Zdzisław Henneberg shot down one He 111 and one Bf 109; P/O Witold Łokuciewski, one Dornier and one Bf 109; F/O Ludwik Paszkiewicz, one Bf 110; P/O Jan Zumbach, one Bf 109; Sgt Josef František, two Bf 109s and one He 111; Sgt Michał Brzezowski, two He 111s; Sgt Eugeniusz Szaposznikow, two Bf 110s, and Sgt Stefan Wojtowicz, one Bf 109. Wojtowicz, fighting with six Messerschmitts, damaged another Bf 109, before his Hurricane was hit and set on fire. The aircraft crashed in a chalk pit at Westerham, killing its pilot. F/O Arsen Cebrzyński was shot down almost at the same time and crashed at Pembury; mortally wounded, he died eight days later.

Polish pilots in British squadrons added five and half more victories to the Polish score: F/O Marian Duryasz, P/O Witold Głowacki, P/O Władysław Różycki, P/O Henryk Szczęsny and Sgt Jan Budziński, one aircraft each, and P/O Stefan Witorzeńc, half of an aircraft. The Air Ministry put the number of Germans destroyed at 93, of which 84 were credited to the RAF and nine to AA batteries, with the Polish "bag" of 19 1/2 representing 23 percent of the fighter total.

For the next three days Luftwaffe operations, affected by adverse weather, were at a reduced scale, though the night raids on London continued. No.303 Sqdn flew a few routine patrols without making contact with the enemy. With London now established as the main target of the German offensive, Fighter Command moved some of its squadrons from the less threatened parts of the country to reinforce No.11 and No.12 Groups, the latter providing reserve strength and relief for the former. For the pilots of No.302 Sqdn still stationed at Leconfield this was the break they were waiting for. Distinguished combatants of the Polish and French campaigns, they had spent most of their time on uneventful patrols over the east coast, with little opportunity for combat.

A shiver of excitement ran through the Squadron when on 12 September at 14:00 hours an order arrived, requesting the immediate departure of a detachment of 12 of its Hurricanes to RAF Duxford in No.12 Group. The rest of the pilots were to stay at Leconfield and maintain one section at "readiness" for defence of the airfield and convoy protection duties.

The fighters took off for Duxford at 15:00 hours, and in the late afternoon carried out one patrol over the London area, flying back to Leconfield in the evening. The following morning the aircraft returned to Duxford and before evening No.302 Sqdn was instructed to send essential servicing personnel to Duxford. No sorties were flown on that day. Two take-offs from Duxford the following afternoon were uneventful, as the enemy did not venture beyond the English coast, as was a convoy protection sortie flown by the Leconfield section.

Sgt Josef František, a Czechoslovak pilot of No.303 Squadron, standing by V7289 RF•S; Northolt, September 1940. With 17 confirmed air victories to his credit he became the RAF's highest-scoring fighter ace of the Battle of Britain. (IPMS)

On 14 September Hitler reviewed the progress of the war and blamed the weather for the lack of air superiority over Britain, reiterating that this must be achieved for a successful invasion, the preparations for which were otherwise complete. He added that four to five days of good weather were needed to eliminate the RAF – the same period of time as that given by Göring as necessary before Adler Tag five weeks earlier! He postponed the operation "Seelöwe" ("Sealion") once again by three days, until 17 September, hoping for a decisive result from the Luftwaffe by then. On the 14th only 1,493 fighter pilots, 94 of them Poles, stood between the survival of England and the final German victory in Europe.

15 September was destined to become the climax of the Battle of Britain and brought about a spectacular defeat of the German bomber force which changed the course of the war. With the greatly improved weather, the Luftwaffe wanted a repeat of 7 September and set out to deliver a final blow to

London as a lead-in to the invasion. The first big bomber formation began to assemble before 11:00 hours in full view of British radar, picking up its massive fighter escort over North-East France, in direct contradiction to the earlier practice in which the escort joined the bombers en route. The Germans were eventually intercepted by 22 RAF fighter squadrons out of 24 scrambled to meet them, with nothing left in reserve. Winston Churchill, making one of his periodic visits to No.11 Group Operations Room at Uxbridge, witnessed the dramatic event and was greatly impressed with the way in which this critical situation was handled.

No.303's 12 Hurricanes left Northolt at 11:20 hours and followed No.229 Sqdn, both being vectored to South London. They sighted one enemy formation at a distance, which escaped, and then another one approaching from the south. The squadron turned, but only one section managed to catch the bombers near the coast, the rest heavily engaging swarms of Messerschmitts of the rear guard which were already under attack from Spitfires. The battle developed into a series of individual duels. F/O Henneberg destroyed one Bf 109 and one Do 17 bomber, and P/O Ferić, P/O Łokuciewski, F/O Paszkiewicz, F/O Pisarek, P/O Zumbach and Sgt Wojciechowski claimed one Bf 109 each. In addition, Sgt Wojciechowski and Sgt Andruszkow shared one Dornier bomber and Sgt František accounted for one Bf 110. P/O Łokuciewski was wounded in the leg in an exchange of fire and his aircraft was damaged.

No.302 took off at the same time as part of Leigh-Mallory's big Duxford Wing (two Spitfires and three Hurricane squadrons), led by the legendary legless S/Ldr Douglas Bader, who only a few hours earlier assured the Poles, "You'll soon be back in Warsaw." The Wing arrived over London from the North and spotted the bombers far away to the West, which moments later began a wide sweep, turning towards them under attack from other fighters. No.302 pilots, led by S/Ldr Satchell at the head of 'A' Flight and with F/Lt Tadeusz Chłopik leading 'B' Flight, hear over R/T Bader's cry, "Weigh-in, everyone for himself," as the fighters tore into a flank of the raiders, bursting through the close escort. While the Spitfires engaged the Messerschmitts, the Hurricanes struck at Dorniers – which were already dispersing – and completely smashed their formations.

In a frenzied combat No.302 gave a magnificent account of itself. S/Ldr Satchell shot down one Do 17Z and claimed another as "probable" and P/O Stanisław Chałupa achieved the same score; F/Lt Tadeusz Chłopik destroyed one and a half Dorniers and F/Lt Franciszek Jastrzębski the other half; P/O Tadeusz Czerwiński and Sgt Marian Wędzik were credited with one Dornier each, and Sgt Jan Palak and Sgt Antoni Siudak shared the destruction of another. P/O Julian Kowalski also claimed one Dornier, Sgt Palak one Bf 109 and Sgt Siudak one He 111 as "probably destroyed." P/O Jan Maliński reported a further Bf 109 as "damaged." P/O Stanisław Łapka's Hurricane, WX•K (V6569) was badly hit, but the pilot baled out

Skies over the Channel at the height of the Battle in September 1940, with one of the Germans, shot down, falling towards the sea.

unhurt. Four other machines returned damaged, which left the squadron's Duxford detachment with only seven serviceable fighters.

Returning squadrons were immediately refueled and rearmed to be ready for the next challenge, which began to emerge in the form of an immense raid in three waves two hours later. All available fighters were airborne and about 170 Spitfires and Hurricanes engaged the enemy over Kent, with 13 more squadrons, including the Duxford Wing, falling on them over London.

Nine Hurricanes of No.303 Sqdn were scrambled at 14:20 hours to patrol base and after receiving directions sighted about 400 enemy aircraft in tight formation stepped up to 25,000 ft, coming head on from the direction of Gravesend. S/Ldr Kellett led Blue Section into a quarter frontal assault on the bombers which wheeled eastwards. Red and Yellow Sections – led by F/O Urbanowicz – then followed, ploughing through the raiders in order to create the maximum confusion and force the Germans to break ranks and turn back. The whole formation, by then also intercepted by another squadron a few miles to starboard, scattered, some turning eastwards and others southwards. The Bf 109s, presumably low on fuel, seemed reluctant to intervene, but the Bf 110s closed in on the Polish squadron, with a number of Bf 109s hesitantly joining the battle. The squadron was compelled to break up and individual dogfights took place, with some of the fighters escaping in cloud. S/Ldr Kellett got one Dornier as "probable" and later destroyed a Bf 110. His wingmates, F/O Walerian Żak, Blue 2, and Sgt Mirosław Wojciechowski, Blue 3, accounted respectively for one Dornier and one Bf 109. F/O Witold Urbanowicz shot down two Dorniers and P/O Mirosław Ferić, one Bf 110. Sgt Michał Brzezowski went down over Gravesend, crashed in the Estuary and his body was never found. Sgt Tadeusz Andruszkow baled out of his crippled Hurricane RF•M (P3939) over Dartford unhurt and a few hours later described the scene:

> "I have never seen such a party in the air. One just prayed not to get some burning piece of wreckage on his umbrella. There were so many of us going down that I thought they would take us downstairs for a bloody parachute division and shoot us with duck shot or catch us on a halberd when landing. After all, the Home Guard want to have fun too and they robbed museums for weapons. The Huns at any rate did not fire at the parachutes, because too many of them were their own."

Seven Hurricanes returned to base landing between 15:40 and 16:00 hours, five of them (Kellett, Ferić, Urbanowicz, Wojciechowski and Żak) damaged. No.303's score amounted to 15 enemy aircraft destroyed by its Polish pilots plus one destroyed and another "probable" by S/Ldr Kellett, but by the end of the day the squadron's operational strength was down to only eight pilots and seven Hurricanes.[31]

No.302 Sqdn took off again at 14:10 hours with the rest of the Duxford Wing, which was scrambled rather late and approached London when the enemy formation had already been scattered. Over North Weald, as the fighters descended on individual Dorniers, they were suddenly attacked by high-flying Bf 109s, diving on them from the sun.

In confused combat, the squadron fought with exceptional bravery. P/O Edward Pilch chased the bombers and shot down two of them. F/Lt Tadeusz Chłopik's Hurricane WX•E (P2954), leading the second section, was set on fire. The pilot baled out but, apparently hit by the airframe, failed to open his parachute and fell dead at Rawreth (North-East of Southend). Sgt Jan Kowalski, his wingmate, succeeded in destroying one Do 17Z and probably another and survived the Bf 109 attack, nursing his fighter, riddled with 76 bullets, back to base. P/O Włodzimierz Karwowski's machine also returned damaged.[32] S/Ldr Satchell claimed one Bf 109 as "probably destroyed." The squadron's Polish pilots achieved nine confirmed victories on that day and congratulations followed. The signal from AVM Leigh-Mallory read: "Germans have made a great effort today and you have played a notable part in frustrating it – heartiest congratulations to you all on splendid results." Sir Cyril Newall, Chief of the Air Staff, signaled: "Well done No.302 Squadron. Your success yesterday is an outstanding example of the hard fighting which is frustrating the enemy's attack. Keep it up."

Poles in British squadrons accounted for another three Germans: F/O Marian Duryasz and P/O Bolesław Własnowolski, for one Do 17 each, and Sgt Józef Jeka, who chased his prey so close to the ground that his bursts lifted up yards of turf, for a Bf 110.

Upon hurried examination of the day's combat reports, the Air Ministry arrived eventually at the figure of 185 enemy aircraft destroyed – 178 by the RAF and seven by AA fire, of which 27, 15 percent of the fighter total, were credited to the Polish pilots.

After the war it was established that the Luftwaffe in fact lost 56 aircraft, with a further 21 damaged, but with the RAF loss of only 26 fighters and 13 pilots, the balance was still very much in Britain's favor. Although the inflated RAF claim was a great morale booster to the defenders, it was arrived at in good faith. The confused nature of the fighting in those two tremendous air battles over London was such that more than one pilot firing at the same aircraft could be convinced that he was responsible for its destruction.

However, the actual figure was of little importance. While the size of the German loss might have justifiably been a matter of controversy, its substance was uncompromising. Despite its gigantic effort, the Luftwaffe, unable to achieve a bombing concentration even approaching that of 7 September and with its crews deeply shocked by the strength of the defending fighter force, had to accept the defeat of its bomber force and break off the massive daytime raids, this having incalculable consequences for the future course of the war. "Seelöve" had to be abandoned and with it went the prospects of Germany's ultimate victory. The Polish contribution to this victory was disproportionately larger than the number of Polish pilots participating in the Battle.

F/O Witold Urbanowicz, the Polish commander of No.303 Squadron, who was credited with 15 confirmed air victories during the Battle. P3120 RF•A depicted in the background has a diagonal strip on the rear fuselage similar to that seen on V6665, which seems to be partly washed off; Northolt, late September 1940.

[31] As recorded in RAF Form 540 (IPMS, Lot.A V,49/30).

[32] The Duxford Wing lost five aircraft in this engagement, including two from No.310 Czech Squadron which formed part of its strength; several more were damaged but repairable.

Purporting to show a "scramble" of No.303 Sqdn during the Battle of Britain, this photograph was taken at Leconfield in December 1940.

At a conference held the following day Göring decided to change tactics once again and to return to attacks against Fighter Command with smaller forces and with even bigger fighter escorts. The RAF also took stock of its combat successes over the past week and introduced measures to increase the effectiveness of its fighter force even further – such as improved rendezvous procedures, pairing of squadrons at higher altitudes, shortening of patrol lines, and others.

Poor weather on 16 September precluded large-scale air activity, permitting squadrons to recover. No.302 received an order at dawn to reinforce its Duxford detachment with four pilots (S/Ldr Mieczysław Mümler, F/Lt Piotr Łaguna, P/O Stefan Wapniarek and Sgt Antoni Markiewicz) and four aircraft before 08:00 hours, to make up the previous day's losses which had left it with only four serviceable Hurricanes.

Both No.303 and No.302 Squadrons made operational sorties the following afternoon, but only the former, flying with No.1 Canadian Sqdn, met the enemy at 16:00 hours over the Thames Estuary. These were Bf 109s seen at altitudes of up to 27,000 ft and difficult to catch, but Sgt Mirosław Wojciechowski did succeed in bringing one down in flames over the sea. F/Lt James Anderson Thomson joined No.302 Sqdn as a British Deputy Flight Commander.

18 September was more eventful. No.303's 12 Hurricanes, operating with No.229 Sqdn, took off from Northolt at 12:20 hours on their second patrol and spotted a single reconnaissance Do 215 with a large Bf 109 escort south of Biggin Hill. Eight pilots shot at the Dornier and saw its crew bale out and the aircraft crash at Yalding. Sgt Josef František shot down one of the escorting Bf 109s, which fell in the sea. The squadron's Hurricanes were attacked by British fighters by mistake and two were slightly damaged.

C.-in-C. Polish Armed Forces, Gen Sikorski visited the squadron and awarded the Virtuti Militari[33] to S/Ldr Zdzisław Krasnodębski (in hospital), F/Os Zdzisław Henneberg, Wojciech Januszewicz, Wacław Łapkowski, Ludwik Paszkiewicz, Marian Pisarek and Witold Urbanowicz, P/Os Jan Daszewski, Mirosław Ferić, Witold Łokuciewski and Jan Zumbach, and Sgts Josef František, Stanisław Karubin, Eugeniusz Szaposznikow, Stefan Wojtowicz and Kazimierz Wünsche; F/Os Arsen Cebrzyński (killed in action), F/O Zygmunt Wodecki, Sgt Michał Brzezowski (killed in action) and Sgt Jan Rogowski received the Cross of Valour.

For No.302 Sqdn the 18th was a day of an outstanding success. After two uneventful patrols, the squadron was airborne again with the rest of the Duxford Wing at 16:15 hours and met a fast raid by Ju 88s of KG 77 approaching London from the East. Leaving the Wing's two Spitfire squadrons to look after the fighter escort, S/Ldr Bader led his three Hurricane squadrons in an almost vertical dive on the first formation, and the whole raid broke up and turned for home. The Poles attacked the bombers with great determination, and S/Ldr Satchell, leading No.302's Hurricanes together with S/Ldr Mümler, shot down one, while Mümler accounted for another, both pilots identifying the planes as Dorniers. F/Lt Riley claimed one Ju 88 destroyed for certain and one "probably." P/O Włodzimierz Karwowski, P/O Edward Pilch, P/O Stefan Wapniarek and Sgt Edward Paterek added one Ju 88 each to the score; F/Lt Franciszek Jastrzębski and F/O Julian Kowalski claimed one Dornier each as "probable," and F/Lt Farmer one Ju 88 damaged. Sgt Paterek damaged his WX•Z in a forced landing near Danbury, out of fuel, and P/O Wapniarek, lost and also without petrol made a successful landing, both pilots returning to the Squadron the following day. AVM Leigh-Mallory congratulated the Poles again on "the keenness and gallantry displayed in these combats" (reflecting) "great credit on all members of the Squadron. ... " Seven enemy aircraft were credited to the Polish pilots on 16 September, out of 48 recorded by the Air Ministry as destroyed.

The next four days saw only slight Luftwaffe activities. On the 19th, the day Hitler finally decided to put off "Seelöve" indefinitely, No.302's 'B' Flight attacked a solitary Ju 88 and F/O Julian Kowalski finished off the bomber, which crashed

[33] The highest Polish combat distinction, in five classes: Grand Cross with Star, Commander's Cross, Chevalier's Cross, Golden Cross and Silver Cross; the last two awarded for outstanding acts of individual courage.

Gen Sikorski pins the Virtuti Militari Cross on Sgt František during decoration of No.303 pilots at Northolt on 18 September 1940. Sgt Szaposznikow and F/O Pisarek are standing next to František. (IPMS)

P/O Witold Głowacki, No.605 Squadron, by his shot down P3832 UP•P on 24 September 1940. Germans reported him to be dead the same evening.

at Bury St. Edmunds at 11:30 hours. On 21 September F/O Juliusz Topolnicki, flying with No.601 Sqdn, was killed in a take-off accident.

On 23 September the enemy raided England in waves of various sizes. No.303 Sqdn was ordered off at 09:30 hours to lead No.1 Canadian and No.229 Sqdns as a Wing. The squadron, at 22,000 ft, saw about 20 Bf 109s at a very high altitude circling the Thames Estuary, which made no attempt to attack. Then about 12 other Bf 109s came from behind, apparently trying to break up the Wing. Red Section turned to meet them and the Germans fled towards France. F/Lt Kent caught up with one and Sgt Eugeniusz Szaposznikow with another; Kent's victim crashed into the Channel and Szposznikow's went down in flames near the French coast. On the way back Kent fired at a twin-engined aircraft which he thought was a FW 58 and reported it damaged.

On the 24th neither of the Polish squadrons on patrol encountered Germans, but P/O Witold Głowacki of No.605 Sqdn, taking off from Croydon at 15:41 hours as Yellow 2 with P/O Ian Muirhead, Yellow 1, and ordered to the Beachy Head area, spotted a Dornier in the clouds, which proceeded eastwards. Muirhead attacked from starboard, quarter astern and Głowacki followed up from the port quarter. The Dornier jettisoned its bombs over the sea; Muirhead closed up to 100 yards and set fire to the bomber's port engine; and then, after Głowacki's repeated bursts, the German plane, with both engines in flames, lost height and crashed into the Channel off Cap Gris Nez. Both Hurricanes were attacked by Bf 109s which suddenly appeared overhead, and dropped down to sea level and crossed the French coast at Albemeuse. Muirhead managed to shake off the Messerschmitts, climbed into cloud and returned to Croydon. Głowacki, flying UP•P (P3832), was shot down and crashed. Both pilots were credited with 1/2 Do 215 each.

P/O Głowacki's shot down Hurricane, which broke into three parts during the crash landing.

Several photographs, taken by a German war correspondent, showed Głowacki with a bandaged head, standing beside his wrecked Hurricane. He did not look to be seriously wounded, yet his death was recorded a few hours later, on 24 September. He was born in Berlin (on 13 October 1913) of Polish parentage and spoke perfect German – he came to Poland with his family after World War I. It seems very likely that he was executed by his German captors as a "traitor."

The following day two Polish pilots in British squadrons – P/O Jerzy Jankiewicz and P/O Władysław Różycki – accounted for two more German aircraft destroyed for certain and F/O Tadeusz Nowierski claimed one "probable." No.302's Duxford detachment was ordered at 18:00 hours to return to Leconfield. The Hurricanes departed almost immediately, leaving behind one unserviceable fighter to be ferried by P/O Włodzimierz Karwowski the next morning, followed by the servicing party.

With London no longer the principal target of the daytime operations the Luftwaffe relocated its forces along the French coast and mounted a few attacks, mainly against the British aviation industry (Bristol, Southampton, Yeovil).

On 26 September at 14:30 hours HM King George VI arrived at Northolt and toured the Station, visiting the squadrons, including No.303. At 16:10 hours No.303's 12 Hurricanes were "scrambled," leading No.229 Sqdn to patrol Guildford. They were then ordered to Portsmouth and spotted a formation of about 50 He 111s with a heavy Bf 109 escort far to starboard and above, bombing the east bank of Southampton

Water. Their target was the Spitfire factory at Woolston. The King was invited to listen to the squadrons' progress from 'Z' Sector Operations Room and returned to the airfield after the combat to watch the aircraft land and talk to the pilots. He signed No.303 Squadron's History Book as a memento of his visit.

On this occasion No.303 caught up with the enemy rather late, as it turned sharply North of Hamble, and No.229 Sqdn was too far behind to be of much help. S/Ldr Kellett led Red Section against the Messerschmitts, but at first they did not react and the squadron attacked the bombers. The Poles heard an order "All Apany aircraft pancake," which was already known to them from previous experience, and ignored it. When the Bf 109 decided to intervene, some Hurricanes became engaged in a dogfight, while others chased the Heinkels all the way to France. Several pilots claimed destruction of enemy aircraft and eventually nine He 111s – with one more damaged – and four Bf 109s were credited to the squadron as follows: two each to P/O Jan Zumbach and Sgt Josef František, and one – and another "damaged" – to P/O Walerian Żak; one each to S/Ldr Kellett, F/Lt Forbes, F/O Witold Urbanowicz, and F/O Ludwik Paszkiewicz, and half each to P/O Bogdan Grzeszczak, Sgts Tadeusz Andruszkow, Marian Bełc and Jan Kowalski.

However, these claims were later proved to be grossly exaggerated, with every pilot apparently firing at one and the same aircraft and reporting it as shot down. The captured Luftwaffe documentation recorded the total loss of only one He 111 and 40 percent damage to another among the Heinkels of KG 55 involved in the Southampton raid (with only two more Heinkels damaged on landing after combat missions on that day), and one Bf 109. Three Hurricanes of No.303 Sqdn suffered damage: F/Lt Forbes returned to base with the port wing tip of his V7465 missing, and F/O Januszewicz and Sgt Bełc made successful forced landings. Polish pilots in British squadrons claimed: Sgt Józef Jeka, two He 111s, and P/O Franciszek Surma, one Bf 109.

The last Luftwaffe effort to mount a truly massive attack during the Battle came on 27th September, with London and Bristol as the main targets. The approaching formations were met by powerful fighter forces which badly mauled the raiders and forced them to retreat in confusion. The few enemy aircraft that managed to reach their objectives had to drop their bombs at random. For No.303 Sqdn this was one of the days of the most intensive fighting and heaviest casualties.

According to RAF Form 540,[34] the squadron took off at 09:00 hours and, operating jointly with No.1 Canadian Sqdn, was sent South and intercepted an enemy raid composed of He 111s and Bf 109s in the Horsham area at 15,000 ft. As on the previous day, just before engaging the enemy, the pilots heard on the R/T an order "Apany Leader pancake." The Poles attacked the bombers from astern and were themselves assaulted by the Messerschmitt fighters, which fled quickly in disorder. In the meantime the Heinkels had wheeled southwards and No.303 attacked the bombers from astern again, with other Hurricanes coming into the battle. Then its pilots sighted about 40 Do 17s approaching head-on in a single line astern with a

HM King George VI begins the inspection of No.303 "Warsaw Kościuszko's" Fighter Squadron after his arrival at Northolt on 26 September 1940. The unit's Polish C.O., F/O Urbanowicz, is standing on the King's right. No.303's Hurricane in the background shows the squadron's badge in a new position, under the radio mast, for the first time. (IPMS)

The King signs his autograph in No.303 Squadron's commemorative history book during his visit on 26 September. (IPMS)

The King shakes hands with No.303 pilots on 26 September. The officers in the first row, wearing life-saving jackets, are, left to right, F/O Urbanowicz, F/Lt Forbes (RAF), C.O. of 'B' Flight, F/O Paszkiewicz and F/O Żak. Paszkiewicz, with six air victories to his credit, was killed in action the following day. (IPMS)

[34] IPMS, Lot. A V,49/30 (p. 14-15, renumbered 36-37). The summary report from that action as recorded on this Form wrongly identified Ju 88s of KG 77 as He 111s and Bf 110s as Do 17s, but in No.303's Combat Diary (A V,49/6) the Bf 110s were correctly recorded as such while in the case of Ju 88s the same error was repeated.

Bf 109 fighter screen above. The Dorniers formed the familiar defensive circle and the Messerschmitts, the similar one above. Only the Polish Commander of the squadron, F/O Urbanowicz, went after the Dorniers and their escort, which also received attention from two other Hurricane squadrons, and he was credited with shooting down one Do 17[35] and one fighter. The rest of the Poles stayed with the Heinkels and in the increasingly chaotic pursuit of the enemy did not follow down the individual Germans claimed as destroyed; it was believed that almost all fell on land. These were credited as follows: one He 111[36] each to F/Lt Forbes, F/O Żak, P/O Ferić and Sgt František, and one "damaged" He 111, to Sgt Kowalski; one Bf 109 each to F/O Grzeszczak, F/O Henneberg, P/O Ferić and P/O Zumbach; and one Bf 110, to Sgt František. Two Polish pilots were shot down and killed: F/O Ludwik Paszkiewicz, flying RF•M (L1696), crashed at Borough Green[37] and Sgt Tadeusz Andruszkow, at Cowden. F/O Żak's Hurricane was set alight and crashed at Stoke D'Abernon, but the pilot managed to bale out with burns to his face and hands.

No.303 Sqdn, ordered off again at 15:15 hours, could muster only six serviceable Hurricanes and these were sent due East and encountered 15 Ju 88s and 30 Bf 109s over London. The Poles, led by F/Lt Kent and F/O Urbanowicz, went after the bombers, which immediately turned tail and fled South. Because of the high speed of the Ju 88s, the Hurricanes caught up with them only at Hastings, over the coast. Only three fighters managed to come into firing position. F/O Urbanowicz struck a Ju 88, the crew of which baled out, landing in the sea. He then attacked a second Junkers, which fell into the Channel in flames. F/Lt Kent set the port engine of another bomber on fire and Urbanowicz saw it crash into the water. Sgt Szaposznikow was about to fire at a Ju 88 when he was attacked by a passing Bf 109. He managed to evade the Messerschmitt and followed it up in a dive, giving it two bursts; the enemy turned on its side and dropped into the sea. F/Lt Kent's RF•J (V6684) returned damaged in an exchange of fire.

In the Day's fighting the Polish pilots of No.303 Sqdn accounted for 13 enemy aircraft (six Bf 109s, five bombers and two Bf 110s) reckoned to be shot down for certain. Poles in British squadrons added one Bf 110, by Sgt Jan Budziński, and Bf 110, by Sgt Józef Jeka, to the Polish score. This represented 11 percent of the overall Air Ministry total of 133 German aircraft regarded as destroyed. The actual Luftwaffe losses amounted to 55 aircraft lost and 12 damaged.

During the following two days the air operations were largely subdued. P/O Władysław Różycki of No.238 Sqdn added one more enemy aircraft, a Bf 110, to Polish victories, and P/O Janusz Żurakowski was credited with another Bf 110 as "probably destroyed." On 30 September a number of smaller attacks and one large raid developed, employing a variety of tactics to confuse the defences.

A close-up of the RF•B's (probably V7182) fuselage shows to advantage the new position to which No.303's Kościuszko badge was moved in the last weeks of September 1940. (K. Chołoniewski)

No.303's 12 Hurricanes, which took off on their third patrol on the 30th at 13:15 hours as the third squadron of the Northolt Wing (with No.1 Canadian and No.229 Sqdns), were vectored towards Dungeness and met about 30 Dorniers with a very strong Bf 109 escort. As the first two squadrons did not attack, No.303 broke away upwards and pursued the enemy over the Channel. Only 'A' Flight made contact with the Germans, 'B' Flight being too late. While Red Section fought an inconclusive combat with the Messerschmitts, F/O Witold Urbanowicz, leading Yellow Section, chased and destroyed two Bf 109s near the French coast and one Dornier (presumably belonging to KG 3) over France. Sgt Stanisław Karubin, Yellow 3, shot down one Bf 109 into the Channel off Beachy Head, and P/O Jerzy Radomski, Yellow 2, accounted for 1/2 Do 17Z (shared with a British pilot), but had to crash land his damaged Hurricane at Lydd; he was unhurt.

As soon as the Polish fighters were refueled and rearmed, ten were scrambled again at 16:40 hours to patrol base with the other two squadrons of the Wing which had taken off 15 minutes earlier. Climbing at full boost the flights became separated and were unable to join the Wing. 'A' Flight was surprised and dispersed by a strong force of Bf 109s (between 150 and 200 Bf 109s and 110s were then in the area), one of which fell victim to F/O Urbanowicz's guns. Cloud cover saved the Flight from disaster. Meanwhile, 'B' Flight made towards bombers near Croydon. Sgt Bełc attacked Bf 109s firing at a Hurricane pilot who had baled out and was descending on a parachute, and rejoined the flight after seeing him safe. The remaining four 'B' Flight fighters were pounced upon by Messerschmitts before they could reach the bombers. Sgt František broke off upwards to cover the Flight from the enemy assault. He got a Bf 109 on the turn in his gun sight and destroyed it. Then, together with six other British fighters, he fought a running battle with several Messerschmitts and claimed one more "probably destroyed." He escaped with great difficulty in cloud. Other pilots also came through what was an extremely hazardous situation and the Squadron suffered no

[35] Correctly recorded in No.303's Combat Diary as one Bf 110.

[36] In fact, as already explained above, these were Ju 88s of KG 77.

[37] Major parts recovered by Malling Aircraft Archaeologists, including the Merlin engine and propeller. One propeller blade presented to the Sikorski Museum in London.

casualties. F/O Tadeusz Nowierski, whose No.609 Sqdn was also involved in this action, destroyed one Bf 109, increasing the Polish score on that day to 7 1/2 enemy aircraft. This constituted 15 percent of 49 Germans counted by the Air Ministry as "destroyed for certain." The Luftwaffe recorded a total loss of 48 aircraft, a number almost equal to the British claim.

The 4th phase of the Battle, at the time appearing as only a modest success for the British air defences, was in fact of immense consequence for the future of the Free World. At its beginning the Luftwaffe was on the verge of establishing lasting air superiority over Southern England, the pre-condition of the "Seelöve" invasion. By its end the Luftwaffe, defeated by the narrowest of margins, had lost its strategic purpose and Fighter Command re-established its firm control of the British skies. The magnitude of the Polish contribution to this victory expressed itself in 112 enemy aircraft credited to the Polish pilots (see Appendix 15) out of 846 deemed by the Air Ministry as "destroyed for certain" (the actual German losses amounted to 436 aircraft totally lost and 161 damaged).

The Polish score represented 13.2 percent of this overall British assessment, which also included the enemy aircraft shot down by AA batteries, and rose to well over 14 percent of the Fighter Command "bag." No.303 Sqdn was credited with an astonishing score of 89 1/2 Germans shot down – 79 1/2 by its Polish pilots and 10 by the British – or 10.5 percent of the total reckoned as "destroyed for certain," the highest among all the squadrons taking part in this phase of the Battle. No.302 Sqdn also achieved a notable success during its fortnight of operations from Duxford.

The Poles paid a heavy price: 11 pilots killed and five seriously wounded. Two more were posted away. With 19 new Polish fighter pilots joining combat squadrons during the phase (see Appendix 14), the number of Polish participants of the Battle on 30 September amounted to 92.

(5.) 5th Phase – 1 to 31 October

At the beginning of October German tactics underwent a dramatic change. With the actual loss of some 1,650 aircraft and well over 2,000 of the most experienced flying crew members in less than three months, and with the British fighter defences stronger than at the start of the Battle, Göring embarked on the use of small, fast fighter-bomber formations, flying at high altitudes, to continue daytime attacks on Britain. Such tactics avoided further bomber losses, but did little significant damage, not being able to achieve in most cases more than a nuisance value. Immune to interception,[38] intensive all-night bomber raids, which reached their peak in October,[39] were to make up for the very limited bomb-carrying capabilities of the fighter-bombers, but although they did caused terrible damage

[38] Night-fighter operations were then still in their infancy.

[39] The tonnage of High Explosive bombs dropped on Britain in October amounted to over 9,000 tons, the highest monthly total in the Battle of Britain. In addition over 7,000 tons of Incendiaries were dropped.

A posed photograph, purporting to show No.303 pilots returning from a successful fighter sortie of the Battle. Left to right, in the front row, are: P/O Ferić, F/Lt Kent, C.O. of 'A' Flight, F/O Grzeszczak, P/O Łukuciewski, F/O Henneberg, Sgt Rogowski, and Sgt Szaposznikow. (IPMS)

to the cities and civilian population, they did not pose a major military threat.

The "hit-and-run" fighter-bomber attacks were difficult to counter and the number of interceptions fell to a disappointingly low level, although – in marked contrast to the preceding phases – the actual Luftwaffe losses were about 25 percent higher than the confirmed victories claimed by the RAF.

On 1 October No.303's nine Hurricanes were "scrambled" together with No.1 Canadian Sqdn at 13:15 hours. On patrol over Brooklands several groups of Bf 109s were spotted at up to 20,000 ft, but neither of the two squadrons attacked. F/Lt Kent broke away alone and dived to investigate an enemy formation below, which he thought were bombers. These were in fact also Bf 109s and before he could reach them, he was attacked by a number of other Messerschmitts. He set one on fire and hit the engine of another, which disappeared into the clouds trailing smoke (later credited to him as "probable"). He managed to disengage and landed at Northolt after the rest of the squadron.

During the next three days several patrols were flown, but no encounters were reported. No.303 Sqdn made contact with the enemy again on 5 October, after take-off at 11:05 hours (12 Hurricanes) and rendezvous with No.1 Canadian Sqdn. Flying southwards they met German fighters over Rochester which descended from upper cloud layers. About 60 Bf 109s and 40 friendly fighters became locked in running dogfights, rolling towards the Channel. Between Ashford and Lympne the squadron came across 18 Bf 110 fighter-bombers of Erpr. Gr. 210,[40] sent to destroy the West Malling fighter airfield, which formed a defensive circle, protected from the top by a large number of Bf 109s. The Poles struck at the circle head-on and below, and broke it up in one of their most daring and spectacular attacks in the Battle of Britain, contemptuous of the threat from the Bf 109s, which shot down in flames and killed F/O Wojciech Januszewicz in a desperate effort to shield their charges. Two Bf 110s fell victim to the Polish onslaught almost immediately over Kent; one of them was the leading machine of the new Commander of the unit, Oblt Werner Weimann, reported "missing" in this action. The others were chased over the sea, sustaining further casualties. Halfway across the Channel the squadron was ordered home.

The officially confirmed No.303 score comprised: four and a half Bf 110s destroyed – credited one each to F/O Zdzisław Henneberg, F/O Marian Pisarek, P/O Mirosław Ferić and Sgt Marian Bełc, and the half (shared with a Spitfire), to Sgt Antoni Siudak; two Bf 110s "damaged" – to F/O Pisarek and Sgt Jan Palak; four Bf 109s destroyed – credited two to Sgt Siudak, and one each to Sgt Stanisław Karubin and Sgt Palak; and one Bf 109 "damaged" – to S/Ldr Kellett. Poles in British squadrons added one and a half more Germans destroyed: P/O Ludwik Martel of No.603 Sqdn, a Bf 109, and P/O Henryk Szczęsny of No.74 Sqdn, half of a Do 17, swelling the Polish "bag" to 10 victories or 43 percent of the total of 23 enemy aircraft credited to the RAF that day.

On 6th October at noon a lone Ju 88 of KG 30 penetrated over Northolt and dropped a few bombs, one of which fell between two hangars and blew to pieces a taxiing Hurricane, the RF•D P3120, killing its pilot, Sgt Antoni Siudak.[41]

On 7 October No.303 Sqdn fought its last engagement in the Battle of Britain on its second patrol at 13:15 hours. Operating at full strength (12 aircraft) with No.1 Canadian Sqdn, it was directed to the Kenley-Brooklands area at 20,000 ft. Over South London both squadrons were attacked by about 50 Bf 109s, which dived on them from a great height in flights of five and then flew away southwards. The squadron evaded the Messerschmitts and attempted to overtake them without success. No.303 then shadowed another formation of yellow-nosed Bf 109s flying westwards, which turned to attack them. In the individual dogfights and subsequent pursuit towards the Channel Sgt Bełc destroyed one Bf 109 in the Redhill area, Sgt Szaposznikow shot down one and probably another over East Sussex and F/O Pisarek, one off the Kent coast. P/O Bogusław Mierzwa crashed and caught fire after force-landing his damaged Hurricane on a decoy airfield at Borstal. Later in the afternoon, Sgt Józef Jeka of No.238 Sqdn destroyed one Ju 88 and Sgt Jan Budziński of No.605 Sqdn, 1/2 of a Bf 109; P/O Zbigniew Nosowicz claimed one enemy "probable" and one "damaged."

A routine patrol by No.303's 12 Hurricanes on 8 October at 09:00 hours resulted in the death of Sgt Josef František, the most acclaimed Czech pilot, whose aircraft, RF•R (R4175), crashed at Cuddington Way, Ewell, Surrey from an unknown cause. One of a few Czech pilots who joined the Polish LW after escaping from Czechoslovakia in March 1939, František shared the fate of the Polish airmen in evacuating to Rumania; he then made his way to France and joined No.303 Sqdn on its formation. He was an individualistic loner and frequently broke away from the rest of the squadron to hunt on his own. With 17 victories officially credited to him during the Battle of Britain,[42] he was the RAF's top-scoring fighter ace of the Battle. He was awarded the Virtuti Militari, the Polish Cross of Valor and three bars, and the Czech Military Cross, and was promoted posthumously to the rank of S-Lt. His only British award was the DFM.

During the following three days No.303 Sqdn flew a number of patrols, but made no contact with the enemy, the last of these – its 35th combat mission from Northolt – being made at noon on 11 October. At 15:40 hours on the 11th its 18 aircraft "left Northolt for Leconfield in order to recuperate after taking a very strenuous and successful part in the Battle of Britain. They were replaced by No.302 Polish Squadron."[43]

[40] Erprobungs Gruppe 210 (Test Wing 210) was a highly specialized crack unit, equipped with the long-range fighter-bomber Bf 110C and 110D variants.

[41] The bomber was then chased by No.229 Sqdn and crashed in Surrey.

[42] František's service register number, 793451, was within the Polish register block, and on pages of this work his victories are all included in the Polish score of No.303 Sqdn.

[43] From the Combat Diary, Royal Air Force Northolt.

V7503 RF•U, in which F/O Pisarek shot down a Bf 109 and damaged another on 5 October and destroyed a further Bf 109 two days later. Photographed at Leconfield in December 1940, the aircraft displays a newly introduced sky-blue band on its rear fuselage. The Kościuszko badge appears to be missing from the white background disc aft of the cockpit.

On the 12th and 13th No.303 Sqdn was on "release" to enjoy a complete rest. Its first operational flight – the 36th – was made from Leconfield on 14 October and until the end of the month the Squadron logged ten missions from this airfield, most of which were three aircraft patrols. No contact with the enemy was reported. On 27 October in the early evening two Heinkels sneaked in undetected over Leconfield and bombed the airfield, mortally wounding AC1 Antoni Rossochacki who was working on one of the squadron's Hurricanes.

No.302 Sqdn spent the period from 26 September to 10 October at Leconfield largely resting and training new pilots posted to it to replace the losses sustained during its operations from Duxford. It was expecting to return to Duxford early in October but on the 9th it received an order to move to Northolt on the 11th. The squadron's aircraft landed at Northolt on 11 October at 13:00 hours and the transport party followed them later in the afternoon. No.302 Sqdn flew its first operational mission from Northolt on 12 October. The Squadron took off at full strength together with No.615 Sqdn at 17:30 hours and spotted the enemy, who was already under attack by several squadrons. The Squadron stayed at 25,000 ft to provide the top cover to the friendly fighters and did not engage the Germans.

After four uneventful patrols over the next two days, No.302 came to grips with the enemy on the 15th. "Scrambled" as part of the Northolt Wing (with Nos.229 and 615 Sqdns) at 09:30 hours, it was ordered to proceed to the Thames Estuary at 30,000 ft. A formation of about 20 Bf 109s flying towards London, passed above No.302, which was still climbing. S/Ldr Satchell wheeled to chase the Messerschmitts and unduly stretched the Squadron. P/O Jan Maliński's engine seized up in the process and he was forced to land at Wrotham. At this point a second, larger group of Bf 109s approached and the first group turned back, making a head-on attack on the Hurricanes. The Germans, having on their side a height advantage, better aircraft and superiority in numbers, shot down in flames Sgt Marian Wędzik's WX•R (P2752). The pilot baled out with burns over Chatham. S/Ldr Satchell's WX•L (P3812) was damaged and force-landed at Slough (No.229 and 615 Sqdn also suffered casualties and lost three aircraft). Although the outcome of the battle was not considered to be a success, No.302 Sqdn achieved two victories: F/Lt Riley and P/O Wacław Król were credited with one Bf 109 each. Polish pilots in British squadrons, F/O Tadeusz Nowierski and P/O Bolesław Własnowolski, added two more Bf 109s destroyed during the noon engagements.

The following day Sgt Wilhelm Korsarz hit a balloon cable in bad visibility, losing 14 inches from his starboard wing, but managed to land successfully at Heston. During a morning patrol on 17 October Sgt Jerzy Załuski lost bearings and, running out of fuel, made a catastrophic landing at Colliers End – his aircraft fell into a ditch and overturned, killing the pilot. F/Lt Jan Czerny and P/O Stefan Kleczkowski also had to force-land on that day, but escaped injury.

18 October was one of the saddest days in No.302's history. After take-off from Northolt at 15:06 hours in bad weather conditions, the squadron, led by F/Lt Thomson and S/Ldr Mümler, Red Section, and with P/O Wczelik, F/O Carter and F/Lt Jastrzębski leading Yellow, Blue and Green sections respectively, climbed above clouds and became detached from the rest of the wing, which remained under the cloud layer. The Poles sighted a Ju 88 bomber and a large Bf 109 escort. Yellow and Green Section dived to attack the bomber and Red and Blue Sections provided the top cover, but the Messerschmitts vanished into cloud. The Junkers was hit and two of its crew were seen to bale out, but as the bomber disappeared in cloud, it was credited – to Sgt Eugeniusz Nowakiewicz – only as "probably destroyed." No German aircraft were deemed by the RAF to be "destroyed for certain" on that day, although in fact the Luftwaffe lost at least eight aircraft as the result of combat.

The squadron began to descend into low cloud and fog and got lost over the Surrey Hills. F/O Carter spotted the Kempton Park Race Course and instructed Blue Section to crash-land. He baled out at a very low altitude, but apparently in agitation detached his parachute and was killed on impact with the ground. F/O Jan Borowski hit the ground at 200 mph; his body was cut to pieces by the wreckage and was unrecognizable. Only P/O Bronisław Bernaś from Blue Section succeeded in force-landing successfully. P/O Stefan Wapniarek and P/O Aleksiej Żukowski, both from Green Section, were also killed, the former attempting to land at Nutwood Farm, Thames Ditton, and the latter, at Harp Farm near Detling, his body not being found until four days later. Investigations of the disaster lasted a few days and replacements were posted to 'B' Flight. The squadron re-commenced combat flying on 24 October, when two uneventful patrols by two and six aircraft were made.

On 25 October during morning patrol No.302 Sqdn, operating at full strength, saw a large number of Bf 109s at a considerable distance over the sea. F/Lt Franciszek Jastrzębski's section attempted to give chase, but Sgt Antoni Markiewicz soon rejoined the squadron and P/O Bronisław Bernaś, also unable to keep pace with Jastrzębski's faster machine, turned back over the Channel. Only F/Lt Jastrzębski managed to engage the enemy, but he was apparently shot down – his WX•V (V7593) was last seen gliding towards the enemy coast – fell in the sea and drowned. P/O Stanisław Piątkowski, flying with No.79 Sqdn, was also killed on that day, crashing near Carew Cheriton from unknown causes after a routine patrol. P/O Martel, of No.603 Sqdn, was shot down in combat with Bf 109s over Hastings and crashed his damaged Spitfire in a forced landing; the wounded pilot never returned to the Battle.

On 26 October the Germans intensified their attempts to keep up daylight pressure using high-flying fighter sweeps, but better coordinated RAF defences were ready to meet them. Nos.302 and 229 Sqdns, after an abortive take-off at about 10:30 hours, were "scrambled" from Northolt again half an hour later and spotted the raiders over the Channel, who promptly formed the familiar defensive circle. No.302 broke up the circle and chased the Messerschmitts to the French coast, catching up with them as they were preparing to land in the Boulogne area (No.229 Sqdn detached over the Channel, went after, and shot down a He 59 seaplane). S/Ldr Satchell, F/Lt Thomson, Sgt Korsarz and Sgt Markiewicz fired a number of bursts from a considerable distance, and Satchell and Markiewicz claimed one Bf 109 each as "probable." P/O Stefan Kleczkowski force-landed near Sittingbourne out of fuel. He was slightly injured and his aircraft slightly damaged. During the afternoon patrol the squadron pursued the enemy beyond Marlborough, Wiltshire, and had to land at Yatesbury to refuel. The aircraft returned to Northolt the following morning. During the next two days five uneventful patrols were flown.

29 October saw the last significant enemy daylight activity in the Battle of Britain and the first one in which the Italian Regia Aeronautica took part,[44] but Fighter Command tactics worked well and 19 raiders were actually shot down (33 were claimed at the time as destroyed). No.302 Sqdn was "scrambled" twice that day but made no contact with the enemy. During the afternoon sortie at about 15:30 hours, F/Lt Thomson flew from behind into the path of F/Lt Jan Czerny's Hurricane. Thomson's tail was severed by Czerny's airscrew. Thomson baled out, but Czerny was unable either to open his cockpit or release his undercarriage and with his engine stopped and shattered airscrew, he made a masterly dead-stick belly landing at White Waltham.

Polish pilots in British squadrons were credited with 2 1/2 confirmed victories: P/O Tadeusz Kumiega of No.17 Sqdn shared a Bf 109 of JG 26 with Sgt Hogg; Sgt Michał Maciejowski of No.249 Sqdn brought down a Bf 109 of LG 2 which attacked the North Weald airfield; and P/O Tadeusz Nowak of No.253 Sqdn claimed a Do 17. P/O Franciszek Surma of No.257 Sqdn was shot down in combat with Bf 109s near Matching, but baled out unhurt.

Bad weather during the next two days severely curtailed flying. On the 30th Sgt Michał Maciejowski ran out of fuel after a duel with Bf 109s over the Channel and made a successful forced landing in East Sussex. The following day Sgt Bronisław Malinowski of No.43 Sqdn force-landed at Chirnside due to engine failure.

During the final, 5th phase of the Battle, with the employment of Messerschmitts as the main striking force, the Hurricane fighters – flown by both Polish squadrons and the majority of Polish pilots in the British squadrons – were at a considerable disadvantage. Their inferior performance did not permit them to deal effectively with the Bf 109s, which, freed from the constraints of bomber escort, could evade engagement at will. Out of 260 enemy aircraft reckoned by the RAF as "destroyed for certain" during this period, 20 (7.7 percent) were credited to the Poles. The actual Luftwaffe losses – which were much higher – amounted to 325 destroyed and 163 damaged.[45] Polish losses included eight pilots killed and two wounded – and one other pilot was posted away, but with 20 new arrivals making their operational debut in the Battle, the total number of Polish fighter pilots rose to 101 by the end of the phase.

(6.) Assessments and Conclusions

Soon after the Battle had ended the RAF began an analysis of all available evidence relating to its course and results, completing the essential work in the spring of 1942. After disallowing and amending[46] a few claims which contained the el-

[44] 15 Fiat BR 20 bombers, escorted by three times as many Fiat C.R.42 biplane fighters, attacked Ramsgate.

[45] In the less continuous fighting by smaller numbers of aircraft – and in particular in fighter-versus-fighter combat – the results of engagements were difficult to establish and in many cases the "kills" could not be confirmed.

[46] Some enemy aircraft initially credited as "damaged" were reclassified as "probable" and vice versa.

ement of reasonable doubt, the RAF estimated that 2,692 enemy aircraft were destroyed during the period of the Battle of Britain, 10 July to 31 October 1940. The Polish score, officially credited by Fighter Command HQ to the Polish pilots[47] in the Polish and British squadrons – and endorsed by the PAF HQ as correct – amounted to 203 1/2 enemy aircraft "destroyed for certain"[48] (7.56 percent of the overall figure), with a further 35 "probable" and 36 "damaged." Of these 110-9-6 were accounted for by the Poles in No.303 (Polish) Sqdn, 16-10-1, by the Poles in No.302 (Polish) Sqdn and 77 1/2-16- 29, by Polish pilots in British squadrons. No.303 (Polish) Sqdn, whose overall score in the Battle, after the addition of 16-4-3 aircraft attributed to its British pilots, amounted to 126-13-9 (4.7 percent of the overall RAF score), and became the highest scoring unit among the 66 squadrons of Fighter Command participating in the combat. The British pilots in No.302 (Polish) Sqdn increased its score by a further 5-5-1 enemy aircraft, bringing it to a total of 21-15-2 (For details of these victories see Appendix 15).

The RAF victory claims had already given rise to controversy during the Battle, with some neutral observers querying the publicly announced British assessments of German losses, and they became the topic of an involved debate in the immediate post-war years, when secret Luftwaffe records became available to the Allies. The vexed question of the size of German losses distracted for some time attention from the nature of the British victory. To the fighter squadrons and their pilots, who bore the burden of the deadly struggle and fought to the point of total exhaustion, the enemy loss figures were the measure of their pride and personal achievement not to be belittled. To the post-war pundits, a cool comparison of the RAF claims with the Luftwaffe records became a newsworthy revelation.

In the spring of 1947, as the result of a study of the captured German files, the RAF officially reassessed the Luftwaffe operational losses during the Battle at 1,733 aircraft destroyed and 643 damaged, or 2,376 put out of action. War communiqués issued at the time by the German High Command admitted the loss of only 896 aircraft. The Luftwaffe's own victory figures were much more exaggerated than the British ones, claiming the destruction of 3,058 British planes, as compared with the actual RAF loss of about 1,300 aircraft, of which about 1,000 belonged to Fighter Command.[49]

The retrospectively revised RAF figures of 1,733 enemy aircraft destroyed and 643 damaged received endorsement by the British Government. They were given by Philip Noel-Baker, the then Secretary of State for Air, to the House of Commons on 14 May 1947 in reply to parliamentary questions. He explained that: "the figures of the German High Command

[47] These included Sgt Josef František, a Czech pilot on the Polish roster of No.303 Sqdn, whose personal score was 17-1-0.

[48] This total did not include F/O Franciszek Gruszka's victory on 18 August 1940, which was never put on file in any of the RAF wartime records.

[49] The official Fighter Command figure was 915 aircraft, but in some cases the fighters listed by squadrons as destroyed escaped inclusion in the RAF summary loss records. "The Battle of Britain Then and Now," Fifth Edition (1989), p.707, puts Fighter Command losses at 1,023 aircraft. In addition some 300 planes were lost by Bomber and Coastal Commands.

With 126 enemy aircraft officially credited as destroyed, No.303 Squadron was the highest-scoring fighter unit in the RAF during the Battle. The "destroyed aircraft" tally was recorded cartoon-fashion on one of its Hurricanes.

showed that during the opening and concluding phases of the Battle, while the numbers engaged were relatively small and the fighting less continuous and intense, the losses actually inflicted on the enemy were higher than the numbers claimed by the RAF. When very large forces were in action and when the battle raged without respite for many days, the estimates were well above the losses which the Luftwaffe sustained." He added that "... this retrospective correction of claims which were honestly put forward did nothing to diminish the achievements or to dim the glory of the men who fought so bravely against great odds."

The corrected RAF figures were subsequently queried by different historians, whose individual interpretations of the German tabulations resulted in assessments of German losses varying from 1.385 to 1,887 aircraft.[50] While it would be reasonable to assume that a few of the destroyed aircraft might

[50] Wood and Dempster in "The Narrow Margin" (Arrow Edition, 1969), Appendix 18, put the number of German aircraft "destroyed due to enemy action" at 1,385, with a further 404 "destroyed on operations but not due to enemy action" – the latter figure being inordinately high and not born out by the main text of the book, where the Luftwaffe daily losses attributed to the RAF seem to add up to about 1,680 aircraft. On the other hand, "The Battle of Britain Then and Now" (Fifth Edition, 1989), p. 707, puts the number of Luftwaffe aircraft lost at 1,887.

Polish mechanics of No.303 Squadron, who did so much to make the unit the most successful one in the Battle, pose beside the victory-marked Hurricane. (IPMS)

have escaped inclusion in the Luftwaffe loss totals and the real numbers might have been somewhat higher, in many instances it was difficult to establish with certainty which of the German losses were due to enemy action and the figure of 1,733 embraced some of these doubtful incidents. On balance, this figure appears to be the most realistic and authoritative one and is accepted as such by the author of this work. The Polish score of 203 1/2 enemy aircraft destroyed, forming part of the original RAF figure of 2,692, should therefore be reduced in the same proportion, to 131 aircraft.[51]

According to the official Fighter Command criteria, 2,927 aircrew, including a small number of gunners in the Defiant and Blenheim squadrons, participated in the Battle.[52] These included 145 Polish pilots (see Appendix 14) – the largest national group beside the United Kingdom and Commonwealth pilots – 87 Czechoslovakians, 29 Belgians, 13 Frenchmen, 11 Americans and nine Irishmen. The average availability of pilots in Fighter Command during the duration of the Battle amounted to about 1,450, of which about 75 were Polish (the average for the whole Battle). During the crucial 3rd and 4th phases of the Battle the average number of Polish pilots participating in combat exceeded 90. However, almost three quarters of all Polish fighter pilots served with the units of No.11 Group, which provided the mainstay of the British fighter defenses and carried the heaviest burden of the fighting, and at the peak of the Battle the Poles constituted over 10 percent of the Group's pilot strength.

In terms of human loss, over 500 aircrew of Fighter Command were killed, including 29 Poles, and about 720 of Bomber Command and 280 of Coastal Command, bringing the RAF total to over 1,500 killed.[53] The estimated Luftwaffe losses amounted to about 2,660 aircrew killed. The balance sheet was in favor of the RAF in both equipment and personnel losses.

But the "numbers game," which preoccupied aviation journalists for many years, was only a side issue, for the Battle was not about the size of the losses, but about survival of Britain and with it, of the Free World. The Luftwaffe mounted its most ambitious offensive effort of the war, having on average about 1,900 aircraft available for the attack on one front throughout the 114 days of the Battle. Its warplanes flew about 60,000 sorties, an average of about 526 sorties per day, amounting to one sortie per aircraft every three and half days.

This figure, formidable as it was, was bettered three times over by Fighter Command, which, having at its disposal on average 720 fighter planes, logged almost 66,400 sorties, a daily average of 582 sorties, meaning one operational flight per aircraft every one and a quarter day. The tremendous effort of the British fighter force, whose effectiveness was greatly enhanced by a sophisticated and well-running warning and guiding system, based on a new, revolutionary instrument of air defence, the radar chain,[54] saved the day.

The final conclusions from the Battle are as follows:

- Fighter Command prevented the all-powerful German military machine from achieving its specified war objective for the first time since the beginning of war and smashed the myth of its "invincibility." This is the undisputed nature of the RAF victory, which changed the course of war.
- After a very precarious and evenly balanced struggle, the RAF subdued Germany's finest weapon, the Luftwaffe, which never fully regained its former strength and thus was never again able to mount an air offensive of such magnitude. By contrast, the RAF, which asserted its control of the British skies, emerged from Battle stronger than at its beginning[55] and continued to build up its strength month after month.
- The margin of the RAF victory was extremely slim, but its consequences were incalculable. Poles made a vital contribution to the outcome of the Battle on two counts: first, by strengthening the squadrons at the time when Fighter Command reached crisis point in pilot strength; secondly, by their share of victories in the decisive 3rd and 4th phases of the Battle, which amounted to over 10 percent of all enemy aircraft destroyed in both these phases and to over 13 percent in the latter phase alone. The first point was best acknowledged by the Rt.Hon. Viscount Thurso, Secretary of State for Air, 1940-45, who said " ... Our shortage of trained pilots would have made it impossible to man the squadrons which were required to defeat the German Air Force and so win the Battle of Britain, if the gallant airmen of Poland had not leapt into the breach and fought with valour alongside their British comrades."
- The number of Poles who took part in the Battle was small as compared with the overall Fighter Command strength – amounting to only between five and six percent of the total – but they represented the elite of the pre-war Polish fighter force. Their experience, fighting skills, tactics and exceptionally developed powers of observation, which allowed them to see enemy aircraft first – a fact

[51] Because of a disbandment of the PAF this was never officially done and almost all Polish publications began to relate the figure of 203 1/2 aircraft originally credited to the Poles to the revised RAF figure of 1,733 without any explanation.

[52] Those awarded the Battle of Britain clasp for having flown at least one authorized operational sortie with a unit of Fighter Command during the period 10 July to 31 October 1940 – as per Kenneth G. Wynn "Men of the Battle of Britain." Several more pilots were on the strength of Fighter Command squadrons, but for one reason or another did not make any operational flights.

[53] The Roll of Honor in the Battle of Britain Memorial Chapel in Westminster Abbey, compiled in 1947, comprises 496 aircrew of Fighter Command, 718 of Bomber Command and 280 of Coastal Command – a total of 1,494 killed; Wood and Dempster in "The Narrow Margin" (1969), Appendix 23, put Fighter Command casualties at 507 killed, while "The Battle of Britain Then and Now" (1989), p.707, at 537 killed, and Kenneth G. Wynn, ibid (1989), at 544 killed.

[54] At the beginning of the Battle there were 21 CH (Chain Home) and 30 CHL (Chain Home Low – to detect low-flying aircraft) radar stations fully operational and these provided long-range "pictures" of all enemy air movements, invaluable to defence.

[55] By the beginning of November 1940 12 new fighter squadrons were formed, although 22 of the existing squadrons were temporarily classified as 'C' squadrons, to rest and recoup.

acknowledged by several British pilots flying with the Poles – made them outstandingly effective. The total Polish score in the Battle: 203 1/2 enemy aircraft credited to 145 Polish participants of the Battle, represented 1.4 enemy aircraft destroyed for every Polish pilot engaged in the combat. In the words of Air Chief Marshal Sir Hugh Dowding, "The first Polish Squadron (No.303) in No.11 Group, during the course of a month,[56] shot down more Germans than any British unit in the same period ... " Despite joining combat only in the middle of the Battle, No.303 Sqdn finished it with 126 accredited victories – the highest score of enemy aircraft reckoned as destroyed by any RAF squadron in the Battle.

- Poland's part in the Battle can not be measured only by statistics. During that turning point of history, Polish airmen were the first and the most numerous among the few Allies – apart from the Commonwealth nations – who fought on the side of Britain. It was Poland's moral, as well as physical, support which made such a vital contribution to the victory when the future of Britain and of the whole Free World hung in the balance. Whether that contribution was the decisive factor in tipping the scales of victory the RAF way, will always remain a matter of speculation.

3. After-the-Battle Operations, November-December 1940

British historians generally accept that the Battle of Britain had ended on 31 October and although such a neatly defined date is rather artificial, by then the twin-engined bombers had almost vanished from the daylight skies[1] and the intensity of harassing fighter-bomber attacks significantly diminished. Nevertheless, the fighting continued throughout November and December 1940 and its scale was still such that the Luftwaffe lost 379 aircraft – albeit a considerable proportion of this loss was sustained at night through the AA fire – and Fighter Command 186.

However, after the defeat by day, the Luftwaffe was still prepared to continue with major offensive operations at night, in order to compensate for its failure by day and so as not to lose face by withdrawing from the air battle altogether. The devastating raid on Coventry on the night of 14/15 November by 437 aircraft, which resulted in colossal damage to its densely populated urban areas, opened a new phase of the night Blitz with the aim of destroying British cities and industrial centers, and of undermining the morale of the British people. This was to last until June 1941, the beginning of the attack on the Soviet Union.

In the autumn of 1940 Fighter Command was unprepared for night operations and almost completely impotent to intercept enemy raids. This permitted the German night bombers to attack Britain with minimal losses[2] and only the introduction of airborne radar, the development of which took some time, began to result in a slow increase in the night interception rate in the early months of 1941. The shortfall contributed to the replacement of Dowding as Commander-in-Chief of Fighter Command[3] by ACM William Sholto Douglas on 25 November 1940.

No.302 Sqdn continued its operations from Northolt until 23 November 1940. On 1 November at 10:00 hours the Squadron was "released" for three hours and joined by crews of Nos.300 and 301 (Polish) Bomber Squadrons for an official visit by Gen Sikorski and Sir Archibald Sinclair, the Secretary of State for Air. ACM Sir Charles Portal, the Chief of the Air Staff, ACM Sir Richard Peirse, Commander-in-Chief of Bomber Command, and Col Iwaszkiewicz, the Deputy Inspector of the PAF, were also present. The squadron flew two patrols the same afternoon.

During the following three weeks bad weather severely curtailed operations, but when flying was possible, the squadron made up to four patrols a day. These were largely uneventful, except for 5 and 9 November. On the 5th, during the day's second patrol at 14:20 hours, No.302's Hurricanes were suddenly attacked from above by enemy fighters (reported to be the He 113s) in the Maidstone area. Two Polish pilots opened fire and Sgt Wilhelm Korsarz reported one German "destroyed," but his uncorroborated claim was rejected by the RAF.

On the 8th, No.302 patrolling with No.615 Sqdn over the Croydon area at 13:30 hours, saw a large number of Bf 109s 5,000 ft above. The enemy dived down on No.302's last section, composed of Antoni Wczelik, Eugeniusz Nowakiewicz and Wilhelm Korsarz, and Korsarz's P3538 was set alight and crashed near Mayfield (Sussex). The pilot was killed. A few minutes later the two other fighters were damaged and forced to land near Detling. Nowakiewicz could not release his undercarriage and crashed.

On 23 November No.302 Sqdn moved to Westhampnett. It was put on the daily State of Readiness as "Available" the following day at 13:00 hours. Rain and fog hampered air activities and in the last week of November only a few patrols were made. At that time relations between S/Ldr Satchell, the British officers in the squadron and the Polish Commander, S/Ldr Mümler, strained for some time, reached a new low. Complaints concerning the behavior of an individual RAF officer in particular, led to a Polish request for his transfer to another unit. Satchell refused to investigate the matter fully and on 6 December Fighter Command HQ decided to post S/Ldr Mieczysław Mümler, F/Lt Jan Czerny, F/O Antoni Wczelik and F/O Władysław Goettel to other squadrons.[4] The follow-

[56] Reference to September, the first month of the squadron's operations

[1] A notable exception was the major daylight raid – the first and the last one – by the Regia Aeronautica on 11 November, which resulted in the humiliating defeat of an Italian formation approaching Harwich.

[2] During the four-month period, Nov. 1940 to Feb. 1941, the British defences destroyed 75 enemy aircraft out of over 12,000 sorties flown by the Luftwaffe. Approximately two-thirds of these were claimed by the AA guns.

[3] Although ACM Dowding received the GCB, his retirement to a relatively obscure appointment in the United States was regarded by many as a scandal. In May 1943 he received a Barony.

[4] IPMS, Lot. A V,48/8 (also other documents on file A V,48).

No.303 Squadron, during respite and training at Leconfield in December 1940, with N2460 RF•D first in the line and N2661 RF•J and P3585 RF•C behind.

ing day Mümler handed over the command of No.302 Sqdn to Piotr Łaguna.

Bad weather resulted in limited activities in December and the only encounter with the enemy took place on the 10th. Red Section, led by F/Lt John F. Finnis (recently posted to No. 302 Sqdn)[5] and patrolling south of Bognor at 10:10 hours, was asked to investigate an unidentified aircraft, which turned out to be a Ju 88. Finnis attacked it twice, and P/O Stanisław Łapka (Red 2) and Sgt Antoni Łysek (Red 3) also fired at the bomber during a chase over the sea. The Junkers began to smoke and lose height, but deteriorating weather forced the section to turn back. Finnis was credited with one Ju 88 "probably destroyed."

Relations with the British officers still remained tense and an announcement on 20 December that they would all be shortly leaving the squadron was received by the Poles with relief. A farewell party for the departing British was given on 31 December – their last day in No.302 Sqdn – and on 1 January 1941 S/Ldr (Cpt) Piotr Łaguna assumed overall command of the unit. Despite the weather, No.302 Sqdn logged over 500 sorties in the November-December period, increasing its 1940 operational effort to about 120 missions, totaling 923 sorties in 1,016 hours flying time.

No.303 Squadron, operating from Leconfield, carried out 66 operational missions in November and 30 in December. These were mostly Sector patrols by three-aircraft sections, which on some occasions extended to a wider territory, including the York area. A few times the squadron was engaged at full strength on convoy protection duties and once – on 27 December – on a bomber escort duty. Its pilots reported sightings of enemy aircraft only three times during this whole period, but made no firing contact as each time the Germans were far away and escaped immediately.

No.303's 1940 operational effort amounted to a total of 141 missions, comprising 1,049 sorties in 1,086 hours flying time. On 15 December at Leconfield AM Sholto Douglas decorated F/O (Lt) Witold Urbanowicz, F/O (Lt) Zdzisław Henneberg, P/O (S-Lt) Mirosław Ferić and P/O (S-Lt) Jan Zumbach with DFCs and handed to No.303's Commander a DFC for the family of F/O (Lt) Ludwik Paszkiewicz, killed in action on 27 September. Gen Sikorski's order of 21 December awarded Virtuti Militari Silver Crosses to P/O (S-Lt) Tadeusz Nowak and – posthumously – to Sgt (Sgt) Antoni Siudak, and 25 Crosses for Valour to other pilots of the Squadron.

No.306 "Toruń" Fighter Squadron, which arrived from Church Fenton at the RAF Station Ternhill (in No.9 Group) with its eight serviceable Hurricanes on 7 November, became operational the following day. The Squadron was allocated provisional and improvised quarters on the very congested airfield and found working conditions rather difficult. Its first three-aircraft section – led by F/Lt Tennant, with S/Ldr Tadeusz Rolski and Sgt Bolesław Turzański as his No.2 and No.3 respectively[6] – was at the "'Available" state since dawn on 8 November and took off on the squadron's first operational mission at 11:15 hours, to patrol the Liverpool area. It made no contact with the enemy.

During the next few days No.306 Sqdn flew several uneventful patrols and on the 13th collected seven new Hurricanes with metal wings. On the same day its third patrol – composed of F/Lt Kennard, P/O Bohdan Bielkiewicz and P/O Edward Jankowski and scrambled at 16:05 hours – fought the squadron's first combat 10 miles West of Worcester. The pilots fired at a He 111, which began to smoke but escaped into cloud, and all three were credited with one Heinkel "damaged."

Later in November a few pilots were posted to other units, but at the same time F/Lt (Cpt) Stefan Witorzeńc arrived to take command of 'A' Flight and P/Os (S-Lts) Aleksander Pietruszka, Karol Pniak and Władysław Różycki joined the squadron, whose equipment was brought up to the full War Establishment. By 22 November No.306 had on strength four British and 29 Polish pilots and 20 Hurricanes: 16 Initial Equipment, two Immediate Reserve and two under repair.

On 26 November S/Ldr Scott, leading a section on patrol, was involved in a mid-air collision with an Anson soon after take-off (he escaped unhurt), but two of his wing mates,

[5] During the Battle of Britain Finnis flew with No.1 and No.229 Squadrons at Northolt, where he learned to work with Polish pilots.

[6] Polish pilots of No.306 Sqdn were prevented for some weeks from leading the patrol sections, because of an unfortunate operation by No.312 (Czech) Sqdn – in the same No.9 Group – on 13 October, which became completely lost, attacked RAF Blenheims and shot one down (IPMS, Lot A V,50/3 & John Rawlings "Fighter Squadrons of the RAF," p. 397).

Witorzeńc and Bielkiewicz, managed to pull clear at the last moment. Six days later, Scott was injured in another accident and went to hospital. The following day (2 December) S/Ldr Denys E. Gillam took over the British command of the squadron.

Numerous patrols flown in the second half of November did not lead to contacts with the enemy and the squadron's activities slackened in December due to very poor weather. In better conditions up to four missions a day were flown, but again, without result. On 8 December P/O (S-Lt) Kazimierz Rutkowski was injured in a landing accident and his Hurricane (P3938) destroyed. On 24 December F/Lt Tennant was killed, crashing almost vertically because of lack of visibility in fog. By the end of 1940, No.306 Sqdn had made 55 missions, logging 169 sorties in a total flying time of 143 hours.

No.308 "Kraków" Fighter Squadron waited impatiently at Baginton to become operational, but the reduced Luftwaffe daylight activities removed the urgency of adding new fighter squadrons to the line.

On 24 November at 14:30 hours No.308's nine Hurricanes, led by W/Cdr J.O.W. Oliver, took off to practice formation flying. About half an hour later over the center of Coventry the pilots noticed condensation trails high above and the leading Blue Section decided to investigate. P/O Władysław Grudziński, Blue 2, had to retire because of lack of oxygen as W/Cdr Oliver, Blue 1, and Sgt Mieczysław Parafiński climbed to 26,000 ft to begin chase. The enemy, a Ju 88, headed due North, and then turned 180 degrees to head due South. Parafiński, who was in a considerably faster Hurricane than Oliver's old, early production model, saw the Junkers drawing nearer very slowly and gave two short bursts from a long distance. The German slowed down. Parafiński gained height and, being on his right side, began the attack from above at 60 degrees. He clearly saw a yellow letter H next to the German crosses. The Junkers fired back, hitting the Hurricane's right wing.

Parafiński, then over Cheltenham, attacked with a series of shots from the front at a sharp angle. Smoke began to pour from the Ju 88's right engine and some debris fell from the aircraft. The Pole rode determinedly on the tail of the enemy plane, which went into a glide and exploded on impact.[7] The Ju 88A-5, 4U+HL (c/n 0451), belonging to 3(F)/123, crashed at Manor House, Coates, soon after 16:00 hours, killing its crew. Sgt Parafiński was credited with No.308's first victory.

A week later No.308 Sqdn became operational. Operational Order No.1 requested one section to be at the "Available" state during the hours of daylight, from 08:00 hours on Sunday, December 1st. Sections were to be provided by 'A' and 'B' Flights alternately. Operational Order No.2, of 12 December, specified that four sections, two from each flight, be at "Readiness." The Polish command of the squadron passed

On 15 December 1940 AM W. Sholto Douglas came to Leconfield to award the first DFCs to Polish pilots and to congratulate them personally on their victories in the Battle of Britain. Here AM Sholto Douglas inspects airmen of No.303 Squadron upon his arrival. (IPMS)

AM Sholto Douglas decorates P/O Ferić with the DFC; 15 December 1940, Leconfield. (IPMS)

The first four Polish recipients of the DFC: F/O Urbanowicz, P/O Zumbach, P/O Ferić and F/O Henneberg, on 15 December 1940. The fifth DFC was awarded posthumously to P/O Paszkiewicz.

to S/Ldr (Mjr) Jerzy Orzechowski on 8 December. The bad December weather greatly affected operational activity and during that month the squadron logged only 11 sorties in five and half hours flying time.

No.307 "Lwów Eagle-Owls" Night Fighter Squadron at Jurby (Isle of Man) was to be temporarily used for daylight

[7] Reports of W/Cdr Oliver, Officer-in-Charge Training, No.9 Group, and Sgt. Parafiński (IPMS, Lot A V,52/1b-1).

Hurricanes of No.306 "Toruń" Fighter Squadron take off from Ternhill after the unit became operational on 8 November 1940. (IPMS)

Three Hurricanes of No.306 Squadron coming to land at Ternhill in November 1940. (IPMS)

V6986 UZ•U at Ternhill, November 1940. (IPMS)

operations, because of the doubtful suitability of its Defiant fighters for night work. An order from HQ No.9 Group requested one section of Defiants to be kept at 15 minutes availability from dawn to dusk from 4 December onwards, for convoy protection duties within a 30-mile radius of the Isle of Man.

Red Section – the crews: P/O (S-Lt) Jan Buczyński and Sgt (Cpl) Kazimierz Frąckiewicz in L7020, Sgt Ludwik Mirończuk and Sgt Stefan Ostolski in N1683, and Sgt Witold Dukszto and Sgt Stanisław Jarzębowski in N1696 – took off on the squadron's first mission on 8 December at 15:13 hours, to patrol over Ramsey Bay. Other patrols followed and in December No.307 made altogether 16 missions, amounting to 47 sorties in 44 1/2 hours flying time. Two Defiants were involved in accidents.

In 1940 the Polish fighter squadrons carried out altogether almost 240 missions, comprising 2,199 sorties in 2,295 hours flying time. Although in the last two months of 1940 several Poles serving with British fighter squadrons during the Battle of Britain obtained transfers to Polish units, their departures were partly compensated by new Polish pilots joining the RAF squadrons. The Poles who remained in the British units added 9 2/3-0-2 enemy aircraft to the Polish bag, to mention only P/O (Lt) Henryk Szczęsny, flying with No.74 Sqdn, who shot down two more Bf 109s (on 1 and 5 December) and damaged another (on 2 December), P/O (S-Lt) Marian Chełmecki, flying with No.17 Sqdn, who was credited with the destruction of a Ju 87 and a Bf 109 (on 8 and 11 November respectively) and shared a Do 17 "damaged" (on 9 November), F/Sgt (Sgt) Michał Maciejowski, flying with No.249 Sqdn, who accounted for two Bf 109s destroyed (on 7 November and 5 December) and P/O (S-Lt) Karol Pniak, flying with No.257 Sqdn, who destroyed 1 1/2 Italian Fiat BR 20 bomber aircraft on 11 November. The overall Polish score at the end of 1940 stood at 214 1/6-35-38 2/3.

Apart from Poles in the RAF fighter units, two Polish pilots, P/O Ryszard Dyrgała and P/O Zbigniew Wysiekierski (the latter nicknamed "Whisky") flew unarmed Spitfires in No.1 Photographic Reconnaissance Unit on highly secret high-altitude, strategic reconnaissance missions over occupied Europe and Germany in the later part of 1940. The latter pilot discovered in April 1941 the absence of the battleship Bismark from its mooring in a Norwegian fjord, which started a search for the mighty vessel, thus leading to its eventual destruction.

4. Expansion and Operations of the Polish Fighter Force 1941-1943

In the winter of 1940/41 Fighter Command recovered from the tremendous exertion and losses of the Battle of Britain and by the end of December mustered six Groups, between them deploying 71 squadrons with 1,467 fighters in strength. The majority of these were new aircraft, including the improved Spitfire II and Hurricane II, delivered since October. The build-up of the fighter force to 75 squadrons, including three new Polish ones, continued into 1941.

V7118 UZ•V manhandled by ground crew at Ternhill, November 1940. (IPMS)

After the defensive victory in the Battle, the revitalized Fighter Command, anxious to look for new successes, began to move to the offensive with shallow penetrations into France and the Low Countries. Sweeps across the Channel, code named "Rhubarbs" were the first step in the changing nature of Command's operations.[1] These were straightforward low-level attacks by fighter aircraft against enemy airfields and on ground targets of opportunity as a sideline. They were soon complemented by a more complex form of attack, joint sorties by bombers and fighters coded "Circuses," against short-range or "fringe" targets, which made a greater impression on the Germans. The "Rhubarbs and "Circuses" added up to some 2,700 sorties between January and June 1941, in which Fighter Command lost 51 pilots and claimed 44 victories.[2] Their effects were marginal.

With about 2/3 of the Luftwaffe's strength committed from June 1941 to an offensive against the Soviet Union and because of the consequential political pressure, the RAF was obliged to step up "Circuses" in the second half of 1941 in order to damage "valuable military objectives," to destroy enemy fighters and to force the Germans to maintain a substantial fighter presence in Northern France. In the autumn of 1941 the initial forms of the fighter offensive were supplemented by more ambitious operations: "Rhubarbs" – by large-scale roving fighter attacks coded "Rodeos," which were aimed at the establishment of local air superiority; and "Circuses" – by "Ramrods," which were aimed at the destruction of specific targets by fighter-escorted bombers.

The RAF paid a heavy price for these activities, which was compounded by the emergence of an excellent new German fighter, the Focke-Wulf Fw 190 in September 1941, superior on all counts to the RAF's latest Spitfire V. In 1941 Fighter Command lost over 460 aircraft and the RAF claimed 752 enemy planes shot down, but the actual combat balance sheet was quite different. German losses on the Western front amounted only to some 170 aircraft, of which about a quarter could not be attributed to RAF action.

Dramatic changes in the strategic situation in 1941 reached their climax in December, when the Japanese attack on Pearl Harbour drew America into the war. The vast military strength of both the Soviet Union and the United States was now lined up against the Germans and their Axis partners.

The British fighter offensive was greatly expanded in 1942. Indicative of its scale was the period from March to June 1942, in which no less than 22,000 fighter sorties were flown across the Channel (bombers contributed a further 700 sorties). A major air operation in support of the Dieppe raid in August 1942 (Operation "Jubilee") was the culminating point of the year's fighter activities, one of the objectives of which was to draw the strength of Luftflotte 3 over the area and destroy it. In this the RAF failed – for the loss of 108 aircraft it succeeded in shooting down only 48 Germans and damaging another 24 – about nine percent of the Luftflotte's combat force (as compared with the contemporary RAF claim of 87 enemy aircraft destroyed, 45 probably destroyed and 143 damaged).

In overall terms Fighter Command's losses in 1942 amounted to 915 aircraft, and although the Luftwaffe did not allow the RAF challenge to go unheeded, it was forced on to the defensive. Having at the end of 1942 only some 3,950 first-line aircraft facing almost three times as many Soviet, British and American warplanes in three war theaters – the Eastern Front, in the Mediterranean and in the West (as compared with about 3,370 combat aircraft in September 1939, most of which were deployed on the Polish Front against less than 400 of LW's machines) – the Luftwaffe became massively overextended.

By 1943 the RAF fighter offensive was overshadowed by Bomber Command's night and USAAF's daylight offensives against the Reich, which paved the way for invasion. With the basic plan for the invasion of France set into motion by July 1943, the RAF fighter force was regrouped to meet the new task of preparation and tactical support for the invasion armies and the name Fighter Command disappeared from the order of battle in November 1943.

The Polish fighter force then numbered seven daylight squadrons, three of which – Nos.315, 316, and 317 – were established early in 1941. This expansion led to the formation of three Polish fighter wings and facilitated organizational changes which gave the PAF a wider autonomy.

(1.) Formation of New Polish Fighter Squadrons Nos.315 "Dęblin," 316 "Warsaw" and 317 "Wilno"

An Air Ministry order of 6 January 1941 authorized formation of a new Polish fighter squadron, No.315, to be set up at RAF Acklington (within No.13 Group). Two days later the Inspector General of the PAF instructed the Polish Depot in Blackpool to select personnel for the unit and entrusted its command to S/Ldr (Cpt) Stanisław Pietraszkiewicz. The squadron, which included former members of the CWL Dęblin Training Centre, was named No.315 Dywizjon Myśliwski Dębliński ("Dęblin" Fighter Squadron) and adopted the badge of 112th Eskadra of 1st (Warsaw) Air Reg. It was assigned the code letters PK.

The squadron was formed on 21 January with the arrival of Pietraszkiewicz and the first Polish officers at Acklington. Other Polish personnel followed in the next few days. S/Ldr H.D. Cooke came to the station as the British Commander of the squadron on 25 January and he was soon joined by F/Lt Davy and F/Lt Edy, appointed British Commanders of its two flights. Living conditions at Acklington were rather difficult and the airfield was waterlogged, with two runways still under construction.

---

[1] Although several small-scale sweeps were flown by the RAF fighters across the Channel in the second half of 1940, they were defensive in character – their objective being to inhibit German preparations for Operation "Sealion."

[2] The actual German losses were less than 20 aircraft destroyed.

Ground training activities began on 26 January and the first Hurricane (L1740) was received on 3 February. Four further Hurricanes were taken on charge by the 8th and the first familiarization flights were made on that day. Over 40 flights were logged in the next eight days, but snow falls interrupted training. By the beginning of March No.315 Sqdn had a full complement of Hurricanes[3] and commenced combat proficiency training. However, its fighters, all of an early production series and with many hours of combat flying behind them, required excessive maintenance and their serviceability was restricted even further by a lack of spares. Heavy snow in the second week of March stopped flying.

On 11 March 1941 No.315 Sqdn received notice that it was to be moved to Speke (Liverpool) on the 13th. The air party left Acklington on the 13th in the morning and the squadron's 16 Hurricanes, led by S/Ldr Cooke, took off for Speke at 11:45 hours. Two Hurricanes followed the next day and two unserviceable ones remained at Acklington. No.315 Sqdn HQ re-opened at Speke on the 15th. The squadron was reorganized and became operational on 18 March. In addition to three British officers and nine other ranks, it had on strength 24 Polish pilots and 165 ground personnel. Its pilots then comprised:[4] S/Ldr H.D. Cooke, British C.O. of the squadron, S/Ldr (Cpt) Stanisław Pietraszkiewicz, Polish C.O. of the squadron; F/Lt Thomas Humphrey Davy and F/Lt (Cpt) Władysław Szczęśniewski, Commanders of 'A' Flight, F/Lt Allen Laird Edy and F/Lt (Lt) Władysław Szulkowski, Commanders of 'B' Flight; F/Os (Lts) Zbigniew Czaykowski, Bernard Groszewski, Marian Pisarek, F/O (S-Lt) Kazimierz Woliński, P/Os (S-Lt) Józef Czachowski, Jerzy Czerniak, Eugeniusz Fiedorczuk, Władysław Grudziński, Tadeusz Hoyden, Franciszek Kornicki, Marian Łukaszewicz, Włodzimierz Miksa and Tadeusz Nowak; Sgt (Sgt Mjr) Józef Kania, Sgts (Sgts) Jan Kowalski, Edward Paterek, Sgt (Cpl C-Off) Marek Słoński-Ostoja, Sgts (Cpls) Jan Adamiak, Tadeusz Krieger, Bolesław Turzański and Piotr Zaniewski. F/O (Cpt) Stanisław Luranc was the squadron's Engineering Officer.

Two more Polish fighter squadrons, Nos.316 and 317, began to organize in Blackpool in the first half of February 1941[5] to the orders of the Air Ministry and Gen Sikorski. F/L (Cpt) Juliusz Frey, charged with the formation of No.316 Sqdn and formally appointed S/Ldr and its Polish Commander on 22 February, was instructed to proceed to RAF Pembrey (in No.10 Group), where the unit, known as No.316 Dywizjon Myśliwski Warszawski ("Warsaw" Fighter Squadron) was formed on 23 February. With its personnel largely coming from the former IV/1 Warsaw Fighter Dyon, No.316 Sqdn continued the traditions of the Dyon and adopted the badge of its 113th Eakadra. The code letters SZ were allocated to the squadron.

A Hurricane P2827 PK•K of No.315 Squadron at Speke in early April 1941. Note the painted over, old PAF chessboard under the cockpit hood, which is replaced by a larger new chessboard forward of the windscreen. (IPMS)

F/Lt C.J. Donovan, promoted to S/Ldr, joined No.316 Sqdn as its British C.O. on 2 March, but his posting to the squadron proved particularly unfortunate. His thoroughly negative attitude towards Poles alienated relations and embarrassed the Group's HQ.

On 24 February the squadron received the first of its 18 Hurricanes and ground and flying training began. The severe winter weather intervened and by the end of the month only eight flights were made. The training programme was greatly intensified in March and 145 practice flights in 217 flying hours were logged during that month,[6] but the squadron suffered considerable personnel shortages and S/Ldr Frey's attempts to bring it to its nominal strength met at first with little success. Despite this, No.316 Sqdn was made operational at Pembrey on 25 March, having on strength one British and 16 Polish pilots, as follows:[7] S/Ldr C.J. Donovan, British C.O. of the squadron, S/Ldr (Cpt) Juliusz Frey, Polish C.O. of the squadron, F/Lt(Cpt) Wieńczysław Barański, Commander of 'A' Flight, F/Lt(Cpt) Wacław Wilczewski, Commander of 'B' Flight; F/Os (Lts) Aleksander Gabszewicz, Roman Nartowicz, Zbigniew Nosowicz, Tadeusz Nowierski, Bronisław Mickiewicz, Tadeusz Sawicz, P/Os (S-Lts) Bohdan Anders, Franciszek Kozłowski, Bolesław Rychlicki, Stefan Stegman, Sgt (Sgt) Władysław Kiedrzyński, Sgt (Cpl C-Off) Romuald Gadus and Sgt (Cpl) Włodzimierz Mudry. F/O (Lt) Edward Christman was the squadron's Engineering Officer.

By the end of March 1941 No.316 Sqdn's ground personnel numbered 148 and five more Polish pilots arrived in Pembrey. These were: P/O (Lt) Dariusz Gozdecki, P/O (S-Lt) Bernard Buchwald, Sgt (Cpl C-Off) Jerzy Szymankiewicz and Sgts (LAC C-Offs) Tadeusz Dobrucki and Eugeniusz Gottowt.

An Air Ministry order of 15 February 1941 requested the formation of "No.317 (Polish) (Fighter) Squadron ... at Acklington commencing 19 February."[8] This was followed by the PAF Inspectorate's order of 20th February. The unit was

[3] As per War Establishment of a Fighter Squadron – (Hurricane) No.War/FC/227, later amended to No.War/FC/231 (IPMS, Lot.A V,53/1a-2).
[4] No.315 Sqdn's Operational Diary (IPMS, Lot A,V53/1c-1).
[5] Initially to War Establishment No.War/FC/227.
[6] No.316 Sqdn's Operational Diary (IPMS, Lot A V,54/3).
[7] Ibid, and Squadron's Personnel File 1941 (IPMS, Lot A V, 54/1a-1).
[8] Order LM 7074/D of O (IPMS, Lot A V, 55/1a-2).

formed on 22 February under the Polish command of S/Ldr (Cpt) Stanisław Brzezina, who arrived at Acklington (in No.13 Group) four days later, preceded by the advance Polish party of four officers and four other ranks. S/Ldr Christopher John Mount was appointed the British Commander of the squadron, which took the name No.317 Dywizjon Myśliwski Wileński ("Wilno" Fighter Squadron) and continued the traditions of III/5 Fighter Dyon, adopting its "Cormorant" badge. It was given the code letters JH.

The first Hurricanes were already allocated to the squadron by 22 February, but its beginnings – on the unfinished airfield temporarily immobilized by heavy snowfalls which caused two Hurricanes to crash on delivery by ferry pilots – proved very difficult. Acute shortages of flying and ground crews magnified the problems – by the middle of March 1941 it had in strength only 15 Polish pilots, five administration officers and 70 ground personnel, a shortfall of 14 pilots and over 100 other ranks. These circumstances adversely affected the training program. Practice flights on Hurricanes did not start until 11 March and by the 16th 15 flying hours were logged.[9]

In the second half of March 13 more pilots arrived at Acklington. Ground crew manning also improved and training accelerated rapidly, amounting to almost 140 flying hours. The very caring attitude of S/Ldr Mount created an excellent atmosphere and helped to overcome various difficulties. The squadron was declared operational on 23 March, keeping one of its Flights on "Readiness." At the end of March 1941 its flying personnel comprised:[10] S/Ldr Christopher John Mount, British C.O. of the squadron, S/Ldr (Cpt) Stanisław Brzezina, Polish C.O. of the squadron; 'A' Flight: F/Lt(Cpt) Antoni Wczelik, Flight's Commander, F/Os (Lts) Marian Trzebiński, Jan Wiśniewski, P/Os (S-Lts) Stanisław Bochniak, Stanisław Juszczak, Tadeusz Kratke, Tadeusz Kumiega, Ludwik Martel, Marian Sikorski, Jerzy Solak, Karol Wójcik and Sgts (LACs) Józef Biel and Michał Maciejowski; 'B' Flight: F/O (Lt) Henryk Szczęsny, Flight's Commander, F/O (Lt) Paweł Niemiec, P/Os (S-Lts) Stanisław Czarnecki, Władysław Drecki, Tadeusz Koc, Stanisław Łukaszewicz, Andrzej Malarowski, Tadeusz Stabrowski, Tadeusz Szumowski, Sgts (Cpls) Stanisław Brzeski, Bronisław Kościk and Leon Malinowski.

F/Lt (Cpt) Adam Kowalczyk and F/O (Lt) Piort Ostaszewski became Operational Officers. F/O (Cpt) Władysław Pomaski was the squadron's Engineering Officer.

(2.) Fighter Offensive over France and the Dieppe Expedition

Polish fighter squadrons began to participate in Fighter Command's offensive across the Channel from the very beginning and in increasing strength. With little experience of offensive warfare, they quickly learned new methods of operations under very different conditions from those of the Battle of Britain. Flights at naught feet became the order of the day and baling out usually meant spending the rest of the war in a prison camp.

With air activities greatly restricted by the winter weather, No.302 "Poznań" Squadron – at Westhampnett – began 1941 with a few routine sector patrols and engaged in exercises which were to prepare it for Mosquito (Rhubarb)[11] and Circus operations. These included low-level flights by two-aircraft sections over the sea and rendezvous with bombers. On 10 January the squadron's 12 Hurricanes, led by F/Lt Julian Kowalski and flying with No.610 Sqdn, made their first sweep over the French coast. No.302 received a special commendation from No.11

[9] Brzezina's Report for the period 1-15 March 1941 (IPMS, Lot. A V,55/1c-1).

[10] Brzezina's Report for the period 15-31 March 1941 (A V,55/1c-1).

[11] Rhubarbs were initially called Mosquitoes, but this name was dropped a few weeks later to avoid any confusion with the Mosquito aircraft.

A Hurricane of No.317 Squadron coming to land at Acklington, circa March 1941.

F/O Henryk Szczęsny, C.O. of No.317's 'B' Flight, points to his name "Hesio" (diminutive of Henryk) painted on his Hurricane. (IPMS)

No.302's pilots at the dispersal hut at Westhampnett early in 1941, with Sgt Nowakiewicz sitting next to the doorway and a WX•A Hurricane visible in the background. (T. Kopański)

Group HQ for the greatest number of operational hours flown and the shortest scramble time in the adverse January weather.

On 5 February 1941 No.302 Sqdn took part for the first time in a Circus operation, protecting – together with five other fighter squadrons – 12 Blenheims which attacked St. Omer-Longueness aerodrome. Five days later a two-aircraft section took off on the first Rhubarb, but had to return because of cloudless sky over the Channel.[12] Routine patrols were also flown over the South Coast, and on 16 February P/O Edward Pilch and Sgt Marian Wędzik intercepted a Ju 88 over Bognor and were credited with its destruction. Four days later Pilch, testing a newly overhauled Hurricane I WX•P (R2687), dived out of control and was killed.

At the beginning of March No.302 Sqdn began to convert to the more powerful Hurricane IIA and twelve-gun IIB and by 18 March had surrendered all its earlier Hurricane Is. These were brand new machines and in the first half of March the squadron made about 80 flights to run-in their Merlin XX engines, in addition to 37 operational missions. On 4 March a three-aircraft section still flying the Hurricanes I engaged three Ju 88s and each of its pilots – F/Lt Julian Kowalski in WX•C V6744, P/O Jan Maliński in •B V6860 and Sgt Antoni Łysek in •A P3207 – were credited with one Ju 88 damaged. On the 13th three pilots – P/O Wacław Król, in the lead, P/O Bronisław Bernaś and P/O Marceli Neyder – fought the first encounter in the new Hurricanes II[13] and were credited with one Ju 88 damaged over Selsey Bill, but the Luftwaffe records confirm that the attacked aircraft, the reconnaissance Ju 88A-5 7A+LM, belonging to 4(F)/121, crashed mid-Channel at 16:40 hours and its entire crew was missing. In the second half of March the squadron carried out 35 patrols and on the 28th its section successfully attacked another Ju 88 over the Channel off Beechy Head, its destruction being credited jointly to P/O Władysław Kamiński and Sgt Antoni Łysek.

April was bustling with patrol activities and the squadron, which moved to Kenley on 7 April, carried out its first Sphere – a high altitude sweep above cloud over enemy territory – but reported no encounters with the Germans. A major combat with enemy fighters took place on 8 May. Squadron's 'B' Flight was scrambled from Kenley around noon to patrol over the Maidstone area and, vectored South-East, engaged the new Bf 109Fs between Ashford and Tenterden in a surprise attack from cloud. F/O Zygmunt Kinel, P/O Wacław Król and Sgt Eugeniusz Nowakiewicz were credited with the destruction of one enemy fighter each, and one of these, the Bf 109F-2 3+ (c/n 5467) from 1/JG 3, crashed at Tenterden at 12:45 hours. Sgt Marian Rytka accounted for one more Bf 109 as "probable."

A second encounter with Bf 109s on the same day was recorded around 18:00 hours over Dungeness and P/O Zbigniew Wróblewski reported one Bf 109 approaching Sevenoaks as probably destroyed. Two days later, in another fight with Bf 109s over Ashford, P/O Marceli Neyder damaged one enemy fighter. The last combat from Kenley was fought on 21 May at 17:30 hours, with Rytka – just promoted to P/O (S-Lt) – shooting down a Bf 109F and P/O Wróblewski claiming another as "probable."

On 23 May S/Ldr (Cpt) Piotr Łaguna was promoted to Mjr and a few days later posted away to take command of No.1 Polish Fighter Wing. On 28 May 1941 No.302 "Poznań" Squadron moved to Jurby (Isle of Man) for rest, changing places

Sgt Rytka, P/O Neyder, Sgt Beda, P/O Gnyś and F/O Duryasz of No.302 Squadron, sitting on the wing of WX•V at Westhampnett early 1941. (W.Gnyś)

[12] Rhubarb orders made a partial cloud cover an essential pre-condition.
[13] Z2485 •U, Z2423 •V and Z2350 •W.

An eight-gun Hurricane IIA Z2668 WX•H No.302 Squadron, bearing a small PAF chessboard under the cockpit enclosure, at Westhampnett, March 1941.

with No.312 (Czech) (Fighter) Squadron and leaving its Hurricane IIA and IIB fighters at Kenley for the Czechs. At Jurby No302 took on charge 20 Hurricane Is – 17 of them operational – which formerly belonged to the Czech squadron. On 29 May F/Lt (Cpt) Stefan Witorzeńc became No.302's Acting C.O. and was officially promoted to Acting S/Ldr on 15 June. Operating from Jurby during the next ten weeks – with a four-aircraft detachment going to Valley in Anglesey in the middle of June – the squadron was engaged on generally uneventful patrols over the Irish Sea and on convoy protection duties. On 7 August No.302 was transferred to Church Stanton (No.10 Group) and reverted to Hurricanes IIAs and IIBs.

No. 303 "Kościuszko" Squadron, which returned to Northolt on 3 January 1941, flew its first operational sortie in the new year on the 9th. This was a routine sector patrol and similar ones followed. Rumoured conversion to Spitfires soon materialized and on 22 January the squadron received its first six Spitfire Is.[14] On the same day the pilots made their last

[14] N3108, N3285, P9555, R6972, R6975 and X7704, ferried from Hornchurch.

A twelve-gun Hurricane IIB, Z3095 WX•N, with a parachute harness on its port wing, at Readiness at Westhampnett in March 1941. A small PAF chessboard under the cockpit enclosure is well visible.

Hurricane sorties on their first Rhubarb mission, spearheading Polish participation in this type of action. Seven Hurricanes attacked in sections Le Touquet, Crecy, Desvres and Auxi-le-Chateau, damaging two Bf 109s on the ground and a convoy of military lorries. F/O Wacław Łapkowski flew into an overhead electricity cable, 25 yards of which wrapped itself around his Hurricane's nose; his friends regarded such a low form of sabotage as beneath the dignity of a fighter pilot. P/O Wiktor Strzemborz went for a gun post and held on so long that he knocked his wingtip off against the sandbags, even though bullets were hitting his windscreen. The operation was regarded as very successful.

During the next ten days the pilots were familiarizing themselves with the Spitfires and their next combat mission, Circus No.1, was flown on 2 February 1941. Twelve Spitfires flew escort – with five other fighter squadrons – to five Blenheim bombers raiding the Boulogne docks. By the end of the month three more Circuses comprising 32 aircraft sorties, were logged, in addition to a few routine patrols. On 3 March No.303 was rearmed with Spitfire IIA fighters and during March flew two Circuses and two offensive sweeps as well as some convoy patrols. A few enemy fighters were sighted, but no contacts were reported.

An interesting photograph of a dismantled ex-No.302's Z3675 WX•B, seen late in 1941 at Sverdlovsk in Russia, in the ground of the evacuated Zhukovsky's Air Academy. Z3675 was crossed off of the strength of No.302 on 18 July 1941 and delivered to the USSR in the autumn of 1941 in a consignment of Hurricanes as part of the British aid program. (T. Kopański)

Col Merian C. Cooper, the founder of the original Kościuszko Squadron, visited No.303 "Warsaw Kościuszko's" Squadron at Northolt on 14 March 1941. Left to right, in the first row, are: F/O Kołaczkowski, F/O Henneberg, G/Cpt Kwieciński, P/O Ferić, Col Cooper (in civilian clothes), F/O Urbanowicz and F/O Grzeszczak. The Spitfire in the background displays the Kościuszko badge under the radio mast. (IPMS)

President Władysław Raczkiewicz, accompanied by the American Ambassador to Poland, Mr A. Drexel Biddle (standing on President's right), arrived at Northolt on 4 April 1941 to visit No.1 Polish Fighter Wing. He was welcomed on the airfield by AM Sholto Douglas and AVM T. Leigh-Mallory, AOC No.11 Group (on the extreme right). (IPMS)

Ambassador Drexel Biddle chats with the pilots of No.303 Squadron at Northolt, while President Raczkiewicz (with his back to the camera) talks to F/O Henneberg. (IPMS)

On 1 April 1941 No.1 Polish Fighter Wing was ordered to be formed at Northolt, initially under the dual command of W/Cdr John Kent and W/Cdr (Cpt) Witold Urbanowicz. The Wing at first comprised Nos.303 and 306 Squadrons, which were to be strengthened later by No.308 Sqdn. No.306 "Toruń" Squadron joined No.303 at Northolt on 2 April after limited winter activity from Ternhill, which included night patrols over the Birmingham area. These commenced in February and claimed their victims. On 13 February two Hurricanes were wrecked while night landing in fog – P/O (S-Lt) Bohdan Bielkiewicz hit trees and was killed instantly and Sgt (Cpl) Wawrzyniec Jasiński survived the crash but suffered serious injuries. At Northolt No 306 began conversion to Hurricane II fighters and retraining for its tasks – night interception over London and

One of the Spitfire IIAs newly delivered to No.303 Squadron, sporting the unit's badge in a new position, under the windscreen, as seen during President Raczkiewicz's visit to Northolt on 4 April 1941.

A Spitfire IIA RF•J, carrying the presentation name "Garfield Weston Ltd" on the forward fuselage and the Kościuszko badge under the windscreen, takes off from Northolt during the President's visit. Note the light-colored disc under the radio mast, where the badge was originally painted. (IPMS)

A Spitfire IIA P8041 RF•F, with unidentified presentation markings on the forward fuselage, Northolt, 4 April 1941.

A Spitfire IIA P8361 RF•R "Krakatao," funded by the Dutch East Indies; Northolt, 4 April 1941. (IPMS)

Circuses and Rhubarbs. The squadron resumed operations after a fortnight.

Offensive operations intensified in April with a Rhubarb on 12 April. In the late afternoon No.303's three two-aircraft sections, briefly operating from Hawkings, attacked dispersed aircraft at Crecy and Le Touquet, starting fires on the latter airfield. S/Ldr (Cpt) Zdzisław Henneberg's Spitfire, RF•P (P8029), damaged by ground fire, ditched in the Channel and its pilot was never found. F/Lt (Lt) Tadeusz Arentowicz became the squadron's Acting C.O. He was succeeded by S/Ldr (Lt) Wacław Łapkowski on 5 May 1941.

On 16 April all Northolt squadrons were engaged in Circus No.7 – including for the first time No.306's Hurricane IIAs, providing close escort to bombers on the Berck-sur-Mer raid. No.303's Spitfire IIAs, flying high cover, fought with Bf 109s over the Channel and P/O (S-Lt) Bogusław Mierzwa and P/O (S-Lt) Mieczysław Waszkiewicz were shot down and killed off Dungeness. Four days later Squadron's six Spitfires on a Sphere avenged the losses. In a running battle with Bf 109s, which started over Le Touquet at about 11:00 hours, P/O Witold Łokuciewski shot down one German fighter and P/O Jan Daszewski, whose guns partly froze, claimed one "probable": On the way back Sgt Jan Palak damaged a solitary Ju 88 over the Channel. On 29 April No.306's 12 Hurricane IIs, sent on patrol over Maidstone, were attacked by passing Bf 109Fs, which retired to France.

With the departure of No.601 Sqdn from Northolt on 2 May, only the two Polish squadrons forming No.1 Polish Fighter Wing remained at the Station, which became largely Polish in character. In the first half of the month No.303 Sqdn flew defensive patrols, some of which involved inconclusive skirmishes with Bf 109s over Biggin Hill, Maidstone, and Dungeness. On 11 May No.306 Sqdn was engaged in night combat over London during a very heavy attack by more than 500 bombers – the last big Luftwaffe raid on London of the war – and achieved a notable success. F/O Władysław Nowak destroyed a He 111 from a range of 50 to 20 yards, which finally crashed in the Channel and F/O Gerard Ranoszek damaged a Ju 88.

A Rhubarb flown by No.303's three two-aircraft sections on 15 May was particularly rewarding. Blue Section – F/Lt Jerzy Jankiewicz and Sgt Wacław Giermer – attacked a Ju 52 transport flying at a low altitude near Aire and sent it to the ground in flames. The Junkers – c/n 5595, of Transportstaffel FI/Flg Kps – was carrying Gen Obst Ulrich Grauert, Air Corps Commander, and his staff, all of whom were killed. White Section – P/O Bołesław Drobiński and Sgt Marian Bełc – strafed airfields in the St. Omer-St. Inglevert area, hitting a taxiing Ju 52, and later a German convoy in the Channel. One of the vessels, 305-ton VP 1805 (ex-French "Senateur Louis Brindeau"),[15] was set on fire by Drobiński. Green Section – P/O Stefan Paderewski and Sgt Mieczysław Popek – attacked six Bf 109s in blast shelters and fired at a group of soldiers.[16]

In the middle of May No.303 Sqdn converted to the cannon-armed Spitfire IIBs and No.306's equipment was augmented by the twelve-gun Hurricane IIBs. On 19 May No.306's 12 Hurricanes, which included some Mk IIBs, were caught unawares by Bf 109s while changing formation over Maidstone and lost four of their numbers – three pilots baled out successfully, but P/O (Lt) Jan Czepiewski was posted missing and his body was never found. On 21 May both squadrons took part in Circus No.10 (No.303 Sqdn for the first time in Spitfire IIBs), and three days later No.303 flew an uneventful Sphere. On 1 June W/Cdr (Mjr) Piotr Łaguna took over the command of No.1 Polish Fighter Wing from W/Cdr Urbanowicz.

A Spitfire IIB P8385 RF•A "Impregnable," with the colourful "Pluto" character on the engine cowling. This aircraft was usually flown by F/O Ferić, but the photograph shows P/O Stanisław Juszczak (killed in training on 11 August 1941) standing by the engine; Northolt, the middle of 1941. (IPMS)

The Polish pilots began their June activities with a sweep on the 4th, during which P/O Bronisław Kłosin shot down a Bf 109 with eight rounds from his Spitfire IIB's cannon but, hit himself, had to make a crash landing in Kent, and S/Ldr Wacław Łapkowski damaged another Messerschmitt. In the course of the following ten days No.303 and 306 Sqdns made one Sphere each and the former Squadron took part in Circus 12.

It is ironic that while the Russians dismissed all the West's warnings about an impending German invasion as "hostile propaganda" and neglected to take any precautionary steps against such an eventuality,[17] the RAF dramatically stepped up its offensive operations against the Luftwaffe in France in anticipation of the German attack in the East six days before it was launched. Altogether 15 Circuses were made between 16 and 30 of June – as compared with 12 in the whole preceding period from the beginning of 1941 – and No.1 Polish Fighter Wing at Northolt made a vital contribution to this effort.

On 16 June, during Circus 14, P/O Kazimierz Rutkowski of No.306 Sqdn damaged a Bf 109 near Boulogne. In the early hours on the following day four Hurricane IIBs from No.306 Sqdn attacked Etaples airfield, and later, at 18:45 hours the

[15] According to John R. Foreman, who also established that the convoy included two Flak (AA) ships (Nos.54 and 55), a minesweeper (MS 3611, ex-Dutch), and a tug.

[16] Reports in the Combat Diary, Royal Air Force Northolt.

[17] On 14 June the Kremlin broadcasted to the world that the rumors of a German attack on Russia were an "obvious absurdity."

whole Squadron took off from North Weald on Circus 13. Numerous combats developed with Bf 109s and three Messerschmitts were destroyed – one each by F/O Jerzy Słoński-Ostoja, Sgt Józef Jeka and Sgt Leon Kosmowski, who also claimed another one damaged – with no loss of either fighters or the escorted Blenheims. At the same time No.303 Sqdn made a sweep over Cap Gris Nez and came to grips with Bf 109s. F/O Wojciech Kołaczkowski destroyed one German and F/O Wiktor Strzembosz claimed another "probable," but the former pilot's aircraft was hit and crash landed near Manston. Four more Germans fell victim to No.303 Spitfires' cannon during Circus 15 on 18 June with no loss to the squadron, the victors being P/O Bolesław Drobiński, 2 Bf 109Es, and S/Ldr Wacław Łapkowski and F/O Witold Łokuciewski, one Bf 109F each. The next day No.303 Sqdn provided top cover to bombers on Circus 17.

In the last ten days of June, 11 Circuses (Nos.16 and 18 to 27 inclusive) and one Rhubarb were flown by Polish fighter squadrons, and several air battles were successfully fought. On 24 June No.308 "Kraków" Squadron, arriving from Chilbolton, joined No.1 Polish Fighter Wing at Northolt, bringing its strength to three squadrons.

In the winter and spring months of 1941 No.308 Sqdn, operating from Baginton, was involved in defensive day and night patrols over the Midlands and in intensive combat training which cost the lives of three pilots – F/Lt (Cpt) Witalis Nikonow and P/O (S-Lt) Jerzy Wolski (on 11 January), and Sgt (Sgt) Mieczysław Parafiński (on 25 February). On 26 March, a section consisting of P/O Wacław Bożek, P/O Franciszek Surma and Sgt Jan Kremski, still in Hurricane Is, claimed a Ju 88 as "probable" South of Coventry. By the end of March the squadron had converted to Spitfire Is and six weeks later to the superior Spitfire IIAs. In a few encounters in April and May P/O Tadeusz Stabrowski reported two combats with Ju 88s and was credited with 1/2 damaged on each occasion. On 1 June the squadron was transferred to Chilbolton near Andover, within No.10 Group, from where it flew several patrol missions in the Southampton-Portsmouth area and the Channel. On 23 June S/Ldr (Cpt) Marian Pisarek became the squadron's new C.O., the change in command coinciding with the move to Northolt. No.308 Sqdn participated with the rest of No.1 Polish Fighter Wing in Circus 21 on 24 June in the late afternoon, the day of its arrival at Northolt.

From 21 June to the end of the month Polish pilots of No.1 Polish Fighter Wing were credited with a score 23-8-3, with No.303 Sqdn in the lead with 21-6-1 enemy aircraft, and all losses – all Bf 109s – were inflicted over French northern coastal areas. The victors were:

- No.303 Sqdn: P/O Bolesław Gładych 3-1-0, S/Ldr Wacław Łapkowski 3-0-0, P/O Bolesław Drobiński 3-0-0, P/O Władysław Drecki 2-0-0, Sgt Marian Bełc 2-0-0, F/Lt Jerzy Jankiewicz 1-1-0, F/O Witold Łokuciewski 1-1-0, Sgt Mieczysław Adamek 1-1-0, F/O Mirosław Ferić 1-0-0, F/O Aleksander Wróblewski 1-0-0, P/O Wiktor Strzembosz 1-0-0, Sgt Mirosław Wojciechowski 1-0-0, Sgt Kazimierz Wünsche 1-0-0, P/O Jan Bondar 0-1-0, Sgt Jan Szlagowski 0-1-0, F/Lt Tadeusz Arentowicz 0-0-1;

Spitfire IIAs of No.308 Squadron, being refueled at Northolt in the summer of 1941. The aircraft display large PAF chessboards on their noses, which were introduced at that time as standard on fighters of No.1 Polish Fighter Wing, except for No.306 Sqdn.

Hurricanes of No.315 "Dęblin" Fighter Squadron, preparing for a take-off at Speke on 22 April 1941.

- No.306 Sqdn: Sgt Leon Kosmowski 1-0-0, Sgt Wawrzyniec Jasiński 0-1-0, P/O Zdzisław Langhamer 0-0-1, Sgt Józef Jeka 0-0-1;
- No.308 Sqdn: F/O Stanisław Wielgus 1-0-0, Sgt Władysław Majchrzyk 0-1-0.

In addition four Bf 109s were claimed as destroyed on the ground and a few others damaged.

Polish casualties included three pilots killed – W/Cdr (Mjr) Piotr Łaguna (on 27 June – he was succeeded as Commander of No.1 Polish Fighter Wing by W/Cdr (Cpt) Tadeusz Rolski), P/O (S-Lt) Stefan Paderewski (No.303 Sqdn, on 25 June), and P/O (S-Lt) Józef Bondar (No.303 Sqdn, on 28 June) one pilot – P/O Józef Żulikowski (No.306 Sqdn) – missing and made PoW, and one – P/O Bolesław Gładych (No.303 Sqdn) – badly injured; five Spitfires were lost and two damaged. The Combat Diary, Royal Air Force Northolt, recorded, "the days were reminiscent of the Battle of Britain days."

In July 1941 Circus operations were expanded even further. In the first 12 days of that month No.1 Polish Fighter Wing took part in 19 Circuses in various strength, No.303 Sqdn being engaged in 15, No.306 (in Hurricanes until 11 July), in seven, and No.308, in 13.

No.308 gave an excellent account of itself in its first major battle with the enemy on 2 July, when a large formation of Bf 109Fs and 109E fighters attempted to prevent Blenheims from bombing a power station at Lille (Circus 29). Five Messerschmitts were claimed by the squadron as destroyed, with a further two "probable" and one "damaged," for the loss of one Polish pilot killed and one missing (taken PoW) and two Spitfires shot down. No.303 added two more Bf 109s destroyed, one "probable" and one "damaged," but its own casualties comprised two pilots killed, including its C.O., S/Ldr Łapkowski, and two fighters lost. One of these was flown by Sgt Ryszard Górecki, who crashed into the Channel and survived for 72 hours without food or water in a dinghy. In addition, Sgt Wojciechowski was injured, but managed to land at Martlesham and climb out of a cockpit which was dripping with blood; he collapsed walking towards Sick Quarters.[18]

Four-engined Stirling heavy bombers joined Circuses for the first time on 5 July, raiding Lille, with No.308 Sqdn providing top cover. The following day No.303 twice provided Stirlings with a fighter escort and during the second mission (Circus 35), acting together with Nos.306 and 308 Sqdns which

[18] Combat Diary, Royal Air Force, Northolt.

Arming of a Hurricane of No.315 Squadron at Speke before a combat sortie in the early spring of 1941. Note the exhaust anti-glare panel and the PAF chessboard below.

No.315's Hurricanes operating from Speke in the defence of Liverpool in the early spring of 1941. (J. Rutkiewicz)

A Spitfire IIA, P8528 PK•J ("Janka"), newly received by No.315 Squadron at Northolt in July 1941. (J. Rutkiewicz)

served as close escort, prevented Bf 109s from attacking the bombers. On 11 July No.306 Sqdn converted to Spitfire IIBs, flying its first mission on the type on Circus 45 the same afternoon. In the early part of July the wing's squadrons also flew a few sorties providing protection to Lysanders on air/sea rescue missions.

On 13 July No.303 Sqdn and No.315 Sqdn changed places, with the latter squadron arriving from Speke to join No.1 Polish Fighter Wing and No.303 moving to Speke for respite and leaving its Spitfire IIBs at Northolt. No.315 "Dęblin" Squadron, which became operational at Speke on 18 March, was initially engaged in sector patrol and convoy protection duties over the Irish Sea. Its early operations were marred by a collision between two Hurricanes on convoy patrol on 27 March in which both pilots, F/Lt (Lt) Władysław Szulkowski and Sgt (Sgt) Edward Paterek, were killed. This work was generally uneventful and the first encounter with the enemy took place only on 25 May, when a Ju 88 was intercepted over Ruthin with no visible effect. In May-June most pilots underwent night flying training. At Northolt the squadron received Spitfire IIA and IIB fighters and achieved combat readiness on Spitfires at the beginning of August.

In the second half of July, with bad weather curtailing air activities during the later part of the month, No.1 Polish Fighter Wing was engaged in eight Circuses in two-squadron strength (Nos.306 and 308), in one Circus with one squadron (No.306), and in two sweeps. Battles with large formations of Messerschmitts were fought, developing into individual duels, which added further victories to the formidable Polish score, but also resulted in heavy casualties. No. 308 distinguished itself during a sweep on 17 July in a perilous combat with three times its number of Bf 109s, which attacked the returning squadron over the St. Omer-Gravelines area. In the uneven fight, the Poles repelled the Messerschmitts, shooting down three at the cost of two pilots killed and one missing (made PoW), and three Spitfires destroyed. The squadron was engaged in another heavy battle and tough dog-fighting on 22 July near Guines and claimed this time four victories and one German as "probable" for a loss of two pilots killed. Two more Bf 109s were destroyed on the ground. The following day No.306 Sqdn bore the brunt of the fighting on Circus 59, bringing down two Germans and damaging a further two, but its own casualties amounted to two pilots killed and one badly burned, as well as five Spitfires destroyed. Anothere encounter with Bf 109s was fought on the 24th.

During July No.1 Polish Fighter Wing – excluding S/Ldr Kent – accounted for 29-8 1/2-5 enemy aircraft, with No.308 Sqdn in the lead with a score 20-4-1, No.303, with 5-3-1, and No.306, with 3-1/2-2. The individual victories were credited as follows:

- W/Cdr Tadeusz Rolski, Wing's C.O., 1-1-1;
- No.308 Sqdn: S/Ldr Marian Pisarek 2-1-0; P/O Jan 16

The individual letter 'J' on P8528 being turned into the girl's name "Janka"; Northolt, July 1941. (J. Rutkiewicz)

Kremski, P/O Witold Retinger, and Sgt Jerzy Zieliński 2-0-0 each. F/O Feliks Szyszka and Sgt Stanisław Widlarz 1-1-0 each; P/O Bruno Kudrewicz 1-0-1; F/Lt Władysław Chciuk, F/Lt Stefan Janus, F/O Franciszek Surma, F/O Stanisław Wielgus, P/O Władysław Bożek, P/O Jan Szyszka, Sgt Tadeusz Hegenbarth, Sgt Tadeusz Schiele and the whole Squadron jointly 1-0-0 each, and Sgt Stefan Krzyżagórski 0-1-0;

- No.303 Sqdn: F/O Jan Zumbach 1-1-0, P/O Bolesław Drobiński 1-1/3-0; F/O Wojciech Kołaczkowski, Sgt Mieczysław Adamek and Sgt Wacław Giermer, 1-0-0 each; F/O Witold Łokuciewski 0-1-0; F/Lt Jerzy Jankiewicz and Sgt Mieczysław Bełza 0-1/3-0 each; and P/O Maciej Lipiński 0-0-1;
- No.306 Sqdn: F/O Stanisław Skalski, F/O Jerzy Słoński-Ostoja and Sgt Aleksander Franczak 1-0-0; P/O Leon Jaugsch 0-1/2-0, and Władysław Nowak 0-0-2.

Polish casualties comprised:

- No.308 Sqdn: six pilots killed – F/Lt (Lt) Erwin Kawnik (02/07/41), F/O (Lt) Stanisław Wielgus (12/07/41), P/O (Lt) Władysław Bożek (22/07/41), P/O (Lt) Mirosław Orzechowski (22/07/41), P/O (S-Lt) Jan Szyszka (17/07/41), Sgt (LAC C-Off) Tadeusz Hegenbarth (17/07/41), and four missing and made PoW – F/Lt Władysław Chciuk, F/Lt Stefan Maciejewski, Sgt Józef Biel and Sgt Paweł Kowala; and ten Spitfires lost;
- No.303Sqdn: three pilots killed – S/Ldr (Cpt) Tadeusz Arentowicz (08/07/41), S/Ldr (Mjr) Wacław Łapkowski (02/07/41), and F/O (Lt) Wiktor Strzembosz (08/07/41), and one badly injured – Sgt Mirosław Wojciechowski; three Spitfires lost;
- No.306 Sqdn: two pilots killed – P/O (Lt) Janusz Kosmowski (23/07/41) and P/O (S-Lt) Jan Maras (23/07/41), and one badly injured – F/Lt Władysław Nowak; five Spitfires lost;
- No.315 Sqdn: one pilot killed – P/O (Lt) Józef Czachowski (24/07/41, flying individually his first sortie from Northolt with No.308 Sqdn); one Spitfire lost.

The weather interrupted flying during the first few days in August 1941, but on 7 August No.1 Polish Fighter Wing took part in Circus 67 with 36 Spitfires from all three squadrons, including 'A' and 'B' Flights of No.315 on their first operational mission from Northolt. During the day's second operation in the late afternoon, No.308 Sqdn lost two pilots in a surprise attack by Bf 109s diving from the sun. Two days later the wing's squadrons fought two major engagements in the vicinity of St. Omer, in the morning and in the afternoon, and in both of them No.315 Sqdn played the leading part in the combat. Chaotic duels made it difficult to establish how many Bf 109s were actually destroyed, but at the end of the day the Polish pilots were credited with a score 2-3-4 enemy aircraft. All of these, apart from one damaged, were attributed to No.315 Sqdn, which lost two pilots killed.

On 14 August all three squadrons took off to rendezvous with bombers, but since these had to turn back owing to cloud conditions, the fighters proceeded to sweep over France and engaged a strong force of Bf 109s near the French coast. Numerous dog fights developed. No.306, flying top cover, lost three pilots, including its C.O., S/Ldr Jerzy Zaremba (who was succeeded by S/Ldr (Lt) Jerzy Slonski-Ostoja), and claimed two Germans destroyed and one "probable." No.308 accounted for four enemies shot down and two more damaged, for a loss of one pilot, while No.315 Sqdn achieved a resounding victory, with eight Germans destroyed, one "probable" and one damaged, without a loss.

P7855 PK•K ("Krysia"), transferred to No.315 Squadron from the East India Squadron and still without the PAF chessboards; July 1941. (J. Rutkiewicz)

The same PK•K ("Krysia"), with the inscription "East India Squadron" still visible under the cockpit, featuring the large PAF chessboards of No.1 Polish Fighter Wing on the nose. (J. Rutkiewicz)

16 August was another day of intensive combat. The wing was engaged on Circus 74, with No.306 flying top cover. This time the Squadron tore the Germans to pieces in a brilliantly executed diving attack on 15 Messerschmitts over St. Omer. Six Bf 109s were claimed destroyed and one damaged for a loss of one pilot shot down and made PoW. No.315 added one more German to the Polish score.

In the second half of August – in addition to the customary Circuses – the wing's operations included the first Roadstead[19] on the 17th and some Rhubarbs and sweeps. Several clashes with the enemy took place and, as the month drew to its close, another major battle was fought on 29 August (Circus 88), resulting in three Germans destroyed and four damaged for a loss of two pilots killed, including S/Ldr Słoński-Ostoja – who was succeeded in turn by S/Ldr (Mjr) Antoni Wczelik – and one made PoW. The operation "Illyad" – a pro-

[19] Fighter protected dive-bombing or low-level attacks on enemy ships.

A damaged Spitfire IIB, P8342 UZ•N (funded by the Dutch East Indies and still carrying its original "Ceram" markings on the front fuselage) of No.306 Squadron, which made an emergency landing at Biggin Hill after a victorious battle with a Bf 109F on 29 August 1941. The aircraft was flown by Sgt Marcin Machowiak. The squadron retained its pattern of markings, with the unit badges on the aircraft noses and chessboards on the rear fuselages.

tective patrol over the Navy-escorted merchant vessels – was made by the wing on 31 August.

Out of 29 Circuses mounted by the RAF in August, No.1 Polish Fighter Wing took part in 18 – in all but three of them at full three-squadron strength. The wing's pilots were credited with 34-6-12 enemy aircraft, with No.315 Sqdn in the lead with a score 17-4-5, No.306, with 11-1-1, No.308, with 6-0-6 and Wing's C.O. with 0-1-0. The individual credits were:

- W/Cdr Tadeusz Rolski, Wing's C.O., 0-1-0;
- No.315 Sqdn: Sgt Aleksander Chudek 3-0-1, F/O Jan Falkowski 3-0-0, F/Lt Bronisław Mickiewicz and P/O Władysław Grudziński 2-0-0 each; F/Lt Władysław Szczęśniewski and P/O Józef Gil 1-1-1 each; S/Ldr Stanisław Pietraszkiewicz 1-1-0; F/O Tadeusz Nowak, Sgt Stanisław Blok, Sgt Michał Cwynar and Sgt Tadeusz Krieger 1-0-0 each; Sgt Eugeniusz Malczewski 0-1-0; P/O Eugeniusz Fiedorczuk and P/O Kazimierz Woliński 0-0-1 each;
- No.306 Sqdn: S/Ldr Jerzy Słoński-Ostoja and F/Lt Stanisław Skalski 2-0-0 each; Sgt Henryk Pietrzak 1-0-1; P/O Zdzisław Langhamer, P/O Zdzisław Radomski, Sgt Wawrzyniec Jasiński, Sgt Józef Jeka, Sgt Marcin Machowiak and Sgt Otton Pudrycki 1-0-0 each; Sgt Stanisław Wieprzkowicz 0-1-0;
- No.308 Sqdn: P/O Witold Retinger 1-0-2, F/Lt Marian Wesołowski and P/O Jerzy Zbierzchowski 1-0-1 each; S/Ldr Marian Pisarek, F/O Feliks Szyszka and P/O Adolf Pietrasiak 1-0-0 each; F/Lt Stefan Janus and Sgt Jerzy Zieliński 0-0-1 each.

Polish casualties comprised:
- No.315 Sqdn: one pilot killed – F/O (Lt) Jerzy Czerniak (09/08/41), and two missing and made PoW – F/Lt Bronisław Mickiewicz and Sgt Andrzej Niewiara; three Spitfires lost;
- No.306 Sqdn: four pilots killed – S/Ldr (Cpt) Jerzy Zaremba (14/08/41), S/Ldr (Cpt) Jerzy Słoński-Ostoja (29/08/41), P/O (S-Lt) Wiesław Choms (14/08/41) and Sgt (Cpl) Stanisław Zięba, who, severely injured, was made PoW, but died of his wounds nine days later (23/08/41), and one pilot missing and made PoW – Sgt (Sgt) Aleksander Franczak; one pilot – P/O (S-Lt) Zdzisław Radomski, hit by cannon shell was badly injured (27/08/41) and had to have his arm amputated; six Spitfires lost;
- No.308 Sqdn: four pilots killed[20] – P/O (S-Lt) Jerzy Bettcher (29/08/41), P/O (S-Lt) Jan Kremski (14/08/41), P/O (Lt) Leonard Stąpel (07/08/41) and Sgt (Cpl) Zdzisław Brózda (07/08/41), and one missing – Sgt (Sgt) Adolf Pietrasiak;[21] six Spitfires lost.

[20] A fifth pilot, F/O (S-Lt) Tadeusz Blach, was killed in an accident on 09/08/41.

[21] Shot down over France on 19/08/41, he evaded capture, reached Spain and eventually returned to the Squadron.

The wing's activities in September 1941 were limited by adverse weather to only eight days of operations in strength, but the continued ferocity of the fighting and the appearance of new aircraft on both sides provided plenty of excitement.

On 4 September the Poles, still flying Spitfire II fighters (on Circus 93), clashed with Bf 109s over the French coast and shot down one German and damaged another for a loss of one aircraft and one pilot, who was made PoW. At the beginning of the month all three squadrons started converting to the superior Spitfire VB and although the first mission in the new type, planned for 11 September, had to be aborted due to cloud conditions, the wing's 37 Spitfire VBs – including one of the Command's aircraft flown by W/Cdr Rolski – made an offensive sweep on 16 September and fought with a formation of Messerschmitts, claiming five Germans destroyed, two "probable" and one damaged; two Spitfire VBs and two pilots were lost. In another encounter the following day P/O Czesław Budzałak of No.308 Sqdn was killed in a collision with a Bf 109F during a dog fight, after which both aircraft went straight down. Two further Messerschmitts and one damaged were added to the Polish score, but one more of the wing's pilots did not return.

Three Bf 109s were claimed by No.308 Sqdn over the Rouen area on 20 September with no losses, but on the next Circus (101), on 21 September, the wing was involved in a desperate combat with 18 Bf 109s on the way to France, which were joined by several new radial-engined enemy fighters, variously identified by the Polish pilots as Blochs, Curtiss Hawks or even Macchi C.200s. Four of these were reckoned destroyed, by S/Ldr Pietraszkiewicz (two) – who himself was shot down and made PoW – F/O Falkowski and P/O Wandzilak, in addition to five Bf 109s brought down, two "probables" and one damaged for a loss of one Polish pilot killed. The Polish combat reports created a sensation in Intelligence circles and were discussed for many days.[22] It was later realized that the enemy aircraft involved were the new Fw 190s.

The second encounter with Fw 190s was reported on 27 September (during Circus 103), with one claimed destroyed by Sgt Chudek. Five Bf 109s were also shot down, with a further two "probables" and four damaged. Polish casualties were one pilot and a Spitfire VB.

The Wing's September 1941 score amounted to 28-6-6, No.308 being the most successful with a score 16-3-3 enemy aircraft. No.315, with a score 8-2-1 and No.306, with 4-0-2, followed. The individual victories were:

- W/Cdr Tadeusz Rolski, Wing's C.O., 0-1-0;
- No.308 Sqdn: P/O Jerzy Popławski 4-0-0; F/O Franciszek Surma 2-2-0, S/Ldr Marian Pisarek 2-0-0; Sgt Stefan Krzyżagórski 1-0-1; F/Lt Stefan Janus, F/Lt Marian Wesołowski, P/O Czesław Budzałek, P/O Olgierd Iliński, P/O Edmund Krawczyński, P/O Stanisław Wandzilak and

[22] Combat Diary, Royal Air Force Northolt.

Sgt Ernest Watolski 1-0-0 each; Sgt Jerzy Zieliński 0-1-1; and Sgt Stanisław Piątkowski 0-0-1;
- No.315 Sqdn: S/Ldr Stanisław Pietraszkiewicz 2-0-1; F/O Jan Falkowski and Sgt Aleksander Chudek 2-0-0 each; Sgt Stanisław Blok 1-1-0; F/Lt Zbigniew Czaykowski 1-0-0, and Sgt Michał Cwynar 0-1-0;
- No.306 Sqdn: F/Lt Stanisław Skalski 2-0-0; P/O Ryszard Malczewski and Sgt Grzegorz Sołogub 1-0-0 each; S/Ldr Antoni Wczelik and Sgt Zygmunt Rozworski 0-0-1 each.
Polish casualties included:
- No.308 Sqdn: one pilot killed – P/O (S-Lt) Czesław Budzałek (17/09/41), and two made PoW – Sgt Henryk Kowalski and Sgt Ernest Watolski; three Spitfires lost;
- No.315 Sqdn: two pilots killed – F/O (Lt) Tadeusz Nowak (21/09/41) and F/O (Lt) Kazimierz Woliński (16/09/41), and one made PoW – S/Ldr Stanisław Pietraszkiewicz (he was succeeded as Squadron's C.O. by S/Ldr Szczęśniewski); three Spitfires lost;
- No.306 Sqdn: one pilot killed – Sgt (Cpl) Stanisław Wieprzkowicz (16/09/41), and one made PoW – P/O (S-Lt) Czesław Daszuta; two Spitfires lost.

The last quarter of 1941 saw a gradual change of emphasis from Circuses to fighter ground attacks – Rhubarbs, Rodeos and Ramrods – and despite the autumn weather, October was still eventful. The composition of No.1 Polish Fighter Wing changed on 7 October, when No. 306 Sqdn went to Speke for rest, replacing No.303 Sqdn, which came back to the front line. While at Speke, No.303 received old Hurricane Is with which it began routine patrols and convoy protection on 21 July, its last Hurricane sortie being flown on 24 August. Two days later the squadron commenced operations on Spitfire I fighters and

Sgt Tadeusz Schiele and his mechanic (No.308 Squadron) sitting on the engine of a Spitfire VB AB968 ZF•H (known as Haberbusch, as in the renowned Polish beer "Haberbusch & Schiele"), which bears Schiele's personal "The Great Bear" marking and large PAF chessboards; Northolt, the autumn of 1941. Schiele destroyed a Bf 109 and probably another in this aircraft on 24 October 1941. (K. Chołoniewski)

F/O Jan Falkowski, newly-appointed C.O. of No.315's 'A' Flight (on 22 September 1941), in a Spitfire VB embellished with his personal marking.

A Spitfire VB PK•T (No.315 Squadron), successfully nursed back to Northolt by Sgt Edward Jaworski after receiving a direct cannon hit in the starboard wingtip in a battle over St. Omer on 21 October 1941. Note the smaller PAF chessboard.

made its last mission on these machines on 1 October. Altogether, the squadron logged 58 operational flights at Speke, most of them in one-section strength, and left without aircraft. Upon its arrival at Northolt No.303 received Spitfire VBs.

No.303, flying Spitfire VBs, took part in wing operations for the first time on 12 October (Circus 107) and claimed one Bf 109 destroyed, No.308 accounting for two further Germans. The following day the squadron recorded its first encounter with Fw 190s and damaged one, while No.308 brought down three more Messerschmitts. In the later part of the month two Rhubarbs and two Rodeos were flown and heavy fighting was reported, which resulted in several victories, with No.315 Sqdn achieving a notable score 4-4-1 in an air battle over St. Omer on 21 October.

Offensive activities continued in November and December 1941 whenever opportunity arose, and the wing provided cover and support for Hurricane bombers (Hurribombers) on a few occasions. Alcohol distilleries were among the targets and one of them was set alight at Beaumont on 15 November by Polish pilots, who, mindful of a wastage of precious liquid, attacked it with heavy hearts. Most casualties were due to greatly strengthened German AA defences and for No.315 Sqdn 23 November was the blackest day in its history with five pilots killed on a single mission when they ran into a very heavy concentration of ground fire near Gravelines on Low Ramrod 12. On 12 December No.308 Sqdn was moved North (to Woodvale) for respite, and No.316 Sqdn was transferred from No.2 Polish Fighter Wing to Northolt to take its place. In November S/Ldr (Lt) Wojciech Kołaczkowski took over the command of No.303 Sqdn and S/Ldr (Mjr) Stefan Janus, the command of No.315 Sqdn. In December S/Ldr (Cpt) Marian Wesołowski became C.O. of No.308 Sqdn.

Wing's balance sheet for the last quarter of 1941 showed a score 24-9-5 for a loss of eight pilots and ten aircraft. No.308 Sqdn was in the lead again with a score 9-1-2, followed by No.315 Sqdn, with 8-6-2 and No.303 Sqdn, with 7-2-1. The credited victories were:

- No.308 Sqdn: F/Lt Stefan Janus 2-0-0; Sgt Tadeusz Schiele 1-1-0; P/O Jerzy Popławski 1-0-1; S/Ldr Marian Pisarek, F/O Franciszek Surma, P/O Jan Jakubowski, P/O Witold Retinger and Sgt Stanisław Piątkowski 1-0-0 each; P/O Tadeusz Stabrowski 0-0-1;
- No.315 Sqdn: P/O Włodzimierz Miksa 1-1-1; F/Lt Zbigniew Czaykowski 1-1-0; S/Ldr Stefan Janus, F/Lt Jan Falkowski, P/O Zygmunt Drybański, P/O Stanisław Łukaszewicz, Sgt Aleksander Chudek and Sgt Ryszard Malczewski 1-0-0 each; P/O Eugeniusz Fiedorczuk, P/O Józef Gil, P/O Władysław Zając and Sgt Zygmunt Gruszczyński 0-1-0 each; P/O Konrad Sembrowicz 0-0-1;

The poignant record of high casualties suffered by No.315 Squadron – the unit's personnel board with crossed off names of the pilots who failed to return from operations in the closing months of 1941 (Pilot's nicknames appear on the right of the list).

Sgt Bronisław Kościk, No.317 "Wilno" Fighter Squadron, entering his Hurricane at Acklington at the end of March 1941. (IPMS)

P/O Stanisław Bochniak (with an accordion) and P/O Ludwik ("Zośka") Martel with No.317's mascot "Lipa" at Acklington in the Spring of 1941. (S. Bochniak/W. Matusiak)

- No.303 Sqdn: F/O Jan Zumbach 2-0-1; Sgt Mieczysław Adamek 2-0-0; F/O Zygmunt Bieńkowski, P/O Bolesław Gładych and Sgt Marian Bełc 1-0-0 each; P/O Bolesław Drobiński and P/O Eugeniusz Horbaczewski 0-1-0 each.

The casualties comprised:
- No.308 Sqdn: one pilot killed – F/O (Lt) Franciszek Surma (08/11/41); one Spitfire lost;
- No.315 Sqdn: six pilots killed – F/O (Lt) Bernard Groszewski (08/12/41), F/O (Lt) Władysław Grudziński (23/11/41), P/O (Lt) Jan Grzech (23/11/41), P/O (S-Lt) Marian Łukaszewicz (23/11/41), Sgt (Cpl C-Off) Grzegorz Kosmalski (23/11/41), and Sgt (Cpl) Marian Staliński (23/11/41), and one made PoW – P/O Eugeniusz Fiedorczuk; eight Spitfires lost;
- No.303 Sqdn: one Spitfire lost.

No.317's 'A' Flight at readiness by a Hurricane II at Exeter in the early summer of 1941. Note the PAF chessboards in a new position, on the Hurricane's nose, with the inscription "Poland," introduced on aircraft of No.2 Polish Fighter Wing. (S. Bochniak/W. Matusiak)

No.2 Polish Fighter Wing began to organize at Exeter early in August 1941 and was formed on 18 August with the appointment of its Polish Commander W/Cdr (Mjr) Stanisław Brzezina, who initially shared the command with W/Cdr Blake. The wing comprised No.317 Sqdn at Exeter and No.302 and 316 Sqdns at Church Stanton, and its prime task was to provide daytime defence of the South Coast ports and the city of Exeter. No.307 (Polish) (Night Fighter) Squadron was charged with similar duties at night.

Pilots of No.317 Squadron, left to right: P/O Szulc, P/O Kratke, S/Ldr Brzezina, F/O Szczęsny and Sgt Kościk, by a Hurricane IIB at Exeter (the summer of 1941). (S. Bochniak/W. Matusiak)

No.317 "Wilno" Squadron moved to Exeter on 21 July 1941. No.317 had become operational on 23 March 1941 at Acklington – at first with one flight composed of three two-aircraft sections – and started routine patrol work. On 23 April the squadron's second flight reached operational status and six days later No.317 moved to RAF Ouston, near Newcastle upon Tyne. Besides patrol duties by day and on some occasions also by night, No.317 was involved in convoy protection sorties, which in the middle of 1941 became its main task. Its first clash with the enemy came on 2 June in the evening, when its eight aircraft, from both flights, were engaged on convoy patrol off Tynemouth. Black Section of 'B' Flight was vectored to investigate an unidentified aircraft, which turned out to be a Ju 88. In a lightning engagement, in which the Junkers took evasive action in cloud, F/O Paweł Niemiec and Sgt Tadeusz Baranowski fired good bursts, with Baranowski almost ramming the German. The Ju 88A-5, M2+DL, from 3/106, crashed in the sea, with two killed and two taken prisoner.

Towards the end of June No.317 Sqdn moved via Colerne to Fairwood Common. From the beginning of June until its departure from Fairwood Common the Squadron logged 201 sorties in 214 flying hours on convoy patrols from Ouston and Fairwood Common. On 10 July its Hurricane Is participated for the first time in escorting Blenheims on a Le Havre raid and during this operation S/Ldr Stanisław Brzezina destroyed a Bf 109 and F/Lt Henryk Szczęsny and Sgt Stanisław Brzeski jointly, another; Sgt Jan Malinowski damaged a further Messerschmitt. Four days later No.317 achieved its next victory, when F/Lt Szczęsny and Sgt Brzeski sighted a Ju 88 at

F/Lt Jerzy Jankiewicz (killed in action 25 May 1942) by a Hurricane of No.316 "Warsaw" Fighter Squadron during his visit to the unit in April 1941. Note the squadron badge under the cockpit. (T. Kopański)

sea level about to bomb a convoy south off Worms Head. They dived head on and fired short bursts, pursuing the Junkers over the convoy to which no damage was done. The German jettisoned his load and fell into the sea.

On 17 July No.317 Sqdn began to receive its first Hurricane IIs and in three days completed its re-equipment with the new model at Fairwood Common, arriving in Exeter with 16 Hurricane IIs and two Hurricane Is. Its first mission on Hurricane IIs, a convoy patrol, was flown from Exeter on 23 July, and the following day its fighters provided escort to bombers raiding Brest.

No.316 "Warsaw" Squadron moved to Church Stanton on 2 August. The squadron, which became operational on 25 March 1941 at Pembrey, was engaged mainly on shipping protection duties and achieved its first victory on 1 April in the late afternoon, when F/O Aleksander Gabszewicz (flying in V7000) and P/O Bohdan Anders (in V6635) encountered three Heinkels approaching a convoy off Linney Head and shot down one bomber in flames and attacked another before the Germans escaped in cloud. The He 111H-5, 1G+LH, from I/KG 27, crashed at Ballyristeen, Wexford, in Ireland and its crew was interned. The second victory followed on 9 April. F/O Tadeusz Sawicz sighted a low-flying He 111, coming out of the mist to bomb a convoy off St. Ann's Head and dived on it. The bomber's port engine burst into flames and its crew baled out and was taken prisoner, except for a gunner, who was killed. Later operations from Pembrey were largely uneventful despite their intensity – by the time of its move to Colerne on 18 June, the squadron had made 996 sorties in 1,420 flying hours on convoy patrols alone. On 2 June No.316 Sqdn suffered a sad casualty when P/O (S-Lt) Bohdan Anders, flying a Magister trainer, ran into barrage balloons over Newport and was killed.

On its arrival at Colerne No.316 Sqdn was rearmed with the Hurricane IIA fighters, which were later supplemented by the twelve-gun Hurricane IIBs, and the formal passing of the overall command of the squadron from S/Ldr Donovan to its Polish C.O., S/Ldr (Cpt) Juliusz Frey, coincided with the move.[23] At Colerne the squadron, while continuing with convoy patrol work, began exercises which were to prepare it for offensive operations across the Channel. Its first such mission – a close bomber escort to Le Havre in which No.317 Sqdn also took part – was made on 10 July, and the second, in which the squadron formed a part of the "first cover wing" – also with No.317 Sqdn – took it over Brest on 24 July. In an encounter with enemy fighters during the latter operation, S/Ldr Frey and F/O Gabszewicz destroyed jointly a Bf 109 and

A Hurricane II JH•R (No.317 Sqdn) being serviced in a blast pen at Exeter in the summer of 1941. (S. Bochniak/W. Matusiak)

Gabszewicz also damaged another. Earlier, on 14 July, F/O Włodzimierz Klawe was credited with a "probable" He 111 over a convoy off Lynton. During July the squadron also made some operational flights from Exeter, Warmwell, and Ibsley, and while based at Colerne logged 91 sorties in 127 flying hours on convoy duties.

At Church Stanton F/Lt (Cpt) Wacław Wilczewski succeeded S/Ldr Frey as No.316's acting Commander on 10 August, and the squadron began to operate more frequently with No.317 Sqdn. Both squadrons flying together took part in offensive sweeps over France on 17 and 26 August. At the same time both squadrons were still individually involved in convoy protection duties. On 20 August S/Ldr (Cpt) Henryk Szczęsny became No.317's new C.O.

No.302 "Poznań" Sqdn was transferred from the Isle of Man to Church Stanton on 7 August without aircraft and upon its arrival received Hurricane IIA and IIB fighters. The squadron completed the composition of the newly organized No.2 Polish Fighter Wing, whose three squadrons were all now equipped with Hurricane IIs.

The wing made its first operation at three-squadron strength on 4 September, taking off from Warmwell under the command of S/Ldr Stefan Witorzeńc and flying high escort for six Blenheim bombers on Gudgeon VI – an attack on ships and docks at Cherbourg. No.302 Sqdn engaged Messerschmitts over the target area, with S/Ldr Witorzeńc accounting for one Bf 109 destroyed and P/O Kazimierz Sporny, for one "probable." During September and the early part of October the squadrons flew a few more offensive sorties, some of them from Bolt Head – a primitive airfield on the Devon coast, providing a good jumping-off point for strikes across the Channel. Early in October No.302 Sqdn moved to Harrowbeer.

In the middle of October 1941 No.317 Sqdn at Exeter and No.316 at Church Stanton began to convert to Spitfire VBs and received at first a few Spitfire IIs for practice flying. No.317 took delivery of the first four Mk VBs on 13 October and

[23] This date of Donovan's departure is incorrectly challenged by Bernard Buchwald – a former pilot of No.316 Sqdn – in his book "316 Warszawski Dywizjon Myśliwski" (MON, Warsaw, 1989), in which he maintains that Donovan "was absent from the squadron already since mid-April." In fact, on 17 April Donovan suddenly refused to recognize the F/O ranks of No.316's Polish officers and this led to a drawn and unpleasant disagreement which had to be referred to No.10 Group HQ. It appears that later in May Donovan's command activities were restricted, but according to official records (e.g. No.316 Operational Diary, vol. 23 Feb. to 11 Dec. 1941, p.123 – IPMS, Lot.A V,54/3), he was not removed from the post until 22 July 1941.

No.316 received on the same day three Spitfire IIs[24] and these were followed two days later by the first Mk VBs. The two units completed re-equipment with Spitfire Vs before the end of October. At the same time No.302 Sqdn went to Ibsley where it was similarly equipped with Spitfire VBs (with some Spitfire IIs for training). The squadron returned to Harrowbeer on 5 November.

A major operation by No.2 Polish Fighter Wing, with the three Spitfire squadrons taking part, was carried out on 8 November. The Wing, led by W/Cdr Stanisław Brzezina was sent to Redhill, and took off from there just before 11:00 hours to act as escort cover on Circus 110. Fw 190s and Bf 109s clashed with the formation after bombs were dropped on a factory in Lille and many combats took place. Sgt Bronisław Kościk attacked an Fw 190 at 80 yards range and sent it straight to the ground, Sgt Stanisław Brzeski accounted for a Bf 109F destroyed and an Fw 190 damaged, and F/O Tadeusz Koc for one Bf 109F "probable" – all from No.317 Sqdn, which suffered no casualties. S/Ldr Wacław Wilczewski, of 316 Sqdn, was shot down and made PoW (he was succeeded by S/Ldr (Cpt) Aleksander Gabszewicz as Squadron C.O.), and so was P/O Zbigniew Gutowski of No.302 Sqdn.

On 11 November, Poland's Independence Day, President Raczkiewicz visited Exeter and held an investiture. S/Ldr Henryk Szczęsny was decorated with the Virtuti Militari and the Cross of Valor, and F/Lt Józef Brzeziński, F/Os Ludwik Martel, Paweł Niemiec, Piotr Ostaszewski, Jerzy Solak and Marian Trzebiński, and Sgts Tadeusz Baranowski, Stanisław Brzeski, Marian Domagała, and Henryk Trybulec, with the Crosses of Valour. S/Ldr Stefan Witorzeńc was promoted to Mjr – and on 28 November became W/Cdr and succeeded Brzezina as Commander of No.2 Polish Fighter Wing, the command of No.302 Sqdn passing to S/Ldr (Cpt) Julian Kowalski; F/Lt (Cpt) Józef Brzeziński was appointed S/Ldr and took over the command of No.317 Sqdn from Szczęsny.

Little flying was possible in the second half of November and only a few sector and convoy patrols were made. These cost the life of P/O (S-Lt) Karol Wójcik of No.317 Sqdn, who crashed on patrol duty in bad visibility on 29 November.

Operations were similarly restricted in December, except for three days. On 6 December, Green Section of 'B' Flight, led by Sgt Stanisław Brzeski on an exercise flight, was vectored to Bolt Head at 11:45 hours and sighted a Ju 88 near Plymouth. The enemy adopted a violent evasive action, seeking safety in cloud, but Brzeski stuck to it, firing cannon and machine guns and breaking off only 30 yards behind. The German – Ju 88D-2 4U+CH from 1(F)/123 – began to break up and hit the sea. Its crew perished. Less than half an hour later, three sections of No.317 Sqdn went on a Rhubarb and set on fire two specified targets (depots and storage tanks) in Northern France.

[24] P7738, P7893, and R6960 – some publications list these incorrectly as operational aircraft.

A Hurricane IIB SZ•S (No.316 Squadron), serviced at Church Stanton in the late summer of 1941. The aircraft carries the unit badge under the radio mast and the PAF chessboards and inscription "Poland" on the engine cowling. (T. Kopański)

On 12 December the composition of the wing changed, with No.316 Sqdn departing for Northolt and No.306 Sqdn arriving at Church Stanton – after two months rest at Speke – to take its place. On leaving Speke, No.306 handed over its Spitfire Is and IIs to Woodvale, where they were passed on to No.308 Sqdn. At Church Stanton No.306 Sqdn received Spitfire VBs and began Sector Recognition flights on 13 December, and convoy patrols four days later.

On 18 December No.2 Polish Fighter Wing was involved in Operation Veracity I – the first major assault (by 18 Stirlings, 18 Halifaxes, and 11 Manchesters) on German cruisers and docks at Brest. The Wing was to cover the return of the bombers by patrolling 80 miles off Bolt Head. No.317 Sqdn lost two aircraft on take off, when F/O (Lt) Marian Sikorski stopped on the runway owing to ground mist and F/O (S-Lt) Jerzy Solak, commencing his run, hit the standing Spitfire and killed its pilot, receiving himself slight injuries. No.317's Red Section (four aircraft) and No.306's Yellow and Blue Sections (five aircraft) provided the lower cover, and No.302 (11 aircraft), the upper cover.

The lower cover, led by W/Cdr Witorzeńc, spotted 15 bombers being attacked by Bf 109s and dived on the Germans who were taken by surprise and broke formation. Individual combats followed and the Poles shot down four enemies and damaged one – F/O Kazimierz Rutkowski and Sgt Leon Kosmowski of No.306 and F/O Tadeusz Kumiega and P/O Zbigniew Janicki of No.317 accounted for one Bf 109 each, and S/Ldr Józef Brzeziński of No.317, for a further one damaged. Several other pilots fired at the Messerschmitts but examinations of camera-gun films proved inconclusive. One bomber crashed in the sea and No.317's Green Section protected the rescue operation.

Veracity II followed on 30 December and No.306 (13 aircraft) and No.317 (12 aircraft), led by W/Cdr Stefan Witorzeńc, again took part, operating from Bolt Head as the first cover wing on patrol 10 miles North of Brest. No.302 (12 aircraft), taking off from Harrowbeer and led by S/Ldr Julian Kowalski,

Bird's-eye view of a Spitfire VB AD140 JH•H (No.317 Sqdn); Exeter, late 1941.

provided close escort to the bombers. Flying in a loose line abreast, the cover wing saw ahead the bombers on the point of being attacked by a large force of Messerschmitts and, in a precisely executed plan of engagement, drew the Germans into combat, while No.302 Sqdn struck from above. Not a single enemy fighter was able to attack the bombers and the fierce combat resulted in a resounding Polish victory. No.302 Sqdn claimed three Bf 109s destroyed – one each by F/O Czesław Główczyński, F/O Kazimierz Sporny and P/O Marian Rytka, two "probables" – by F/O Alojzy Rodziewicz and Sgt Bronisław Malinowski, and one damaged – by P/O Ryszard Budrewicz; No.306 Sqdn, three Bf 109s destroyed – one each by S/Ldr Antoni Wczelik, F/Lt Kazimierz Rutkowski and Sgt Jan Śmigielski, two "probables" – by P/O Grzegorz Sołogub and P/O Sergiusz Szpakowicz, and one damaged – by P/O Zdzisław Langhamer; No.317 Sqdn, two Bf 109s destroyed – by F/Sgt Mirosław Maciejowski. Polish casualties comprised one pilot of No.306 Sqdn – F/Lt (Cpt) Stanisław Zieliński – in killed and one pilot of No.317 Sqdn – Sgt Tadeusz Baranowski – made PoW; two aircraft were lost.

With the formation of new Polish squadrons, the number of Poles serving with British squadrons gradually decreased in 1941, dropping to only about a dozen by the end of the year. The Polish pilots in these units added 17 more enemy aircraft destroyed and a few further "probable" and damaged to the Polish score. Most successful among them was Sgt Adolf Pietrasiak, who, flying with No.92 Sqdn, achieved 6 confirmed victories in July 1941 – all Bf 109s – before moving to No.308 Sqdn. Polish losses included F/Lt Stanisław Z. Król, serving with No.74 Sqdn, who was shot down in combat with Bf 109s over France on 2 July 1941 and made PoW.[25]

In the spring of 1941 an agreement was reached between Gen Ujejski (the Polish Air Force) and Gen Janoušek (the Czechoslovak Air Force) on interchange of fighter pilots between the Polish and Czechoslovak squadrons in order to promote better links between the two air forces. Air Ministry permission was obtained for the proposed exchange of one pilot between No.303 (Polish) and No.312 (Czechoslovak) Sqdns and one between No.306 (Polish) and No.310 (Czechoslovak) Sqdns.

The Czechs asked for F/O (S-Lt) Jan Zumbach (No.303 Sqdn) and P/O (S-Lt) Władysław Różycki (No.306 Sqdn), but because of shortages in the Polish squadrons, the PAF HQ was unable to spare these officers and eventually only one exchange took place. On 15 May 1941 F/O (Cpt) J. Sudený from No.310 Sqdn was assigned to No.306 Sqdn for one month and flew 15 operational sorties with the Polish unit, and on 31 May 1941 P/O (S-Lt) Wilhelm Śniechowski, a Polish pilot who had served with No.32 (RAF) Sqdn since October 1940, was posted to No.310 Sqdn. Śniechowski stayed with No.310 Sqdn until his death on a training flight on 8 December 1941, when his Spitfire IIA (NN•W, serial AD468) went into a spin and crashed at Dyce airfield (Scotland).[26]

The Polish contribution to the RAF fighter offensive in 1941 was very formidable. Out of seven Polish fighter squadrons, six were in the front line – operating within No.11 and No.10 Groups – for most of the year. No.1 Polish Fighter Wing took part in almost three-quarters of all Circuses flown by Fighter Command. A great effort was also put into shipping protection operations, particularly by Nos.316 and 317 Sqdns which alone flew more than 1,700 aircraft sorties in about 2,400 hours in the nine month period (April-December 1941), with No.317 achieving a record with 558 sorties on convoy duties in the space of 32 days (in August-September).

The RAF overall claim for 1941 comprised 752 enemy aircraft deemed destroyed, 328 "probable" and 421 damaged. Of these 198-52-57 1/2, representing 26 percent of the Germans destroyed, were credited to Polish pilots. Most of these victories were earned in the demanding fighter-versus-fighter combat. In addition four aircraft were destroyed and three damaged on the ground. No.308 Sqdn, with 52-10-13 enemies to its credit, became not only the most successful Polish squadron, but also achieved the year's highest score in Fighter Command. Although the RAF figures proved later to be grossly exaggerated, and the actual Luftwaffe losses were less than a quarter of the RAF assessment, this did not alter the share of the Polish contribution.

The Polish victories were hard-won and their casualties heavy. Forty-one pilots were killed in action, including a Wing Commander and three Squadron Leaders, and three more lost their lives in accidents on operational flights; 15 pilots were missing and made PoW, including two Squadron Leaders, and several others were wounded, four of them seriously. One Hurricane and over 70 Spitfires were lost.

In terms of operational effort, in 1941 the Polish fighter squadrons (including No.307 Night-fighter Squadron) put up 13,032 aircraft sorties in 16,722 flying hours, with No.316 Sqdn in the lead with 2,460 sorties in 3,428 hours (for further details see Appendix 22).

The RAF daytime offensive, which began with tentative low-level ground attacks by a few fighters (Rhubarbs) and grew in strength in the later part of 1941 involving small numbers of bombers escorted by large fighter formations (Circuses), entered 1942 with gradually increasing emphasis on bomber operations across the Channel. These were rapidly boosted by the arrival of the USAAF 8th Air Force in Great Britain in July 1942.

The first half of 1942 saw – besides the usual Circuses – mounting numbers of Ramrods, which, although generally similar to Circuses, included more substantial bomber forces, and of Rodeos, which were designed to lure the Luftwaffe fighters into a battle of attrition and thereby assure local air superiority

No.303's F/Lt Witold ("Tolo") Łokuciewski, C.O. of 'A' Flight, S/Ldr Wojciech Kołaczkowski and F/Lt Walerian Żak, C.O. of 'B' Flight, by Kołaczkowski's Spitfire VB "Wojtek" (the name, diminutive for Wojciech, appearing on the engine cowling); Northolt, March 1942. (W. Kołaczkowski)

[25] F/Lt Król took part in the mass escape of Allied airmen from the Stalag Luft III at Sagan on 24 March 1944, but was caught and executed by the Gestapo three weeks later.

[26] Gen Ujejski's letter of 21 May 1941 and other information on the subject were kindly supplied to the author by Zdenek Hurt from Prague.

A close-up of the cockpit of No.303's Spitfire VB AA882 RF•G, which was funded by the Borough of Wanstead & Woodford; Northolt, the early spring 1942. (IPMS)

Another donated Spitfire VB, "Hendon Lamb," depicted at Northolt at the same time. F/O T.B. Simm, a RAF pilot briefly serving with No.303 Sqdn, is sitting in the cockpit. (IPMS)

preparation for a planned probing invasion exercise on the coast of France in the summer of 1942. The RAF operations associated with the landing provided the high point of air activities in 1942 and combat over Dieppe on 19 August was one of the RAF's greatest air battles of the war, at least in terms of numbers of aircraft involved. Five Polish fighter squadrons made an outstanding contribution to this operation and gave an excellent account of themselves in the fighting.

Little flying was possible in the early months of 1942 owing to the very sharp winter. January was a very dead month and activities were somewhat revived in February. No.1 Polish Fighter Wing, at Northolt, played a small part in searching for the German battleships (Operation Fuller), which successfully slipped out of Brest and sailed through the Channel to Norwegian waters on 12 February. The following day, during a sweep in search of shipping Sgt (Sgt) Jan Musiał, of No.316 Sqdn, was killed in a mid-air collision and two Spitfires were lost. No.316 Sqdn suffered further casualties in the late part of the month: on the 20th Sgt (LAC) Walenty Kopa, hit by AA fire, ditched in the sea and was never found, and eight days later, in attacks on shipping at Ostend (Ramrod 13), F/O Józef Górski and P/O (S-Lt) Tadeusz Dobrut-Dobrucki were lost in similar circumstances.

Four Circuses (112-115) – the last one combined with Ramrod – were made in the first half of March 1942, two being led by W/Cdr Tadeusz Rolski and two by S/Ldr Wojciech Kołaczkowski of No.303 Sqdn The wing's 38 Spitfires, escorting Bostons over Hazebrouck on 13 March (Circus 114), fought with Messerschmitts, and S/Ldr Kołaczkowski (No.303), F/O Bolesław Drobiński (No.303) and F/O Mieczysław Wyszkowski (No.316), accounted for one Bf 109 each and F/O Maciej Lipiński (No.303), for one more "probable." F/Lt Witold Łokuciewski (No.303) was shot down and made PoW and F/O Eugeniusz Horbaczewski's aircraft was damaged by enemy fighters. Three sweeps and a Ramrod were carried out later in March, with a rendezvous by the wing's three Polish squadrons over London's Hyde Park to put up a show for the public. Two days later, during a brief skirmish of No.303 Sqdn with Fw 190s, P/O Antoni Kolubiński damaged one German, but F/O (Lt) Aleksander Wróblewski was shot down and drowned. A larger encounter with Fw 190s took place on 27 March, when the Wing acted as forward support to Ramrod 18, attacking targets in Ostend. No.316 Sqdn, flying on top, dived on the Germans and chased them off, damaging four – credited one each to S/Ldr Aleksander Gabszewicz, F/O Maciej Piotrowski, F/O Tadeusz Szumowski and Sgt Marian Żółciński

A Vic of three Spitfire VBs of No.317 Squadron in flight in the early months of 1942. (IPMS)

– without loss. F/Lt Zbigniew Czaykowski (No.315) was credited with one Fw 190 "probable."[27]

Operations of No.2 Polish Fighter Wing, with HQ at Exeter, were similarly curtailed by the winter weather, but F/Lt Piotr Ozyra and F/O Stanisław Łukaszewicz of No.317 Sqdn managed to destroy a marauding Ju 88 on convoy patrol on 11 January 1942. No.302 Sqdn suffered two casualties on a shipping reconnaissance mission in the Brest area on 26 January, when F/Lt (Cpt) Bronisław Kosiński and P/O (Lt) Zdzisław Bednarski were shot down and killed by German fighters. In addition to convoy sorties, the Wing's squadrons patrolled the Exeter and Bolt Head areas and in March bomber escorts were also flown. On 7 March No.317 lost F/Sgt (Sgt) Bolesław Kościk, whose Spitfire developed engine trouble on a defensive patrol and crashed near Marlborough. On 15 March, on escort to Bostons attacking enemy shipping in the Channel (Roadstead 12), No.306 and 317 Sqdns ran low on fuel and disaster struck the latter unit, which missed its airfield in poor visibility. Ten of its 12 aircraft crashed in forced landings and S/Ldr (Cpt) Józef Brzeziński was killed and three pilots injured, command of the squadron passing to F/Lt (Lt) Piotr Ozyra.

For No.308 Sqdn, sent to Woodvale (No.9 Group) for rest and provided there mainly with Spitfire IIs, the beginning of 1942 was marred by the death of S/Ldr (Cpt) Marian Wesołowski in a mid-air collision during a night exercise flight on 9 January. He was succeeded by S/Ldr (Cpt) Tadeusz Nowierski. From the middle of February the squadron began to receive Spitfire VBs. In the first quarter of 1942 No.308 flew mostly uneventful patrols over the Liverpool-Preston area, including some at night, and over convoys docking at Liverpool. Two encounters with single Ju 88s were reported and in the second of these, on 21 February, Sgt Stanisław Zieliński, flying a Spitfire VB, claimed one Junkers "damaged" off Barrow-in- Furness.

On 1 April 1942 both Polish wings changed their composition. No.315 Sqdn left Northolt for Woodvale for rest, and No.317 Sqdn, from No.2 Wing was transferred to No.1 Wing at Northolt, to replace it. At the same time No.308 was moved from Woodvale to Exeter, to replace No.317 Sqdn in No.2 Wing.

In April the air offensive was resumed in earnest and considerable fighting took place. W/Cdr Tadeusz Rolski left No.1 Polish Fighter Wing to join the operational Staff at No.11 Group and W/Cdr (Mjr) Marian Pisarek took over, but he held the command only briefly, being killed in action ten days later. Heston was opened as a Northolt satellite airfield and No.316 Sqdn was transferred there on 22 April.

After Ramrod 19 on 1 April, flown by No.303 and 316 Sqdns, the Wing, led by W/Cdr Rolski, took part in Circus 119 on 4 April at full strength, including, for the first time, No.317 Sqdn The Poles met two formations of Fw 190s; the first engaged No.303 while the second attacked the other two squadrons and attempted to get to the bombers. In the dog fight near St. Omer the wing destroyed three Germans with a further three "probables," but lost two pilots killed – one each from No.303 and 316 Sqdns – and one made PoW. Two Rodeos followed and on the second, on 10 April, a short but effective battle was

[27] Combat Diary, Royal Air Force Northolt, and RAF Form 540.

Polish sailors by AB895 PK•X during a visit to No.315 Squadron at Woodvale in April 1942.

fought, bringing a notable success to No.316 Sqdn whose score amounted to 4-4-3 Fw 190s for no loss. No.317 Sqdn added 0-1-1 enemy fighters to the victory.

During eight days, starting with 12 April, the wing made 11 operations – Circuses, Rodeos and Ramrods, led by W/Cdr Rolski, S/Ldr Kołaczkowski or S/Ldr Ozyra – which totaled 384 sorties and included hard-fought engagements, resulting in casualties to both sides. After a brief interlude, activities of similar intensity were conducted in the last week of April, with W/Cdr Pisarek leading the wing into action for the first time on the 24th, on Circus 132. On 29 April, on Circus 145, the Wing was attacked by an overwhelming force of Fw 190s and suffered the sad loss of W/Cdr Pisarek and S/Ldr Ozyra. F/O Tadeusz Koc (No.317), fighting with seven Focke-Wulfs, fired at one head-on up to 50 yards and flew through the explosion and debris of his prey, receiving some damage to his aircraft. F/Lt (Lt) Stanisław Skalski, a few days later promoted to S/Ldr, became the new C.O. of No.317 Sqdn.

The wing's record – which was also the Northolt Sector's record – in April showed 26 operations, totaling 899 sorties, with a balance sheet of 24-13-11, with No.316 Sqdn in the lead with a score of 11-5-7. Its casualties were nine pilots lost (killed or made PoW) and 11 Spitfire VBs destroyed. The individually credited victories were:

- W/Cdr Tadeusz Rolski, Wing's C.O., 0-2-0;
- No.316 Sqdn: S/Ldr Aleksander Gabszewicz 3-0-0, Sgt Tadeusz Szymański 2-1-0, P/O Mieczysław Wyszkowski 2-0-1, F/O Józef Dec 1-1-0, F/Lt Stanisław Skalski 1-0-1; F/O Tadeusz Szumowski and Sgt Karol Sumara 1-0-0 each; F/O Antoni Cholajda and P/O Jerzy Szymankiewicz 0-1-1 each; Sgt Marian Żółciński 0-1-0; F/O Jerzy Radomski, Sgt Tadeusz Góra and Sgt Władysław Kiedrzyński 0-0-1 each;
- No.317 Sqdn: F/O Tadeusz Koc 2 1/2-0-1, Sgt Adam Kolczyński 1-1-0, F/Lt Marian Duryasz, Sgt Stanisław Brzeski and Sgt Ignacy Olszewski 1-0-0 each; F/O Zygmunt Słomski 1/2-0-1/2; F/O Stanisław Łukaszewicz 0-1-0; F/O Jerzy Mencel and F/O Janusz Walawski 0-0-1 each; F/O Tadeusz Kumiega 0-0-1/2;
- No.303 Sqdn: F/Lt Zbigniew Kustrzyński and F/O Eugeniusz Horbaczewski 2-0-0 each; F/Sgt Mieczysław Popek and Sgt Mieczysław Adamek 1-0-0 each; F/Lt Jan Zumbach, F/O Maciej Lipiński, P/O Antoni Głowacki and P/O Stanisław Socha 0-1-0 each.

Polish casualties comprised:
- W/Cdr (Mjr) Marian Pisarek, killed 29/04/42; one aircraft lost;
- No.316 Sqdn: one pilot killed – F/O (Lt) Jan Muszel (04/04/42), and one made PoW – F/O Bernard Buchwald; two aircraft lost;
- No.317 Sqdn: three pilots killed – S/Ldr (Lt) Piotr Ozyra (29/04/42), P/O (S-Lt) Zbigniew Borusiewicz (28/04/42) and Sgt (LAC) Edward Olszewski (14/04/42); three aircraft lost;
- No.303 Sqdn: one pilot killed – F/Lt (Lt) Jan Daszewski (04/04/42), and two made PoW – F/Lt Zbigniew Kustrzyński and F/O Stefan Kołodyński; five aircraft lost.

No.303's AB824 RF•S, shot down by Fw 190s on Circus 119, on 4 April 1942, being retrieved by the Germans. Its pilot, F/Lt Zbigniew Kustrzyński, was taken prisoner. (T. Kopański)

Spitfire VB AD130 SZ•E (No.316 Sqdn), shot down over France on 12 April 1942. The aircraft was flown by F/O Bernard Buchwald who was taken prisoner. (M. Payne/W. Matusiak)

Pilots of No.303 Sqdn, the victors of the Fighter Gunnery Contest, run by No.11 Group on 11 April 1942. The Cup, awarded to the unit a month later, is held by F/O Bronisław Kłosin, with F/Lt Bieńkowski and F/Lt Zumbach standing on his sides. No.316 and 315 (Polish) Squadrons were placed second and third respectively among the 22 competing units.

RF
M
AR335

RF
M
AR335

In spite of its extensive operational commitments, No.11 Group organized a target shooting contest, which was held on 11 April with all the Group's 22 squadrons taking part. The results astounded Fighter Command. The first three places were taken by the Poles, No. 303 scoring 808 points, No.316, 432 points and No.315, 183 points. The best RAF squadron came fourth with 150 points. A message of congratulations from AVM Sir Trafford Leigh-Mallory, Air Officer Commanding No.11 Group, was received by the three Polish squadrons. The accurate shooting of the Poles revealed by the contest was one of the reasons behind the exceptionally high numbers of enemy aircraft they destroyed in combat.

On 6 May W/Cdr (Mjr) Stefan Janus arrived at Northolt to take charge of No.1 Wing and S/Ldr (Cpt) Tadeusz Nowierski became his No.2. These appointments led to other personnel changes: S/Ldr (Mjr) Mieczysław Wiórkiewicz replaced Janus as C.O. of No. 315 Sqdn and S/Ldr (Cpt) Feliks Szyszka took over command of No.308 Sqdn from Nowierski. The following day, with S/Ldr (Cpt) Wojciech Kołaczkowski resting after completion of his tour of duty, S/Ldr (Cpt) Walerian Żak became the new C.O. of No.303 Sqdn, only to be moved less than a fortnight later to assume command of No.308 Sqdn after the death of S/Ldr Szyszka on 17 May. S/Ldr (Lt) Jan Zumbach followed Żak as the next Commander of No.303 Sqdn.

The first week of May 1942 saw intensive operations continuing, with five Rodeos and four Circuses put up by the wing – all at three-squadron strength – in the first seven days of the month. Enemy fighters were engaged on three occasions. On 3 May, the wing – acting with two other wings as a diversion to the main operation (repeat of Circus 145) – engaged Fw 190s which attacked No.303 in the Boulogne-Calais area and shot down Sgt Mirosław Cybulski, who was made PoW. In the ensuing combat P/O Czesław Jaworowski destroyed one enemy, F/Lt Stanisław Skalski claimed one "probable" and Sgt Władysław Kiedrzyński, one damaged – all from No.316 Sqdn.

On 5 May, on Rodeo 19, No.316 Sqdn's pilots scored again, Sgt Kiedrzyński accounting for one Fw 190 shot down and S/Ldr Gabszewicz, leading the wing, for another "probable" over Boulogne. The following day, on Rodeo 21, No.303 Sqdn lost one pilot, F/O (Lt) Witold Domański, who dived into the sea, apparently unconscious through the loss of his oxygen supply. Later on the same day, when the wing flew escort to Boston bombers (Circus 160), No.316 Sqdn dispersed the attacking Fw 190s and probably destroyed one, which was jointly credited to S/Ldr Aleksander Gabszewicz, F/O Jerzy Radomski and F/O Władysław Walendowski.

On 7 May No.1 Wing was reinforced by No.302 Sqdn, transferred to Heston from No.2 Wing as part of preparations for the anticipated Dieppe landing, but bad weather in the rest of May reduced activity to a few uneventful missions. The month ended with only one brief, but sharp, combat with the enemy on 31 May, which resulted in one pilot, Sgt (LAC) Jan Gałązka from No.302 Sqdn, shot down and killed, and one Fw

This page and opposite: A sequence of photographs of a Spitfire VB AR335 RF•M being prepared for a sortie at Northolt in April 1942. The PAF chessboards, then newly introduced on No.303's aircraft noses, are well visible on the fighter.

Pilots of No.303 Squadron, with the unit's departing C.O., S/Ldr Kołaczkowski in the middle, flanked by F/Lt Zumbach, the newly appointed C.O. of 'A' Flight on his right, and F/Lt Żak, on his left; Northolt, early May 1942.

Above left: Left to right: F/Lt Zumbach, S/Ldr Żak, a new C.O. of No.303 Sqdn, and F/Lt Bieńkowski, a new C.O. of 'B' Flight, standing by the new Spitfire VB "Wojtek" BL670 RF•K ("Krysia") at Northolt in early May 1942 (the aircraft carried two names). (IPMS)

Above right: Zumbach's personal marking "Donald Duck," as it first appeared on his Spitfire VB BM144 RF•D in the spring of 1942, with the baseball bat down.

Right: Briefing of a pilot before a flight at Northolt in May 1942 beside No.303's Spitfire BL670 RF•K (note a larger size of the inscription "Wojtek" on the engine cowling as compared with that appearing on the earlier aircraft). Affectionately known as "Ever Ready "Krysia"," it eventually accounted for four Germans destroyed for certain.

Two above and right: W3970 JH•Y, displaying unusually small RAF cockades on the fuselage and wings, which appeared briefly on some of No.317's Spitfire VBs at a time of the introduction of the new style RAF roundels in May 1942.

190 claimed damaged by F/O Maciej Lipiński from No.303 Sqdn.

During the 16 operations carried out in May 1942, the last of them at four-squadron strength, the wing scored 2-3-2 victories, all but one damaged aircraft being credited to No.316 Sqdn. Its own casualties amounted to 3 pilots and 3 aircraft lost in action. Non-operational losses included F/O (Lt) Sylwester Godlewski from No.302 Sqdn, who crashed from a high altitude, presumably starved of oxygen.

June 1942 was dominated by preparations for the impending combined operation at Dieppe. With the realization that the Fw 190 fighter was superior to the Spitfire V on every count, squadrons were instructed to avoid engaging the enemy in combat, but to keep him constantly on the jump, pending re-equipment with the Spitfire IX. In the first week of the month eight major operations were carried out, six of them at four-squadron strength, and a few minor ones.

On 3 June No.316 Sqdn fended off 15 Fw 190s threatening Bostons over Le Havre on Circus 184; Sgt Tadeusz Góra claimed one German destroyed and P/O Mieczysław Wyszkowski, one "probable." Two days later larger encounters were fought over a widespread area between Caen and Le Havre on Circus 188, and three Fw 190s were destroyed by No.303 Sqdn, F/O Bolesław Gładych, P/O Marian Szelestowski and Sgt Józef Stasik being the victors, and one was destroyed and another "probable" by No.316 Sqdn, these being credited to F/Lt Wacław Król and P/O Kazimierz Samofał respectively. P/O (S-Lt) Antoni Łysek from No.302 Sqdn was shot down and killed. On 6 June S/Ldr (Cpt) Janusz Żurakowski took over command of No.316 Sqdn from S/Ldr Gabszewicz. During the rest of the month activities were greatly reduced and no contact with the enemy was made. Some "Feint Rodeos" were flown – one together with a Special Defiant Echelon – to confuse the enemy radar system.

In the middle of June a further re-alignment of forces took place. On 15 June No.303 Sqdn went to Kirton-in-Lindsey for a well deserved rest, changing places with No.306 Sqdn which came to Northolt. Croydon and Redhill were earmarked as operational airfields for the Dieppe operation and the squadrons of No.1 Wing from Northolt and Heston were originally timed to move to Croydon on 16 June. No.308 Sqdn was to come to Redhill on the same day to boost the strength of No.1 Wing to five squadrons. However, these moves were postponed until 30 June. Croydon was not well suited for Spitfire units and three aircraft crashed on landing. The squadrons at Croydon and Redhill remained under the operational control of Northolt.

Until early May 1942 No.306 and 308 Sqdns formed a part of No.2 (Exeter) Wing, which in April, operating at three-squadron strength, was increasingly engaged in Circuses, Rodeos and fighter sweeps, at first giving support to No.11 Group and later participating in those put up by No.10 Group. In addition to these exhausting missions, which were flown from various airfields, such as Tangmere, Ford, Redhill and others, the wing's squadrons carried on with defensive patrols of the Exeter area (including some at night) and convoy protection duties. A number of ferocious engagements were fought.

On 14 April, on Circus 123, No.306 Sqdn flying top cover to 12 Bostons was suddenly attacked on the way back over the sea by Messerschmitts, which shot down and killed two pilots: S/Ldr (Mjr) Antoni Wczelik and F/O (Lt) Edward Jankowski. In the battle that followed F/Lt Józef Żulikowski[28] and F/O Adam Fliśnik destroyed one Bf 109 each. S/Ldr (Cpt) Tadeusz Czerwiński succeeded Wczellik as the new C.O. of No.306.

[28] Shot down over France on 28 June 1941, Żulikowski managed to escape soon after capture with the help of the French resistance movement and returned to England (via Spain) to rejoin his squadron.

No.303's Spitfire Vs at Kirton-in-Lindsey in the summer of 1942, with BM531 RF•V in the foreground. (IPMS)

A Vic of three No.303's Spitfire Vs about to take off from Kirton-in-Lindsey, with AB183 RF•A nearest, AR371 RF•B behind, and Zumbach's BM144 RF•D "Donald Duck" in the lead.

No.302 "Poznań" Fighter Squadron reinforced No.1 Polish Fighter Wing at Heston in May 1942. Depicted on the ground is WX•N.

Right: Spitfire VB W3902 WX•T showing to advantage the wing detail and cannon and wheel blisters; Heston, May 1942.

Below: A fitter disconnects the trolley accumulator as the engine of W3902 WX•T fires in preparation for a sortie from Heston in May 1942.

W3902 WX•T rolls to the take-off point; Heston, May 1942.

On 16 April, flying with Bostons to Le Havre on Ramrod 20, P/O Grzegorz Sołogub of No.306 claimed two Bf 109s and P/O Brunon Kudrewicz of No.308 a further Bf 109. The following day, on No.10 Group's Circus 2, No.302 Sqdn destroyed one Bf 109 and damaged another near Cherbourg – the credits going to P/O Marian Rytka and P/O Erazm Wardziński respectively – and Sgt Zdzisław Horn of No.306 brought down one more Messerschmitt in that same area.

On 22 April F/O Tadeusz Stabrowski and P/O Stanisław Madej of No.308 Sqdn damaged a Ju 88 on a defensive patrol, and on 3 May F/O Stabrowski and F/Sgt Władysław Majchrzyk damaged another Junkers and silenced its dorsal gunner, but after a long chase to the Dutch coast they had to turn back because of approaching nightfall. A few days earlier, on 25 April, No.2 Wing provided escort to Bostons on No. 10 Group's Circuses 3 and 4. The following day, while on Rodeo sorties overt Cap Gris Nez, No.302 and 306 Sqdns were in combat with Fw 190s – S/Ldr Tadeusz Czerwiński of No.306 shot down one German, and S/Ldr Julian Kowalski of No.302 was credited with one damaged. F/O (Lt) Adam Fliśnik of No.306 was killed in this action. On 30 April the wing escorted bombers over Morlaix on No.10 Group's Circus 5.

On 3 May 1942, No.306 Sqdn was transferred north, to Kirton-in-Lindsey airfield within No.12 Group. On 7 May No.302 Sqdn went to Heston to reinforce No.1 Wing, and No.308 Sqdn departed to Hutton Cranswick in Yorkshire, also within No.12 Group. At the same time No.2 Polish Fighter Wing was temporarily disbanded. Thereafter until joining No.1 Wing, both No.306 and 308 Sqdns continued to be chiefly engaged on sector patrol and convoy protection duties and participated in offensive operations across the Channel only occasionally. During the night of 30/31 May both squadrons were sent on patrols over the North Sea to provide extended cover to the bombers returning from the first 1,000-aircraft raid on Germany. Earlier, on 17 May, No.308 lost its C.O., S/Ldr (Cpt) Feliks Szyszka, when Sgt (LAC) Witold Knott collided with his Spitfire over Lockington on a training flight, with both pilots being killed. S/Ldr Walerian Żak took over command of the squadron.

Unusual camouflage scheme, applied to No.302's WX•G during the period of preparations for Operation "Jubilee" – the Dieppe Raid.

No.1 Wing's squadrons waited at Croydon and Redhill for a week, but the expected Dieppe operation did not take place because of unsuitable weather. Apart from one Roadstead by 12 Spitfires, only formation exercises kept the pilots busy and during one of these, on 6 July, No.308 Sqdn suffered further casualties in a mid-air collision in which P/O (S-Lt) Julian Kawczyński was killed and Sgt (Sgt) Witold Herbst seriously injured. While at Croydon, the Poles began to look for ways to facilitate instant recognition of "friendly" aircraft and came up with the idea of painting the noses of their fighters with white stripes. These were experimentally applied to W/Cdr Witorzeńc's Spitfire VB WX•C (AA853). A few other

No.302's Spitfire VB W3934 WX•B, with S/Ldr Julian Kowalski in the cockpit, depicted at Croydon on 5 July 1942, during the fighter concentration for the aborted attempt at Operation "Jubilee." (T. Kopański)

Above and below: Spitfire VB AA853 WX•C, the personal mount of W/Cdr Stefan Witorzeńc, as seen on 6 July 1942 at Croydon, with the white identification stripes painted on the cowling.

machines were marked with chalk in preparation for painting, but the idea was abandoned, apparently because of lack of official RAF approval.[29] On 7 July, all five squadrons returned to their former airfields, including No.308, which was withdrawn from the Wing and went back to Hutton Cranswick.

During the rest of July, poor weather conditions and two other factors – waiting for Spitfire IX replacements and absorbing USAAF units into the pattern of operations – restricted air activities. No.1 Wing took part in only eight major missions and reported combats with enemy fighters on two occasions. On 15 July six Spitfires of No.317 Sqdn were "scrambled" from Northolt to intercept five Fw 190s which were attacking rescue launches in the Channel, and one Fw 190 was destroyed jointly by F/O Jerzy Mencel and F/O Jerzy Szymankiewicz, one was claimed "probable" by P/O Jerzy Zbrożek and another one damaged by F/O Stanisław Łukaszewicz.

A larger air battle was fought with about 20 Fw 190s over the Channel near St. Omer on 26 July (Rodeo 81), bringing the wing a substantial victory with no losses. W/Cdr Stefan Janus,

[29] The similar idea – white and black stripes on the fuselages – was generally adopted by RAF for Operation "Overlord" in 1944.

Left: Another Spitfire (probably WX•N) being prepared for the painting of the "friendly" identification strip; Croydon, 6 July 1942. (IPMS) Right: Mechanic Wacław Szarejko in the cockpit of F/O Bochniak's Spitfire VB JH•B "Hala" – the girl's name appearing on a number of Bochniak's aircraft – in the summer of 1942. No.317's badge under the radio mast is prominently visible. (S. Bochniak/W. Matusiak)

two No.302 pilots – S/Ldr Julian Kowalski and F/O Alojzy Rodziewicz – and one No.317 pilot – P/O Stanisław Brzeski – brought down one Fw 190 each, and two more enemies were probably destroyed – by No.302's P/O Marian Wędzik and No.317's Sgt Władysław Grobelny – and one damaged – by No.302's Sgt Hipolit Mikusek. However, intensive and increasingly accurate AA fire resulted in a number of casualties. P/O Eugeniusz Nowakiewicz of No.302, attacking with F/O Bolesław Gładych a German army camp near Boulogne, was shot down on 23 July and at first evaded capture, but he was later detained by Gestapo and after a prolonged and brutal interrogation, eventually went to a PoW camp. F/O Zygmunt Słomski of No.317 and P/O Jan Pentz of No.306 were brought down over France on 29 and 30 July respectively and were taken prisoner. F/O Tadeusz Kratke of No.317 parachuted into the sea on 31 July and was picked up by a launch. Other aircraft were damaged, including F/O Witold Łanowski's Spitfire, which had its rudder nearly destroyed but managed to return.

On 30 July No.308 Sqdn returned to Heston and rejoined No.1 Wing, changing places with No.316 Sqdn, which on the same day left for Hutton Cranswick to recoup after seven and a half months of intensive operations within No.11 Group. No.308 Sqdn ended its period of rather monotonous work from Hutton Cranswick with a probable destruction of a Ju 88 60 miles out to sea by F/O Julian Żuromski and P/O Adam Habela on 24 July.

In the first half of August 1942, with squadrons still waiting for the Combined Operations against Dieppe, activities were extremely slight and No.1 Wing as a formation was not involved in any action. Its units were warned to be ready for immediate moves similar to those that took place at the end of June, but unfortunate experiences at Croydon led to the decision not to use that airfield after all and the wing remained in situ. Only Redhill was to be retained and on 16 August No.303 Sqdn moved from Kirton-in-Lindsey to that airfield to increase the strength of No.1 Wing to five squadrons once again.

During its seven weeks at Kirton-in-Lindsey, No.303 was engaged in routine patrol work and met the enemy twice. On 25 June F/O Bolesław Gładych and F/O Olgierd Sobiecki damaged a Ju 88 east of Grimsby. On 3 July early in the evening two sections were "scrambled" to intercept a low-flying enemy. In the encounter two Ju 88A-4 bombers, the M2+BK and M2+KK, were shot down and crashed in Lincolnshire, one crew being killed and the other captured. The two Ju 88s were credited jointly to four pilots: P/O Tadeusz Kołecki (in AR318), F/Sgt Mieczysław Popek (in BL670), F/Sgt Kazimierz Wünsche (in AB151) and Sgt Aleksander Rokitnicki (in W3948).

In July No.303 Kościuszko's Squadron was joined at Kirton-in-Lindsey by the USAAF's 94th Combat Squadron, flying P-38 Lightning fighters which, after its arrival from America was temporarily assigned to No.12 Group. The traditional links of No.303 with the United States – revived in March 1941 when Col Merian C. Cooper, who had commanded the Kościuszko Squadron in 1920, visited the squadron in England – led to the immediate establishment of cordial relations with the American airmen. No.303 pilots, having unsurpassed experience in fighting the Luftwaffe, were quickly involved in helping the 94th Squadron with its training program (and also, equally enthusiastically, with its off duty activities). The 94th

A skillful extermination of two Ju 88s of 2/KüFlGr 106 by four pilots of No.303 Squadron on 3 July 1942 turned out to be of special significance. Three of the aircrew of Ju 88 M2+KK, the 178th aircraft to be officially destroyed by the squadron, wore the Iron Crosses with the date "1939," for the campaign in Poland, and the victors got one of the Crosses as a trophy. Here F/Sgt Popek shows his prey to Sgt Bondarczuk.

Left: The fin of the Squadron's 178th victim admired by No.303's airmen by Zumbach's BM144 RF•D which bears the modified first image of "Donald Duck," with the raised baseball bat. The five pilots around the fin are, left to right: Sgt Rokitnicki, F/Sgt Wünsche, F/Sgt Popek, F/O Horbaczewski and F/O Lipiński. Right: The fin of M2+KK displayed on Zumbach's Spitfire, with "Misia," the squadron's mascot, sitting by the windscreen. (IPMS)

Squadron flew seven of its first operational sorties over France together with No.303, which the American pilots considered to be "the best damn fighting squadron in the world." On 1 September, a USAAF delegation came to No.303's 1942 Feastday Celebration during which the roll of the dead was read, beginning with the American airmen who lost their lives in the defence of Poland against the Bolsheviks in 1919-1920, and Col Cooper, then Chief of Staff of the China Air Task Force, sent a moving letter to the Polish pilots on this occasion. No.303's representative gave the US Ambassador to the Polish Government in London, Anthony Drexel Biddle, the Honorary Badge of the squadron.

On 18 August 1942 at 16:00 hours a decision was finally taken to carry out Operation "Jubilee," the landing of Commando forces on the Dieppe beaches, the following day. The objective was to silence the local coastal gun batteries, to destroy the enemy radar and other military installations in and around Dieppe up to 5 miles inland, and to re-embark the ground force and withdraw across the Channel. The immediate air support, under the control of No.11 Group, was to be provided by 58 fighter squadrons, including five of the PAF (this force being composed of Spitfire, Typhoon and Beaufighter cover squadrons, and six Hurricane and two Hurribomber close-support squadrons), seven (four Mustang and three smoke- laying Blenheim) Army Cooperation squadrons, two Bomber Command Boston squadrons and three USAAF Boston intruder squadrons.

Air operations on 19 August began at 04:20 hours with Bostons, Hurricanes and Spitfires taking off to attack gun emplacements and to cover the landing ground forces. No.1 Wing carried out eight missions during the day, each of its five squadrons – Nos.306 and 317 from Northolt, Nos.302 and 308 from Heston, and No.303 from Redhill – flying four sorties.

During the wing's first operation, between 05:00 and 07:15 hours, 50 Spitfires of No.302, 308, 306 and 317 Sqdns, led by S/Ldr Tadeusz Nowierski and G/Cpt Stefan Pawlikowski, patrolled off Dieppe at 6, to 11,000 ft. In a brief hit – and – run attack by German fighters on No.317 Sqdn F/O Marian Cholewka was wounded by a cannon shell which shattered his right arm. Losing and regaining consciousness through loss of blood, he returned from France. Lowering the wheels with his left hand and holding the stick between his knees, he locked the undercarriage with his foot and made a perfect landing at Lympne. He taxied the aircraft to the perimeter, switched off the engine and blacked out. He was lifted from the cockpit and taken to hospital.

Left: F/Sgt Popek, sitting in the cockpit of Zumbach's "Donald Duck," with the squadron's war trophy, the 1939 Iron Cross, attached to the Spitfire's fuselage just above Zumbach's personal score, as it then was: 11 destroyed (the white-outlined crosses), 2 "probables" (red-outlined crosses) and 2 damaged (crosses without outline), this being later amended to 3 "probables" and 1 damaged. Right: No.303's Spitfire VBs preparing to get underway during the period preceding the Dieppe raid. (IPMS)

Step up port echelon of No.303's Spitfire VBs on an operation preceding to the Dieppe raid, with BM540 RF•L nearest.

The same six Spitfires climbing above a cloud carpet. (IPMS)

Five of 303's Spitfire VBs, with S/Ldr Zumbach in BM144 RF•D in the lead. (IPMS)

Going to battle: No.303's Spitfires, led by S/Ldr Zumbach, in a tight formation above cloud in the weeks immediately before the Dieppe raid. Note "Donald Duck" with the raised bat on the C.O.'s engine cowling.

Prolonged and sharp engagements with enemy aircraft were fought during the wing's second mission between 09:30 and 11:45 hours, which was flown by 26 Spitfires of No.303 and 317 Sqdns. W/Cdr Janus and S/Ldr Nowierski leading the wing dived on and dispersed a formation of Do 217 and Ju 88 bombers, which were about to attack the shipping. The Germans jettisoned their bombs and fled towards France in disorder. No.303, seeing Fw 190s attacking the beaches and then joining the combat over the ships, split off its outside sections to deal with the bombers, while the rest remained to engage the fighters. No.317 was below the bombers and, alerted by W/Cdr Janus, climbed towards them and attacked the Fw 190s coming to their rescue. The squadron then patrolled in loose fours over the ships, fighting individual duels with various enemy aircraft. The Polish Spitfires maintained an effective umbrella for the battle below and achieved an impressive score 7-5-3 for no losses.

Three more take offs were made in the late morning and at midday – one by No.302 and 308 Sqdns together and two by No.306 Sqdn alone – to escort bombers to the operational areas, but these were uneventful. On the sixth mission, between 12:45 and 14:25 hours, 47 Spitfires of No.302, 303, 308 and 317 Sqdns intercepted about 20 Fw 190s which were trying to dive-bomb the returning convoys (two Biggin Hill squadrons were also detailed to the area). The enemy was successfully engaged by No. 303 Sqdn, which destroyed 2 Fw 190s.

The next mission was flown by No.1 Wing, led by W/Cdr Janus, at its full five-squadron strength, with 62 Spitfires taking off at 15:30 hours to patrol over the returning convoy – Nos.302 and 308 above cloud and the remainder below. No.303, led by S/Ldr Kołaczkowski, saw two He 111s about to begin a bombing run over the ships and immediately attacked the enemy, forcing the bombers to make for home and eventually destroying both Heinkels in spite of intervention by Fw 190s. No.317 joined in No.303's action and both Squadrons fought several successful combats with further enemy bombers and fighters under the fighter umbrella provided by the rest of the wing, and added six more Germans destroyed for certain to the overall Polish score. The final operation of the day at 18:30 hours was made in the rapidly deteriorating weather by 12 Spitfires of No.303 Sqdn, with nothing to report.[30]

Although the Dieppe Raid as a whole was largely a fiasco, with the landing force suffering extremely heavy casualties and not reaching the specified objectives – its only value being a lesson in combined operations and in assault techniques on Hitler's Atlantic Wall – the air cooperation was judged to be "faultless" and enabled the ground troops to fight completely unmolested from the air. The RAF's unquestionable success in affording this protection was paid for by fairly heavy losses, which amounted to 106 aircraft – 88 of them fighters – and 71 pilots killed or missing.

303's Spitfire VBs being serviced at Kirton-in-Lindsey on 15 August 1942, a day before the squadron's departure to Redhill to take part in Operation "Jubilee," which was eventually launched in the early hours of the 19th. (IPMS)

No.1 (Polish) Wing emerged from Operation "Jubilee" as the most successful unit, with very few casualties, indicating careful planning of sorties, a high degree of operational professionalism and excellent teamwork. The wing, which flew 245 sorties out of 2,399 put up by the RAF, was officially credited with the score 15 1/2-5-3 enemy aircraft for the loss of only two pilots, and the RAF assessment of operations concluded that this was due to: economizing on petrol and thus giving the pilots every chance to engage and take avoiding action when necessary; excellent contact and cooperation between sections in each squadron, which always effectively covered each other; good adherence to the allotted patrol areas by the pilots and good reassembly practices, with the maintenance of height always correctly observed; good withdrawal procedures on W/Cdr's orders, with the relieving squadrons effectively taking up their positions.[31]

Although No.1 Wing constituted about 10 percent of the fighter force deployed, 15 1/2 confirmed victories credited to it represented 17 percent of the 91 enemy aircraft reckoned by the RAF as destroyed for certain[32] and the Polish casualties were less than 3 percent of the total losses.

The individually attributed scores were (types other than fighters indicated in brackets):

- S/Ldr Tadeusz Nowierski, Wing's Command, 0-0-2 (Do 217s);
- No.303 Sqdn: P/O Stanisław Socha 2 (1 Ju 88)-0-0; S/Ldr Jan Zumbach 1 1/3 (l/3 He 111)-1-0; F/O Eugeniusz Horbaczewski, P/O Tadeusz Kołecki and Sgt Józef Stasik 1-0-0 each (Kołecki 1 Ju 88); F/Sgt Mieczysław Popek -1/2-0-0; P/O Antoni Głowacki and Sgt Wacław Giermer 1/3 (He-111)-1-0 each; Sgt Aleksander Rokitnicki and Sgt Alojzy Rutecki 1/3 (He 111)-0-0 each; F/L Janusz Marciniak and Sgt Józef Karczmarz 0-1-0 each;
- No.317 Sqdn: P/O Michał Maciejowski 2 1/2 (1 Ju 88

[30] The whole account of the day's operations is based on Combat Diary, Royal Air Force Northolt.

[31] Ibid.

[32] The actual Luftwaffe losses turned out to be 48 aircraft totally destroyed and a further 24 damaged.

and 1/2 Do 217)-0-0; F/Lt Kazimierz Rutkowski 2 (1 Do 217 and 1 Ju 88)-0-0; P/O Stanisław Brzeski 1(He 111)-0-0; P/O Stanisław Łukaszewicz 1-0-0; Sgt Adam Kolczyński 1/2 (Do 217)-0-0; F/Sgt Kazimierz Sztramko 1/3(He 111)-0-0; and Sgt Władysław Pawłowski 0-0-1.
The casualties comprised:
- No.303 Sqdn: one pilot, P/O (Lt) Adam Damm, killed, and one aircraft lost;
- No.306 Sqdn: one pilot, F/O (Lt) Emil Landsman, missing and made PoW, and one aircraft lost.

BS456 UZ•Z, one of the Spitfire IXCs delivered to No-306 Squadron at Northolt in September 1942.

Another Polish casualty of the Dieppe Raid was F/O (Lt) Antoni Waltoś, who, flying with No.87 Sqdn in Hurricane IIC, was killed during a ground attack sortie on 19 August.

No.302, 306 and 308 Sqdns had no conclusive combat with the enemy but did excellent work protecting the other two squadrons of the wing and other RAF aircraft which operated in the area.

On 20 August No.303 Sqdn was withdrawn from No.1 Wing and returned to Kirton-in-Lindsey. The wing, operating at four-squadron strength, carried out a major mission in the late afternoon on that day, providing a diversion for Circus 206, and fought with Fw 190s, one of which was shot down by F/O Florian Martini of No.317 Sqdn. A combined Rhubarb by No.302 and 306 Sqdns against St. Omer airfield followed two days later, but this resulted in heavy casualties. No.306, in a daring attack, lost four pilots to intensive fire from ground defences, including its S/Ldr (Cpt) Tadeusz Czerwiński, with command of the squadron passing to S/Ldr Kazimierz Rutkowski. The other casualties included F/O (Lt) Witold Szyszkowski and Sgt (Sgt) Zdziaław Horn, who were also killed, and F/Sgt (Sgt) Brunon Kroczyński, who was made PoW.

On 24 August two Spitfire VBs from No.308 Sqdn were "scrambled" to intercept a single enemy aircraft flying at a very high altitude. This was the first encounter of Polish pilots with the Ju 86P reconnaissance bombers, which began to appear

BS459 UZ•T, another Spitfire IXC of No.306. Standing by the aircrew is F/O Sergiusz Szpakowicz. Six pilots on the aircraft are, in the top row: F/O Władysław Potocki, F/O Zygmunt Jeliński, F/O Bogdan Arct and F/Sgt Edward Martens. Sitting on the wing are F/O Zdzisław Langhamer and P/O Grzegorz Sołogub; Northolt, October 1942.

over Britain during the latter part of the month. The Spitfires were not able to reach the necessary height to attack. On 25 August S/Ldr (Cpt) Julian Kowalski passed the command of No.302 Sqdn to his successor, S/Ldr (Cpt) Stanisław Łapka.

In the last week of August the Wing took part in three more Circuses and a few other operations, but these were uneventful.

(3.) Maintaining the Offensive in the Quest for Air Superiority

A great effort was made to keep the offensive going throughout the inclement autumn and winter months of 1942, the escalating strength and range of the operations partly offsetting their reduced frequency.

The weather was good in the early part of September, but broke up around the 10th day of the month, bringing further air activities almost to a halt. The composition of No.1 Wing changed again on 5 September, when No.317 Sqdn was replaced at Northolt by No.315 Sqdn, which arrived from Woodvale after a comparatively restful five-month recuperation period.

While at Woodvale, No.315, engaged mainly on defensive patrols and shipping protection duties, reported only a few contacts with the enemy and these resulted in 1/2 Ju 88 damaged by F/O Konrad Stembrowicz on 3 May 1942 near Heysham (the other 1/2 being credited to P/O Lamerton of No.452 RAAF Sqdn), and 1/2 Ju 88 damaged by Sgt Jerzy Malec on 14 August over the sea near Barrow (shared with No.41 Sqdn). On 23 August one of No.315's sections, together with Spitfires from No.152 and 504 Sqdns, took part in a fight with Ju 88D-1 4U+KH of 1(F)/123, which was eventually destroyed, crashing near Tramore (County Waterford) in Ireland, but F/O (Lt) Bolesław Sawiak was shot down by the Ju 88's return fire and, mortally wounded, died in hospital.

Between 1 and 8 September No.1 Wing participated in five major operations. On 7 September, acting as the middle wing of the main diversion (No.302, 306 and 308 Sqdns, led by S/Ldr Nowierski – 37 Spitfires in all) on Circus 217, the Poles were jumped by several Fw 190s, which attacked No.302 Sqdn out of the sun. P/O Tadeusz Zdzitowiecki had to bale out and was picked up by an Air/Sea Rescue launch, and Sgt Ignacy Czajka was wounded by a cannon shell in his right arm, but managed to return and make a belly landing in Kent. In the ensuing combat 1 Fw 190 was destroyed by P/O Czesław Śnieć, 1 probably destroyed by S/Ldr Stanisław Łapka and 1 damaged by P/O Eugeniusz Ebenrytter, all of No.302 Sqdn.

The following day, on Ramrod 37, the wing, at four-squadron strength, including No.315, escorted Bostons to Le Havre and fought a sharp combat with Fw 190s near the target and on the way back. One German was damaged by S/Ldr Mieczysław Wiórkiewicz (No.315), but P/O (S-Lt) Jerzy Urbański was killed and Sgt (LAC) Hipolit Mikusek shot down and made PoW, both pilots of No.302 Sqdn.

W/Cdr Stefan Janus, C.O. of No. 1 Polish Fighter Wing, in the cockpit of a new Spitfire IXC of No.306 Squadron, Northolt, late 1942.

Above: The year 1942 ended triumphantly for the PAF in general and for No.306 Squadron in particular. On 31 December 1942 its pilots fought a victorious battle with Fw 190s and P/O Henryk Pietrzak brought down the 500th enemy aircraft to be officially credited to the PAF in Great Britain as "destroyed for certain." F/O Langhamer claimed the 501st. The jubilant Pietrzak is seen here in the cockpit of his Spitfire IX. Below: P/O Pietrzak carried shoulder-high by his comrades after his historic victory, Northolt, 31 December 1942.

A few minor sorties included an attempted interception of a lone Ju 86P by two Spitfire VBs of No.302 on 5 September. The enemy continued to use these high-altitude reconnaissance bombers for isolated harassment raids with impunity in the early part of the month, until a Special Spitfire Flight, formed at Northolt with Spitfire VIs, VIIs and the specially adapted Spitfire IXs and capable of operating at heights of up to 45,000 ft, began to chase them off.

In the later part of 1942 squadron markings on Spitfire Vs of No.317 Squadron were moved to a new position, under the windscreen, as shown in this photograph of JH•M. (IPMS)

No.1 Wing's only sortie later in the month was made on 24 September by three brand new Spitfire IXCs of No.306 Sqdn, which was among the first fighter squadrons to receive this superior model and completed conversion to it two days later. The Spitfire IXCs were sent to intercept Fw 190s in the Dungeness area on a practice flight, but the Germans immediately turned for home and no claims were made. The last days of the month the Squadron spent at Ipswich on armament training. On 25 September S/Ldr (Cpt) Tadeusz Sawicz succeeded S/Ldr (Mjr) Wiórkiewicz as the new C.O. of No.315 Sqdn.

In October 1942 offensive activities were carried out whenever the weather permitted, and the Polish squadrons of No.1 Wing were involved in nine major operations. These included escort to large daylight raids by USAAF Fortresses and Liberators on targets in France. The first such mission was made by eight Spitfire IXs of No.306 Sqdn, led by S/Ldr Tadeusz Nowierski, which took off from Biggin Hill on 2 October at 14:50 hours to act as close escort to the Fortresses on the first part of Circus 221. On the way back the squadron was engaged with Fw 190s and F/O Władysław Walendowski destroyed one enemy aircraft. The rear support to Circus 221 was provided by 37 Spitfire VBs of No.1 Wing, led by W/Cdr Stefan Janus.

On 9 October 12 of No.306's Spitfire IXs and two RAF Spitfire IX squadrons led by Nowierski as Wing Leader, escorted 117 USAAF "heavies" in an attack on Lille (Circus 224). No.306 fought a battle with German fighters and F/Lt Józef Gil was credited with one German destroyed and one damaged, F/Lt Stanisław Kędzierski and P/O Henryk Pietrzak, with

F/Lt Zygmunt Bieńkowski, taxiing in his Spitfire VB BM144 RF•H ("Halszka") (formerly RF•D) at Kirton-in-Lindsey on 1 September (No.303's feastday) 1942. (IPMS)

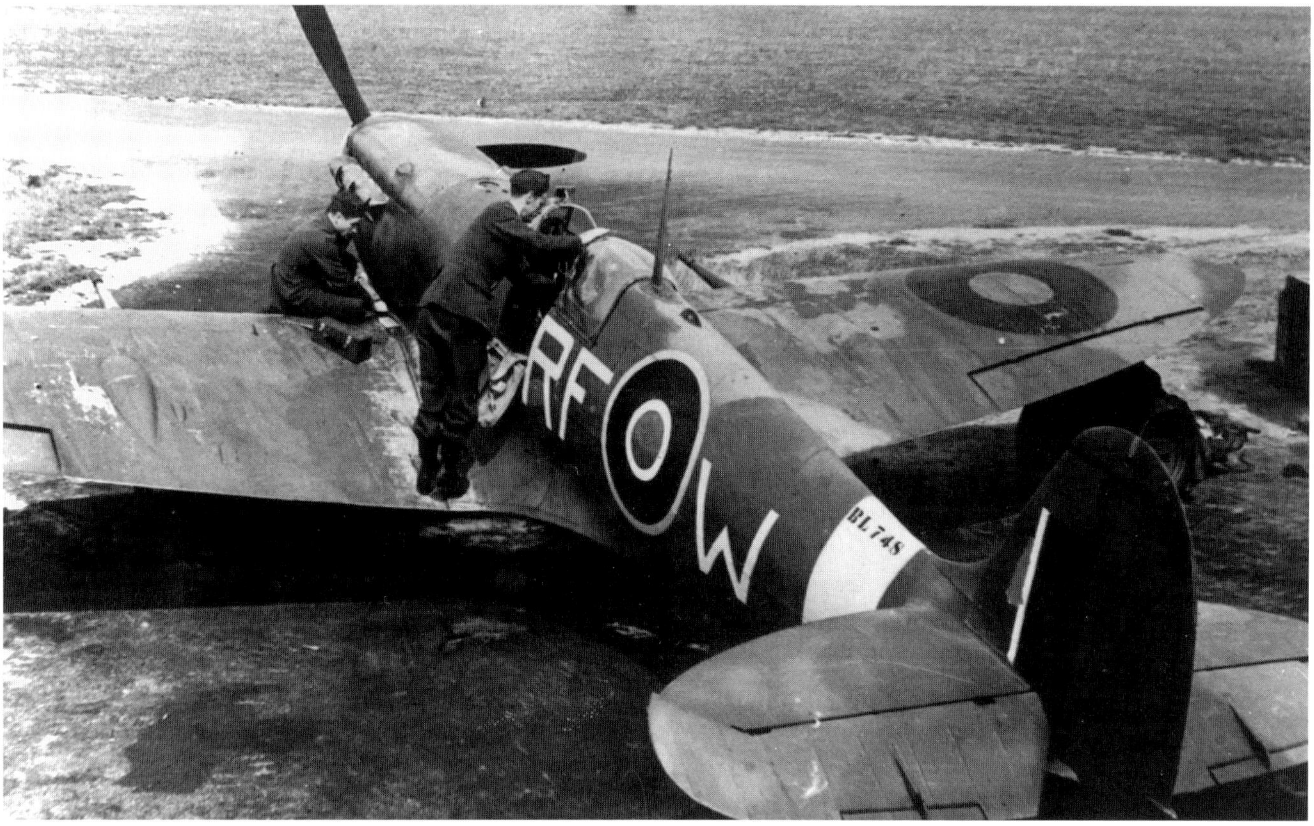

A Spitfire VB BL748 RF•W at Kirton-in-Lindsey on 1 September 1942.

one destroyed each and F/Lt Józef Żulikowski, with one "probable." Sgt (LAC) Marian Kordasiewicz, wounded in the fight, lost his life in a crash landing near Deal. The American gunners, as yet unable to distinguish between friend and foe, persistently fired at the close escort fighters, causing some damage to the Spitfires. The rest of No.1 Wing, 34 Spitfire VBs led by W/Cdr Janus, acted as the main diversion, but saw no action. Two days later an uneventful Rodeo (101) was flown by the wing, and on 12 October the wing's 36 Spitfire VBs patrolled over Channel shipping. The wing took off at three-squadron strength in another Circus on 16 October, but the operation was abandoned midway because of the bad weather.

On 23 October Spitfire IXs of No.306 Sqdn, together with two other Spitfire IX squadrons, with Nowierski as the Leader, acted as the rear support to Fortresses returning from Lannion, while 37 Spitfire VBs from the wing, led by Janus, provided similar cover for Fortresses coming back from Maupertus. No action was reported by either formation. The wing's three Spitfire VB squadrons were again called upon to fly the target cover to Bostons on Circus 232 on 25 October, but the bombers were held back owing to the weather and the mission was aborted. On 29 October No.308 Sqdn, which had already gained experience in night work during its stay in Exeter, was moved from Heston to Northolt to strengthen the London night defences

S/Ldr Zumbach in the cockpit of his new mount, EP594 RF•D, displaying a different "Donald Duck" with the raised baseball bat and his new score, 13 crosses for the destroyed aircraft (in fact it was 12 1/2) and five for "probable" and damaged; Kirton-in-Lindsey, 1 September 1942.

President Raczkiewicz visited No.303 Squadron during its 1942 feastday celebrations. He is seen here decorating F/O Horbaczewski. Lined up for the decoration next to Horbaczewski are: F/Lt Marciniak and F/O Majewski.

against the possibility of increased Luftwaffe activity. The month's operations ended on the 31st, with two sections of No.315 Sqdn attacking locomotives in the Abbeville-Dieppe area.

Persistent fog restricted No.1 Wing's activities in November to seven major operations, four at four-squadron and three at three-squadron strength, during which no contact was made with the enemy. In addition, the squadrons, operating mainly in sections, carried out ten individual Rhubarbs, which totaled 32 sorties and resulted in the destruction of an alcohol distillery and four locomotives, with seven further locomotives and a tug reported damaged. Own casualties amounted to three pilots killed: Sgt (Cpl) Zygmunt Rozworski (No.306 Sqdn), who, hit by AA fire, perished in the sea on 8 November; P/O (S-Lt) Wiesław Benoit (No.306 Sqdn), who flew into barrage balloons near Woolwich in bad visibility on 18 November; and P/O (S-Lt) Bogdan Mijakowski (No.308 Sqdn), who was shot down into the sea by intense AA fire on 26 November. No.315 Sqdn completed conversion to the Spitfire IX at Northolt an 28 November, becoming the second PAF squadron to be so equipped.

Despite the poor December weather, offensive action was continued on every possible occasion. The wing acted as escort to the USAAF Fortresses and Liberators on four Circuses (241 to 244) and also flew a number of Rodeos and Rhubarbs,

In November 1942 the third basic edition of the "Donald Duck" image – initially without an egg, which was soon added to it – emerged on Zumbach's EP594 RF•D, as depicted by this photograph which was taken before Poland's Independence Day, 11th November, in 1942. The last cross of the latest "destroyed" score received the correct inscription 1/3; the C.O.'s banneret appeared under the windscreen. (IPMS)

but the only significant encounter with the enemy took place on 31 December, on Rodeo 140, when 25 Spitfire IXs of No.306 and 315 Sqdns, led by W/Cdr Stefan Janus, met 13 Fw 190s over Berck. No.306 Sqdn attacked while No.315 provided top cover. P/O Henryk Pietrzak brought down one Focke-Wulf north-east of Le Crotoy at about 14:40 hours. F/O Zdzisław Langhamer accounted for one German destroyed and one dam

A group of No.303 pilots by the same aircraft during the celebration of Poland's Independence Day; Kirton-in-Lindsey, 11 November 1942. Left to right are: P/O Kołecki, F/Sgt Adamek (on the wing), F/O Kobyliński, P/O Podobiński (partly obscured), F/Lt Bieńkowski, W/Cdr Witorzeńc, P/O Głowacki (standing by the windscreen), F/Sgt Popek, S/Ldr Zumbach, Sgt Palak, Sgt Stasik, F/O Majewski, Sgt Karczmarz and F/O Zarębski. (IPMS)

aged soon afterwards; P/O Korneliusz Górniak and P/O Władysław Szajda claimed one "probable" each and W/O Jan Śmigielski, one damaged, but the success of No.306 Sqdn was paid for by the lives of two of its own pilots: F/Lt (Lt) Józef Gil, who led the attack, and P/O (S-Lt) Leon Kosmowski.[33]

Pietrzak's victory on the last day of 1942 was a momentous event for the PAF, since his victim was the 500th enemy aircraft to be officially credited to the Polish Fighter Force in Great Britain as "destroyed for certain." Langhamer's was the 501st. An Air Ministry communiqué made special mention of the achievement of No.306 Sqdn and this prompted the British press to voice once again their admiration for the Polish airmen and their fighting spirit, The Times of 1 January 1943 calling them "the tool of justice." Pietrzak was awarded a silver shield trophy which was founded by Polish pilots to mark the 500th victory and both Pietrzak and Langhamer were ceremonially decorated with the Crosses of Valour by President Raczkiewicz in the presence of representatives of the Polish Government, RAF, Army, Navy and all Polish squadrons.

No.2 Polish Fighter Wing was reformed at Kirton-in-Lindsey on 25 September 1942 and W/Cdr (Cpt) Wojciech Kołaczkowski was appointed its new C.O. The wing comprised No.303, 316 and 317 Sqdns, the first two within No.12 Group, and the last named, at Woodvale, within No.9 Group.

Its squadrons continued to operate individually for some time, their activities consisting largely of training replacement pilots and extending the tactical skills of their experienced pilots, with special emphasis on battlefield support and night combat. Operational tasks were performed on a limited scale and these included defensive patrols, convoy protection duties and occasional participation in offensive operations.

No.303 Sqdn was sent back to Kirton-in-Lindsey immediately after Operation "Jubilee" and on 16 September took part together with No.316 in the first Circus to be mounted by No.12 Group. Both units took off from Coltishall and acted as rear support to Boston bombers. No.303 flew another Circus on 9 October, this time taking off from Eastchurch, and on 8 November supplemented fighters of No.11 Group on Circus 235. No combat with the enemy was reported during any of these missions. In addition, two Rhubarbs were carried out by single sections in November. On 1 December 1942 S/Ldr (Cpt) Jan Zumbach passed the command of No.303 Sqdn to S/Ldr (Lt) Witomir Bieńkowski. On 15 December the squadron lost one of its pilots, F/O (S-Lt) Ryszard Stankiewicz, who crashed on a training flight in bad weather.

No.316 Sqdn's respite period on Hutton Cranswick airfield was marred by accidents. On 28 August P/O (S-Lt)

[33] Leon Kosmowski was promoted from Sgt to the officer's rank in June 1942.

No.303's artist was busy to decorate several Spitfires with various Disney characters in preparation for the 1942 Independence Day festivities and portrayed himself doing so in the Squadron's Commemorative Book. (IPMS)

One of the more interesting among these colorful markings was the cheeky impression of "Pee-wee" ("Pop-eye's" son). (IPMS)

Two other examples of the artist's handiwork; Kirton-in-Lindsey, November 1942. (IPMS)

Władysław Balon hit the water during a training flight over the sea and was killed. On 20 November F/O (Lt) Ryszard Stróżak, newly posted to No.65 Sqdn, perished in a crash landing in dense fog near Pooley Bridge, Penrith. The following day four of No.316's Spitfire VCs flew a Rhubarb, attacking ground targets in Holland – the unit's only sortie of this type in the last quarter of 1942. On 22 November Sgt (LAC) Tadeusz Orzechowski lost his life in a midair collision with a Beaufighter on night exercises. On 12 December F/O (Lt) Edward Kowalski insisted on flying back to Hutton Cranswick despite the warnings and became another casualty of the weather when his Spitfire disintegrated on impact with the ground in the landing approach.

On 20 December two squadrons of No.2 Wing – Nos.303 and 316 – set out from West Malling to provide a diversion for Fortresses attacking Romilly-sur-Seine on Circus 244, No.1 Wing flying escort to the bombers.

No.317 Sqdn spent the last quarter of 1942 recuperating at Woodvale. Its sections were "scrambled" on a few occasions to fly defensive patrols, but the squadron's main activities involved intensive training and in particular close-support practice and participation in various army cooperation exercises.

In 1942 the Polish fighter force underwent various organizational changes and made significant progress towards greater autonomy, with Polish officers assuming higher operational command responsibilities. G/Cpt Mieczysław Mümler became Station Commander at Northolt on 25 September and W/Cdr Stanisław Brzezina, Station Commander at Heston, Northolt's satellite airfield, on 6 October. The Northolt Sector airfields were in the thick of combat activities and these appointments added a new dimension to the PAF's stature.

During 1942 the Polish fighter squadrons were credited with the destruction of 90 enemy aircraft, with 36 further "probable" and 43 damaged. No.303 Sqdn, with the score 21-10-3, was the most successful among the Polish units. The overall operational effort of the Polish fighter force in 1942 amounted to 10,390 sorties in 15,365 flying hours,[34] with No.302 in the lead with 1,583 sorties in 2,233 hours (for further details see Appendix 22).

Although the greatly diminished Luftwaffe operations in the West resulted in a smaller number of air victories, the Polish fighter squadrons gained an excellent reputation among the bomber crews for the effectiveness of the cover provided during escort duties. Poles developed a special technique, perfected during "Operation Jubilee." One squadron acted as a decoy, flying closely behind and somewhat below the bombers, its pilots behaving like the rawest novices, wobbling about and almost colliding. Another squadron lurked well above in the sun, waiting until the Germans pounced on the apparently easy prey. The "novices" would then suddenly learn how to fly and the top squadron would dive on the unsuspecting enemy from the rear, both units engaging the Germans in a perfectly coordinated manoeuver and preventing them from attacking the bombers.

Later, in the closing months of 1942, the Poles perfected another method of protecting the American "heavies." RAF fighter cover consisted at first of escort fighters flying above and behind the bombers, but the B-17 Flying Fortresses and B-24 Liberators were most vulnerable to head-on attacks, which the British tactics would not prevent. No.1 Polish Fighter Wing worked out a way of combating this threat by positioning one squadron high above and well in front of the bomber formation, one directly above, and one below and behind it. The high cover engaged the oncoming enemy fighters and frustrated their attack, while the rest of the escort dealt with the dispersed Germans. The tactics proved highly effective and reduced the American losses, and the USAAF asked the RAF to employ similar methods. The Poles also took particularly good care of lame bombers, damaged by the enemy, staying with them until they reached England or affording protection to their dinghies when they ditched in the Channel, and this endeared them to the American crews.

In addition to the intense offensive and escort operations, the squadrons continued with shipping protection duties, which accounted for about 1,700 aircraft sorties in almost 2,100 flying hours in 1942.

A few Polish pilots still flew in British squadrons. The most prominent among them was S/Ldr (Mjr) Jerzy Jankiewicz, appointed C.O. of No.222 Sqdn on 22 May 1942, the first Pole to command a British unit. Three days later W/Cdr Francis D.S. Scott-Malden asked Jankiewicz to lead the whole North Weald Wing on Rodeo 51. In a difficult encounter with German fighters west of Dunkirk, Jankiewicz, flying Spitfire AD233, was shot down and drowned. His body was never found.

On 10 February 1942, ten Polish pilots, temporarily posted at their own request – out of 22 applicants – from the Aircraft Delivery Unit of the Polish Detachment RAF Takoradi, in North Africa, to No.112 "Shark" Fighter Squadron, began operational flying on the Libyan front. The group, which passed through the OTU in RAF Middle East Command, comprised: F/Lt Feliks Gazda, its Commander, F/O Witold Jander, F/O Feliks Knoll, F/O Czesław Matusiak, P/O Henryk Knapik and five NCOs, Sgt Józef Derma, Witold Giedroyć, Stanisław Grondowski, Jerzy Różański and Zbigniew Urbańczyk. Originally Sgt Jan Mikulski was also included, but he was posted back to ferry work in January 1942. P/O Władysław Włodarczyk was appointed as interpreter. The Poles operated from Gambut and Gazala and flew the P-40 Kittyhawks.

F/O (Lt) Matusiak was killed on 12 February, when his Kittyhawk (AK584) stalled after take off on a practice flight at Gambut. Three days later Otto Schulz, a German pilot from II/JG 27, reported seeing the Polish markings on some of the Kittyhawks over the Libyan Desert. On 21 February F/O Jander and Sgt Derma were shot down during combat with Bf 109s

[34] All these totals include No.307 Night-Fighter Squadron.

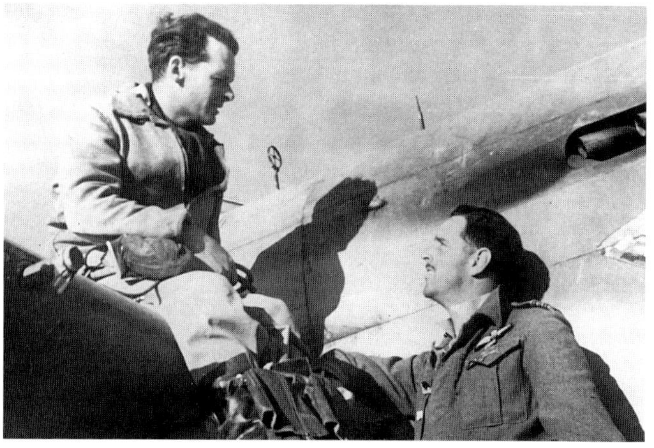

Sgt Jerzy Różański, sitting on the wing of a Kittyhawk I (P-40D) in a talk with S/Ldr Clive R. Caldwell, C.O. No.112 "Shark" Fighter Squadron, at Gambut Main in February 1942. (Crown Copyright, Imperial War Museum, Ref.No.2431).

S/Ldr Caldwell, behind the wheel of a Morris 15, with the Polish pilots of No.112 "Shark" Squadron, left to right F/Lt Gazda, F/O Jander, F/O Matusiak (killed in training 12 February 1942), F/O Knoll (killed in action 10 July 1942), P/O Knapik, Sgt Różański and Sgt Urbańczyk; Gambut Main, early February 1942. (Crown Copyright, Imperial War Museum, Ref. No.2422)

and crash landed. Jander, wounded in the leg, had to go to a hospital. On 13 March two Kittyhawks, one of them piloted by Sgt Różański, became detached from the rest of the Squadron and were brought down in a surprise attack by a Bf 109 flown by Otto Schulz, Różański sustaining some injuries. The following day the Poles recorded their first – and the only – success, when Sgt Urbańczyk, flying with S/Ldr Clive Robertson Caldwell, fought a victorious engagement with enemy fighters near Gazala, both pilots being credited with 1/2 Bf 109 each. On 15 March No.112 Sqdn was withdrawn for rest. Sgt Giedroyć returned earlier to ferry duties, but F/O Jander rejoined the group, which in the first half of April numbered eight.

However, it was realized from the beginning that such a large number of operationally inexperienced pilots could not be absorbed into any one squadron without affecting its combat efficiency, and although the Poles performed well and were highly regarded, their poor command of English and physical condition of some of the older pilots added to the problem. When the squadron returned to combat duties in the middle of April 1942, only Jander, Knapik and Knoll stayed on its opera-

A Curtiss Kittyhawk I (P-40D) AK772 GA•Y "London Pride" of No.112 "Shark" Squadron, flown by F/Lt Gazda, Sgt Derma and Sgt Różański in March 1942. (Crown Copyright, Imperial War Museum, Ref.No.2595)

tional strength. The five remaining Polish pilots were posted away from the unit at the beginning of May. F/O (Lt) Feliks Knoll was killed in combat with Bf 109s over Mersa Matruh on 10 July 1942. Jander flew with the squadron until the end of July, and Knapik some weeks longer.

A much more ambitious and eminently successful involvement of Polish pilots in North African air operations came some six months later with the organization of the Polish Fighting Team.[35]

In 1943 the Allies strived to establish and maintain air supremacy over the Continent, a precondition of any invasion attempt, and penetrated enemy air space in greater strength and deeper than ever before. The duration of fighter sorties increased on average by some 20 percent (as compared with 1942) and by the spring the more aggressive Ramrods began to supersede Circuses. At the same time, the modest Fighter Command "private venture" Rhubarbs almost disappeared from operational schedules.

In the first quarter of 1943 the weather was generally better than could be expected, and in January No.1 Wing carried out 12 major operations – eight Circuses and four Rodeos – and one Rhubarb, one defensive patrol and one Air/Sea Rescue patrol, but was engaged in air combat only once. On 21 January the wing flew Circus 252 at midday and Rodeo 156 in the afternoon. The latter operation was carried out jointly with No.303 and 316 Sqdns of No.2 Wing, which moved from Kirton-in-Lindsey and Hutton Cranswick, their respective airfields, for the day to Northolt for that purpose. 22 Spitfire IXs of No.306 and 315 Sqdns and three Spitfire IXs of No.1 Wing HQ – G/Cpt Stefan Pawlikowski, W/Cdr Stefan Janus, and S/Ldr Henryk Szczęsny – and 24 Spitfire VBs of No.2 Wing left Northolt at 14:25 hours and crossed the French coast near Berck. Flying North, the formation attacked about 20 Fw 190s, but because of 10/10 cloud cover results of the battle were difficult to determine. Three enemy aircraft were credited as "probable" and one as damaged to No.306 pilots – to F/O Sergiusz Szpakowicz, one "probable" and one damaged, and to S/Ldr Kazimierz Rutkowski and Sgt Herbert Szwaba, one "probable" each – and two as "probable" to No.316 pilots – one each to P/O Jerzy Szymankiewicz and F/Sgt Paweł Gallus – with no losses. No.316 was led by S/Ldr (Cpt) Marian Trzebiński, who took over command of the squadron from S/Ldr Żurakowski on 1 January 1943.

On 26 January, misfortune struck No.1 Wing on Circus 256. W/O (Sgt Mjr) Wawrzyniec Jasiński (No.306 Sqdn) collided with W/Cdr (Mjr) Janus over the Channel and was killed. Janus, on his last sortie as Wing Leader before going on a well-earned rest – with W/Cdr (Cpt) Wojciech Kołaczkowski already appointed Leader of No.1 Wing and the change-over arranged for that afternoon – managed to bale out, but he was picked up by the enemy and made PoW. W/Cdr (Cpt) Aleksander Gabszewicz succeeded Kołaczkowski as Commander of No.2 Wing.

No.316's Spitfire VB SZ•N, damaged in landing on the snow-bound Hutton Cranswick aerodrome in the winter of 1942/43. The inscription "Nigeria," indicating the previous user of the aircraft, is still visible forward of the windscreen. (K.Chołoniewski)

Composition of both Polish fighter wings changed on 1 February, with No.303 Sqdn coming via Northolt to Heston to replace No.302 Sqdn in No.1 Wing and No.302 going to Kirton-in-Lindsey, to No.2 Wing. On 13 February No.317, arriving from Woodvale, joined No.302 at Kirton-in-Lindsey. This move brought all three squadrons of No.2 Wing closer together, to the benefit of joint operations.

On 3 February No.308 Sqdn put up a magnificent show defending 12 Venturas on Circus 258. The bombing of airfields in France was abandoned owing to bad weather and when other escort squadrons (from Hornchurch) were recalled 11 Spitfires of No.308 and G/Cpt Mieczysław Mümler (flying a 12th Spitfire) were left to protect the bombers alone. The formation was jumped by some 30 Fw 190s over the Courtrai area and the Poles took the Germans on in a desperate attempt to frustrate the attack on the Venturas. The Fw 190s pursued their prey halfway across the Channel, but succeeded in shooting down only one bomber and damaging another, which crash landed on return. Aptly enough, one of the machines that returned safely carried on the fuselage a motto "Heaven can wait." However, No.308 paid a heavy price – four of its aircraft were lost and four pilots missing: F/O (Lt) Jan Wiejski was killed, F/O (Lt) Jerzy Zbierzchowski and F/O (S-Lt) Jan Okrój were made PoW and F/Lt (Lt) Tadeusz Koc baled out over Dunkirk, but avoided capture. Helped by the French, Koc reached Gibraltar as early as 12 February, and to everyone's astonishment walked into Northolt Officer's Mess on 21 February. Results of the battle were again difficult to establish. S/Ldr Walerian Żak and F/Lt Tadeusz Koc were credited with one "probable" Fw 190 each, and G/Cpt Mieczysław Mümler with one damaged.

No.306 and 315 Sqdns, led by W/Cdr Wojciech Kołaczkowski and S/Ldr Henryk Szczęsny, acted as target support of 2nd Part of Circus 258 the same afternoon and No.315 successfully beat off attacking Fw 190s. Francis Stanley

[35] Its story is related beginning on page 261.

"Gabby" Gabreski, USAAF pilot born of Polish parents in Oil City, Pennsylvania, who later became one of the greatest American fighter aces of World War II with 28 confirmed air victories to his credit,[36] distinguished himself in this engagement.

Gabreski, posted to No.56 Fighter Wing USAAF which came to England towards the end of 1942, was temporarily detached to No.1 Polish Fighter Wing for operational training and from the beginning of 1943 flew with No.315 Sqdn in Spitfire IXs. He learned fast and developed what he called "the Polish eagle eyes" – the exceptional ability to see the enemy in the far distance, for which the Poles became famous within the RAF. In a running battle with several Fw 190 formations over the Calais area on 3 February, P/O Stanisław Blok and Gabreski became separated from the rest of the squadron. The starboard wing of Blok's Spitfire PK•B (BS409) was seriously damaged by enemy cannon shell, but in spite of this he was able to shoot down one of his attackers. He had more Germans on his tail, when Gabreski came to his rescue. Blok, finding his damaged aircraft difficult to fly, wanted to bale out, but was unable to do so. Gabreski stayed with Blok and they both had to fight their way out from France with some 20 Fw 190s in hot pursuit until they crossed the English coast. Blok made a perfectly good landing at Manston. During the same operation P/O Michał Cwynar (in EN123), also of No.315 Sqdn, set one Fw 190 alight and, in his own words, "continued to fire at him to make certain the pilot did not bale out."[37] No aircraft were lost. Cpt Gabreski was later decorated by Gen Sikorski with the Polish Cross of Valour for his actions and he received the Polish pilot's wings.

On 7 February No.308 Sqdn flew a patrol over the train in which Winston Churchill traveled to London after a visit to North Africa. With S/Ldr (Cpt) Walarian Żak going for rest on 12 February, F/Lt (Lt) Franciszek Kornicki took over command of No.308 Sqdn as Acting S/Ldr and held it for three weeks before being succeeded by S/Ldr (Cpt) Paweł Niemiec.

A large German vessel was discovered in Boulogne Harbor on 12 February and was subsequently attacked (Circus 262). The ship moved to Dunkirk and on 15 February Circus 266 was mounted to bomb it, with No.306 and 308 Sqdns, led by W/Cdr Kołaczkowski, acting as first fighter echelon, and No.303 and 315 Sqdns, led by S/Ldr Szczęsny, as target support. Nos.306 and 308 Sqdns engaged Fw 190s, S/Ldr Kazimierz Rutkowski and W/O Jan Śmigielski of the former unit destroying one German each, and W/O Stanisław Piętkowski of the latter unit, probably destroying a further one. During February the squadrons of No.1 Wing assisted in 11 Circuses and one Roadstead, and in addition carried out two Rodeos.

Francis S. "Gabby" Gabreski emerging from the cockpit of No.315's Spitfire IX during his operational practice with the squadron early in 1943. (J. Koniarek)

In March 1943, 13 major offensive operations were carried out. This apart, at the beginning of the month No.303 Sqdn went for nine days to Debden, to take part in Exercise "Spartan," a large-scale rehearsal for the landing in France by combined forces, during which the squadron simulated the "enemy" aircraft.

F/O Stanisław Blok by his Spitfire IXC BS409 PK•B (No.315 Sqdn), severely damaged on Circus 258, flown on 3 February 1943.

[36] His father, Stanisław Gabryszewski, settled in the USA in 1909, arriving from the Lublin area and married a Polish woman from the same region. Gabreski achieved his first air victory on 24 August 1943 over France. He shot down 28 Germans in air combat; after adding three more enemy aircraft destroyed on the ground, he was credited by the USAAF with 31 Germans destroyed. He later fought in Korea in F-86 Sabres as a wing commander and destroyed 6 1/2 MiG-15s.

[37] Quoted from Combat Diary, Royal Air Force Northolt.

On 11 March F/O (Lt) Tadeusz Stabrowski of No.308 Sqdn was hit by AA fire while attacking ground targets in France. The pilot ditched in the Channel, but was never found. On the same day No.306 changed places with No.316, which came to Northolt, to No.1 Wing, from Hutton Cranswick (No.2 Wing) without aircraft and upon its arrival converted to Spitfire IXs. No.306 moved to Hutton Cranswick where it was equipped with Spitfire Vs.

The following day, on the 12th, No.315 and 316 Sqdns, both flying Spitfire IXs, took part in Ramrod 42, acting as escort to a box of 70 American Fortresses which attacked Rouen marshaling yards. Near Le Havre and again near Etretat a few Fw 190s and Bf 109s endeavored to attack the Fortresses, but were driven off by No.316, F/O Czesław Jaworowski destroying one Bf 109 and P/O Władysław Kiedrzyński, one Fw 190. On 13 March, on a very similar operation against Amiens marshaling yards (Ramrod 43), the same two squadrons, led as before by W/Cdr Kołaczkowski and S/Ldr Szczęsny, fought a bitter combat with enemy fighters. F/O Leopold Zakrzewski (No.316) brought down one Fw 190, and F/O Michał Najbicz (No.315) accounted for one "probable" Bf 109, but this time the squadrons lost two aircraft: F/Lt (Lt) Franciszek Kozłowski (No.316) was killed and F/O (Lt) Brunon Semmerling (No.315) was posted missing. The latter pilot baled out slightly wounded but evaded capture and, helped by the French Resistance, reached Spain, returning to the Squadron on 11 May.

In the second half of March, as the two Spitfire IX squadrons of No.1 Wing continued with bomber escort duties, No.303 and 308 Sqdns, flying Spitfire Vs, made a few sweeps over France and during one of these, on 20 March, Sgt Stanisław Górny (No.303) was wounded in his left arm by ground fire and crash landed at Lympne. On 26 March No.303 Sqdn went for 12 days to the Martlesham Heath armament school for firing practice.

In the second quarter of 1943 activities of No.1 Wing intensified considerably and in April 19 offensive operations – almost all of them at two-squadron strength, involving Spitfire IXs of No.315 and 316 Sqdns, with W/Cdr Wojciech Kołaczkowski as Wing Leader – and 22 defensive patrols were put up. The patrols were flown individually by all four squadrons of the wing or their sections. In addition, No.308 Sqdn was involved in four special "RAG" Exercises, by single aircraft or the whole squadron, during which one or two pilots carried on R/T messages or conversations in Russian using different intonations to simulate more "Russian" pilots. A ground controller also spoke Russian. This was intended to mislead the Germans and make them believe that Russian pilots were operating from England. In the middle of the month command of No.315 Sqdn passed from S/Ldr (Cpt) Tadeusz Sawicz to S/Ldr (Cpt) Jerzy Popławski.

Several encounters with enemy fighters were fought. On 3 April W/Cdr Kołaczkowski damaged one Fw 190 over the Channel, but the Poles, running very low on fuel, could not pursue the escaping enemy. The next day a major battle took

An extremely interesting photograph of what appears to be a Spitfire VIII in Polish markings. The fighter, featuring the extended-span, "pointed" wingtips, was among the Spitfire IXs received by No.316 Squadron at Northolt in March 1943. The pilots in front of the aircraft include F/Lt Gnyś (standing fourth from the left). (W.Gnyś/J. Lansdale)

place on Ramrod 51 when No.315 and 316 Sqdns, acting as withdrawal cover to 70 Fortresses, came to grips with Fw 190s. F/Lt (Lt) Marian Łukaszewicz and F/O (Lt) Przemysław Panek collided in mid-air near Rouen while turning to look for the enemy and were both killed. S/Ldr Henryk Szczęsny (Sector HQ), leading the attack, destroyed two Fw 190s and F/Lt Tadeusz Andersz (No.315), one. Two more were damaged and credited one each to S/Ldr Tadeusz Sawicz and F/Sgt Mieczysław Matus (both No.315). Three Spitfire IXs were shot down and three pilots lost – Sgt (Cpl) Tadeusz Ostrowski (No.315) was killed, and S/Ldr Szczęsny and F/O Bolesław Gęca were made PoW.[38]

On 16 April W/Cdr Kołaczkowski, flying some distance away from the rest of the wing, attacked and damaged a single Fw 190 that was shadowing the Polish squadrons. Four days later the wing bounced six Fw 190s which were flying ahead and F/O Ryszard Dubielecki shot down one and S/Ldr Jerzy Popławski and Sgt Tadeusz Jankowski damaged one each. Bad weather in the last week of the month interrupted activities. No.308 Sqdn, posted to No.2 Wing for rest, left No.1 Wing for Church Fenton on 29 April.

Although the scale of operational flying in May 1943 remained similar – except for a marked decline in the number of defensive patrols – heavy encounters with the enemy became more frequent. On 4 May No.315 and 316 Sqdns, led by W/Cdr Wojciech Kołaczkowski, flew first fighter cover to the Fortresses bombing the Ford Works in Antwerp (Ramrod 68) and repelled protracted attacks by German fighters. One Fw 190 was probably destroyed and one damaged by F/O Michał Maciejowski (No.316) and a further one damaged by F/O Ryszard Dubielecki (No.315). Two fighters were lost – F/Lt (Lt) Maciej Lipiński (No.315) was killed and F/O (Lt) Tomasz Łęgowski (No.316) was made PoW.

[38] Combat Diary, Royal Air Force Northolt. F/O Gęca managed to escape and return to England; from May 1944 he flew operationally with No.306 Sqdn.

On 13 May the wing's two Spitfire IX squadrons fought a bitter combat with German fighters, which continually attacked 80 Fortresses dispatched to bomb the Potez factory at Meaulte (Ramrod 71). The Poles observed for the first time a Fw 190 dropping light bombs on the Fortresses. One Bf 109 was credited as "probable" to W/Cdr Kołaczkowski and another to P/O Tadeusz Góra (No.316), who brought back exceptionally good camera-gun pictures of his target. Two Fw 190s were damaged, one each by F/O Stanisław Blok (No.315) and S/Ldr Zygmunt Bieńkowski (No.303 Sqdn, who on this occasion flew with No.316 Sqdn). P/O (S-Lt) Piotr Kuryłłowicz (No.315) was shot down and made PoW. Some of the Spitfires, including Bieńkowski's, were damaged by the Fortresses' defensive fire.

P/O Antoni Cholajda (No.316 Sqdn) resting on the wing of his Spitfire IX which bears his personal "dancing girl" marking and one German cross for an aircraft "probably destroyed" under an unreadable inscription under the windscreen; Northolt, the spring of 1943. (K. Chołoniewski)

The following day an equally determined battle was fought with about 50 Fw 190s attacking Fortresses on Ramrod 73. One Fw 190 was probably destroyed by F/O Stefan Karnkowski (No.316) over the target – Courtrai airfield – and four were deemed damaged – two by F/O Michał Najbicz (No.315) and one each by F/O Czesław Jaworowski (No.316) and F/O Tadeusz Żurakowski (No.315). No casualties were suffered. On 15 May 23 Spitfire IXs (from No.315 and 316 Sqdns and including G/Cpt Pawlikowski, Fighter Command) and 12 Spitfire Vs (No.303 Sqdn), assisted in Circus 297 as high cover. In combat near the target area, Caen airfield, F/O Stanisław Blok destroyed one Fw 190 and F/O Ryszard Dubielecki damaged another, but the Polish fighter force sustained one of its most painful losses – G/Cpt (Col) Stefan Pawlikowski was killed in this action. Pawlikowski belonged among the most distinguished and respected Polish aviation officers of the old generation, with unsurpassed combat experience.[39] The senior Polish Liaison Officer at Fighter Command HQ since November 1940, he strived to secure the greatest measure of independence for the Polish units within the Command – a goal which he largely achieved. Despite his position and age he insisted on flying operationally on several occasions. He was posthumously promoted to the rank of Brig General. Sgt Piotr Lewandowski (No.315) was also shot down on the same sortie and made PoW.

[39] He began his pilot's career in France in 1917, flew with the SPA.96 Escadrille on the Western Front in 1918, joined the Polish Aviation of Haller's Army and fought in the Russo-Polish War of 1919-20, held various command posts in the LW in the inter-war era and commanded the Pursuit Brigade in September 1939.

BS451 RF•M, one of the Spitfire IXs newly received by No.303 Squadron upon its return to Northolt on 1 June 1943.

On 17 May Spitfire IXs of No.1 Wing on Rodeo 217 clashed with enemy fighters again, this time over Amiens. Several pilots engaged the Germans, but only three were credited with successes – F/O Andrzej Próchnicki (No.316), with one Fw 190 probably destroyed, S/Ldr Janusz Żurakowski (flying with No.316), with one Bf 109 damaged and F/Lt Władysław Zając (No.315), with one Fw 190 damaged. On the same day Gen Montgomery arrived at Northolt and spent a few days at the Station. A number of uneventful operations, involving all three squadrons of No.1 Wing, followed in the later part of the month, except for Circus 305 on 28 May, during which four Bf 109s dived on the starboard section of No. 315 Sqdn and, shooting as they passed, shot down and killed Sgt (Sgt C-Off) Jan Rogowski.

June 1943 began with a considerable reorganization of the Polish fighter wings, which brought about changes of airfields by four squadrons on the first day of the month. No.316 Sqdn, the only Polish unit of No.1 Wing unaffected by these moves, was joined at Northolt by No.303 Sqdn arriving from Heston and converting to Spitfire IXs on the same day. No.315 was posted to No.2 Wing and went to Hutton Cranswick where it converted to a later variant of Spitfire V, the clip-winged LF VB and VC, better adapted for low-level ground-attack work. At the same time two Spitfire V squadrons left No.2 Wing for Heston, to bring the strength of No.1 Wing to four squadrons. These were No.302 from Kirton-in-Lindsey and No.317, which spent the whole of May at Martlesham Heath. No.2 Wing – comprising No.315 Sqdn at Hutton Cranswick, No.308 at Church Fenton and No.306, which on 30 May moved to Catterick – thus returned to its earlier three-squadron strength.

In the first quarter of 1943 the three squadrons of No.2 Wing spent most of their time training new pilots, extending their night-flying capabilities and perfecting tactical skills needed for combined operations. In February the squadrons of the wing began to work together and maintained alternate "Readiness" periods. In March defensive patrols became more frequent.

Offensive operational sorties by No.2 Wing began towards the end of March 1943, at first at two-squadron strength, with Nos.302 and 317 taking part. No. 306 Sqdn, which replaced No.316 in the wing on 11 March, converted to the low- altitude Spitfire LF VB and VC and had a short rest at Hutton Cranswick. On 17 March F/Lt (Cpt) Włodzimierz Karwowski took over command of No.306 as Acting S/Ldr from S/Ldr (Cpt) Kazimierz Rutkowski.

On 28 March 25 Spitfire Vs of Nos.302 and 317, taking off from Coltishall and led by W/Cdr Aleksander Gabszewicz, acted as rear support to 24 Venturas on No.12 Group's Ramrod 3, on their way back from Rotterdam. Two similar operations, involving take offs from Coltishall and protection of bombers returning from Rotterdam, were carried out by both squadrons the following day. No contact with the enemy was made on any of these missions.

The next four Ramrods were flown in April, two of them (Ramrods 6 and 7) on the 4th day. The first of these, escorting 105 American "heavies," was uneventful. On the second, about ten minutes after the wing had met the returning bombers off the Dutch Coast, 15 Fw 190s endeavoured to attack the formation. W/Cdr Gabszewicz engaged the enemy fighters and shot down one which was seen to hit the sea. A few other pilots fired but saw no results and the Germans fled. On 16 April No.302 and 317 Sqdns, led by W/Cdr Gabszewicz, went to Coltishall to act as back cover to Venturas attacking the Haarlem railway workshops on Ramrod 8. The following day No.306 and No.317 Sqdns assisted in Ramrod 9, acting as forward support to 65 Fortresses which attacked Bremen and Wilhelmshaven. The same afternoon 12 Spitfires of No.317 led by F/Lt Zbigniew Janicki took part in a Circus.

In the middle of April F/Lt (Lt) Zbigniew Wróblewski succeeded S/Ldr (Cpt) Zbigniew Czaykowski as Commander of No.317 Sqdn, becoming Acting S/Ldr. Three weeks later Wróblewski passed command of the squadron to S/Ldr (Lt) Franciszek Kornicki. A number of accidents in April 1943 cost the lives of three pilots: on 5 April Sgt (LAC) Brunon Makowski and F/Sgt (Cpl) Piotr Hanzelka (both from No.317), returning from Coltishall to Kirton-in-Lindsey in a Tiger Moth, were caught on take-off by a gust of wind, which caused the aircraft to crash, and on 24 April F/Lt (Lt) Józef Dec (No.302) was killed in a mid-air collision with another aircraft which landed safely.

On 29 April No.317 moved to Martlesham Heath and received the low-altitude Spitfire LF VB and VC. The squadron stayed at Martlesham throughout May, dividing its time between Sector patrols, convoy protection sorties and firing practice. On 15 May in a skirmish with Fw 190s off Orford Ness P/O Werner Kirchner probably destroyed one German and Sgt Tadeusz Tamowicz damaged another. During May No.302 and No.306 Sqdns took part in five Ramrods, but apart from combat with Fw 190s west of the Hague on 2 May, which resulted in the destruction of one enemy fighter by F/O Stanisław Brzeski (No.302), these were largely uneventful. On 16 May S/Ldr (Cpt) Wieńczysław Barański succeeded S/Ldr (Cpt) Stanisław Łapka as Commander of No.302 Sqdn. No.308 Sqdn, which with its arrival in Church Fenton on 29 April belonged administratively to No.2 Wing, was engaged on defensive patrols, shipping duties and night-flying exercises. On 14 May the squadron suffered the loss of two pilots, F/O (Lt) Adam Habela and Sgt (LAC) Jan Osoba, killed in a mid-air collision during training. On 16 May S/Ldr (Cpt) Paweł Niemiec passed command of No.308 to S/Ldr (Cpt) Józef Żulikowski.

The June 1943 activities of the Polish fighter force were affected by organizational changes, the second stage of which took place in the later part of the month. In the case of No.1 Wing, its two Spitfire IX squadrons – Nos.303 and 316 – continued to be heavily involved in the fighting, putting up 19 major offensive operations and seven defensive patrols during June. The two Spitfire V squadrons – Nos.302 and 317 at Heston

The "Aerial Post" Spitfire LF VB, EN919 JH•J, on a visit to No.315 Squadron at Hutton Cranswick. The aircraft is covered with greetings and messages from pilots of No.317 Squadron to their colleagues in No.315 Squadron.

– which stayed with the wing only for the first three weeks of June, flew fewer missions.

On 5 June No.302 Sqdn, already renowned for its effective close cover, was called upon to escort a Liberator in which Winston Churchill was returning from his visit to North Africa on the final stage of its journey to Northolt. At the end of their mission the pilots were ordered to maintain a low level patrol over the airfield until the cars of the VIP party had left Northolt. In the following two weeks No. 302 took part in two Ramrods, one Circus and one Rodeo, totaling 47 sorties, and No.317, in the same Circus and Rodeo.

On the first of these operations, Ramrod 86 on 10 June, 12 Spitfire Vs of No.302 Sqdn, led by F/Lt Bolesław Gładych, flew close escort to five Mitchells attacking a power station at Langebrugge in the Ghent area, and the Poles achieved a notable victory. Mjr Tadeusz Kwiatkowski – then F/O – flying in 'B' Flight, recalled the action:

"... As the bombers crossed the Belgian coast still gaining height, we climbed up to them to take our positions. At about 12,000 ft, half way towards the target, the Mitchells turned east, setting course for Langebrugge and two Spitfire IX squadrons of the top cover went missing. A few minutes later we were attacked head-on by some 20 red-nosed Focke-Wulfs which set on fire the port engine of one of the bombers. All our sections engaged the Germans, blocking their further attempts to get at the Mitchells. In a furious running battle with the dispersed enemy the squadron claimed several Fw 190s without loss ... Two days later we received a letter from the Mitchells' crews thanking us for the magnificent protection which probably saved their lives and congratulating us for our fighting efficiency."

The battle was fought between the area south of Knocke and Ostende. Polish victories were confirmed as follows: F/Lt Bolesław Gładych, flying in AR377, F/O Czesław Śnieć in EN865, P/O Jan Krajewski in BM651 and Sgt Jerzy Krzysztofiński in EN828, one Fw 190 destroyed each, F/O Bolesław Kaczmarek in W3765, one Fw 190 probably destroyed and Sgt Eustachy Łucyszyn in BM272, one damaged.

Nos.303 and 316 flew several Ramrods and some Circuses and Rodeos. On 11 June their 20 Spitfire IXs engaged 25 Bf 109s over the Beauvais area and F/O Michał Maciejowski (No.316) destroyed one. Three days later Sgt (LAC) Józef Dąbrowski of No.303 was killed on a training flight. Both squadrons were in combat again on the 17th; F/O Andrzej Próchnicki (No.316) damaged one Fw 190 and F/Lt (Lt) Stefan Stegman (No.316) was shot down and killed. The heaviest fighting of the month occurred on 24 June on Ramrod 102, when the Spitfire IX squadrons of No.1 Wing sighted two formations of enemy fighters. As they turned towards the larger one (20 plus aircraft), they were attacked from the sun by a third pack of Fw 190s. In the ensuing dog-fights Sgt Czesław Bartłomiejczyk (No.316) destroyed one Fw 190, and F/Lt Antoni Kolubiński (No.303) and F/O Józef Sobolewski damaged one each, but No.303 lost two pilots, P/O Józef Karczmarz and P/O Janusz Kobyliński, who were made PoW. A further casualty was sustained on 26 June (on Ramrod 108), when one of the escorting Spitfire IXs of No.316 Sqdn, piloted by Sgt (Cpl) Feliks Grzywacz, dived vertically from 30,000 ft for no apparent reason, an oxygen failure being suspected.[40]

Squadrons No.306 and 315 of No.2 Wing carried out a small number of offensive operations in June 1943. These included assistance to Ramrods 99 and 103 by No.11 Group, in which units of No.1 Wing also took part, and in one Ramrod and one Roadstead by No.12 Group, but no contacts with the enemy were reported.

The struggle for greater independence for the PAF and preparatory realignment of the PAF units to the likely needs of the emerging invasion plans brought about a further regrouping of the Polish fighter force beginning in the summer of 1943.

On 20 June W/Cdr (Cpt) Wojciech Kołaczkowski was posted to HQ No. 11 Group and W/Cdr (Cpt) Aleksander Gabszewicz took over command of No.1 Wing. W/Cdr (Cpt) Tadeusz Nowierski was given command of No.2 Wing at the same time. The two Spitfire V squadrons, Nos.302 and 317, were withdrawn from No.1 Wing and posted to Perranporth (in No.10 Group) on 20 and 21 June, where No.3 Polish Fighter Wing was about to be organized. The formation of this wing, also referred to as the Perranporth Wing, was completed on 29 June, with the appointment of W/Cdr (Cpt) Walerian Żak as its Commander. Both squadrons began operations almost immediately after their arrival and during the last week of June, in addition to convoy protection patrols, they carried out one fighter sweep and four escort missions over France. On the first of these, on the 24th, the Poles fought a desperate battle with Fw 190s north of Brest. No.302 pilots F/O Stanisław Brzeski (in EE662) destroyed one German and F/Sgt Władysław Giertkierewicz (in EP558) probably destroyed another. No.317 pilots F/O Janusz Walawski (in AB198) damaged two and F/O Marian Ciach damaged one, but Sgt (LAC) Edward Kostański (in EP360) crashed into the sea and was killed. In another engagement in the same area on 30 June, Sgt Konstanty Makosz (in ER516) was shot down and made PoW.

Final preparations of the Polish fighter force for "Operation Overlord" and its role in the invasion of Northern Europe are covered in Part III, Chapter I, of this work.

(4.) Operational Activities of No.307 "Lwów Eagle-Owls" Night Fighter Squadron

No.307 "Lwów Eagle-Owls" Night-fighter Squadron at Jurby entered the year 1941 continuing with daylight convoy protection duties and with its future role not yet clearly defined. Confirmation that the squadron would be switched to night operations as originally intended came on 6 January, when HQ No.9 Group advised that a detachment of five improved Defiants, fitted with new-type VHF R/T radios, would operate from Squires Gate from 9 January, providing night defence for the Liverpool area. The first five improved Defiants[41] were received at Jurby on 6 and 7 January and the PAF white-and-red chessboards were painted on the fuselages of the aircraft, under the pilot cockpits, for the first time. The detachment, led by S/Ldrs Tomlinson and Grodzicki, landed at Squires Gate, near Blackpool, on 9 January at noon, as ordered.

The crews were unfamiliar with the VHF radios and made their first flights from the new airfield the same afternoon to try the new equipment and acquaint themselves with the Sector. A two-aircraft section, with crews S/Ldr Stanisław Grodzicki, pilot, and F/O Lech Karwowski, air gunner, and F/O Roman Smok, pilot, and Sgt Zygfryd Blachowski, air gunner, was put on "Readiness" on 9 January at 17:30 hours and was "scrambled" to patrol over Formby just over an hour later. On landing in poor visibility F/O Smok ran into an unmarked deep ditch in the airfield and damaged his Defiant (N3315). A second "scramble" was ordered just before midnight, the first crew remaining the same and the second comprising Sgt Antoni Joda, pilot, and Sgt Wiktor Gandurski, air gunner. Sgt Joda had difficulties with maintaining radio contact with the ground control and went missing. Next morning reports were received that an aircraft was in the sea off Barmouth upside down and half submerged. This was Joda's Defiant (N3401); both crew members were killed.

These experiences seemed to indicate that the detachment's crews were diverted from daylight to night operations too suddenly and without adequate preparation. Although two more patrols were carried out within the next few days – both by single aircraft, piloted by S/Ldr Tomlinson – the crews were primarily engaged in intensive night-flying training, which was frequently interrupted by the bad January weather.

On 12 January notification was received that the whole squadron was to be provided with the VHF radio-equipped Defiants and would move to Squires Gate on 20 January. During the following few days the squadron returned most of its existing Defiants for modification and took on charge ten new Defiants, but the move from Jurby to Squires Gate had to be postponed because of snow and gale force winds. On 23 January, on the express orders of No.9 Group, a section of three Defiants attempted to reach Squires Gate, but only P/O Eugeniusz Przysiecki completed the move without mishap. F/O Jerzy Antonowicz force-landed in a field near Southport and subsequently continued to the new airfield and Sgt Jan Mikszo crash landed near Ormskirk, damaging his Defiant. The rest of the squadron finally reached Squires Gate on 27 January, with its last six Defiants flying in and ground crews arriving in Harrow transport aircraft. The move was completed on 1 February. Owing to a shortage of accommodations at the base, the personnel were dispersed to various hotels, making its control rather difficult. On 29 January two Battles[42] were allocated to the squadron to facilitate training.

[40] Combat Diary, Royal Air Force Northolt.

[41] Serial numbers N3315, N3435, N1809, N3391 and N3401.

[42] Serial numbers L5126 and P6685.

A post-war (1972) photograph of the restored Defiant I N1671, which was among the first to be delivered to No.307 "Lwów Eagle-Owl" Night-Fighter Squadron in September 1940.

On 3 February the whole squadron was put on "Readiness" and was required to keep half its strength at that state by day and another half by night. Intensive training was to continue at the same time, but the inclement weather enforced the grounding of aircraft. Training by day and night could begin only on 6 February and was interrupted again four days later by a return of wintry squalls.

On 9 February in the evening three Defiants were "scrambled" individually to patrol over the base and the Formby area and one crew flew close to an enemy raider, but was unable to see him. Another patrol was flown by two aircraft on the night of 15/16 February, when an enemy bomber dropped some flares in the vicinity of the airfield. After landing P/O Antoni Alexandrowicz collided during taxiing with a roller, which was left very close to the runway (the crew was unhurt). On 25 February F/O Jerzy Antonowicz, starting a take-off run on night patrol, hit a lorry on the runway, killing its RAF driver. The pilot managed to retain control of the damaged aircraft and the crew was not hurt.

N1671, as it is preserved in the Hendon RAF Museum. The Defiant carries now the code EW•D (the individual D letter being assumed) and No.307's squadron markings, which in fact never appeared on the unit's aircraft until the middle of 1945.

A number of other accidents occurred on training flights. The most serious of these took place in the early afternoon on 17 February, when a Defiant (N3314), piloted by Sgt Kazimierz Bocheński and with air gunner Sgt Kazimierz Frąckiewicz, left a formation of three aircraft and began to execute low altitude aerobatics. The machine hit the ground almost vertically, killing the crew. In the middle of the month most of the British personnel were posted away from the squadron. During February only six aircraft sorties were logged by the squadron in night operations, in 4 hours 50 min flying time, and none by day, as compared with 10 hours 10 min operational flying by day and 4 hours 5 min by night in January.[43]

By early March 1941 differences of opinion between S/Ldr Grodzicki – who thought that, with operational duties and extensive training running concurrently, too much was expected of No.307's crews – and No.9 Group – who pointed to a large number of accidents and problems with radio communications, arising from language difficulties, and suggested that the Poles might not be well suited for night operations – became more intense. At the same time the crews grew more and more frustrated by the lack of any success with night interceptions.

On 7 March No.9 Group decided to limit the operational duties of the squadron to six aircraft, which were to be kept on "Readiness" only during the hours of darkness, in order to make more time available for practice flying. Training for "Shortshrift" – the night fighting tactics involving five aircraft formed one above the other with 1,000 ft vertical separation and the lowest flying at 17,000 ft – which began towards the end of February, was stepped up and cooperation with searchlights was also expanded. G/Cpt Pawlikowski, Polish Liaison Officer HQ Fighter Command, visited the squadron to help to resolve its difficulties.

On the night of 12/13 March seven patrol sorties were flown along the Formby-Denbigh line ("George South") and at about 22:00 hours the first night encounter with the enemy was fought by a No.307 crew, Sgt Franciszek Jankowiak and Sgt Jerzy Karais. The pilot saw to port a He 111 silhouetted against white clouds, flying below and in front. The Defiant (N3739) dived below the Heinkel and made a 3/4 attack. The gunner fired a burst from 80 yards and the pilot observed his bullets enter the center part of the bomber's fuselage. The German opened fire at the same time, but missed the Defiant. The Poles broke away to renew their attack, but the enemy could no longer be found. Sgt Jankowiak was credited with one He 111 damaged.[44] Other crews also reported brief sightings of enemy bombers but lost them when manoeuvering to the firing position.

News of the first encounter traveled among the squadron's personnel with the speed of wildfire and some crews on "Release" arrived at the airfield on foot in the middle of the night, begging to fly. The dampened spirits were instantly revived.

The following night the whole squadron was put on "Readiness" in expectation of increased activities by the Luftwaffe and three Hurricanes from No.229 Sqdn arrived to cooperate with the Defiants. In addition to continuous patrols, a "Shortshrift" was also to be carried out over Liverpool. Around midnight, F/O Maksymilian Lewandowski and Sgt Marian Niewolski, participating in the "Shortshrift," spotted a twin-engined enemy aircraft. The air gunner opened fire, hitting the nose and starboard engine of the bomber. The German turned violently to port and dived straight down. The gunner fired another burst from 70 yards and the enemy vanished from sight. The bomber, initially credited to F/O Lewandowski as "probable," was later reclassified as damaged.

Bad weather stopped flying for several nights. On 18 March S/Ldr Tomlinson was posted away from No.307 and S/Ldr Grodzicki took over overall control of the squadron. On 26 March No.307 was transferred to No.10 Group and moved to Colerne, a new airfield, not yet fully finished. The Group's attitude towards the squadron was far more accommodating. First daylight Sector reconnaissance flights from Colerne were made on 28 March and continued for a number of days. The prevailing weather prevented night flying until early April.

Night patrols and intensive night training began on 6 April 1941. A Do 17 was seen two nights later, but as soon as the Defiant's gunner opened fire, the enemy escaped into the clouds. On 10 April in the afternoon, Sgt (Cpl) Maksymilian Frychel and LAC (Cpl) Jan Dziubek, a radio mechanic, returning from a radio test, were killed when their Defiant lost speed and plunged to the ground on the landing approach. The following night one crew fired at a He 111 without result.

The squadron's first confirmed victory came on 12 April at 01:35 hours, after Sgts Franciszek Jankowiak and Józef Lipiński were vectored towards an enemy aircraft. The pilot recognized a He 111 about 1,000 ft in front, flying just above the clouds in the opposite direction. The Defiant EW•C (N3315) turned sharply to the left and circled the bomber at 40 yards distance. The air gunner fired several bursts and the crew saw

Sgts Jankowiak (pilot) and Lipiński (gunner), who achieved No.307's first confirmed victory on 12 April 1941, depicted by a Defiant I N3437 EW•K at Exeter. (IPMS)

[43] The above details and figures come mainly from No.307's Operational Diary (IPMS, Lot A V,51/3b) and the RAF Forms 540 (A V,51/6).

[44] Intelligence Combat Report, Form F (IPMS, Lot. A V,51/3b). Although the Defiant carried no pilot's guns and only the air gunner could fire, all the victory credits were recorded against the pilot's name.

pieces of metal flying off the Heinkel and a number of small explosions. The raider dived steeply and was lost in the clouds. Reports soon verified that the He 111P-2, belonging to 9/KG 27, code letters 1G+HT (c/n 2002), crashed and exploded at Lydlinch, near Stalbridge, killing its entire crew. A part of its wing-spar was later nailed to the door of 'A' Flight's hut as a trophy.

The next night, the crew of F/Lt Mieczysław Jakszewicz and Sgt Józef Starosta engaged single Heinkels on four different occasions during one patrol, firing a total of 1,780 rounds. They saw no conclusive results and filed no claims. They thought it likely that the first three combats involved the same bomber, which they kept losing in the darkness. On the night of 15/16 April F/O Maksymilian Lewandowski and Sgt Mieczysław Zakrocki engaged a He 111, rapidly firing several bursts. The German lost height and smoke poured from one of its engines. The Defiant followed it down to 5,000 ft, expending all ammunition and the crew was credited with one Heinkel damaged.

In the second half of April more night patrols were made, but these were largely uneventful. On 24 April Sgt Witold Dukszto and Sgt Jan Banyś had to abandon their Defiant during an operational sortie owing to engine failure. Both parachuted to safety.

On 26 April the squadron moved to Exeter, where it was destined to stay for the next two years. Operational flights from the new airfield began the following day. In April 1941, No.307 carried out 56 night and seven day sorties in 59 h 40 min and 8 h 55 min respectively. Five replacement T-serial Defiants were received.

On the night of 4/5 May the crew of P/O Eugeniusz Przysiecki and Sgt Jan Woźny, who twice encountered the enemy and attempted to engage, experienced difficulties in finding the airfield, which became shrouded in fog and ran out of fuel in its vicinity. Both crew members had to jump to safety. On 8 May Flight 'B' was detached to Pembrey. The same night the Exeter airfield was severely bombed and a number of the Polish personnel were buried under rubble. All were recovered without serious injuries.

On the night of 11/12 May the airfield suffered even heavier attacks and No.307 was involved in combat. The squadron's first patrol landed before the Germans struck. A two-aircraft section was ordered to "scramble" on a second sortie just as the second wave of bombers began to release their load. Only one crew, Sgt Jerzy Malinowski and Sgt Stanisław Jarzembowski, managed to take off (at about 3:00 hours) as the bombs started to fall, the other crew being unable to reach their aircraft. Flying at about 1,000 ft Malinowski sighted a He 111 against a background of clouds. After a long chase the Defiant (N3403) made a beam attack, the gunner hitting the central part of the fuselage. The bomber turned in towards the Defiant and dived gradually as the gunner fired again. The Heinkel rolled on its back and went vertically into the sea. This was the squadron's second confirmed victory. Both crew members were later decorated with the Virtuti Military Silver Crosses. During the bombardment one Polish Defiant was destroyed and two others and a Battle were damaged.

A number of uneventful patrols followed later in May. The Pembrey detachment was recalled to Exeter on 18 May and at the end of May the squadron learned that they were to be rearmed with a new night fighter with forward-firing armament. The news was enthusiastically welcomed. During May the squadron put up 90 night sorties.

Although in June 1941 night patrols continued with only slightly lesser intensity, no contacts with enemy aircraft were reported. By that time the Sector Controllers began to pay less attention to the Defiant squadrons, concentrating their efforts on helping the more efficient Beaufighter and Havoc night-fighters. A total of 69 sorties by night and 17 by day were flown during the month. On 11 June S/Ldr (Cpt) Jerzy Antonowicz took over command of the squadron from S/Ldr Grodzicki.

A similar pattern of activities extended into July 1941, but – with a decline in Luftwaffe operations against Britain because of the attack on the Soviet Union – No.307 reverted in the later part of the month to shipping protection duties, which briefly took precedence over its night work. Cordial relations were established with No.317 which arrived at Exeter on 21 July. In the first half of August, with re-equipment pending with new aircraft, only a handful of sorties were made. The last two patrols by Defiants were flown on the night of 15/16 and the last practice "Cooperation" flights (with Hurricanes), on 26 August.

On 12 August one Blenheim I (L8436) was allocated to the squadron to facilitate conversion to the Bristol Beaufighter IIF twin-engined night-fighters. The first four Beaufighters[45] that were delivered two days later and by 18 August 13 had been taken on charge. During the next eleven days all but two of No.307's Defiants were ferried to various Maintenance Units. The Beaufighter, carrying a forward-firing pilot's armament of four 20-mm Hispano cannon and six 0.303 in. guns, and the 1.5 meter AI (Airborne Interception) Mk IV radar with a range of 4 miles which was operated by the radio navigator/observer, was a formidable weapon, but the aircraft was tricky to handle under certain conditions. It had a tendency to swing on take-off and in the event of one engine cutting suddenly, it was virtually uncontrollable. On landing, its large flaps pulled the aircraft up rapidly, but stability problems were also experienced and led to frequent ground loops (the machine's c.g. being too far aft).

Intensive training on the new type, which involved not only handling of the aircraft but also learning how to use the new, and highly secret, AI radar,[46] began in the middle of August and lasted two months, during which No.307 was non-

[45] Serial numbers R2443, R2446, R2447, R2449.

[46] No.307 was the first non-British unit to operate the AI radar-equipped aircraft and all its personnel were obliged to sign a special secrecy pledge.

operational. A second Blenheim I (L1186) was assigned to the squadron and a number of its air and ground crews were posted to RAF units already operating the Beaufighters for various courses. Several personnel changes also took place.

The first accident involving a Beaufighter happened on 30 August, when one of the Merlin engines cut immediately after take off, making it bank sharply. The pilot managed to bring the aircraft down straight ahead, and although it crashed and burst into flames, its crew, Sgt Jerzy Malinowski and LAC Zygmunt Modro, escaped serious injury. A second accident occurred less than a month later, on 27 September. P/O (S-Lt) Witold Gayzler, flying with P/O (S-Lt) Włodzimierz Pfleger, encountered very low clouds and crashed into a hill. Both crew members were killed instantly. Between 20 August and 30 September 1941 No.307's pilots carried out 396 flights in 382 flying hours and 22 of them completed training on Beaufighters. On 30 September the squadron had on charge 17 Beaufighter IIFs, 2 Blenheim Is, 1 Defiant I and 1 Magister.[47]

No.307 became operational on Beaufighters on 15 October 1941, its first three crews being put on "Readiness" on that day. Operational sorties began the following night. On the night of 23/24 October the crew of S/Ldr (Cpt) Jerzy Antonowicz and F/O (Lt) Lech Karwowski on a patrol south of Exeter, suffered an engine failure. The pilot returned on one engine, but the Beaufighter (T3025) crashed on landing, killing both crew members. F/Lt (Lt) Maksymilian Lewandowski became the Acting C.O. of the squadron, passing over command to S/Ldr (Mjr) Stanisław Brejnak on 16 November.

The first combat of the squadron's Beaufighters with the enemy was fought on 28 October at about 20:00 hours, when the crew of Sgt Bolesław Turzański and Sgt Henryk Ostrowski, radar operator, obtained a radar contact with an enemy raider near Okehampton and fired four bursts from about 80 yards. The enemy rapidly dived into clouds and disappeared. The claim for one enemy "probably destroyed" was eventually rejected as not corroborated. Two nights later P/O Ryszard Zwoliński and P/O Zbyszko Lissowski made a similar contact and briefly fired at a bomber, but made no claim. By the end of October No.307's Beaufighters had carried out 33 night sorties in 72 flying hours. During the month the last Defiant (N3404) left the squadron, and one Oxford and one more Blenheim I were received for training.

Real success came on the night of 1/2 November. Sgt Bolesław Turzański, with Sgt Henryk Ostrowski as radar operator, attacked and shot down two Do 217s in one sortie. Both crashed into the sea, one near Sidmouth and the other off Watchet. One of the aircraft was identified as Do 217E-2, U5+EM (c/n 1123), belonging to 4/KG 2. Several congratulations were received; two most appreciated ones came from AM Sir Sholto Douglas, C.-in-C. Fighter Command and from AVM (Gen) Stanisław Ujejski, Inspector General of the PAF.

During the night of 23/24 November another bomber was claimed by a No.307 crew, this time over Plymouth, in a most

A Beaufighter IIF R2445 EW•A in the late autumn of 1941 at Exeter. (IPMS)

spectacular kill. The victim, a Ju 88A-4, 4D+GS (c/n 3416), falling from a considerable height, crossed the night sky like a flaming torch, visible to thousands of people for miles around, before it crashed at Porsham Farm in Tamerton Foliot. The victors, P/O Antoni Dzięgielewski and P/O Mieczysław Świerz, flying in Beaufighter EW•B (R2440), received hundreds of appreciative letters and some time later a postal order for 2s 6d, with the request that it be handed to the pilot for a drink "as thanks for the greatest spectacle" the sender, a taxi-driver, had ever seen.

Two nights later P/O Marceli Neyder and Sgt Jan Woźny engaged a Ju 88 near Church Stanton and damaged it, and on 28 November at about 20:00 hours Sgts Bolesław Turzański and Henryk Ostrowski added the third air victory to their November score (all achieved in Beaufighter R2379), shooting down a twin-engined German aircraft (type not identified) south of Sidmouth; its fall into the sea was confirmed by witnesses on the ground.

November 1941 was one of the most eventful and successful months in the history of No.307. The monthly score 4-0-1 was a clear indication that the squadron, after protracted and very difficult beginnings, had begun to excel in night combat tactics and mastered the radar interception techniques. During the month 48 night sorties in 85 flying hours (and none by day) were logged. On 30 November No.307's equipment comprised 19 Beaufighter IIFs, 2 Blenheim Is,[48] 1 Oxford and 1 Magister. Its flying personnel totaled 18 officer and 15 NCO pilots and 9 officer and 15 NCO radar operators; ground personnel totaled 13 officers (including 5 British) and 304 other ranks (including 44 British).

December saw less activity because of persistent fog and the patrols that were made were uneventful. In 1941 No.307 Sqdn carried out 534 sorties in 706 flying hours.

The first quarter of 1942 was a depressing one for No.307. Bad weather and very limited night operations by the Luftwaffe resulted in no encounters, and a number of accidents deepened the gloomy atmosphere. The second week of February became known as the "Black Week" of the squadron. On 11 February just before midnight Sgt (Cpl) Kazimierz Thiesler, pilot, and Sgt (LAC) Edward Podgajny, radar operator, returning from a patrol, were killed when their Beaufighter hit a tree during the

[47] No.307's Operational Diary (IPMS, Lot. A V,51/3b).

[48] One Blenheim crashed on a training flight on 21 October 1941.

No.307's Beaufighter IIFs in flight, with R2390 EW•U nearest.

landing approach in fog. Two nights later F/O (Lt) Zdzisław Kurek and F/O (Lt) Józef Skrouba on GCI (Ground Control Interception) practice, died in a crash near the airfield when one of the engines of their aircraft stopped owing to fuel starvation, the pilot apparently forgetting to switch over to a full tank. On 14 February P/O (S-Lt) Stefan Maxymowicz, flying with F/O (Lt) Jerzy Łazarowicz to test Beaufighter R2445 after an engine change, made a low-level pass over the airfield, which was intended "to lift the personnel spirit" after the two earlier accidents, and crashed after catching a tree with the port wingtip.[49] The pilot died instantly and Łazarowicz was very seriously wounded; his leg was amputated, but he eventually recovered. Other less serious accidents during this period resulted in the loss of a further three Beaufighters.

The first half of April 1942 was equally daunting. On 3 April an engine failure immediately after take-off on a night patrol brought about a crash in which F/O (Lt) Grzegorz Bukowiecki and F/O (Lt) Stanisław Puzyna lost their lives. On the night of 12/13 April S/Ldr (Cpt) Roman Smok and P/O (S-Lt) Mieczysław Świerz were killed when their Beaufighter crashed in the sea near Sidmouth on an operational sortie; an engine malfunction was suspected.[50] In the middle of the month the squadron's command asked for its Beaufighter IIs to be replaced by the Beaufighter VI, which was powered by a more reliable Bristol Hercules radials and had improved handling characteristics.

[49] S/Ldr Lewandowski's letter, published by Andrzej Janczak in his article "Czarny tydzień Dywizjonu" (The Black Week of the Squadron") in "Skrzydlata Polska" magazine in 1976.

[50] Although the Rolls-Royce Merlin XX Vee engine was generally very good, the units mass-produced in the Crewe factory by the rapidly expanded work force lacked reliability.

On 23 April the Luftwaffe launched the first of a series of so-called Baedeker night raids[51] against cathedral cities in Britain, to terrorize the British civilian population. In the opening hours of this new Blitz, on 23 April before midnight, the crew Sgt Władysław Illaszewicz and P/O Zbyszko Lissowski attacked and hit a Ju 88, which went down towards the sea and was credited to them as probably destroyed. This, the first combat in 1942, opened a new round of successes for No.307 Sqdn and revived the badly flagging spirit of its personnel.

Two nights later the same crew achieved another "probable," a Ju 88, in the same area south of Sidmouth, and another crew, F/Lt Marceli Neyder and Sgt Jan Woźny, chased a Ju 88 escaping towards the French coast and destroyed it not far from Berok at the last possible moment. The pilots were not allowed to fly farther than the middle of the Channel, so that the AI radar might not fall into enemy hands. The next night the Neyder/Woźny team added one Ju 88 damaged to its score and F/O Mieczysław Pietrzyk and F/O Czesław Krawiecki accounted for another Junkers bomber probably destroyed. Up to ten sorties per night were flown in the last week of April and during the month No.307's crews logged 206 flying hours in night operations (in addition to five hours by day).

The first week of May was equally busy and the night of 3/4 May brought the squadron a spectacular victory – four enemy aircraft destroyed. During a heavy German raid on Exeter three Polish Beaufighters took off soon after 01:00 hours and were vectored by GCI Station Exminster towards the enemy. The crew of Sgt Illaszewicz and P/O Lissowski made contact with a Junkers bomber which was flying very slowly and set it alight with the first burst. The Ju 88A-5 4D+FU (c/n 5169)

[51] Named after the famous German tourist guide books, from which the targets were selected.

from 10/KG 30 crashed in the Topsham Barracks area, Exeter, killing its crew. The same crew was subsequently directed to another raid and after a few minutes' chase the pilot sighted two bombers, while the radar operator had contact on a third. The pilot attacked the leader of this loose formation and closed in firing. The German took evasive action and fired back. A running fight followed until the pursued Ju 88A-5 3Z+HW (c/n 0331) from 12/KG 77 crashed at Colyton at 02:55 hours.

F/O Stanisław Andrzejewski and Sgt Stanisław Kaliszewski found another Junkers, which they engaged and hit. The crew observed the bomber to spiral down in flames and to fall into the sea. The third crew, F/L Marceli Neyder and Sgt Jan Woźny, after a four-minute AI radar chase, spotted the enemy dead ahead and got in a short burst, setting his petrol tank on fire. Their Ju 88 dived steeply and crashed into the sea.

The self-satisfaction of the jubilant crews was somewhat shaken when upon landing they were given the news that the debris from one of their victims fell upon their living quarters in the Topsham Barracks and burnt all their belongings. P/O Zbyszko Lissowski received the DFC after this victory, the first to be awarded in the unit, and several congratulation signals arrived, including one from Sir Archibald Sinclair, the Secretary of State for Air.

This action was the swan song of the Polish Beaufighter IIs. On 1 May 1942 No.307 received its first seven Beaufighter VIFs and completed re-equipment with the new model within a week. The conversion training progressed well, and on 13 May all its Beaufighter VIFs were put on "Readiness." The first three patrols in the new aircraft were made on the night of 19/20 May. All the earlier Beaufighters left the squadron in the second half of May. By the end of the month, altogether 26 operational night sorties were logged on the Beaufighter VIs, but the Luftwaffe activities diminished after 5 May and no contacts with the enemy were reported. On 6 May the Squadron gave a farewell party to No.308 Sqdn, which operated from Exeter for the previous five weeks.

For most of June the crews were involved in training and in routine defensive patrols, which, due to the absence of the enemy in the air, were uneventful. One aircraft crashed and burned out upon landing from a patrol, when one of its tires burst, but the crew escaped injury.

The Germans reappeared again towards the end of the month and in the early hours of 26 June the Beaufighter VIFs achieved their first successes. F/O Karol Rach and F/O Jan Mika gave chase to a Do 217 and destroyed it after firing a few short bursts. Less than half an hour later Sgt Edward Wojczyński and Sgt Emil Słuszkiewicz sighted another Do 217 and saw a violent explosion as the pilot's bullets hit the aircraft. The enemy went down, but was lost in haze and was credited only as damaged. Still later F/O Gerard Ranoszek and Sgt Jerzy Trzaskowski accounted for one more Do 217 damaged, the crew observing many strikes before the bomber vanished into cloud.

The following night F/Sgt Bolesław Turzański and Sgt Henryk Ostrowski made contact with a Do 217 which, after two bursts, caught fire and was seen to crash into the sea. Thus, No.307's June score rose to 2-0-2. During the month 135 night and 2 day sorties were made in 295 hours 15 min and 1 hour 40 min respectively. The duration of some night patrols began to exceed 3 hours.

On the night of 1/2 July P/O Stefan Podgórski and F/Lt Stanisław Sawczyński made contact with a Do 217 and engaged it from about 80 yards, seeing several hits. The enemy

A Beaufighter VIF EL146 EW•V, carrying Gen Sikorski, lands at Exeter on 10 September 1942.

dived into cloud and the crew lost it. They were credited with one Dornier damaged, but it appeared that when the same bomber came out of the cloud, it was fired at by AA batteries which subsequently claimed it destroyed. On the night of 17/18 July one crew made two contacts with the enemy in the course of one patrol, but both were inconclusive, the second being abandoned owing to shortage of petrol. During the same night F/Sgt (Sgt) Władysław Illaszewicz and P/O (S-Lt) Zbyszko Lissowski lost their lives in the sea after the pilot reported a failure of one engine and warned about his "emergency landing" attempt over the R/T. On 30 July P/O Podgórski and F/Lt Sawczyński made successful parachute escapes while on patrol, when their aircraft unaccountably went out of control, the loss of an aileron being suspected. They were picked up by a rescue launch. On 23 July S/Ldr (Cpt) Jan Michałowski took over command of the squadron from S/Ldr (Mjr) Stanisław Brejnak.

On 1 August 1942 in the afternoon a Beaufighter VIF on a test flight crashed and burst into flames after its engines had cut. In addition to the crew, the aircraft was carrying a British passenger, Sgt Ronald Newman of the ATC. F/Sgt Józef Malinowski, pilot, and Newman, were only slightly injured, but Sgt (Cpl) Stanisław Jarzembowski was killed.

On the night of 4/5 August three more confirmed victories were added to No.307's score. F/O Gerard Ranoszek and Sgt Jerzy Trzaskowski, flying in Beaufighter X8005, were directed from the West onto a northbound raid in the Kingsbridge area and obtained contact with a Ju 88. The pilot closed to nearly point-blank range and gave one very short burst with cannon only. The Ju 88A-4 M2+CL, from 3/KüFlGr 106, immediately went down in flames, narrowly missing the Beaufighter and crashed in the sea off Start Point. Within less than 10 minutes the Polish crew found another raider, which turned out to be the Ju 88A-4 M2+HL from the same German unit, and shot it down, the enemy exploding in a ball of fire and falling at Molescombe Farm, Frogmore, near Kingsbridge. Both combats were watched by some coast guards. The third victim was accounted for by F/O Stanisław Szabłowski and Sgt Henryk Gajewski just before 03:00 hours. The pilot, guided to his target, caught a glimpse of the enemy and approached from below, aiming at his bright exhausts, and fired. The bomber's starboard engine exploded and the pilot had to pull up sharply to avoid flying debris. The aircraft, the Ju 88A-4 M2+CK, also from 3/KüFlGr 106, was seen burning on the sea.[52]

Gen Sikorski attending No.307's Feastday celebrations on 10 September 1942.

The remainder of August was quiet with nothing to report, and the next combat took place on the night of 4/5 September, when Sgt Jan Pacholczyk and Sgt Tadeusz Więckowski fired four bursts at a Ju 88, observing strikes on its fuselage before the enemy disappeared into cloud and AI contact with him was lost. The Junkers was credited to Pacholczyk as damaged.

On 10 September, the squadron's Feastday, Gen Sikorski arrived from London in Beaufighter VIF EW•V (EL146) piloted by S/Ldr Michałowski and decorated several of the unit's

[52] Intelligence Combat Reports (IPMS, Lot. A V, 51/3c), RAF Forms 540 (A V,51/6) and "The Blitz Then and Now," Vol.3.

S/Ldr Gerard Ranoszek's Beaufighter VIF X8005 EW•R, with four swastikas (denoting his four air victories) behind the cockpit enclosure. The aircraft carries the new universal camouflage scheme of Dark Green and Medium Sea Gray. (J. Koniarek)

A Beaufighter VIF EL154 EW•Z, at Exeter in the autumn of 1942.

members. The later famous squadron anthem "The Lwów Eagle-Owls," composed by F/O Jan Maliński, was performed for the first time during the festivities, which were attended by a number of Polish Ministers of State, senior RAF officers, members of the Diplomatic Corps and the City Fathers of Exeter.

On the night of 24/25 September No.307 fought its last successful engagement in 1942. S/Ldr Gerard Ranoszek, flying with F/Lt Stanisław Sawczyński, radar operator, in his EW•R (X8005), made contact with a twin-engined aircraft, presumably a Do 217, and attacked. In the exchange of fire, the bomber was seen to be hit and after a running fight one of its engines began to smoke heavily. The pilot closed in for a kill and almost immediately completely lost all his forward vision as oil from the German machine smothered the Beaufighter's windscreen. The Pole began to circle and saw the wreckage on the sea out of his side window. As his attempts to clean the perspex with deicing fluid failed, he returned to base flying blind and managed to land successfully. This was No.307's 17th confirmed victory, with no losses suffered from the enemy in combat.

In the last quarter of 1942 the enemy appeared rarely and the Squadron settled into the old, boring operational routine – nightly defensive patrols, occasional daylight "scrambles" and some convoy protection sorties – with nothing interesting to report. A few accidents occurred, one of them fatal: a crash of a Beaufighter on landing on 8 November resulted in the death of Sgt (LAC) Wacław Gajek, a Polish volunteer from the USA training to become radar operator; the pilot, S/Ldr František Rylp, a Czech temporarily attached to No.307, escaped with minor injuries. On 20 December the squadron began to receive the Mosquito NF II night-fighters and by 24 December had on

The new "arrow-head" radar on No.307's Mosquito NF II trainer, devoid of guns, which was received by the squadron towards the end of 1942. (IPMS)

S/Ldr Ranoszek's new Mosquito NF II night fighter DZ271 EW•R, delivered to No.307 Squadron at the end of 1942.

charge 21 aircraft of this type. The Mosquitoes were equipped with the superior AI Mk V "arrow-head" radars, and their speed and handling characteristics delighted the Polish crews.

During 1942 No.307 Sqdn put up 1,050 sorties – almost twice the 1941 figure – in 2,176 flying hours – more than three times the 1941 figure. To ease a shortage of fully trained Polish flying personnel, particularly radar operators, a few No.307 crews were Anglo-Polish and remained so for some time to come, with the British pilots and operators playing their full part in operational activities of the unit.

The deliveries of Mosquitoes, which besides their excellent night-fighting capabilities were also ideally suited for offensive intruder operations, hinted at a forthcoming widening of the squadron's operational scope, in line with the overall transition of the RAF from the defensive to an offensive posture.

During January 1943 the Mosquitoes gradually superseded the Beaufighters on operations. The first Mosquito sortie was made by S/Ldr Gerard Ranoszek and F/O Czesław Krawiecki (on EW•R DZ271) on 14 January 1943 at 20:00 hours, and the last Beaufighter patrol was flown (in X8148) on the last night of January. S/Ldr (Mjr) Antoni Alexandrowicz, formerly of No.307 Sqdn, and his navigator, F/O (Lt) Zbigniew Domański, both serving with No.219 (RAF) Sqdn, were killed on 14 February, when their Beaufighter flew into dense fog during a low-level practice flight and hit high ground in the Ireleth area (Cumbria).

On the night of 15/16 February F/O Ryszard Zwoliński and Sgt Henryk Gajewski, on a patrol in DZ289, fought No.307's first Mosquito combat which resulted in the probable destruction of a Ju 88. During the rest of February and the following two months the Luftwaffe was largely absent from the scene. This allowed the squadron to begin training for the "Ranger/Intruder" tasks awaiting it in the near future. However, these activities were not free from accidents. The most serious one happened on 21 March, when W/Cdr (Mjr) Jan Michałowski, flying with F/O (Lt) Stanisław Szkop, stalled during a single-engine landing approach after a practice flight and were both killed in the crash. Command of the squadron passed temporarily to S/Ldr (Cpt) Gerard Ranoszek, who was succeeded by S/Ldr (Mjr) Jerzy Orzechowski on 1 April 1943.

On 15 April No.307 Sqdn moved from Exeter to Fairwood Common. The unit left Exeter with much regret, after a very happy association over some two years with the RAF Command and the local population. The squadron flag was presented to the City of Exeter and hung in the Guildhall, "as a lasting link between Exeter and the 307 Lwów Squadron. 'Semper fidelis' appropriately is the motto of both cities."[53]

At Fairwood Common some relaxation of the squadron's night readiness state was to help the unit to proceed with a concentrated program of training for the "Ranger/Intruder" scheme. The radar operators in particular had to be fully retrained to their new radio/navigator role. No.307 was reorganized by the addition of 'C' Flight which had no regular air crews, only a commander and maintenance personnel, and few Mosquitoes from which the AI radar had been removed (for security reasons, to permit them to fly over enemy held territories) and the "GEE Box" navigational system installed. While 'A' and 'B' Flights carried on with their normal operational duties, some of their crews were now and then detailed to fly 'C' Flight's aircraft across the Channel on "Ranger/Intruder" practice. This training was to be fully completed by all crews by 26 June.

Three sorties made on the night of 15/16 May included No.307's first Ranger operation by one crew, F/O Roman Grzanka[54] and F/O Konrad Małuszek, who penetrated into France, but owing to heavy ground mist had no success in finding any targets. The following night two crews carried out the first Intruder operations, taking off from Castle Camps airfield and setting course for the Dutch coast. These were: F/O Jerzy Damsz flying with F/O Witold Sylwestrowicz, and F/O Grzanka with F/O Małuszek. Their sorties coincided with No.617's famous attack on the Moehne and Eder dams and formed part of the large-scale diversion tactics. Two more Rangers were put up on each of the next two nights, involving three new crews. A few locomotives and other targets of opportunity, including searchlights and two control boats at the end of seaplane flare paths, were shot at and hits observed, with one of the boats being claimed as sunk. Some AA fire was experienced and F/O (Lt) Antoni Dzięgielewski and F/O (Lt) Adam Węgiel, radio/navigator, were shot down and killed on the night of 18/19 May. Another crew, F/O (Lt) Jan Fusiński and F/O (Lt) Stefan Tabaczyński, lost their lives on the night of 27/28 May, when attacked by a German nightfighter in the Lahn area (North-West Germany).

During the last week of May night patrols and Rangers continued and by the end of the month five crews were fully trained for the latter operations, but then the training had to be slowed down as further commitments were added to No.307's list of operational duties. In view of increased Luftwaffe activities over Britain and the Bay of Biscay, the squadron was requested to provide three Mosquitoes at Predannack (effective from 4 June) and also to take over from No.264 Sqdn night patrols in the Portreath Sector (Cornwall). This stretched the unit's resources to the limit.

In the first half of June 1943 No.307 began 'Instep' patrols over the Bay of Biscay. Their purpose was explained by the Intelligence Officer: "Ships with foodstuffs, troops and munitions cross the Atlantic. They are ours. The enemy sends out U-boats against these ships. Against these U-boats we send out Wellingtons and Sunderlands. Against these Wellingtons

[53] RAF Form 540.

[54] Grzanka, serving with the 1st Air Reg. in Warsaw as a fighter pilot in the early 1930s, lost one leg in a SPAD 61 crash. His determined efforts brought about his return to flying duties with the LW. He evacuated to France and then to England and became one of the very few one-legged pilots to fly operationally from Great Britain.

A Mosquito NF II HJ932 EW•L, used for intruder operations in the spring of 1943. (W. Sylwestrowicz/J. Koniarek)

and Sunderlands they send out Junkers 88s and Messerschmitt 110s. Against these Ju 88s and Me 110s we send out Beaufighters and Mosquitoes. Against these Beaufighters and Mosquitoes the enemy sends out Focke-Wulfs. Against these Focke-Wulfs we're going to send out Tempests or Mustangs. That's all. Now you know what to look for and to avoid."

On 14 June, during the second 'Instep' patrol, S/Ldr Stanisław Szabłowski, flying with F/Sgt Henryk Gajewski, his radio/navigator, in Mosquito EW•B (HJ648) and leading a formation of four aircraft, spotted five fully surfaced U-boats. He attacked the second U-boat and saw strikes on the conning tower. He transferred his fire to the third boat with a similar result. The cannon of the second Mosquito jammed, and the very heavy combined return fire from the U-boats, which hit Szabłowski's aircraft and stopped one of its engines, forced the remaining two crews to break away. Szabłowski flew the 500 miles back on one engine and made a belly landing at Predannack. Two U-boats, U-68 and U-155, were seriously damaged and had to turn back to Lorient for repairs.

On 19 June three of No.307's Mosquitoes and one from No.410 (RCAF) Sqdn, led by S/Ldr Szabłowski on another 'Instep' from Predannack, sighted a BV 138 twin-boom flying boat at naught feet. As the enemy began to climb towards the clouds, each Mosquito made two firing passes, from stern and from port. The enemy, with one engine smoking, failed to reach the clouds and, attacked again, burst into flames and crashed into the water. The destruction of the BV 138 was credited 1/4 each to the four crews: S/Ldr Szabłowski and F/Sgt Gajewski, F/O Jan Bieńkowski and F/O Czesław Borzemski, F/O Jerzy Pełka and F/Sgt Mieczysław Zakrocki – all from No.307, and F/O E.A. Murray and F/O P.R. Littlewood – Canadians, from No.410 (RCAF) Sqdn.

Routine defensive patrols and Ranger/Intruder sorties during which trains and other ground targets were attacked, were also flown in June. On 27 June F/O (Cpt) Roman Grzanka and F/Sgt (Sgt Mjr) Wacław Oyrzanowski were killed after flight-testing a newly overhauled Mosquito, when one of its engines failed on landing.

By the end of June 1943 ten of No.307's air crews had completed the Ranger/Intruder training program and the remaining crews were close to achieving the required standard. The squadron was now ready for its new offensive role.

(5.) The Polish Fighting Team of the Desert Air Force

In the early part of the war North Africa was the only region where the British Army was engrossed in a serious and sustained combat with the enemy. For the RAF, whose tactics on cooperation with ground forces received very little attention between the wars and were not yet put to test on a battlefield, the Libyan campaign provided practical experience and a proving ground of immeasurable value. The bitter fighting in the desert in 1942 brought about a turning point in British land/air cooperation and knocked into shape the Allied "combined operation" concept of warfare, the main architects of which were ACM Arthur Tedder, Air Officer Commanding-in-Chief, RAF Middle East Command, and AVM Arthur Coningham, Air Officer Commanding, the Western Desert Air Force.

The PAF HQ was keen to give Polish cadres the widest possible experience in air support operations, which it was realized would be of paramount importance in the planned invasion of Europe. Discussions with the RAF in October-November 1942 led to a proposal that a Polish elite fighter team should be sent to North Africa to become proficient in combat tactics specific to that campaign. It was further stipulated that the Polish pilots in the team should be replaced every two to three months, to extend the experience to a larger group. In contrast to the earlier, rather spontaneous and makeshift initiative of the Polish Takoradi Detachment, this plan was well thought out and reflected the long-term objectives of the PAF.

Middle East Command approved the proposals and 15 pilots – ten officers and five NCOs – were selected by the PAF Inspectorate out of 68 who volunteered to the unit, which became officially designated by the PAF the Polska Ekipa Myśliwska, or the Polish Fighting Team (PFT). It is interesting to note, however, that in all the available Air Ministry and RAF communications the unit was referred to as the Polish Combat Team.

Acting W/Cdr (S-Col) Tadeusz Rolski was appointed Liaison Officer and Acting S/Ldr (Cpt) Stanisław Skalski, operational C.O. of the Team, which became later popularly known as Skalski's Circus. Its 14 pilots, in addition to Skalski, comprised: F/Lts (Lts) Eugeniusz Horbaczewski, Wacław Król, and Karol Pniak; F/Os (Lts) Bogdan Arct, Władysław Drecki, Ludwik Martel, Kazimierz Sporny, Mieczysław Wyszkowski; P/O (S-Lt) Jan Kowalski; W/Os (W/Os) Władysław Majchrzyk, Bronisław Malinowski, and Mieczysław Popek; and F/Sgts (Sgt Mjrs) Marcin Machowiak, Kazimierz Sztramko.

The PFT badge, a black "Winged Sphinx" on a white ornamental cross background, was designed by F/O Arct.

On 13 February 1943 the PFT pilots assembled at RAF West Kirby. They were posted to Middle East Command effective from 20 February, and to "Polish Spitfire Flight of No.145 Squadron," with effect from 7 March.[55] No. 145 Sqdn, then based at Bu Grara, Tunisia, was commanded by S/Ldr Lance Wade, a noted American fighter pilot serving with the RAF, and the PFT was attached to it as 'C' Flight for administrative purposes only. The Team was operationally fully independent and received its operational orders directly from No.244 Wing of the Desert Air Force, which provided support for Montgomery's Eighth Army.[56] No.244 Wing was commanded by W/Cdr Ian R. Gleed and after his death in action on 16 April 1943, by Acting W/Cdr Peter Olver, and comprised five Squadrons: Nos.92, 145 and 601 (RAF), No.1 (SAAF) and No.417 (RCAF), all flying Spitfire Vs.

On 24 February the Polish unit sailed from Glasgow on board MS "Letytia" for North Africa and on 4 March arrived at Oran, from where it proceeded by air via Maison Blanche, Castel Benito, El Adem and Cairo to the RAF base at Hazbub. On 13 March the PFT arrived at Bu Grara. Ten Spitfire Vs with which the unit was initially provided carried the same code letters (ZX) as No.145 Sqdn, but instead of individual letters they were given numbers from 1 to 0 and were marked with the PAF chessboard insignia on the forward fuselages forward of the windscreens. The living conditions, in tents, at Bu Grara were very primitive and the location – salt flats near the sea, where the aircraft were prone to subside in patches of soft sand – was far from ideal.

After some practice local reconnaissance flights, the PFT became operational on 17 March, beginning with patrols over Rommel's Afrika Korps' positions on the Mareth Line, which the Eighth Army was preparing to attack. On 20 March, at the beginning of Montgomery's advance, the Poles escorted 12 Mitchells which bombed the Axis-held Gabes airfield and F/Lt Horbaczewski fired the Team's first shots in anger at a Macchi MC 202 Folgore of the Regia Aeronautica, but without apparent results. Two days later F/O Arct had a similarly inconclusive skirmish with a Bf 109, while W/O Majchrzyk suffered some damage from the enemy fighters, but successfully returned to base.

Between 23 and 26 March the Team received ten new Spitfire IXs (numbered 1 to 0) from the first consignment of 24 which had just arrived at Algiers, and became the first unit in North Africa to be fully re-equipped with this superior model. On 28 March came the first success: over Sfax five Spitfire IXs encountered six Ju 88 bombers escorted by ten Bf 109s and in a lightning attack S/Ldr Skalski and F/Lt Horbaczewski accounted for one Ju 88 each, opening the Team's impressive score. Seeing below the streets of Sfax cram-packed with troops, the pilots celebrated the victory with strafing passes which created panic on the ground.

A sand storm briefly affected the Team's readiness and on 2 April only four Spitfires were available for operations. Scrambled at 15:30 hours, they came to grips with 15 Bf 109s north of Gabes and in a bitter combat S/Ldr Skalski, F/Lt Horbaczewski and F/Sgt Machowiak destroyed one German each; the latter pilot and F/O Arct were each also credited with one more Bf 109 damaged. Arct's aircraft was hit, but managed to land at base. Two days later, again over the Gabes area, six Polish Spitfires engaged Ju 88 bombers and their 12-strong Bf 109 escort. S/Ldr Skalski and F/Lt Król shot down one enemy fighter each, while F/O Martel damaged another, sustaining himself a number of hits to no ill-effect.

More offensive sorties followed every day and on 6 April the Team flew six patrols on the last of which they encountered ten Bf 109s over Sfax. F/Lt Horbaczewski shot down one German, but his own EN459 ZX•1 was badly damaged and he force landed at Gabes, which by this time was in Allied hands. F/O Wyszkowski accounted for a Bf 109 probably destroyed and W/O Majchrzyk, for one damaged. The next day five of the Team's Spitfires were in combat and F/O Sporny and W/O Malinowski scored one confirmed victory each. On 8 April W/O Popek damaged a Bf 109. Three days later, following the capture of Gabes and Sfax, No.145 Sqdn moved to La Fauconnerie, a former Luftwaffe base, littered with the battered remains of many Bf 109s and other enemy aircraft and four days later proceeded to Goubrine.

On 18 April, F/O Wyszkowski (in EN355 ZX•O), acting as a decoy to draw enemy fighters while the other Spitfires, flying above waited for their prey, was suddenly attacked from the sun and shot down by passing Messerschmitts, which escaped. Wyszkowski baled out from his blazing aircraft over

[55] Postagram from ACM, AOC.-in-C. RAF Middle East to No.145 Sqdn, dated 13/04/43 (IPMS, Lot. A V,30/1d file).

[56] On 23/02/43 the Desert Air Force came under operational control of the newly established North-West Tactical Air Force, also referred to as the 1st, of Allied, Tactical Air Force.

enemy-held territory, and, injured, was taken PoW. This was the only loss suffered by the Team during its operations in Africa. Later the same day the Polish pilots sighted a formation of Macchi MC 202 Folgores, but these immediately turned back and only F/O Arct managed to fire some good bursts at one, which was credited to him as "probable."

Pilots of the Polish Fighting Team, the legendary Skalski's Circus, by Spitfire IX EN268 ZX•5; standing, left to right: F/O Sporny, F/O Martel, F/Lt Pniak, F/O Drecki, F/Lt Horbeczewski, W/O Majchrzyk, F/Sgt Machowiak, F/Sgt Sztramko, F/Lt Król, S/Ldr Skalski and F/O Wyszkowski; sitting on the ground: P/O Kowalski and W/O Malinowski (F/O Arct took the photograph).

Acting S/Ldr Skalski in the cockpit of his Spitfire IX at Goubrine.

In view of the appalling losses suffered by Ju 52 transports trying to supply the collapsing Afrika Corps, the Luftwaffe put up standing fighter patrols over Cape Bon and on 20 April six Polish Spitfires took part with the rest of No.145 Sqdn in an operation to provide protection to Kittyhawks which were to deal with another Ju 52 convoy. No.145 Sqdn clashed with some 20 Bf 109s and MC 202s south of Pantellaria and in the ensuing battle the Poles, flying top cover, accounted for six of the enemy fighters out of nine destroyed in a combat which lasted barely two minutes. The victors were: Arct, Drecki and Martel, credited with one Bf 109 each, and Król, Majchrzyk and Popek, credited with one MC 202 each. In addition, Martel accounted for one more MC 202 probably destroyed and Drecki, for one Bf 109 damaged. Drecki's aircraft was the only one to be hit by enemy fire, but safely returned home. The following day, flying at low level over Pantellaria, F/Lt Król added another Bf 109 destroyed to his score and F/O Arct, one damaged.

On 22 April the PFT again provided six Spitfires as top cover for 36 Kittyhawks of 7th Wing USAAF, escorted by No. 1 (SAAF) Sqdn. Over the Gulf of Tunis, near the island of Zembra, the Allied fighters encountered 20 giant six-engined Me 323 Gigant transports, each capable of carrying ten tons of supplies, flying with a force of Bf 109 and MC 202 fighters. While the Kittyhawks enjoyed themselves with the Gigants, the Spitfires attended to the more demanding task of dealing with the enemy fighters. The Poles alone destroyed six without loss: Horbaczewski and Sporny shot down two Bf 109s each and Sztramko added one Bf 109 and a MC 202 to the score.

Towards the end of April W/Cdr Mateusz Iżycki, Senior Polish Staff Officer at Middle East Command, visited the Team and over a few days took part in a number of operational sorties. On 27 April W/O Popek accounted for one more MC 202 destroyed and three days later F/Sgt Machowiak damaged a Fw 190, which escaped in cloud. By then the enemy's air force had been largely eliminated from the battle. In the final Allied attack on 6 May the Luftwaffe made its last appearance, only to lose 20 fighters in three hours; one of these was shot down by F/Sgt Sztramko and S/Ldr Skalski added one more damaged to the Team's final account. On the same day the PFT moved to Hergla, from where it flew its last operational sortie on 8 May. The PFT became non-operational on 12 May. Two days later the Tunisian Campaign ended with the final surrender of the Afrika Corps.

The Polish Fighting Team – with only 15 pilots and 10 aircraft – became the highest scoring fighter unit in North Africa in April 1943. During eight weeks of its spectacularly successful operations the Team was officially credited with the score 25-3-9 for the loss of only one pilot (made PoW) and two aircraft.

The unit, usually led into action by S/Ldr Skalski or F/Lt Król, and on a few occasions also by its rising star F/Lt Horbaczewski, flew a total of 539 sorties (excluding W/Cdr Iżycki). Horbaczewski became its highest scoring "ace" (5-0-0), while four of its pilots – Skalski, Król, Sporny and Sztramko

W/Cdr Iżycki at the controls of EN315 ZX•6, about to take off from Goubrine at the end of April 1943.

– achieved three confirmed air victories each, and only two of the Team's pilots did not succeed in recording any results. Skalski's Circus became a legend and further enhanced the high regard which the RAF had for the Polish Air Force.

An expression of this were two cypher messages from ACM Sir William Sholto-Douglas, Air Officer Commanding-in-Chief, Middle East Command[57] – one to Gen Ujejski, Inspector General of the PAF, which read: "I wish to thank you for sending the Polish Combat Team to fight in this Command. They have done magnificent work ... Please convey to all members of the Polish Combat Team my heartiest congratulations on their splendid record in this theatre and my thanks for all they have done to help to bring about our recent victories." – and another, directly to the "Polish Combat Team," sent to S/Ldr Skalski, " ... the record of the Polish Combat Team which flew with the Desert Air Force is regarded with pride. I congratulate you and I am proud that earlier in Great Britain I commanded you when you served with the top-scoring Wing in Fighter Command. Regardless of personal safety, Polish pilots have accomplished their one desire – to knock the Hun out of the sky."

Early in May 1943, when the future of the PFT began to be considered, AVM Harry Broadhurst, AOC RAF Western Desert, wrote to HQ RAF Middle East: "The Polish Combat Team is composed of very experienced fighter pilots, who have done excellent work since they have been attached to the Western Desert. So long as they are employed as a separate group on their own, however, their experience and seniority is largely wasted. Nearly all of them are suitable as flight or squadron commanders and I am fully prepared to consider them for such posts within 244 Wing."[58]

His views were endorsed by ACM Sholto-Douglas, who conveyed them to the Air Ministry and to Gen Ujejski. But such experienced leaders were also needed by the PAF and, after consultations with the PAF Inspectorate, Gen Sikorski agreed to the British proposal only partially, limiting his permission to those pilots who could be immediately offered suitable command posts in British squadrons and who would be prepared to stay in North Africa and hold them for some time.

The PFT was formally disbanded on 22 July 1943 and at the end of that month 11 of its pilots returned to England by air. Three pilots accepted the RAF offer and stayed behind: S/Ldr Stanisław Skalski was appointed C.O. of No.601 Sqdn, becoming the first pilot of the Team to command a RAF squad-

[57] He took over this Command from ACM Tedder on 11 January 1943. Both messages were sent out in July 1943.

[58] All documents quoted above come from IPMS, Lot. A V, 30/1d file.

The Team's "Operation room" at Hergla in May 1943. Standing left to right are: F/O Sporny, W/Cdr Rolski, W/O Popek, F/Lt Pniak, and P/O Kowalski.

A Spitfire HF VIII JH447 UF•7, flown by S/Ldr Skalski, C.O. No. 601 Sqdn, in the Sicilian campaign. On 30 August 1943 Skalski led the squadron in it, providing top cover to some 100 fighter-bombers over the Straits of Messina, prior to the invasion of Italy.

ron – he held the post until December 1943; F/Lt Eugeniusz Horbaczewski was Flight Commander and later C.O. of No.43 Sqdn – he returned to England early in 1944; and F/Lt Władysław M. Drecki, who, posted to No.152 Sqdn at Lentini East (Sicily) on 8 August, was given command of its 'B' Flight.

F/Lt Horbaczewski's first victorious encounter with the enemy in No.43 Sqdn occurred within days of his joining the unit. On 27 July 1943 he led an attack on Messerschmitts which were about to strike at Mustangs escorted by No.43 Sqdn near Messina (Sicily). He and his wingmates, P/O W.H. Reid and F/Sgt T.E. Johnson, engaged a Bf 109, which was shot down. Its destruction was confirmed and credited 1/3 each to the three pilots involved. However, Horbaczewski decided to disclaim his share in the victory in favor of Reid and Johnson, who were subsequently apportioned 1/2 Bf 109 each. Horbaczewski's victories in No.43 Sqdn, as officially recorded on his personal

A Spitfire VC MA289 UM•T of No.152 Squadron, in which F/Lt Drecki shot down Oblt Steffens on 11 September 1943. The photograph was taken at the Lentini East airfield, Sicily, in September. (T. Drecki)

PAF Form A/1 "Register of flights, victories and decorations," comprised: 1 Bf 109 destroyed on 4 September 1943, 1 Fw 190 damaged on 15 September and 2 Fw 190s destroyed on 16 September.

A story of F/Lt Drecki's victory in No.152 Sqdn ran even a stranger course. The squadron moved to Milazzo East airfield on the north coast of Sicily on 6 September 1943 to cover the landing of Allied forces on the Salerno beaches in Italy, which began three days later. On 11 September in the late afternoon a section of the squadron on a Salerno patrol, led by F/Lt Drecki in Spitfire VC MA289 UM•T, met two Bf 109s. One of them was attacking an American Lightning and Drecki got onto its tail and had the satisfaction of seeing the German bale out some 15 miles off Monte Corvino. The Americans picked up the pilot, Oblt Rudolf Steffens of JG 53 (The other Bf 109 was damaged by P/O R.E.J. McDonald). Two days later, on the 13th, the heat of the sun caused a tire of Drecki's Spitfire VC UM•U ES112 to blow out on take-off. He swung into another Spitfire parked too near the runway, his 90-gallon long-range tank exploded and he was killed instantly.[59]

Drecki's last victory – increasing his score to 4-0-1 – though fully documented in No.152 Sqdn records, was apparently never notified to the PAF HQ nor to Fighter Command in England,[60] and therefore escaped inclusion on his personal PAF Form A/1. For the same reason it was omitted from the "Roll of the Polish Fighter Pilots' victories," as finally compiled by the Polish Liaison Officer to HQ Fighter Command in March 1946, which incorrectly listed Drecki's score as 3-0-1.[61]

Apart from the PFT, a few other Polish airmen served with distinction in RAF Mediterranean Air Command in the first half of 1943. These included the Mosquito II crew S/Ldr Jerzy Orzechowski, pilot, and F/Lt Ignacy Szponarowicz, navigator, who flew with No.23 Sqdn based in Malta. F/Lt (Lt) Marceli Neyder, serving as pilot with the same squadron, vanished without trace on a sortie on 21 February 1943 and was presumed dead.

(6.) Assessments and Conclusions

The Polish fighter forces of eight squadrons was the largest of the Allied fighter formations within the RAF and constituted on average about 10 percent of Fighter Command's operational strength throughout the 1941-43 period:

- Polish fighter squadrons were in the forefront of the early fighter offensive. They took part in about 70 percent of all Circus operations, provided almost one third of the fighter support for the Dieppe landing in 1942 and contributed up to one half of fighter strength for a majority of the later Ramrods.
- The Polish Fighter Team in North Africa, led by S/Ldr Skalski, became a legend and its exceptional score was ranked first in the achievements of RAF fighter squadrons in 1943.
- Notwithstanding considerable over-estimation of the numbers of enemy aircraft actually destroyed by the RAF, the Poles were consistently credited with between 15 and 20 percent of the losses inflicted upon the Luftwaffe during the 1941-43 period.
- The British, and later the American, bomber squadrons highly valued the skill and determination of the Polish fighter escorts in preventing German fighters from penetrating the defensive screen. Numerous messages of thanks from bomber crews were the best confirmation of this.
- No.307 Squadron, expanding its night-fighting capabilities with patient dedication, played a vital role in the defence of Exeter and endeared itself to the people of the town, establishing a lasting bond with them.

[59] RAF Forms 540 & 541 (Public Records Office, AIR 27/1025 File) and photocopies of logbooks and diaries of No.152 Sqdn personnel kindly provided by Tomasz Drecki, nephew of F/Lt Drecki.

[60] Drecki was killed in Sicily and No.152 Sqdn belonged to Mediterranean Air Command and in November 1943 moved to India; it was disbanded in Singapore in March 1946.

[61] "Polish Fighter Pilots' achievements (1.9.1939-6.5.1945)," No.FC/S.5/1/AIR/CPLO. INTEL. (IPMS, Lot. A V,1/4-XXIV).

Chapter III

POLISH ARMY COOPERATION ACTIVITIES

1. No.309 "Czerwieńska Land" Army Cooperation Squadron

A Polish army cooperation squadron was organized on the specific insistence of the C.-in-C. Polish Armed Forces. A postagram from the Air Ministry, dated 27 September 1940, authorized the formation of No.309 (Polish) (Army Cooperation) Squadron to War Establishment No.WAR/AC/126 on 7 October 1940, at Renfrew, near Glasgow, within No.22 Group. This Establishment was soon amended by WAR/AC/137, which increased the unit's personnel by duplicating some posts, in line with the practice adopted for other Polish squadrons. The aircraft establishment, comprising 12 Initial Equipment Westland Lysanders, one Percival Proctor and two Tiger Moth trainers, remained unchanged. Code letters AR were assigned to the unit. No.309 Squadron was officially formed on that date at Renfrew and W/Cdr N.W.F. Mason, Commander of the Station, was appointed its British C.O.

Selection of Polish personnel at Blackpool to the unit, which was named No.309 Dywizjon Współpracy "Ziemi Czerwieńskiej" (No.309 "Czerwieńska Land" Army Cooperation Squadron – the name linking it with the former 6th (Lwów) Air Regiment) progressed very slowly (the Polish fighter and bomber squadrons having priority) and W/Cdr (S-Col) Zygmunt Pistl, the Polish Commander of the Squadron, did not arrive at Renfrew Until 20 November. Most of the Polish personnel, which included a few older, higher ranking officers, reached the Station in the middle weeks of November. The first Lysander IIIs were received at about the same time and equipment of No.309 was deemed to be completed on the 23rd of the month. The aircraft were soon marked with large PAF chessboards, which were carried on the fuselage next to the RAF roundels. At first the squadron had no badge, but eventually a badge, consisting of two arrows embellished with a wing and surrounded by a circle, was adopted as the result of a competition (and approved by C.-in-C. Polish Armed Forces in September 1944). The squadron was to be attached to the Polish Army Corps when it became operational and, for operational purposes only, was to be under orders of the Corps.

At the end of November 1940 the flying personnel comprised:

- W/Cdr (S-Col) Zygmunt Pistl, the Polish C.O., S/Ldr (S-Col, Obs.) Władysław Madejski, the Polish Deputy C.O.;
- Flight Commanders: F/Lts (Cpts, Obs.) Aleksander Łukiński and Fidelis Łukasik;
- pilots: F/Os (Lts) Piotr Dunin and Jan Narewski; P/Os (S-Lts) Bronisław Baster, Włodzimierz Berezecki, Janusz Lewkowicz, Ryszard Łopacki, Eugeniusz Pietrzyk, Kornel Stefanus and Roman Suwalski; P/O (Sgt Mjr) Stanisław Sawicki; Sgts (Sgt Mjrs C-Offs) Roman Jarema and Eugeniusz Rajewski; Sgt (Cpl C-Off) Wincenty Miniszewski; Sgts (Cpls) Mieczysław Budziński and Andrzej Piaskowski;
- observers (air gunners): F/Os (Cpts) Tadeusz Kern and Jerzy Wolf; P/Os (Lts) Stanisław Daniel, Tomasz Kasprzyk and Adam Solecki; Sgts (Sgt Mjrs C-Offs) Marian Bujak, Walerian Fuks, Jan Marczewski, Czesław Maślankiewicz and Jan Wasikowski; Sgts (Sgts C-Offs) Aleksander Bączek and Witold Korzeniewicz; Sgt (Cpl C-Off) Józef Kocoń.

P/O (S-Lt) Lucjan Nowiński became Engineering Officer of the unit.[1]

The squadron faced very difficult and dilatory beginnings. While in Poland all the Military Aviation activities were looked upon only as a part – or at best an extension – of Army operations and observers, who were considered extremely important, were extensively trained in the art of "observing" and close cooperation with ground troops. In contrast, the RAF was a totally independent armed force, intended to fight a war in the air on its own terms. In 1939 the concept of cooperation with the Army had been alien to it as to no other air force in the world. Only after Dunkirk's experimental exercises were car-

[1] Compiled from No.309's personnel roll lists (IPMS, Lot. A V,61/3a).

ried out in Northern Ireland (in September 1940) with the object of coordinating the action of forward troops and supporting aircraft, and, largely on the initiative of AM Sir Arthur Barratt, former AOC, British Air Forces in France, Army Cooperation Command was eventually set up on 1 December 1940. But an air of unreality in its role still lingered.[2] It was only in the battles of East Africa and the Desert that successful systems of Army/Air cooperations were evolved and perfected by the British. With establishment of the Command, No.309 was transferred to its No.71 Group.

The Polish army cooperation legacy was apparent from the high caliber of No.309's personnel, who had to reconcile themselves with the fact that the RAF instructors who came to teach them knew very little about army cooperation and nothing about the experience and capabilities of their "pupils" and soon began to learn from those whom they were supposed to "instruct." The army cooperation training syllabus was not yet in existence and the crews kept themselves busy getting acquainted with the Lysanders and putting in as many flying hours as the winter weather permitted. The greatest benefit of the protracted training period was that it gave the Poles the time to gain a fair knowledge of English.

Early in 1941 the Poles went through the gunnery schools at West Freugh and Dumfries and 17 crews completed the courses by the end of February. Cooperation exercises with the 1st Polish Army Corps began and the first instructions for cooperation with the artillery were drafted. Personnel shortages were slowly reduced and by 1 March 1941 the overall strength rose to 38 officers and 301 other ranks. Even so, this represented only some 75 percent of the Establishment.

Intensive night activities of the Luftwaffe in the Glasgow area during the spring months of 1941 affected the squadron. Heavy bombing on the nights of 13/14 and 14/15 March resulted in casualties – three airmen, Antoni Ptaszkowski, Paweł Radke and Zygmunt Sokołowski, were killed and one NCO injured on the first night – and partial destruction of the living quarters. Another attack was made on the airfield on the night of 7/8 April, adding to the damage.

On 14 March six Lysanders of 'A' Flight left for Perth for

[2] John Terraine "The right of the line," p.350-351.

No.309's aircrew climbing into a Lysander III at Renfrew at the beginning of 1941. The aircraft carries large PAF chessboards on the rear fuselage, but no code letters. (W. Korzeniewicz-Kay)

a two-week night flying course and on 1 April 'B' Flight followed suit. At the same time the squadron's Lysanders were gradually fitted with increased armament, their single observer's guns being replaced with twin-guns. There was also much talk about re-equipment with new aircraft and Hurricanes and American Brewsters were briefly mentioned in this context. Notwithstanding all these activities, on 15 March No.309 (its 'B' Flight) held exercises with the 1st Polish Army Corps for the first time. In April the unit began to keep three aircraft ready for local defence purposes, but the Lysander was virtually useless for interception[3] and no such sorties were flown. President Raczkiewicz, accompanied by the US Ambassador, Mr. Drexel Biddle, visited the squadron on 15 April.

During March and April 1941 great emphasis was put upon radio communications and ground support (strafing and bombing)[4] training. On 15 May No.309 moved to Dunino, where its personnel had to be accommodated in tents in a nearby wood and had to cope with a lack of water supply. At the time of the move the squadron's roll comprised 411 officers and other ranks, including the RAF contingent. On arrival at Dunino, the unit began to receive the somewhat improved Lysander IIIAs.

[3] Its only recorded success in this role was the shooting down of a He 111 by a Lysander of the British Air Forces in France in November 1939.

[4] The Lysander could carry a bombload of up to 500 lbs on the undercarriage stub wings and on racks under the rear fuselage.

Left: A Lysander III T1572 in flight over Scotland in the early months of 1941. Right: President Raczkiewicz arrives with the US Ambassador Drexel Biddle to inspect No.309 Squadron at Renfrew on 15 April 1941. The Lysanders are still without code letters. (IPMS)

Squadron's ground personnel by a Lysander III in spring 1941

No.309's Lysander IIIA V9484 AR•M on army exercises.

Large scale exercises with the 1st Polish Army Corps in the middle of May 1941 included reconnaissance of armour movements, maintenance of radio liaison with the ground, and cooperation with artillery and support for the troops on a battlefield. Towards the end of June No.309 exercised with the 51st Highland Division and received a special commendation from Scottish Command HQ for this work. S/Ldr Władysław Madejski was posted away from No.309 and S/Ldr (Mjr, Obs.) Zdzisław Stranz succeeded him as Deputy C.O. of the squadron on 2 July. British duplicate officers began to leave in the middle of the year. Various accidents occurred in the first half of 1941, most of them resulting in only minor damage to aircraft. The most serious one occurred on 18 June, when the Lysander flown by Sgt Zygmunt Kowalczyk, with F/Lt Aleksander Łukiński as observer, caught fire in mid-air and burnt out after crash landing, its crew receiving minor burns.

By 1 August 1941 the squadron had 22 fully trained pilots and 18 observers – one pilot and five observers below Establishment (23 crews envisaged) – and persisting shortages of skilled ground personnel and training commitments allowed it to keep on average only six of its Lysanders in combat condition. This further delayed the unit becoming operational, the operational state requiring at least eight aircraft to be ready for combat (at two hours notice) at all times. In the second half of August No.309 was extensively involved in the major "Rabbit" and "Hare" army exercises of Scottish Command, playing the role of "enemy" aircraft. During these and other similar operations, which included reconnaissance, gas spraying, supply dropping and intensive ground support work, the squadron achieved a high degree of mobility.[5]

In view of the demands put upon No.309 by the British and Polish army commands during the summer and autumn of 1941, at the beginning of October its equipment strength was increased by four more aircraft and consideration was given to the formation of a third ('C') flight. During October the squadron took 1,323 aerial photographs at the request of various army commands in Scotland and on 11 November 1941 the Air Ministry declared it operational. On the 24th of the same month W/Cdr Pistl took over the overall command of No.309 from W/Cdr N.W.F. Mason and on 1 December F/Lt (Cpt, Obs.) Aleksander Łukiński succeeded S/Ldr (Mjr, Obs.) Stranz as Deputy C.O. During December 'A' Flight took part in the exercise "Scorch," operating from Ipswich, and 'B' Flight, in "Stag," operating from Inverness. Organization of the third flight began. At the beginning of 1942 the squadron was commended by Army Cooperation Command HQ for the highest number of hours flown among its squadrons in November and December 1941.

On 13 January 1942 HRH the Duke of Kent paid No.309 a visit. In the first half of the year the squadron's Lysanders continued to provide tactical reconnaissance and widely ranging air support to Polish, British and Norwegian troops in nu-

[5] No.309's half-monthly and monthly reports (IPMS, Lot. A V,61/4).

A Lysander IIIA V9576 AR•N being prepared for a flight.

The same aircraft with the trolley accumulator in the foreground.

Lysander IIIAs involved in army cooperation work with the 1st Polish Army Corps in 1942. (W. Korzeniewicz-Kay)

merous exercises. On 1 February 1942 the third, 'C' Flight, equipped with Lysanders and commanded by F/Lt (Cpt, Obs.) Tadeusz Kern, began work. On 9 February No.309 suffered its first fatality when a Tiger Moth on a liaison flight crashed into the sea, killing its British pilot, Cpt J. Worledge and F/O (Lt) Jerzy Sadowski. On 25 February F/O (Lt) Piotr Dunin, pilot, and F/O (Lt) Jerzy Homan, observer, lost their lives in a Lysander, which dived to the ground out of control owing to a steering failure. No.309 retained its three Lysander flights until June 1942.

In April 1942, an operational order of Scottish Command assigned No.309 Sqdn to the 1st Polish Army Corps. At about the same time, with the upgrading of army cooperation work to new operational demands which included the long-range photo-reconnaissance capability and more effective air support for the army by fast fighter aircraft, the squadron began the long process of transformation into a fighter-reconnaissance unit. To begin with, its 'B' Flight was to convert to the powerful North American P-51 Mustang I fighter-reconnaissance aircraft. The first six pilots – F/Lt Maciej Piotrowski, F/Lt Jerzy Gołko, F/O Jan Bendix, F/O Roman Jarema, F/O Janusz Lewkowicz and F/O Stanisław Sawicki – were sent to OTUs at Old Sarum and Hawarden for training. They were subsequently posted to No.26 Sqdn at Gatwick for combat proficiency practice on Mustangs. F/Lt Piotrowski made the first operational sortie on a Mustang on 21 May 1942, flying a Rhubarb with two British pilots and attacking ground targets in the Le Touquet area. A week later F/O Lewkowicz participated in another Rhubarb with No.26 Sqdn, shooting up a train and other targets of opportunity. F/O Jarema followed suit later.

No.309 received its first two Mustang Is on 7 June and by the end of July its 'B' Flight was completely re-equipped with the new aircraft. Since Dunino was not suitable for Mustangs, the fighters were initially based at the nearby Crail airfield and on 31 July, 'B' Flight, commanded by F/Lt Piotrowski, moved to Dalcross, near Inverness, some 100 miles away. At Dalcross, the Flight went through gunnery and navigational courses and became much talked about on account of a very imaginative flight by a single individual, which notched a place for itself in the annals of air operations as the most enterprising "private venture" sortie of the war. The individual was F/O (Lt) Janusz Lewkowicz, who before the war graduated at the Warsaw Technical University, obtaining a BSc Engineering Diploma.

From the moment of his first flight in a Mustang, on 30 April 1942, Lewkowicz fell in love with the aircraft and became fascinated by its capabilities. He approached the North American representative in London and obtained from him an analysis of the Mustang's fuel consumption at most favorable boost pressure and airscrew-rotation speed settings for various altitudes and speeds. Thinking about a most spectacular way of demonstrating the Mustang's range, he became captivated by the idea of a fighter sweep from Scotland to Norway. He started his own studies and produced petrol consumption/range calculations, graphs and navigational charts. These were put to practical tests during various exercises by No.309 pilots, which fully verified his theoretical findings.

At the end of June Lewkowicz forwarded his proposals for the Norway sweep, backed up by results of his research, to HQ No.71 Group, but received no response of any kind. In a desperate attempt to prove the feasibility of his proposals, he decided to put them to test without authorization. His chance came on 27 September, when he obtained permission for a long-range navigational practice flight over the North Sea from Dalcross (which was some 100 miles further away from Norway than Peterhead, the most suitable airfield for such a sortie). He took off at 10:05 hours, making for Stavanger, Norway, and dropped down to the altitude of naught feet 100 miles from the Norwegian coast. He made a low-level sweep over the Stavanger Bay and the town itself, strafing a few targets of opportunity. He intended to land at Peterhead, but veered off course and landed at Dunino at 13:50 hours.

Meanwhile, there was consternation on the ground when he became "overdue" and upon checking the Operations Book, his entry "Rhubarb to the Stavanger area" was discovered. Upon landing, he was put under house arrest at Dunino, pending investigations. Eight days later he was released and returned to

No.309's Lysander IIIA AR•P, standing beside a Mustang I recce-fighter of 'B' Flight at Dunino in the summer of 1942. (J. Gołko/D. Koniarek)

A Mustang I AG648 AR•E, depicted at Dalcross on 10 October 1942.

flying duties at Dalcross, but he was charged with an unauthorized flight and called to appear before AM Sir Arthur Barratt, AOC.-in-C. Army Cooperation Command. He was first severely reprimanded and then congratulated on a remarkable flight. The range of Mustangs' operational sorties was increased in October 1942 and some British commentators linked this to "a Polish pilot from No.309 Squadron, who flew to Stavanger."

No.309's 'B' Flight was posted to No.35 Wing on 29 November 1942, its seven Mustangs landing at Gatwick on 1 December. The Flight became operational on that day and carried out its first official operational sortie on 5 December. This was a "Popular" operation over the French coast,[6] made by two pilots: F/Lt Maciej Piotrowski and F/Lt Jan Narewski who took oblique photographs of German fortifications. Over the next ten days six more Populars and two Rhubarb-Populars were flown, but the bad weather interrupted further operations and on 29 December 'B' Flight returned to the squadron, then based at Findo Gask, after flying 16 sorties in 15 hours 30 min with No.35 Wing from Gatwick.

In the meantime, the two Lysander-equipped 'A' and 'C' Flights carried on with the routine cooperation activities, exercising with the ground troops. On 26 November No.309 completed its move from Dunino to Findo Gask and during December 1942 took part in exercises with British infantry and the 1st Polish Armored Division.[7]

The winter weather curtailed flying in January 1943, and because of the inability of the Mustangs to operate from the water-logged Findo Gask airfield, the construction of which was not yet finished, 'B' Flight moved to Peterhead on 17 January. Between 23 February and 5 March the flight made 22 convoy protection sorties over the North Sea and rejoined the rest of the squadron at Kirknewton on 10 March 1943.

An Army Cooperation Command directive from 6 February 1943[8] ordered a complete reorganization of No.309 Sqdn as a two-flight all-Mustang unit, to new Establishment No.WAR/AC/179 (12 Initial Equipment Mustangs, plus 2 Immediate Reserve). The pilots and maintenance personnel to man the additional Mustangs were to be drawn from the Lysander flights and the surplus flying and ground personnel were to be disposed of in accordance with PAF requirements to provide the nucleus of a new Army Cooperation (Fighter-Reconnaissance) squadron (No.318) for service with the 2nd Polish Army Corps overseas.

[6] "Popular" operation was the unarmed photo-reconnaissance of the German Atlantic Wall fortifications by pairs of aircraft, during which no attack on enemy ground or air (aircraft) targets was permitted. "Rhubarb-Popular" operation was the armed photo-reconnaissance penetration inland by single aircraft, which included strafing attacks on targets of opportunity.

[7] No.309's monthly reports (IPMS, Lot. A V,61/5).

[8] ACC/S.1071/19/Org. (IPMS, Lot. A V,62/1a).

A unique picture of one of No.309's last Lysander IIIAs V9550 'T', with the AR code letters and large PAF chessboards, removed from the rear fuselage and a new small chessboard added under the windscreen. (T. Kopański)

No.309's Mustang I tactical-reconnaissance fighter undergoing maintenance in the spring of 1943. (T. Kopański)

The new Establishment came into effect on 20 February 1943 and on the 23rd of that month No.309 was reclassified as Tactical-Reconnaissance Squadron. Its 'A' Flight converted to Mustangs and 'C' Flight was disbanded at Findo Gask. On 8 March the unit moved to Kirknewton, leaving its last Lysanders behind, and its 'B' Flight arrived at the same airfield two days later. Earlier, on 14 February 1943, W/Cdr (S-Col) Zygmunt Pistl, who had commanded the squadron since its formation, was recalled to Staff duties and S/Ldr (Mjr) Witold Jacek Piotrowski was appointed its new C.O. The last observers left the unit at the end of April for Staverton, to be retrained as navigators.

At Kirknewton the squadron, forming part of No.32 Wing, underwent intensive training on Mustangs. With the disbandment of Army Cooperation Command and the simultaneous formation of 2nd TAF on 1 June 1943, No.309 was transferred to No.12 Group of Fighter Command, which was to be assigned to 2nd TAF. On 4 June the squadron moved to Snailwell, near Newmarket, a Fighter Command station within the Group, and embarked on a new chapter of operations in a new field.

No.318 "Gdańsk" Fighter-Reconnaissance Squadron was formed on 20 March 1943 at Detling to cooperate with 2nd Polish Army Corps in the Middle East. Its story will be related in Part III, Chapter 4, of this work.

Chapter IV

POLISH BOMBER OPERATIONS

1940-1943

1. Formation and Deployment of the Polish Bomber Squadrons

The principal task envisaged for the RAF in pre-war plans was to wage a strategic air offensive, and Bomber Command, as the chosen instrument, was of fundamental importance to the British conduct of the air war. However, the Command's first operations at the beginning of the war had shown that unprotected bombers could not operate in daylight and that the effectiveness of night bombing with its existing equipment was very poor. During the British battle for survival in the summer of 1940, Bomber Command had to assume temporarily a defensive posture, concentrating its main effort on counter-invasion strikes. In the late months of 1940 the Command began to resume the role for which it had always been intended – the offensive against Germany. Two Polish bomber squadrons entered operations during this difficult transition period and, together with two further squadrons that followed, made up the Polish bomber force which helped the RAF air offensive to gain momentum.

According to the Anglo-Polish Agreement of 25 October 1939, the Polish aviation personnel arriving in Britain were to be used only for bomber duties and two Polish heavy[1] bomber squadrons were to be organized, with two more as reserve. Retraining of bomber crews began in February 1940 and although a number of the new arrivals took an early opportunity to apply for transfer to Fighter Command, the first Polish squadron to be formed on British soil was in Bomber Command.

(1.) No.300 "Mazovian" Bomber Squadron

Air Ministry signal 0/360 dated 28 June 1940 ordered the formation No.300 (Polish) (Bomber) Squadron at Bramcote – within No.6 Bomber Group – on 1 July. Gen Sikorski's authorization issued at about the same time called the unit the "Ist Polish Bomber Squadron." It was to form initially on Fairey Battle medium bombers under the establishment No.WAR/BC/119, with 16 Battles as Initial Equipment and one Battle Trainer. Its personnel was to consist of 20 three-man aircrews and 180 ground crew. The squadron was named No.300 Dywizjon Bombowy Ziemi Mazowieckiej ("Mazovian" Bomber Squadron) and received the code letters BH. It was to convert later to Vickers Wellington heavy bombers. Its badge (not officially authorized by the RAF) represented the coat of arms of the Mazovian Dukes with the Mazovian Eagle and the British Lion on its white fields, topped by a crown bearing the Roman numeral CCC (300).

On 26 and 27 June the Polish Training Unit of No.18 OTU was moved from Hucknall to RAF Bramcote (still in the final construction stages) to provide initial personnel for the Squadron and the first of the Fairey Battles arrived on the 26th.[2] In contrast to the Polish fighter units, which were at first commanded jointly by British and Polish officers, the bomber squadrons were commanded from the beginning by Polish officers, though with British officers in an advisory role. W/Cdr (S-Col) Wacław Makowski, the former C.O. of the Polish Training Unit at Hucknall, was appointed commander of the squadron and W/Cdr K.P. Lewis, its RAF advisor.

On 1 July the PAF flag was hoisted at Bramcote for the first time and a few more Battles were received by the squadron. The aircrews were organized into two Flights; 'A' Flight, under S/Ldr (Cpt) Stanisław Cwynar, and 'B' Flight, under F/Lt (Cpt) Mieczysław Pronaszko. However, Cwynar soon became Deputy C.O. of the squadron and was succeeded as commander of 'A' Flight by F/Lt (Cpt) Stefan Kryński, while Pronaszko was posted to No.304 Sqdn and was succeeded by F/O (Cpt) Romuald Suliński. Engineering Officer, F/O (Cpt) Mikołaj Kaczanowski, put in charge of 'S' (Engineering) Flight, was a few weeks later replaced by F/Lt (Cpt) Stanisław Dudziński. P/O (S-Lt) Ludwik Białkowski was the Gunner Officer.

[1] Wellingtons, Whitleys and Hampdens were then counted as "heavies," and Blenheims, Bostons, and Battles, as "medium" bombers. With the later introduction of four-engined "heavies," Wellingtons, Whitleys and Hampdens were re-classified as "mediums."

[2] No.300 Sqdn's Operational Diary, which, together with other documents in IPMS's file Lot. A V, 34/4, are the main source of this chapter.

LOCATIONS OF THE SQUADRONS

BENBECULA	QG	304
BLACKBUSHE	WQ	301
BRAMCOTE	VK	300, 301, 304, 305
CHEDBURGH	WM	301*), 304*)
CHIVENOR	VS	304
DALE	VN	304
DAVIDSTOW MOOR	VS	304
DOCKING	WG	304
FALDINGWORTH	WA	300, 305*)
HARTFORD BRIDGE**	WQ	305
HEMSWELL	WA	300, 301, 305
INGHAM	WA	300, 305
LASHAM	WQ	305
LINDHOLME	WA	304, 305
NORTH WEALD	WM	301*), 304*)
PREDANNACK	VX	304
ST. EVAL	VX	304
SWANTON MORLEY	WG	305
SWINDERBY	WF	300, 301
SYERSTON	WF	304, 305
TALBENNY	VN	304
TEMPSFORD	WL	301
TIREE	QM	304

*) only after the end of war
**) renamed Blackbushe in November 1944

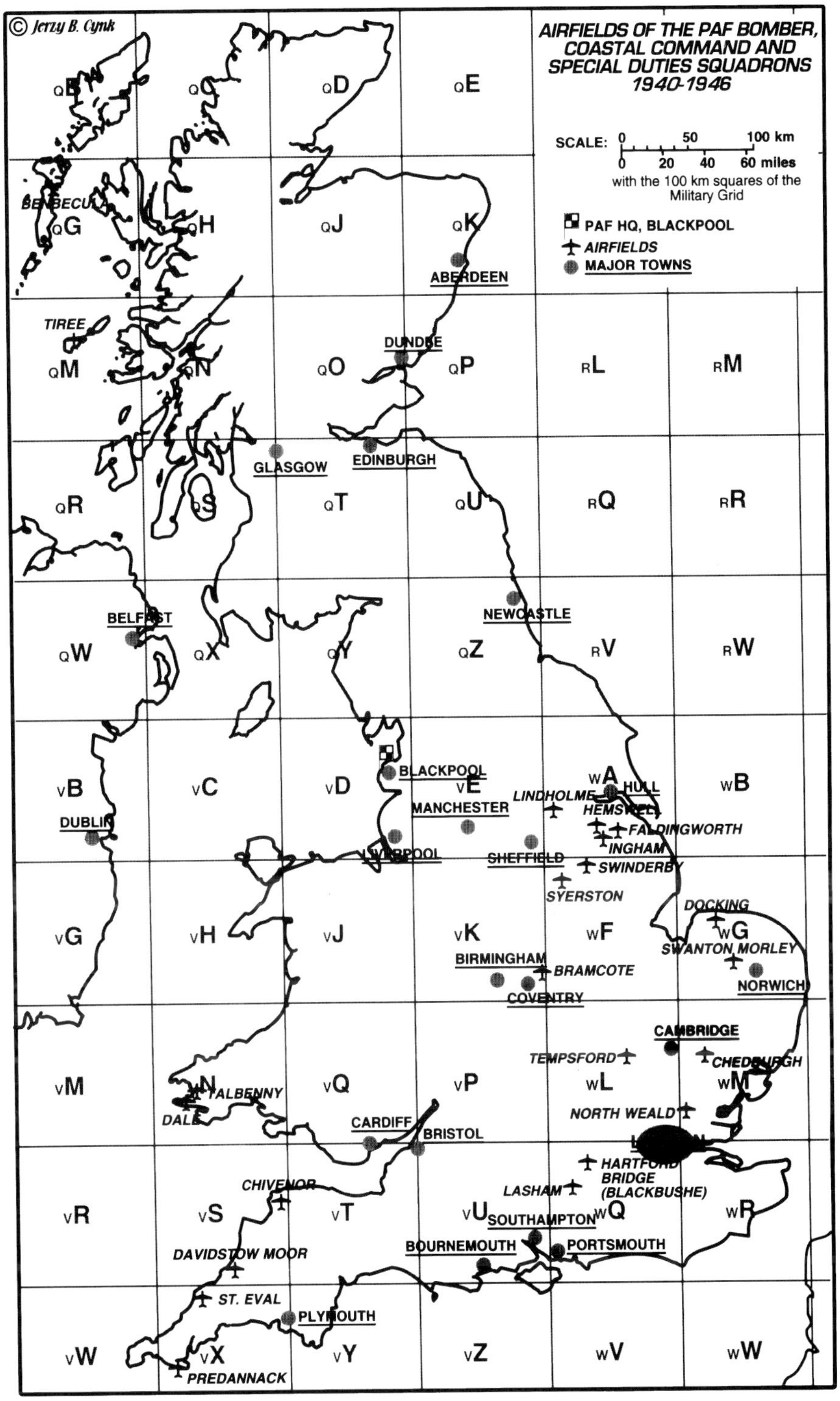

During the next few days the aircrews began daylight training, including navigation and dive-bombing, but because of an initial shortage of servicing tools and certain spare parts only two aircraft were serviceable on average. An Avro Anson was loaned to the unit to help with the training. On 11 July the first night flights were made. In the second half of the month training intensified and the Squadron started to exercise in gunnery and bombing on the Otmoor firing range. On 3 August No.300 and 301 Squadrons were inspected at Bramcote by AM Sir Charles Portal, C.-in-C. Bomber Command, and after a dive-bombing demonstration his staff remarked that the Poles "handled the Battles like Spitfires." Four days later Gen Sikorski paid a call on the squadrons. On 20 August, as Gen Sikorski visited Bramcote for the second time, accompanied by Gen Ujejski, Inspector General of the PAF, and other dignitaries, King George VI arrived unexpectedly at the station. The King inspected the Polish detachments, talked to the crews, watched flypast and bombing demonstrations and signed No.300's Operational Book.

On 22 August No.300 Sqdn moved to Swinderby, a new RAF station in No.1 Bomber Group not yet fully completed, which was to become its home for the next 11 months. Its aircraft landed on the new airfield before nightfall and training was resumed on the 25th. Its flying personnel comprised:

- pilots: S/Ldr (Mjr) Stanisław Cwynar (No.300's Deputy C.O.), F/Lt(Cpt) Stefan Kryński ('A' Flight C.O.), F/O (Cpt) Romuald Suliński ('B' Flight C.O.); F/Os (Lts) Jerzy Antonowicz, Jan Gębicki, Jerzy Gołko, Mieczysław Kałuża, Jan Konarzewski, Jan Michałowski; P/Os (S-Lts) Tadeusz Jasiński, Kazimierz Kula, Zdzisław Kurowski, Marek Rusiecki; Sgt (Sgt Mjr) Henryk Szwedowski; Sgts (Sgts) Czesław Dziekoński, Gerhard Goebel, Józef Kuflik, Marian Kostecki, Piotr Nowakowski, Alojzy Ratajczak, Kazimierz Szymanowski, and Sgt (LAC) Józef Kazimierczak;
- observers: F/Os (Lts) Aleksander Bujalski, Tadeusz Chrostowski, Stanisław Firley-Bielański, Tomasz Łubieszko, Adam Męciński, Roman Miarczyński, Włodzimierz Pęski, Marian Piotrowski, Władysław Rogalski, Ignacy Szponarowicz; P/Os (S-Lts) Czesław Dej, Nikodem Koziński, Zenon Luba, Józef Morawiak, Zygmunt Pluta, Antoni Pułczyński, Karol Szymański; Sgts (Sgts) Edward Morawa, Marian Przybylski and Bolesław Sobieszczuk;
- gunners: Sgts (Sgts) Jan Artymiuk, Tadeusz Bączak, Jan Bieżuński, Tadeusz Egierski, Marian Gędziorowski, Władysław Graczyk, Otton Herman, Feliks Jezierski, Alfons Kowalski, Henryk Kudełko, Władysław Łapot, Albin Socha, Tadeusz Swidziński, Tadeusz Szmajdowicz, Marian Sztul and Władysław Urbanowicz.

During August the Squadron made 458 training flights. To increase the psychological effect of their attacks, the Poles decided to fit wailing sirens to their Battles. It was all unofficial, but Robey of Lincoln made them and they were installed on the aircraft without any fuss. The whole thing caused concern only when the Poles made unannounced, dummy dive-bomb attacks with the sirens on against No.1 Group HQ, whose ruffled staff hurriedly sought shelter under office desks and tables in the belief that enemy bombers were overhead.

On 7 September the British Chiefs of Staff issued the code word "Cromwell." This meant that invasion was imminent and the station went on Alert State one. Six crews – three from No.300 and three from No.301 Sqdns – were put on stand-by and 12 Battles were bombed-up. The squadron became operational on 12 September.

(2.) No.301 "Pomeranian" Bomber Squadron

An Air Ministry order of 17 July asked for No.301 (Polish) (Bomber) Squadron to be formed at Bramcote on 22 July under the same establishment as No.300 Sqdn. The Polish affirmation of the order referred to the unit as "IInd Polish Bomber Squadron." W/Cdr (S-Col) Roman Rudkowski, appointed its C.O., and S/Ldr (Mjr) Stefan Floryanowicz, its Deputy C.O., arrived at Bramcote on the 19th. During the following few days they were joined by several Polish officers and most of the airmen assigned to the squadron, and by S/Ldr C.G. Skinner, the RAF adviser.

Three more British officers and some 30 airmen were initially attached to the squadron to help with its organization and training. The squadron was named No.301 Dywizjon Bombowy Ziemi Pomorskiej ("Pomeranian" Bomber Squadron) and the code letters GR were allocated to it. An unofficial badge – the red Pomeranian Griffin on a white pentagon shield, closely resembling that of the pre-war 41st Eskadra – was carried on its aircraft from the middle of 1941 until late 1942. Later a different badge – a decrescent forming part of a circle with a

A DH 95 Flamingo with Gen Sikorski on board lands at Bramcote on 7 August 1940. (IPMS)

S-Col Luziński greets Gen Sikorski as he steps out of the Flamingo to inspect No.300 and 301 Bomber Squadrons at Bramcote on 7 August 1940. Gen Kalkus stands on Sikorski's left. (IPMS)

An aircrew of No.300 Squadron climbs into the cockpits of Battle L5427 BH•E at Bramcote on 7 August 1940.

flying bat superimposed, reminiscent of that of the 64th Eskadra – was proposed. Disbandment of the squadron early in 1943 prevented its approval by the C.-in-C. Polish Armed Forces. When the squadron was reformed in November 1944, a new circular badge, combining elements of the previous two – the decrescent and the Pomeranian Griffin – with the Piast[3] Eagle and the Warsaw Siren emblem was designed. This was approved and the name "Obrońców Warszawy" ("Defenders of Warsaw") was added to the official title of the squadron. The squadron received its first four Battles on 26 July and began training flights the same day. By the end of the month more Battle bombers and one Battle dual-control trainer arrived, and an Avro Anson was loaned to the unit. On 8 August No.301 suffered its first fatality, when F/O (Lt) Dominik Fengler crashed in Battle L5597 soon after a night take-off. During the month a number of aircrews completed advanced training and ground crew training proceeded equally well. The flying personnel (as of 21 August) comprised:[4]

- pilots: S/Ldr (Mjr) Stefan Floryanowicz (No.301's Deputy C.O.), F/Lt (Cpt) Stanisław Brejnak ('A' Flight C.O.), F/Lt (Cpt) Witold Jacek Piotrowski ('B' Flight C.O.), F/O (Lt) Wacław Butkiewicz; P/Os (S-Lts) Leopold Mielecki, Bolesław Murawski, Marian Olszyna, Teofil Pożyczka, Józef Waroński, Józef Wójcik; Sgts (Sgts) Tadeusz Bojakowski, Roman Bońkowski, Stanisław Jensen, Stanisław Kłosowski, Kazimierz Lenczowski, Antoni Palul, Zygmunt Popławski, Ignacy Radzymiński, Adam Weiss and Franciszek Zaremba;

[3] The first Polish royal dynasty, the founders of the Polish State.
[4] IPMS, Lot. A V, 35/28.

King George VI, with Ambassador Raczyński and Gen Sikorski to his left, at the beginning of his visit to Bramcote on 20 August 1940. Gen Ujejski walks behind Sikorski. (IPMS)

- observers: F/Os (Lts) Eugeniusz Domański, Stefan Hałłas, Józef Koryciński, Jan Kozak, Stanisław Król, Hieronim Kulbacki, Julian Pałka, Stanisław Rewkowski, Henryk Sawlewicz, Stanisław Starowicz, Leon Szymkiewicz, Antoni Voellnagel, Stanisław Waszkiewicz, Antoni Wronka; P/Os (S-Lts) Bolesław Biliński, Józef Kuliński, Mieczysław Łapa, Marek Wodzicki and Kazimierz Wójcicki;
- gunners: Sgts (Sgts and Cpls) Ignacy Bator, Jan Bujak, Antoni Dydo, Franciszek Florczak, Konstanty Gołębiowski, Jan Hejnowski, Edward Janik, Gustaw Kasianowski, Bernard Kowalski, Czesław Kozłowski, Antoni Lipecki, Tadeusz Madejski, Karol Paliwoda, Zbigniew Pieczyński, Kazimierz Sawicz, Ryszard Tydman, Walenty Wasilewski.

P/O (S-Lt) Narcyz Krupowicz was the Engineering Officer.

No.300's Battles preparing to take off for a flying display during the Royal visit on 20 August 1940.

A Battle Trainer R7399 at Swinderby. The aircraft carries a small PAF chessboard on the fuselage aft of the first cockpit. (W. Sylwestrowicz/J. Koniarek)

The squadron followed No.300 Sqdn to Swinderby on 28 August and its 16 Battles and one Anson landed on the airfield in the afternoon. Because of a shortage of accommodations at the station, the new arrivals were partly billeted at the nearby RAF Winthorpe, Swinderby's satellite airfield, the building of which was not yet completed. Despite the lack of amenities on both airfields and a Luftwaffe attack on the night of 30 August, the two squadrons rapidly completed their preparations for operational duties. After the alert on 7 September, No.301 Sqdn became operational at the same time as No.300, on 12 September.

(3.) No.304 "Silesian" Bomber Squadron

In the middle of August 1940 the Air Ministry and the Inspectorate General of the PAF issued orders authorizing the formation of No.304 (Polish) (Bomber) Squadron ("IIIrd Polish Bomber Squadron") at Bramcote on 22 August, under the establishment No.WAR/BC/142 (basically the same as BC/119 except for the addition of ancillary British personnel), with 16 Battles of Initial Equipment and a Battle Trainer. The Squadron was to be manned by the "French Poles" – the PAF personnel who had arrived in England after the capitulation of France.[5] Most of the officers and airmen selected for it came originally from the pre-war 2nd (Kraków) and 6th (Lwów) Air Regiments. W/Cdr (S-Col) Jan Biały was appointed its Commander, S/Ldr (Mjr) Julian Wojda, Deputy C.O., and S/Ldr F.M. Graham, the British adviser. The squadron adopted the name No.304 Dywizjon Bombowy Ziemi Śląskiej im. Ks. Józefa Poniatowskiego ("Silesian Price Józef Poniatowski's" Bomber Squadron) and was assigned the code letters NZ. Its badge consisted of a falling bomb, with the RAF roundel and the PAF chessboard embellished with stylized wings on its sides and number 304. Following preliminary work on the formation of the unit and allocation of personnel to flights and sections, the first of the 16 Battle bombers and two dual-control Battle Trainers were received on 26 August.

[5] RAF Form 540 (IPMS, Lot. A V,36/12).

First Battle bombers received by No.301 Squadron at Bramcote, with L5392, still without unit code letters, nearest, and L5471 GR•F, behind.

Battle Trainer P6723 NZ•Y of No.304 Squadron, with a small PAF chessboard on the fuselage forward of the letter Y. (IPMS)

During the next few days these were prepared for training. A Miles Magister and later an Avro Anson were loaned to the squadron. At the end of August several "French Poles" were posted to Hucknall for courses in their trade and were replaced by Poles trained in Britain. The strength of personnel as of 1 September was 31 Polish Officers and 154 airmen, plus five British officers and 34 airmen. Five more Polish officers joined the squadron in September and one British officer was posted away. The flying personnel then comprised:[6]

- pilots: F/Lt (Cpt) Jan Gaździk ('B' Flight C.O.), F/Os (Lts) Jan Błażejewski, Rudolf Christman, Kazimierz Czetowicz, Bronisław Kuszczyński, Mieczysław Lewandowski, Aleksander Onoszko, Jan Waroczewski; P/Os (S-Lts) Zenon Horski, Jerzy Iszkowski, Antoni Sym; Sgt (Sgt Mjr) Władysław Żołnowski, Sgts (Sgts) Stefan Gajewski, Stanisław Tofin; Sgts (Cpls) Piotr Głydziak, Michał Leszkiewicz. Stanisław Płonczyński, Mikołaj Zykow; Sgts (LACs) Wacław Przedsławski, Zygmunt Witkowski and Konrad Ziółkowski;
- observers: S/Ldr (Mjr) Julian Wojda (No.304's Deputy C.O.), F/Lt (Cpt) Jan Buczma ('A' Flight C.O.); F/Lts (Cpts) Stefan Alberti, Jerzy Kranc, Henryk Łepkowski, Edward Stańczuk; F/Os (Lts) Mieczysław Daab, Zbigniew Gałczyński, Zbigniew Jaroszyński, Marian Kostuch, Janusz Kurek, Bolesław Nowicki, Zygfryd Stępień, Józef Zawadzki; P/Os (Lts) Zygmunt Sokołowski, Cezary Wieczorek; P/Os (S-Lts) Stanisław Duchnicki, Adam Gisman, Zdzisław Kaniewski, Feliks Sobieralski and Stanisław Wójcik;
- gunners: Sgt.(Sgt) Józef Apanasewicz, Sgt (Cpl) Wojciech Lichota; Sgts (LACs) Henryk Glucklich, Zdzisław Kaliciński, Eugeniusz Krzeczewski, Henryk Kwiatkowski, Konrad Paszkiewicz and Kazimierz Szewczyk.

F/O (Lt) Wacław Pijanowski was the Engineering Officer.

Training on Battles began on 6 September, but the following day, due to the invasion alert, six Battles were prepared for operations and training was temporarily halted, and later continued on a greatly reduced scale. Towards the end of the month No.304 took part in Army cooperation exercises, carrying out dummy attacks on infantry. A surprise, low-level raid on the airfield by a single Ju 88 on 26 September damaged one Battle. A total of 286 flying hours – all by day – were recorded during the month. In October all aircrews had completed initial training, but advanced operational training was greatly hampered by rain, low clouds and mist which prevailed throughout the month. In the middle of October official notification was received that the squadron was to convert to a heavy bomber unit with eight Initial Equipment Wellington aircraft,[7] to be set up under establishment No.WAR/BC/150. This was to be expanded later to 16 Initial Equipment Wellingtons, under a revised establishment BC/154.[8]

The first two Wellingtons, Mk.IA trainers (N2899 and N2989), were delivered on 1 November and the first bomber, a day later. By the end of the month the squadron had in strength seven Wellington IC bombers and two IA trainers and all op

[6] No.304 Polish Bomber Squadron, List of Personnel, 20 September 1940 (IPMS, Lot. A V,36/2a).

[7] Air Ministry letter B.165612/40/E.4(b) of 14 October 1940 and Bomber Command letter BC/17739/E of 20 October (IPMS, Lot. A V,37/1a).

[8] Dated 1 November 1940 and superseding the earlier establishment No.WAR/BC/116 for a bomber squadron with 16 I.E. Wellingtons. At this stage two bombers adapted for training were included in the 16.

erational Battles were disposed of.[9] Continued bad weather restricted training in the air during November to 195 flying hours (all by day), 93 of them on Wellingtons. Ground training proceeded well. On 26 November an advance party went to RAF Syerston to prepare reception of the squadron. At about the same time F/Lt (Cpt) Mikołaj Kaczanowski, former Engineering Officer of No.300 Sqdn, was appointed Technical Officer of the station. The main party moved to the new airfield – in No.6 Bomber Group – on 1 December, but the bad weather delayed the arrival of its aircraft until the 4th. On 19 December W/Cdr Biały, a highly experienced officer, Commander of the IInd Light Bomber Dyon of the Bomber Brigade in September 1939, relinquished command of No.304 Sqdn because of misunderstandings with his British adviser. He was succeeded by Acting W/Cdr (S-Col) Piotr Dudziński, who assumed command on the 20th.

The hard winter of 1940/41 handicapped training – only 251 flying hours were logged in January and February 1941 – and this, combined with the change over to Wellingtons, considerably delayed the whole air training program. Advanced training, including night and instrument flying, bombing practice and tactical exercises, began in earnest in March; 240 flying hours by day and 11 hours 50 min by night were recorded during the month. On 15 April the squadron suffered its first casualties when its Wellington R1212 crashed on landing approach due to engine malfunction. F/O (Lt) Rudolf Christman, Sgt (C-Off) Wiesław Pietruszewski and Sgt (Cpl) Antoni Berger were killed, and three other members of the crew were wounded, two of them seriously. No.304 Sqdn reached operational status on 25 April 1941.

(4.) No.305 "Wielkopolan" Bomber Squadron

No.305 (Polish) (Bomber) Squadron ("IVth Polish Bomber Squadron") was formed at Bramcote on 29 August 1940, one week after No.304 and under the same establishment. The two squadrons filled the places vacated by the move of No.300 and 301 Sqdns. The squadron's development and training proceeded in parallel with that of No.304. Its personnel were recruited mainly from among the "French Poles" and included a large number of men originating from the pre-war 3rd (Poznań) and 5th (Lida) Air Regiments. W/Cdr (S-Col) Jan Jankowski was appointed Commander of the squadron, S/Ldr (Mjr) Kazimierz Kielich, Deputy C.O., and W/Cdr J.K.M. Drysdale, the British adviser. The squadron assumed the name No.305 Dywizjon Bombowy Ziemi Wielkopolskiej im. Marszałka Józefa Piłsudskiego ("Wielkopolan" Marshal Józef Piłsudski's Bomber Squadron). Its 'A' flight was unofficially named the "Lida" Flight and 'B' Flight, the "Poznań" Flight. The code letters SM were allocated to the squadron. The main motif of its badge was the wing of the Polish Winged Hussars with the letter P, marshal's mace and number 305 superimposed; a section of the PAF chessboard and the RAF roundel attached to its left bottom corner completed the design.

The squadron expected a lengthy formation period because of considerable personnel shortages in general and the complete lack of British-trained crews in particular. Courses at Benson and Hucknall were to provide a long-term solution. At the beginning of September its personnel numbered 32 officers and 156 other ranks, excluding the ancillary British staff, and six Battle bombers, two Battle trainers and a Miles Magister were received. The bombers were prepared for operations on the 7th due to the invasion alert. By the end of the month the squadron had in strength 16 Battle bombers. Conversion to Wellingtons began in the first half of November, with delivery of the first aircraft, Mk.IA trainer (P2531) on the 5th. The first three Mk.IC bombers of the eight initially expected, arrived on the 22nd. With the subsequent increase in strength to 16 heavy bombers, six further Mk.ICs were taken on charge by the 30th, and the squadron's C.O. made urgent appeals to Blackpool for more personnel.[10] The flying crew, as of 30 November, comprised:[11]

- pilots: F/Lt (Mjr) Kazimierz Kielich (No.305's Deputy C.O.), F/Lt (Mjr) Bogdan Kleczyński ('A' ("Lida") Flight C.O.), F/Lt (Cpt) Szczepan Ścibior ('B' ("Poznań") Flight C.O.) F/Os (Cpts) Tadeusz Stefanicki, Karol Zaucha; F/Os (Lts) Władysław Cichowski, Henryk Śliwiński; P/Os (Lts) Jerzy Bzowski, Zenon Kurzempa; P/Os (S-Lts) Eugeniusz Ebenrytter, Mieczysław Jonikas, Antoni Kleniewski, Józef Lach, Józef Nogal, Edward Zakrzewski; Sgt (Sgt C-Off) Jan Ostrowski; Sgts (Sgts) Stefan Hildebrandt, Tomasz Kasprzyk, Marian Kleinschmidt, Ludwik Mołata, Kazimierz Mruk, Julian Pieniążek, Stefan Tomicki, Jan Trembaczowski; Sgt (Cpl) Wacław Januszkiewicz; Sgts (LACs) Mieczysław Balcerowicz, Stanisław Dusza, Bogdan Engel, Alojzy Łozowicki and Wacław Simiński;
- observers: F/Lt (Cpt) Czesław Korbut, F/O (Cpt) Stanisław Sawczyński, F/Os (Lts) Antoni Lisiński, Alfons Nowak, Kazimierz Nowak, Brunon Okoński, Jerzy Sukiennik, Wacław Ungar; P/O (Cpt) Wacław Malak; P/Os (Lts) Mieczysław Gorzeński, Tadeusz Wianecki; P/Os (S-Lts) Stanisław Barzdo, Jan Horocki, Aleksander Jastrzębski, Stanisław Kowalcze, Bogdan Ławreńczuk, Jan Łucki, Mieczysław Marcola, Tadeusz Martusiewicz, Jan Ochalski, Stanisław Ostaszewski, Fabian Pokorniewski, Jan Pragłowski, Wacław Reda, Edward Ryciak, Zbigniew Siarkowski, Marian Wojtowicz and Zygmunt Zbucki;
- gunners/radio operators and gunners: F/O (Lt) Mieczysław Saferna, P/O (Lt) Zygmunt Policiewicz, P/O (S-Lt) Aleksander Zirkwitz; Sgts (Sgts) Jan Dębiec, Paweł Krenżel, Zdzisław Pisarek, Wacław Ptak, Wacław

[9] RAF Form 540 (IPMS, Lot. A V,36/12) and No.304's Operational Diary (A V,36/8).

[10] IPMS, Lot. A V, 37/1f and A V, 37/19.

[11] IPMS, Lot. A V, 37/1f and A V, 37/52.

Weinberg, Feliks Zemmler; Sgts (Cpls C-Offs) Bronisław Blicharski, Jan Jabłoński, Jan Maćkula; Sgts (Cpls) Leon Hampel, Zbigniew Idzikowski, Wacław Kowalski, Walenty Sieczka; Sgts (LACs C-Offs) Henryk Bogdanowicz, Franciszek Pantkowski; Sgt (LACs) Janusz Barcz, Józef Czech, Ignacy Drewelus, Ignacy Gacoń, Jan Gomuła, Władysław Kordyś, Zdzisław Nowiński, Jan Podziemski, Czesław Tęgowski and Tadeusz Żuk.

F/O (Lt) Stanisław Wesołowski was the Engineering Officer, and F/O (Cpt Obs) Stanisław Poziomek, the Gunner Officer.

By 1 January 1941 the Polish personnel strength amounted to 47 officers and 177 airmen, this representing only about 60 percent of the establishment.

At the beginning of December 1940 the squadron moved together with No.304 to Syerston. On 27 January 1941 Their Majesties the King and Queen visited the station and for both squadrons, still non-operational and frustrated because of the interruption of training by snow and mist, this provided an exciting break in the winter monotony. The airfield remained unserviceable during the greater part of February and only 159 hours were flown that month. With somewhat better weather in March, 380 hours were flown, though only 2 1/2 hours at night. On 29 March W/Cdr Jankowski went on leave and S/Ldr (Mjr) Bogdan Kleczyński, later promoted to W/Cdr, took over command. Advanced training was completed in April and on the 25th No.305 Sqdn became operational.

2. The Polish Contribution to the Bomber Offensive

(1.) Harassing the Invasion Fleet – the Autumn of 1940

The first two Polish bomber squadrons went into action at the height of the German invasion threat. On 12 September 1940 Station Operational Order No.1 was issued to No. 300 and 301 Sqdns at Swinderby, requiring each to put up three aircraft for a night operation "to destroy enemy barges and shipping in Boulogne." Bad weather resulted in a cancellation of the order, which was repeated the following day with the same result. On 14 September the order was issued for the third time and the clear skies allowed six Battles to take off at dusk and carry out the first operational sortie. These were BM•T, •K, and •O, crewed respectively by F/O Suliński (pilot) - F/O Bujalski (obs.) - Sgt Bieżuński (gunner), by F/O Antonowicz-P/O Dej-Sgt Alfons Kowalski[12] and by P/O Jasiński - Sgt Sobieszczuk - Sgt Łapot; and GR•B, •G, •K, crewed respectively by S/Ldr Floryanowicz-F/O Voellnagel - Sgt Bernard Kowalski, by Sgt Jensen - F/O Rewkowski - Sgt Wasilewski and by Sgt Lenczowski - F/O Hałłas - Sgt Kasianowski. Each bomber dropped four 250-lb bombs. The Station's Combat Diary recorded: "The punctuality of their take-offs, and their keenness earned a message of congratulation from AOC No.1 Group. All the aircraft bombed successfully, but were prevented from returning to their base by raid warnings and deteriorating weather."[13] Other signals received later included congratulations from AM Sir Charles Portal.

Queen Elisabeth examines the rear turret of one of the Polish Wellingtons during the Royal visit to RAF Syerston on 27 January 1941. (IPMS)

The toss of a coin was to decide which of the two squadrons would fly the next mission the following night. No.300 was the winner and its six Battles – led by W/Cdr Makowski in BH•O and including aircraft piloted by F/O Gębicki, F/O Kałuża, F/Lt Kryński, P/O Kula and Sgt Kuflik – attacked the same target in Boulogne. Four crews bombed successfully, but Kryński and Kuflik were not able to release their bombs. Six more similar raids, involving high-altitude, as well as low-level and dive bombing techniques, were carried out on every clear night in the second half of September, each squadron taking it in turns to bomb Bolougne and once also Calais, the alternative target. On 18 September, while a large parachute flare was being secured on a Battle of No.301 Sqdn, its detonator was activated and its striker ejected, wounding an armourer-fitter, LAC Stefan Nowak, in the arm. Realizing the danger to the aircraft and maintenance crew working on it, Nowak pulled the flare off its hooks and ran with it away from the bomber for some 16 yards before it exploded in his hands. He sustained further injuries and severe burns. Nowak's bravery earned him

[12] The same order of trades, pilot-observer-gunner, applying throughout.
[13] IPMS, Lot. A V, 35/3a.

Battle L5556 GR•G and some of the aircrew of No.301 Squadron which took part in the first operational mission on Boulogne on the night of 14 September 1940. (IPMS)

the first Polish decoration to be awarded to an airman in Great Britain.

On the night of 24/25 September No.301's Battle L5351, believed to be attacked by an enemy fighter, crashed and exploded in the Thetford area, killing its crew: P/O (S-Lt) Józef Waroński-P/O (S-Lt) Józef Kuliński-Sgt (Sgt) Karol Paliwoda. The last September raid, the following night, was directed against enemy barges and shipping at Ostend. Nine Battles of No.300 Sqdn bombed successfully despite numerous searchlights and intensive AA fire, and returned home safely in deteriorating weather. Earlier, on 23 September, when the previous two-hour readiness was tightened to 20-minute readiness and the station defence plan was put to test, 12 Battles – six from each squadron – took off in just 11 minutes from receipt of the order – a No.1 Group record.

These September operations gave the squadrons their first combat experience, and although generally successful, revealed some shortcomings in night navigation (due to erroneous navigation one of No.301's Battles landed in Scotland!) and the use of radio. A succession of gales, heavy rain and low cloud interrupted bombing activities on 27 September and two days later the station command suspended operational flying for ten days.

Operations were resumed on the night of 9 October for No.301 sqdn and on the night of the 10th for No.300, each unit putting up eight aircraft carrying six 250-lb bombs. Three further raids against the invasion ports were made on 12, 13 and 15 October, the last one marking the end of Polish operations on the Battles. While all the aircraft reportedly found their targets and bombed successfully, various problems were encountered on their return due to a surprise "Red" alert, sudden fog patches over the airfield and other unforeseen weather conditions. Unhappily, two Battles of No.300 Sqdn had to make force landings on 13 October before midnight because of fog and one was damaged and the other, BH•Y (L5499), crashed near Oxton, killing the crew F/O (Lt) Jan Gębicki-Sgt (Sgt) Edward Morawa-Sgt (Sgt) Tadeusz Egierski. A few hours earlier RAF Swinderby was attacked by a single German bomber which destroyed one Battle of No.301 Sqdn and damaged two of No.300. The total Polish combat record on Battles (14 September-15 October 1940) amounted to 15 raids comprising 85 aircraft sorties – eight raids and 46 sorties by No. 300 Sqdn and seven raids and 39 sorties by No.301 – flown on 13 nights (on two occasions both squadrons operated on the same nights); 45 tons of bombs were dropped. All further operational flights on Battles were officially suspended on 18 October.

(2.) Lending Strength to the Thrust into Germany, December 1940-February 1942

Conversion to Wellingtons, under Establishment BC/150, with Initial Equipment of eight bombers, announced in the middle of October 1940, coincided with the end of the Battle of Britain and the return of Bomber Command to offensive operations. The first Wellington – Mk.IA trainer (L7789), later

Painting of the code letters GR•P on a Wellington IC bomber L7819, newly received by No.301 Squadron at Swinderby in November 1940.

brought up to operational standards – arrived at Swinderby on 18 October and appropriately it was delivered by P/O Janas, a Polish pilot of the Air Transport Auxiliary. Other Wellingtons began to follow two days later – including P9214, the first to be taken on charge by No.301 – and by the middle of November each squadron had eight Wellingtons in strength and disposed of all its operational Battles. In view of the subsequent decision to bring up both squadrons to Establishment BC/154, with Initial Equipment of 16 Wellingtons, further aircraft were delivered late in November.

First training flights on Wellingtons, with British instructors, were made on 24 October, but routine training on Battles still continued. On 29 October a wing broke off of Battle L5356 of No.300 Sqdn and the aircraft spun to the ground, killing its crew: Sgt (Sgt) Gerhard Goebel-F/O (Lt) Stanisław Firley-Bielański-Sgt (Sgt) Tadeusz Szmajdowicz. On 1 November six Battles from each squadron flew to Northolt for a review of the PAF squadrons by Gen Sikorski and Sir Archibald Sinclair, the Secretary of State for Air. On the night of 14/15 November, during a Luftwaffe raid on Winthorpe, the Swinderby's satellite airfield, a land mine – one of the first to fall on England – descended on a parachute and blew up a guard hut from which the duty airmen just managed to jump clear into

Sir Archibald Sinclair, the State Secretary for Air, and Gen Sikorski (in the middle of the group) by a Battle bomber, during a review of the PAF squadrons at Northolt on 1 November 1940. (IPMS)

the ditches. Night flights and operational training began on Wellingtons in the second half of November. On the 30th Władysław Raczkiewicz, the President of the Polish Republic, visited the station and stayed overnight.

No.300 Sqdn was ready to resume operations on 12 December 1940, and No.301, nine days later. No.300 received an order for the first Wellington attack on the 18th, but unsuitable weather led to its cancellation and the first Wellington raid was eventually carried out jointly by the two squadrons in the small hours of 22 December. The new type of target – oil tanks at Antwerp – reflected the beginnings of the bomber offensive whose primary objectives included the destruction of oil plants and the morale of the German population.[14] Three Wellingtons from each squadron, starting with No. 301, made the attack. These were GR•M, •R and •B – captained respectively by W/Cdr Rudkowski, F/Lt Piotrowski and Sgt Jensen,[15] and BH•O, •M, and •Q captained by F/O Suliński, F/O Kałuża and P/O Kula. Kałuża lost his way and bombed shipping at Ostend, the alternative target.

The next raid on the Antwerp oil plant was delayed by bad weather until 28 December. It was made after nightfall by six Wellingtons of No.300 Sqdn – BH•A, •F, •K, •B, •T and •Q, captained respectively by F/Lt Kryński, Sgt Nowakowski, S/Ldr Cwynar, Sgt Kuflik, P/O Jasiński and Sgt Kazimierczak – operating for the first time without the moon or a completely clear sky, and all crews reported difficulty in finding the target. The return trip was made in deteriorating visibility and all the landing approaches were hectic. F/Lt (Cpt) Stefan Kryński's bomber, BH•A (R1055), crashed into trees after he had opened up to go around again because of a faltering landing attempt; the pilot was killed instantly, Sgt (Sgt) Henryk Węgrzyn, his front gunner, died in hospital a few hours later, and the rest of the crew, F/O Jerzy Gołko (2nd pilot)-P/O Karol Szymański (obs.)-Sgt Władysław Graczyk (gunner)-Sgt Aleksander Suczyński (gunner), sustained various injuries. Thus, the last Polish bomber raid of 1940 ended on a sad note. No.300 Sqdn recorded in December a total of 212 flying hours 25 min of which 42 hours 10 min were operational, and No.301 Sqdn, 164 hours 15 min of which 13 hours 25 min were operational.

In 1940 the two Polish bomber squadrons made 17 raids totaling 97 aircraft sorties (88 percent against the invasion ports and 12 percent against oil plants) in 368 flying hours and dropped 62 tons of bombs. The operational casualties amounted to two Battles and one Wellington lost and eight aircrew killed.

At the beginning of operations in 1941 tragedy befell two crews. On New Year's Day Polish aircraft were to bomb a German town for the first time. The attack on Bremen was ordered as a reprisal for the indiscriminate dropping of incendiaries on London on the night of 29/30 December, and six Wellingtons from No.301 Sqdn, GR•A, •G, •H, •K, •R, and •M – S/Ldr Floryanowicz's, F/L Brejnak's, Sgt Kłosowski's, P/O Murawski's, Sgt Weiss's and Sgt Mierniczek's crews – took part. The returning bombers encountered heavy snow showers over England and GR•A (T2517) and •K (T2518) crashed on landing. The entire crew of the former – S/Ldr (Mjr) Stefan Floryanowicz (1st pilot)-P/O (S-Lt) Marian Olszyna (2nd pilot)-F/O (Lt) Hieronim Kulbacki (obs.)-Sgt (Sgt) Jan Hejnowski (gunner/radio op.),-Sgts (Cpls) Ludwik Gackowski and Adam Guzowski (gunners); and five crew members of the latter bomber (which was said to be damaged by a night fighter) – P/O (S-Lt) Bolesław Murawski (1st pilot)-P/O (S-Lt) Tadeusz Dziubiński (2nd pilot)-P/O (Lt) Bolesław Sadowski (obs.)-Sgt (Cpl) Kazimierz Sawicz (gunner/radio op.)-Sgt (Cpl) Gustaw Kasianowski (gunner) – lost their lives. Sgt Jan Wancisiewicz, a gunner, was the only survivor. The wintry weather grounded the squadrons for most of January, permitting only one more raid during the month – on the 9th, on an oil plant in Rotterdam – in which four crews from No. 300 Sqdn took part. A Royal visit on the 27th provided a highlight for the Poles at Swinderby in January.

Further operations became possible only in the second half of February. Two bombers from each squadron flew on the 10th to Newton, to join RAF bombers in a night raid on enemy shipping at Boulogne. The attack was repeated five nights later by four Wellingtons from each squadron taking off from Waddington (because of unserviceability of all airfields in No.1 Group), and again by one aircraft from each squadron on the night of the 23rd/24th, taking off from Mildenhall. The bombings were deemed to be highly successful and operations from strange airfields afforded crews excellent experience. Raids on Germany, which began in earnest at the end of February, had a far greater appeal to the Poles. Düsseldorf was named as the target of the raid on 25 February in which two crews from No.300 participated, and the following nigh three bombers from No.300 and six from No.301 Sqdns were sent over Cologne (two also dropped leaflets on Düsseldorf). Both attacks were made from Newton and despite haze over the target area, several fires were started.

On the night of 2/3 March 1941 five Wellingtons from each squadron ventured over the heavily defended Brest to attack a German battleship and all the aircraft bombed the port. F/Lt Suliński showed remarkable skill and daring by carrying on when his throttle control broke just after taking off. He had to fly at practically full throttle the whole time and managed to land with one engine. On 12 March three crews from No.300 and two from No.301 flew to Mildenhall to launch a night attack on Hamburg. A fine night with the full moon made their job easy and plenty of good fires were reported. The next night the two squadrons put up four bombers each, half of them to raid Hamburg and half to attack the Rotterdam oil depot and both Squadron Commanders, W/Cdr Makowski and W/Cdr Rudkowski, took part in the operations. However, three of the Wellingtons were beset with technical troubles – two suffered complete radio failures, and Rudkowski was obliged to return with five 500 lb bombs, to the intense disappointment of the

[14] As per Portal's Directive to Bomber Command of 25 October 1940.
[15] 1st pilots were captains of the aircraft, regardless of their ranks.

crew. Apart from that, both missions were successful. On 18 March, for the first time since 9 January, an operational take-off from Swinderby was made possible by the weather and three aircraft from No.301 were detailed to attack the Rotterdam oil storage area. A very large fire was started, the glow of which could be seen from 60 miles out to sea by the returning crews.[16] On 21 March the squadron suffered three training fatalities when Sgt (Sgt) Kazimierz Lenczowski (pilot), P/O (Lt) Józef Koryciński (observer) and Sgt (Cpl) Bernard Chrzanowski crashed on landing in a Wellington.

An order of 23 March, sending Polish aircraft to bomb Berlin for the first time, was received with great excitement. Eight bombers from No.300 and six from No.301 Sqdns were dispatched to Langham, RAF Bircham Newton Station's satellite airfield, for the raid, but the operation was not free from disappointments. The airfield was small and undulating, and proved very difficult for a night take-off. After two Wellingtons of No.300 Sqdn had been damaged in crashes on take-off (no one was hurt), the remainder were canceled. The four bombers of No.300 that managed to get into the air – BH•O, •B, •K and •U, captained respectively by W/Cdr Makowski, F/Lt Suliński, Sgt Dziekoński and Sgt Hajdukiewicz put up a fine show in spite of difficult weather conditions. Berlin was bombed by the Poles after all, if not by all the aircraft.

On 6th April 1941 No.300 – four aircraft – and No.301 – six aircraft – were detailed for a night attack on German battleships at Brest, taking off from Waddington. Their navigation was perfect, but 10/10th cloud over the port prevented accurate bombing. For the following two nights Kiel was the target. Crews from both squadrons returned from the two missions in high spirits, reporting raging fires all over the city. On 14 April the Poles were ordered to repeat the attack on Brest, this time from Swinderby, and their take-off was witnessed by Gen Sosnkowski, who was visiting the squadrons. Cloudy weather again made bomb-aiming tricky. Two nights later five Wellingtons from each squadron raided Bremen. They all located the target, but the show was generally marred by very poor visibility. As the returning bombers were landing, a German intruder attacked one and dropped a bomb on the base, but did no harm.

On 17 April the Poles – four crews from each squadron – attacked Berlin for the second time and they were all able to report complete success despite the exceptionally heavy anti-aircraft fire. BH•R, Sgt Szymanowski's crew, was caught by a concentration of searchlights and suffered some damage; its observer, F/O Władysław Rogalski, was wounded by a piece of shrapnel in the thigh, but continued his bombardier's duties regardless. During the next few nights Rotterdam, Brest, Le Havre and Kiel were bombed. The last, and the biggest, April operation by the two squadrons, putting up 14 aircraft to attack Mannheim, was flown on the night of the 29th/30th, with no hitches and excellent navigation.[17]

[16] Main sources: the RAF Swinderby's Combat Diary (IPMS, Lot. A V,35/3a) and No.301 Sqdn's Operational Diary (A V,35/6).

The Syerston airfield. photographed by one of No.304's Wellingtons on 13 March 1941. (IPMS)

Aircrew of No.305 Squadron in front of Wellington IC SM•E "Elżunia" at Syerston in the early spring of 1941, in the final stages of training, before the unit became operational (T. Kopański)

On 25 April No.304 and 305 Sqdns at Syerston received their first operational orders – to bomb oil and petrol storage tanks at Rotterdam. Two aircraft from No.304, NZ•E (R1230) and •D (N2852), were crewed respectively by P/O Antoni Sym (1st pilot)-W/Cdr Piotr Dudziński (2nd pilot)-S/Ldr Julian Wojda (observer)-Sgt Leon Hampel (gunner/radio op.) – Sgt Stanisław Białek (gunner)-P/O Stanisław Duchnicki (gunner), and by F/O Kazimierz Czetowicz-P/O Jerzy Iszkowski-P/O Bronisław Lewkowicz-Sgt Józef Apanasewicz-Sgt Alojzy Sankowski-Sgt Władysław Thier.[18]

Three aircraft from No.305, SM•F (R1322), •A (R1016) and •L (R1019), were crewed by S/Ldr Bogdan Kleczyński-P/O Jerzy Bzowski-F/O Stanisław Poziomek-Sgt Jan Dębiec-P/O Mieczysław Marcola-P/O Stanisław Sawczyński, by P/O Mieczysław Jonikas-Sgt Stanisław Dusza-P/O Tadeusz Martusiewicz-Sgt Franciszek Laskowski-P/O Zygmunt Policiewicz-P/O Mieczysław Gorzeński, and by Sgt Jan Trembaczowski-Sgt Ludwik Mołata-P/O Bolesław Jarkowski-

[17] Ibid.

[18] The same order of trades, 1st pilot-2nd pilot-observer-gunner/radio operator-two gunners, applying throughout.

R1348 GR•O of No.301 Squadron, being prepared for a sortie at Swinderby in the spring of 1941. (M.J. Ingham)

Sgt Kazimierz Kratochwil-Sgt Czesław Reng-P/O Józef Suska.

However, the target was obscured by haze and while four aircraft succeeded in bombing the area, P/O Sym's crew was unable to find it and dropped their bombs into the sea. Cloudy and hazy nights led to the cancellation of further operations in April. During April personnel shortages in both squadrons were made good and at the end of the month their individual strength increased to 55 officers and over 270 airmen each (this excluded the British ancillary staff of five officers and over 50 airmen each).

On the night of 2/3 May all four Polish bomber squadrons were in action. The Swinderby squadrons mounted their biggest effort yet, with eight Wellingtons from No.300 and nine from No.301 raiding Hamburg and distinguished by their selection by C.-in-C. Bomber Command to lead the way to the target – as the 'pathfinder' force – to raise the fires which would provide a beacon for the rest of the bombers. The Air Officer Commanding No.1 Group came to see the take-off and to wish the Poles "good luck" in this heavy attack delivered on the Polish National Day. Despite intense searchlight and AA fire the operation was a great success.[19]

The same night four bombers from No.304 and seven from No.305 carried out Operation No.2 – a raid on the docks and the industrial center of Emden. Numerous fires were observed as a result of the attack. One aircraft, unable to locate the primary target owing to cloud, dropped its load on the railway at Groningen. No. 305 suffered its first operational loss – its SM•N (R1214) was shot down by a German fighter over Holland and P/O (Mjr) Wacław Malak, navigator, and P/O (Mjr) Mieczysław Ryszkiewicz, flying as a volunteer gunner, were killed, while the rest of the crew, P/O Józef Nogal, 1st pilot, Sgt Tomasz Kasprzyk, 2nd pilot, P/O Aleksander Jastrzębski and Sgt Tadeusz Żuk were made PoW. With 21 sorties flown in one night, the Polish contribution to the bomber offensive began to assume formidable proportions.

Two nights later No.300 and 301 Sqdns attacked German battle-cruisers at Brest reporting near misses, but the landings were delayed considerably because of the presence of hostile aircraft – nick-named "duty Jerries" – in the area. At the same time Nos.304 and 305 bombed the docks at Le Havre, observing a very large explosion and numerous fires.[20] P/O Iszkowski's W5668 (No.304) caught fire soon after its engines were started and the coolness and disregard for personal safety showed by Cpl Antoni Chojnacki, LAC Mieczysław Czajkowski and AC1 Czesław Prusski of the servicing crew in extinguishing the fire, earned them praise from the C.-in-C. Bomber Command himself. Sorties flown on the nights of 5 and 6 May against Mannheim, Hamburg and Le Havre were hindered by 10/10th cloud over the targets. On the latter night P/O Ryszard Zygmuntowicz's crew, flying GR•K (X3163), had a close encounter with a Bf 110 in the vicinity of Hamburg, but successfully returned to base in the damaged Wellington. No.304 was less lucky – bomber R1443 attacking Le Havre was shot down into the sea and its crew, P/O (Lt) Antoni Sym-Sgt (Sgt Mjr)

[19] The RAF Swinderby's Combat Diary (IPMS, Lot. A V,35/3a).

[20] RAF Forms 540 (IPMS, Lot. A V,36/12 and 37/11).

Władysław Żołnowski-P/O (S-Lt) Feliks Sobieralski-P/O (Lt) Stanisław Duchnicki-Sgt (Sgt) Leon Hampel-Sgt (LAC) Stanisław Białek, all perished.

The night of 8/9 May was again one of great effort, with 27 Polish Wellingtons – 18 from Nos.300 and 301, and nine from Nos.304 and 305 – attacking Bremen as part of a big area bombing operation against Bremen and Hamburg, totaling 362 aircraft. Three bombers did not return. GR•M (R1227) from No.301 was lost, and its crew, Sgt (Sgt) Tadeusz Bojakowski-P/O (S-Lt) Władysław Brzozowski-F/O (Cpt) Stanisław Rewkowski-Sgt (Sgt) Antoni Lipecki-P/O (S-Lt) Ignacy Dudek-Sgt (Sgt) Konstanty Gołębiowski, were all drowned. One bomber (R1473) from No.304, captained by F/O Lynes and with W/Cdr Graham, the squadron's adviser, and the British crew, was destroyed. Another Wellington, SM•F (R1322) from No.305, attacked by fighters over Holland, exploded in mid-air, killing its crew, F/Sgt (Cpl) Jan Dorman-Sgt (Cpl) Zdzisław Gwóźdź-P/O (Cpt) Maciej Socharski-Sgt (LAC) Ludwik Karcz-Sgt (Cpl) Stanisław Pisarski-Sgt (Cpl) Henryk Sikorski. Congratulations were received from the Secretary of State for Air by HQ No.1 Group for its squadrons for the successful night's work, deserving "the highest praise" and revealing "the growing power of Bomber Command." On the two following nights Mannheim and Hamburg were hit again by all four squadrons, while four "freshman" crews from Nos.304 and 305 were sent over Emden. On the night of the 12th/13th three crews from No.304 and four from No.305 were detailed to raid Mannheim. One Wellington returned damaged by an enemy fighter. On the night of 16/17 May 17 bombers from Swinderby and seven from Syerston took part in the area bombing of Cologne. In addition, two "freshmen" crews from No.304 and three from No.305 attacked the docks and shipping at Boulogne.

One of the lighter moments in the life of aircrews of No.304, a "gunnery practice" on a Tiger Moth at Syerston in the middle of 1941. (IPMS)

The persistently rainy weather and gusty winds in the second half of May and the early part of June 1941 brought operations almost to a standstill. On the night of 27/28 May two "freshmen" crews from No.304 and three from No.305 went to bomb Boulogne. The aircraft R1392 of the former squadron was struck in the port engine by AA fire and crashed in Kent; F/O (Lt) Jan Waroczewski (1st pilot), F/O (Lt) Bronisław Kuszczyński (2nd pilot), P/O (Lt) Cezary Wieczorek (obs.) and

No.300's Wellington IC BH•T at Swinderby.

Sgt (LAC) Józef Drożdż (gunner) were killed and two other crew members bailed out.

The next task, an attack on Duisburg, was carried out by six bombers from No.304 and seven from No.305 on the night of 2/3 June. All aircraft reached the target despite the unfavorable weather and intense AA fire. On 11 June Swinderby and Syerstone were visited by AM Sir Richard Peirse, C.-in-C. Bomber Command, accompanied by AVM R.D. Oxland, AOC No.1 Group. HRH the Duke of Kent visited both stations on the 12th and 13th. On the night of 11/12 June No.300 and 301 Sqdns put up 18 bombers for the first operation from Swinderby for nearly a month – an attack on Düsseldorf. Some aircraft experienced engine trouble in the air and were forced to bomb other targets, as well as to jettison bombs in order to gain height. No.300 lost one Wellington, BH•H (W5666) – five members of the crew of F/O Stanisław Sędzik, comprising besides him F/O Walerian Sojka and Sgts Stanisław Kruk-Szuster, Ludwik Maciej and Wacław Weinberg, were made PoW, and Sgt (Cpl) Zygmunt Chowański lost his life. GR•H (T2921) was damaged by an enemy fighter and its rear gunner, Sgt Władysław Langer, was wounded, but the aircraft returned safely.

The following night both squadrons dispatched 21 bombers (of the 22 originally ordered) – the largest number yet – to attack Osnabrück. Serious icing problems were experienced over the North Sea and three Wellingtons were obliged to jettison their loads. GR•K and •O had to turn back, and the former got stuck in soft earth while taxiing after landing and the latter ran into it – both were severely damaged.

The remaining 18 aircraft – and five from No.304 Sqdn which undertook the same task – successfully bombed the target. No.305, which on the 12th suffered the unfortunate loss of SM•K (R1017) on a day training flight with its entire crew, F/Lt (Cpt) Tadeusz Stefanicki-Sgt (Sgt) Kazimierz Mruk-F/O (Lt) Marian Wojtowicz-F/O (Lt) Aleksander Zirkwitz-F/O (S-Lt) Stanisław Kowalcze-Sgt (LAC) Jerzy Krawczyk, when an Oxford trainer collided with it, raided Duisburg with seven Wellingtons the same night. Both Syerston squadrons, flying seven aircraft each, participated in the bombing of the harbor of Brest and the German battleships Gneisenau, Scharnhorst and Prinz Eugen on the night of 13/14 June. Three crews from No.305 – with captains S/Ldr Ścibior, S/Ldr Kielich and Sgt Mołata – followed the earlier heavy strike by the other bombers by a daring low-level attack. Bursts were observed close to the dry dock and the ships, and a large explosion was seen.

Düsseldorf was the target of 18 bombers from No.300 and 301 Sqdns on the night of 15/16 June. The operation was regarded as a success despite the fact that thin cloud covered the town. The next night, a raid on Duisburg called for another major effort and 17 bombers from Swinderby and eight from Syerston took part. F/O Wacław Butkiewicz and his crew (No.301) showed great courage when they were repeatedly fired at by a Bf 110, which was reported damaged, possibly shot down, by the gunners after its third attack. Fires broke out in the rear turret, injuring Sgt Marian Koc, and in the navigator's compartment, but were extinguished and Butkiewicz succeeded in making a safe landing in the damaged aircraft, although the bomb door would not close and the landing gear, flaps and emergency system were out of action. As the summer nights became shorter and the raids longer, part of the flight had to be made in daylight, with take-offs before dusk and landings after daybreak. The brightness of the moon and the Northern Lights, which were brighter than usual during the summer of 1941, posed additional hazards. Interception by the German night fighters became more frequent and effective, and the bomber casualties increased.

On the night of 18/19 June, during an attack on Bremen by 15 Polish bombers from No.300 and 301 Sqdns and five from No.305, three aircraft did not return. No.300 lost BH•M (W5665) and its crew, Sgt (Cpl) Władysław Paleniczek-Sgt (Sgt) Józef Domański-Sgt (Cpl) Walenty Sieczka-Sgt (Cpl) Piotr Bankowski were killed, and F/Lt Wiktor Cebrzyński (obs.) and Sgt Stanisław Winek (gunner) were made PoW. The GR•D (R1365) of No.301 was shot down into the sea and its crew, P/O (S-Lt) Jerzy Krassowski-Sgt (Sgt) Roman Bońkowski-P/O (Lt) Kazimierz Bernaś-Sgt (Sgt) Stanisław Franaszczuk-Sgt (Sgt) Marian Manasiak-Sgt (Sgt) Andrzej Wagner, all perished. No.305 suffered the lost of SM•F (R1696). Its observer, F/O (Cpt) Kazimierz Żerebecki, was killed and the rest of the crew, Sgt Stanisław Lewek-Sgt Edward Ołenyn-F/O Kazimierz Jaklewicz-Sgt Henryk Rogowski-Sgt Mieczysław Dębkowski, were made PoW; Rogowski, mortally wounded, died on the 19th in a German hospital.

The next task, a bombing of the main railway station and industrial center in Cologne, was flown by four Wellingtons from No.304 and four from No.305 on the night of 21/22 June. The following night 12 bombers (of the 13 ordered) from Swinderby raided Bremen. The 13th, GR•R (R1026), lost one of its airscrews on take-off and the engine caught fire. It crashed into trees and P/O (Lt) Stanisław Pietruszka (1st pilot) and Sgt (S-Lt) Czesław Tęgowski (gunner) were killed, Sgts Zygmunt Popławski, Feliks Florczak and Mieczysław Grzymski were wounded, and F/O Henryk Sawlewicz escaped injury. Two nights later nine aircraft from No.304 and 305 Sqdns were sent over Cologne, while two "freshmen" crews from the latter unit – captained by S/Ldr Kielich and P/O Lach – attacked the docks at Boulogne. On the way back, the former aircraft, SM•F (W5723), was hit by the fire from an unidentified ship off Calais and one of its engines stopped and the other began to overheat. It was about to crash on Clacton, when the pilot turned out to sea, presumably to avoid civilian casualties. Most of the crew baled out over the water, but only W/O Zbigniew Idzikowski (observer) and Sgt Tadeusz Frankowski were rescued. S/Ldr (Mjr) Kazimierz Kielich, Sgts (Sgts) Wacław Januszkiewicz (2nd pilot) and Stanisław Witczak, and Sgt (Cpl) Zbigniew Lewoniec were drowned.[21]

---

[21] RAF Form 540 (IPMS, Lot. A V,37/11).

On the night of 25/26 June 15 bombers from No.300 and 301 Sqdns had successfully bombed Bremen and two nights later four aircraft from No.304 and five from No.305 were allotted the same target. The last June mission was flown by 16 Swinderby Wellingtons on the night of 29/30, also against the docks and town of Bremen. One bomber of No.301 (R1373) was shot down by night fighters over Germany and its radio operator, Sgt (Sgt) Hubert Korab-Brzozowski, was killed; the rest of the crew, Sgt Adam Weiss-P/O Leon Kozłowski-P/O Edward Hubicki-Sgt Jan Bujak-Sgt Mieczysław Hasiński, were made PoW. Sgt Nowakowski's crew (No.300 Sqdn) lost their fix on the return flight and had to come down in the North Sea, but they were quickly found and picked up within hours.

By the middle of 1941 combat losses among the most experienced Polish bomber crews were already a matter of concern. Replacements were not immediately available, but were expected to come from No.18 OTU, where (as of 1 July) 101 officers and 159 airmen – including 81 pilots, 59 navigators,[22] 72 radio operators and 48 gunners – were undergoing operational training.

On the night of 2/3 July nine bombers from No.304 and 305 bombed Bremen. The following night 14 Wellingtons from No.300 and 301 hit the same target. Enemy fighters were encountered and shot down one aircraft from each squadron. No.300 lost R1642, its crew F/O (Lt) Kazimierz Kula (1st pilot) and Sgts (Sgts) Władysław Urbanowicz, Otton Herman and Jan Mieczkowski were killed, and F/Lt Witold Krupowicz (2nd pilot) and W/O Marian Przybylski (nav.) were made PoW. Of No.301's GR•M (R1492) crew, G/Cpt (Col) Bolesław Stachoń,[23] the Polish Commander of the Swinderby Station who flew voluntarily on operations as 2nd pilot, was killed, and the rest, F/Lt Wacław Butkiewicz (1st pilot), F/O Julian Pałka, W/O Antoni Dydo, W/O Zbigniew Idzikowski and Sgt Zdzisław Dzięgiel, were made PoW.

On the night of 4/5 July four aircraft from No.304 and six from No.305 attacked very successfully the harbour of Lorient and submarines lying in the docks, and two "freshmen" crews from the former unit were sent to Cherbourg. The next night 11 Wellingtons from Swinderby bombed a power station at Bielefeld, while one "freshmen" crew was dispatched to Rotterdam. During the following five nights similar attacks were carried out against varied targets. Cologne and Osnabrück were hit hardest, but Bielefield, Emden, Brest and Boulogne were also raided, some of them by "freshmen" crews. On the night of the 9th/10th night fighters destroyed SM•G (R1762), killing F/O (Lt) Brunon Okoński, flying as a gunner; the rest of the crew, Sgt Wacław Simiński-Sgt Jan Mikszo-F/O Mieczysław Marcola-Sgt Jan Dębiec-Sgt Mieczysław Minta, were made PoW. On the 10th/11th Sgt Kuflik's crew of No.300, shot down over Germany, was made PoW. On 11 July the following congratulatory message from the C.-in-C. Bomber Command was received by AOC No.1 Group: "With a record month of operations in June behind us and the opening days of July exceeding even those of June in their effort, severity and effectiveness against the enemy, I express to you my wholehearted praise and admiration of the achievements of your Squadrons ... To all those charged with the maintenance of our aircraft I send a particular word of praise. Whether they be in operational units or the OTU's their tireless efforts have maintained a wonderful standard of serviceability."

On the night of 14/15 July another big raid was mounted on Bremen, with 14 Polish bombers from Swinderby and 11 from Syerston taking part. In addition three "fresher" crews bombed Rotterdam. Very heavy AA fire was encountered over Bremen and SM•Q (W5726) was shot down and crashed on the town, killing its crew F/O (Cpt) Jerzy Bzowski-Janota-P/O (S-Lt) Jan Ostrowski-F/O (Lt) Antoni Lisiński-Sgt (Cpl) Stanisław Mitkowski-Sgt (Sgt) Czesław Kaczalski-Sgt (LAC) Antoni Burak. The navigator of another bomber, F/O Riedl, was wounded by a shell splinter. F/O Ostrowski's aircraft (No.304 Sqdn) had to abandon the attempt to reach Bremen owing to failure of the starboard engine, but running on only one engine, he continued to locate a secondary target on the Dutch coast. Trying to make a forced landing at Langholme

Gen Sikorski addresses airmen of the Polish bomber squadrons at Swinderby on 16 July 1941 during the 1941 Polish Air Force Day celebrations. (IPMS)

[22] Earlier usually referred to as 'observers' in most official documents.

[23] Stachoń was the first Polish rotary-wing aircraft pilot, the Commander of the Advanced Flying School of the 4th (Toruń) Air Regiment, and, in September 1939, of the Aviation of Army Pomorze; he was promoted posthumously to the rank of Brig Gen.

airfield in fog, he crashed in a wood. His front and rear gunners, Sgts Paszkiewicz and Gębicki, sustained injuries and the aircraft (R1002) was destroyed.

On 15 July 1941, on the eve of Polish Air Force Day, the Polish Memorial in Newark Cemetery was unveiled by the President of the Polish Republic, Władysław Raczkiewicz, and was blessed by Bishop Gawlina. The ceremony was attended by Gen Sikorski and AVM Ujejski, C.-in-C. PAF. On the 16th, Polish Air Force Day, a High Mass was sung by Bishop Gawlina at Swinderby and the Polish Air Force Color, secretly made in Soviet-occupied Wilno and smuggled to England, was ceremonially presented to the PAF by Gen Sikorski in the presence of President Raczkiewicz and representatives of the Polish and British governments and high ranking officers. No.300 Squadron was the first to receive it, and each Polish squadron was to keep it in turn for a period of three months throughout the war.

Gen Sikorski and ACM Sir Charles Portal salute the Polish Air Force Color after its handing over to No.300 Squadron at Swinderby on the Polish Air Force Day, 16 July 1941. (IPMS)

On 17/18 July all the four squadrons were engaged in a major operation against Cologne, where several large fires were started. Their "fresher" crews were given the customary easier task, the bombing of Rotterdam, but on their return the "freshmen" from No.304, flying in N2840, were shot down near Newark by enemy aircraft, and F/O (Cpt) Bolesław Kuzian-Sgt (S-Lt) Janusz Tomaszewski-F/O (Lt) Bronisław Klatt-Sgt (Sgt) Jan Sylwestrowicz-Sgt (Cpl) Jan Podziemski-Sgt (LAC) Marian Czerniejewski all lost their lives.

On 18 July W/Cdr (Mjr) Stanisław Cwynar took over command of No.300 Squadron from W/Cdr Makowski – promoted to Polish Station Commander at Lindholme. On the same day No.300 and 301 Sqdns moved from Swinderby to Hemswell, 'B' Flight of No.300 being accommodated at Ingham, Hemswell's satellite airfield. F/Lt (Cpt) Stanisław Dudziński was appointed Chief Technical Officer at Hemswell and F/O (Lt) Teodor Piotrowski succeeded him as Engineering Officer of No.300 Sqdn. On the 19th the main and air parties of No.304 Sqdn left Syerston for Lindholme and the next day No.305 followed suit. F/Lt (Cpt) Mikołaj Kaczanowski became Chief Technical Officer at Lindholme. Both new airfields were large and well appointed, and, although like the previous stations they still had no runways, they were well drained and appeared bone-dry as compared with the previous bases. At the same time Nos.300 and 301 at Hemswell and No.305 at Lindholme began to convert to new Wellington variants, the former two to the Mk.IV, which was basically the Mk.IC re-engined with the American 1,200 hp Pratt & Whitney Twin Wasp two-row radials. No.305 re-equipped with the Mk.II, which had 1,145 hp Rolls-Royce Merlin X Vee engines instead of Bristol Pegasus radials.

No.305 Sqdn received its first two Mk.IIs (W5533 and W5557) on the evening of 20 July and took six more on charge in the next two days, completing a smooth conversion to the new mark on 2 August. The unit returned its Wellington IC bombers and two IA trainers (P9216 and P2519) in the first days of August. Nos.300 and 301 on the other hand, which had their first Mk.IVs delivered early in August and began to train on the Mk.IV towards the middle of that month, took several weeks to convert fully to the Pratt & Whitney-powered version. The American engines, to which the maintenance crews were not accustomed, initially lacked spares and tools and posed various "teething" problems. The Mk.IV bombers themselves required several modifications before becoming operationally fit.

Despite the moves and re-equipment with new aircraft, operations from the new airfields began immediately, with all four squadrons flying for the time being Mk.IC bombers. No.300 and 301 Sqdns flew the first mission from Hemswell – a very successful attack on Frankfurt a/M – on the night of 21/22 July. One returning bomber from No.301 was forced to crash land on the beach near Clay because of powerplant failure, but the crew were uninjured. Nos.304 and 305 carried out their first task from Lindholme, also against Frankfurt a/M, the following night and reported a tremendous fire. The next night Nos.300 and 301 were sent to Mannheim, and on the 24th/25th ten bombers from Nos.304 and 305 bombed Emden, but X9620 (No.304) was shot down by AA batteries over Holland and its crew, F/O (Cpt) Leszek Karczewski-Sgt (Cpl) Zygmunt Witkowski-F/O (Cpt) Stanisław Rzepa-Sgt (AC1) Zbigniew Żuwała-Sgt (Cpl) Bolesław Salomon-F/O (Lt) Jan Musiał, were all killed. The last July operation – the bombing of Hamburg – was flown by ten aircraft from Nos.300 and 301. The former unit lost two aircraft, but while P/O Klecha's crew, flying BH•O, was rescued after landing on the sea due to engine trouble, BH•E (X9639) was destroyed by the ground defenses and Sgt (Cpl) Mieczysław Słoma-P/O (Lt) Władysław Jakimowicz-P/O (Lt) Tadeusz Srzednicki-Sgt (AC 1) Andrzej Horak-Sgt (Cpl) Feliks Zemmler perished; only W/O Jan Danielewicz (gunner) survived and was made PoW. A take-off crash on a training flight on the night of 25 July resulted in the death of P/O (S-Lt) Włodzimierz Rebuszyński, pilot of No.301. W/Cdr (Mjr) Witold Jacek Piotrowski succeeded W/Cdr Rudkowski as Commander of No.301 Sqdn on 27 July.

The Polish bomber effort in July 1941 was very formidable; No.300 Sqdn put up 13 operations, totaling 60 sorties,

Left: In the middle of 1941 the red Pomeranian Griffin on a white pentagon shield appeared on Wellingtons of No.301 Squadron. This unofficial badge, similar to that of the pre-war 41st Karaś Eskadra, was carried on the unit's bombers until late 1942. Right: Wellington IC W5690 GR•W, with the Griffin markings on the forward fuselage, depicted at Hemswell in late July 1941.

No. 301, 10 operations with 62 sorties, No.304, 11 operations with 52 sorties and No.305, 11 operations with 48 sorties.

On the night of 2/3 August four Wellington ICs from No.304 attacked Hamburg, while No.305 carried out its first operation on Wellington IIs, sending six of these aircraft, captained by F/O Zakiewski, Sgt Engel, F/O Jonikas, F/O Jankowski, Sgt Płachta and Sgt Mołata, to bomb the docks and shipping at Cherbourg. The crews expressed their complete satisfaction with the performance of their new aircraft. Beginning with 5 August, the squadrons from Lindholme, alternating with those from Hemswell, flew missions on four consecutive nights against targets in Frankfurt a/M, Aachen, Hamburg, and Essen, losing three aircraft. On the 5th Wellington II SM•P (W5593) was shot down over Belgium; F/O (Lt) Jerzy Sukiennik, nav., and gunners F/O (Cpt) Mieczysław Saferna and Sgt (Cpl) Wacław Rybak were killed. S/Ldr Szczepan Ścibior, 1st pilot, was made PoW, but Sgts Stefan Tomicki and Mieczysław Kowalski managed to escape and with help from the Belgians returned to England five months later. The following night BH•M (X9676), attacking Frankfurt, failed to return and its crew, F/L (Lt) Władysław Cichowski-F/O (Lt) Stanisław Jasiński-F/O (Cpt) Tadeusz Chrostowski-F/O (Cpt) Edmund Jura-Sgt (Sgt) Feliks Jezierski-Sgt (Sgt) Czesław Szczukowski, were drowned. Two nights later GR•B (T2625) was destroyed by fighters over Germany; F/O (Lt) Michał Liniewski-F/O (Lt) Jan Wiszniewski-P/O (Lt) Stanisław Cioch-Sgt (Sgt) Edward Kozłowski were killed and two gunners, Sgts Zygmunt Baderski and Józef Gacoń, were made PoW. On 8 August S/Ldr (Mjr) Robert Beill, C.O. of No.304's 'B' Flight, was appointed to command No.305 Sqdn and promoted to W/Cdr, replacing W/Cdr Kleczyński, who had become medically unfit.

The same W5690 GR•W ready to receive its deadly load, Hemswell, late July 1941.

Z8339, SM•N, one of the new Wellington II bombers of No.305 Squadron, operating from Lindholme in autumn 1941.

The weather affected operations in the later part of August. Only single bombers were involved in an action on 12 August. Nos.300 and 301 mounted larger raids on the nights of the 14th, 16th, 18th and 22nd against Hannover, Cologne, Duisburg and Mannheim respectively, while Nos. 304 and 305 put up only two or three aircraft – including "fresher" crews – on these nights, attacking a variety of targets. On the 14th/15th two Wellingtons from No.301 were severely damaged by Bf 110s on returning from Hannover, but thanks to the skills of their pilots, F/O Wójcik and F/O Wolski, they landed safely. Two nights later Wellington II SM•E (W5463), raiding Cologne, was shot down by intense AA fire with the loss of the crew, Sgt (Cpl) Stanisław Przecławski- Sgt (Cpl C-Off) Jan Płachta-F/O (Lt) Alfons Miondlikowski-Sgt (Cpl) Feliks Wardeński-F/Sgt (Cpl) Edward Majewski-Sgt (AC1) Stanisław Stankiewicz. On the 18th, No.300's R1641 crashed during take-off, resulting in the death of both pilots, Sgt (Sgt) Kazimierz Cegliński and Sgt (Cpl) Stanisław Kuropatwa; P/O (Lt) Antoni Paszkowski (nav.), severely wounded, died later in hospital; Sgts Wybraniec, Kordyś and Glucklich sustained minor injuries.

Raids on Cologne by No.300 and 301 Sqdns, and on Le Havre, Ostend and Mannheim by Nos.304 and 305, followed towards the end of the month, and the last August task, the bombing of Essen, was flown by the two Hemswell squadrons, still using the Wellington IC bombers, on the night of 31 August/1 September.

During September 1941 No.300 and 301 Sqdns directed their efforts to convert to the Wellington IV, and this and the bad weather restricted their operational activities, particularly in the second half of the month. Nos.304 and 305 (12 aircraft) were ordered to attack Frankfurt a/M on the night of 2 September. Sgt Mołata's bomber (No.305) was badly damaged by AA fire. The pilot tried to return on one engine, but was forced to ditch in the sea – all the crew were picked up uninjured.

On 7 September six Wellington IIs from No.305 and one "fresher" crew from No.304 bombed Boulogne, and Nos.300 and 301 sent five crews to attack the same target, No.300 using operationally for the first time two Wellington IV bombers – in addition to one Mk.IC – which were captained by P/O Jasiński and Sgt Żalejko. The results of these attacks appeared to be very successful.[24]

A larger operation was put up on the night of the 13th, when 15 aircraft from No.300 and 301 Sqdns were detailed to attack Brest. At the same time six "fresher" crews – from Nos.300, 301, 304 and 305 – bombed Le Havre. Five more "freshers" were sent to Le Havre two nights later and started a very large fire, but Wellington II SM•J (W5526) with a British crew – captained by W/Cdr J.K.M. Drysdale, the RAF adviser to No.305 Sqdn – did not return. A few small raids followed, but operations intensified again only towards the end of September, when on the 26th six bombers from No.305 were sent

[24] No.300 Sqdn's Combat Diaries (IPMS, Lot. A V,35/3a and 34/7).

The Merlin X Vee engines of Z8339 SM•N being serviced in the open at Lindholme in autumn 1941. (IPMS)

A Wellington IC of No.301 Squadron rolls on the grass at Lindholme in the late summer 1941.

to Cologne. Two crashed on return and the crew of SM•G (W5557) suffered casualties; Sgt (Sgt) Eugeniusz Buszko (pilot), Sgt (Cpl) Jerzy Leyche (radio op.) and Sgt (AC1) Wiktor Wasilenko were killed outright, Sgt (Cpl) Tadeusz Korczyk (2nd pilot) died in hospital, and P/O Stanisław Barzdo and Sgt Zdzisław Pisarek were seriously wounded.

On 29 September 11 bombers from No.300 and 301 Sqdns took part in a major operation against Hamburg. At the same time seven "freshmen" crews from both squadrons were detailed to attack Le Havre, but this was canceled owing to unfavourable weather over the port. The following night two bombers from No.304 and four from No.305 raided Hamburg while two aircraft from each squadron were directed to Stettin for the first time. However, only one of them reached the target. Three "freshers" were sent to Le Havre.

Intensive efforts to convert the flying and maintenance crews to Wellington IVs, which were gradually replacing the earlier Mk.IC bombers on operations, continued in the early part of October 1941. Apart from five "fresher" raids on Ostend on the 10th, no missions were flown until the night of the 12th/13th, when the Polish squadrons put up their biggest effort yet in sending 19 bombers from Hemswell and nine from Lindholme over Bremen, and six on smaller raids on Nuremberg, Cologne and Boulogne, attacked by two aircraft each. Sgt Mołata's crew was credited with starting a very large fire in Bremen. All aircraft returned safely. Four nights later seven Wellingtons from No.304 bombed Duisburg, and one "freshmen" crew from No.300 and one from No.305 were sent over Dunkirk, but the latter, flying SM•L (W5579) was shot down and its crew, Sgt (Sgt) Stefan Hildebrandt-F/O (Lt) Marian Kosowski-F/O (Lt) Antoni Bryk-Sgt (Cpl) Jan Hejnowski-P/O (S-Lt) Jan Łucki-Sgt (AC2) Tadeusz Lang, were drowned.

On the 20th 12 aircraft from Nos.304 and 305 were detailed to bomb the docks at Emden and one "fresher" from the latter squadron, Antwerp. NZ•D (N2852) was hit by AA fire and crashed into the sea off the German coast, and Sgt (Sgt) Mikołaj Zykow, P/O (S-Lt) Stanisław Borzęcki-F/O (Lt) Adam Gisman-Sgt (Cpl) Henryk Plis-Sgt (Cpl) Wilhelm Adamik-Sgt (AC1) Ryszard Klimiuk lost their lives. The following night 20 Wellingtons – ten from No.300 and ten from No.301 – successfully bombed Bremen, but GR•W failed to return and its crew, F/Sgt Cieślak-Sgt Borodej-F/O Riedl-Sgt Stalewski-Sgt Młodzik-Sgt Kleeberg, were made PoW.

Another major operation was carried out on the night of 26/27 October, when 19 Wellington IVs from Hemswell and 11 Wellington ICs and IIs from Lindholme were detailed to attack Hamburg. In addition, five "fresher" crews (from Nos.300, 301 and 305) raided Cherbourg. For No.301, all of whose 14 Mk.IV bombers[25] achieved combat readiness only the previous day, this was the first operation to be made entirely on the new bombers. Because of the inability to fix overload tanks to four of their Mk.IVs in time and engine trouble grounding one more aircraft, the squadron could dispatch only five bombers of the ten ordered for the attack on Hamburg. These were GR•B, •A, •J, •H, and •T, captained respectively by Sgt Kłosowski, P/O Liszka, Sgt Bolcewicz, F/Lt Kurzempa and P/O Dobromirski, but •H experienced W/T problems and was recalled to base and •T bombed Emden, the alternative target, because of irregularity in one of its engines.[26] The bombers which did reach the primary target reported causing several large fires. F/O Ladro's crew in Wellington IC W5720 (No.304) suffered engine difficulties and had to ditch in the sea; they were rescued 18 hours later, except for the front gunner, F/O Stefan Stenocki, who drowned.

Apart from a few more minor missions, the docks at Brest were attacked by seven Wellington IVs from No.301 Sqdn on the night of the 29th, thanks to very determined efforts by the Polish mechanics to start the engines, which were badly affected by the freezing temperatures. On 31 October 12 bombers from Nos.304 and 305 were detailed to take part in a raid on Hamburg, but owing to mist results could not be observed.

November operations began on the night of the 1st, with 18 aircraft from No.300 and 301 Sqdns ordered to bomb Kiel, but the mission was hampered by 10/10th cloud, which pre-

[25] Of the 16 Initial Equipment bombers two were still dual control trainers (usually Mk.IAs), but within the following few weeks the existing establishment was revised by the addition of two Initial Reserve bombers, increasing the number to 18, including two for training.

[26] No.301 Sqdn's Combat Diary (IPMS, Lot. A V,35/6).

vented positive location of the target and obliged reliance on radio fixes. On the night of 7/8 November Bomber Command launched one of its largest offensive operations of 1941, which included simultaneous attacks on Boulogne, Ostend, the Ruhr, Cologne, Mannheim and Berlin, and the mining of Oslo Harbour, with 400 aircraft (some 30 of them four-engined Stirlings and Halifaxes) taking part. The four Polish bomber squadrons provided 33 Wellingtons – 17 to bomb Berlin, 15 to bomb Mannheim and one to bomb Boulogne.

The operation, which went ahead in spite of the meteorologists' warnings of convection cloud centered over the North Sea with tops rising to 20,000 ft, freezing temperatures and storms, turned into one of the Command's biggest fiascoes. Severe icing en route caused technical failures in several of the aircraft. Only about 30 percent of all crews claimed to have reached their prime target areas, and that at the price of 37 machines and some 220 experienced men which failed to return. The Berlin attackers suffered most, losing 21 of the 169 bombers dispatched, with very little to show for their sacrifice.

The Poles did well enough by comparison, with eight crews reporting bombing Berlin and nine Mannheim, but four of their aircraft failed to return. No.300 lost two Wellington IVs, BH•A (Z1271) with the crew of Sgt Piotr Nowakowski-P/O Władysław Radwański-F/O Mieczysław Taras-Sgt Henryk Kudełko-Sgt Gustaw Chrzanowski-Sgt Bolesław Iwanowicz, who were made PoW (Tares and Chrzanowski managed to reach the Vichy controlled France and were interned, but the latter eventually died from TB); and BH•U (R1705) with the crew Sgt Kazimierz Sobczak-Sgt Brunon Biliński-F/O Piotr Kowalski-Sgt Jerzy Budzyński-Sgt Stanisław Konarzewski-P/O Zbigniew Groyecki (some of whom were wounded), who shared a similar fate. No.301 lost its Wellington IV GR•Z (Z1277). Its crew, Sgt Henryk Bolcewicz-Sgt Zygmunt Lenczewski-F/O Włodzimierz Kolanowski[27]-W/O Alfred Szczepanowski-Sgt Bolesław Wróblewski and Sgt Tadeusz Moryn, lost on the way back, saw suddenly an airfield in front of them to light up. When they landed, they found out they were in Belgium and surrounded by German guards. Their intact Wellington was repainted in the Luftwaffe markings and used for tests. A similar mistake was made by P/O Tadeusz Blicharz of No.304 Sqdn, flying Wellington IC R1215, after his navigational aids had failed. Running short on fuel, he landed on another lit airfield, but he realized quickly that this was the German-held Brussels aerodrome. He set his aircraft on fire before he and his crew, P/O Wacław Rekszczyc-F/O Jerzy Mondschein[28]-Sgt Józef Jaworosiuk-Sgt Marian Lewandowski-Sgt Stanisław Krzawiecki, were made PoW. Other damaged bombers returned safely. P/O Mieczysław Rzemyk's crew did well in bringing back their GR•D from which they had at one stage nearly baled out due to a port engine fire. The following day reconnaissance sorties were flown over the sea in search of the missing crews.

A No.304 Wellington IC about to be loaded for the next night attack on Germany in the closing weeks of 1941. (IPMS)

On the night of the 9th Nos.300 and 301 each sent six aircraft to bomb Hamburg and very good results were reported. After that both airfields at which the Polish bombers were based remained unserviceable for a fortnight – except for the 15th, when two "freshers" bombed Boulogne. On 12 November W/Cdr Dudziński was posted to Newton as senior Polish officer and handed over command of No.304 Sqdn to S/Ldr Stanisław Poziomek, formerly of No.305 Sqdn, who was promoted to W/Cdr four days later. AVM (Gen) Ujejski, the Polish Inspector General, inspected the Polish squadrons at Hemswell and Lindholme on the 20th and 21st respectively and decorated several officers and airmen with the Cross of the order of Virtuti Militari and the Cross of Valour, awarded on the occasion of Polish Independence Day – 11 November.

On the night of the 25th seven "fresher" crews (from all four squadrons) were detailed to attack Cherbourg. A larger raid, by 15 Wellington IVs from Hemswell, was made on Emden the following night. On the last night of November Emden was bombed in excellent visibility by three "freshers" each from Nos.300 and 301, and Hamburg by five crews from No.305 Sqdn. In the later part of the month a number of officers and airmen from Hemswell were posted to begin four-engined training.

December 1941 was a moderately busy month. On the night of the 7th six Wellington IVs from Hemswell bombed Calais, but the aircraft were heavily engaged by AA batteries, and two – P/O Radecki's BH•R (Z1244) and P/O Boczkowski's BH•W (Z1275) – crashed on landing. The same night ten bombers from Nos.304 and 305 also raided Calais and two from the latter squadron were sent to Aachen, but P/O Skarpetowski, unable to locate Aachen owing to cloud, dropped bombs on the railway near Maastricht, causing an enormous explosion.[29] On 11 December No.301 dispatched three aircraft to attack Cologne, and No.300, three against the same target the following night.

[27] Kolanowski, caught during a mass escape of RAF officers from the Stalag Luft III at Sagan in March 1944, was one of the 53 escapees executed by Gestapo.

[28] Shared the same fate as F/O Kolanowski above.

[29] RAF Form 540 (IPMS, Lot. A V,37/11).

On the 16th six "fresher" crews from No.304, six from No.305 and one each from Nos.300 and 301, were detailed to attack the docks at Ostend. No.304's R1064 crashed into the sea off Margate and its crew, S/Ldr (Cpt) Jan Błażejewski (on his 27th operational flight)-F/O (Lt) Marian Szczodrowski-P/O (Lt) Jan Komłacz-Sgt (AC2) Bogusław Gołąbek-Sgt (Cpl) Kazimierz Suwalski-Sgt (LAC) Hubert Rutkowski, were drowned. Two Wellington IIs from No.305 – P/O Skarpetowski's W5420 and Sgt Gusowski's Z8427 – crashed on landing and four crew members of the latter needed treatment in hospital. On the 17th 21 bombers from Hemswell raided the docks and ships at Brest.

A major operation was put up on 23/24 December against Cologne with 21 aircraft from Hemswell and ten from Lindholme taking part. In addition nine junior crews from all four units were ordered to bomb Ostend, this bringing the total of Polish bombers involved in that night's sorties to 40. SM•J (W5374) crashed and burnt out on the return journey, killing its crew, Sgt (Cpl) Henryk Rozpara-P/O (Lt) Czesław Gołacki-F/O (Cpt) Alfons Nowak-Sgt (Cpl) Eugeniusz Barącz-Sgt (LAC) Zdzisław Kurowski-P/O (S-Lt) Jan Siwiec. P/O Rodecki's BH•R (Z1244) was lost in a crash, but its crew was unhurt. On the 27th 20 Wellingtons from Hemswell attacked Düsseldorf, encountering very strong AA fire and night fighters, and three from No.300 and one from No.301 were damaged to various degrees by the enemy defences and in forced landings.

The last missions of 1941 were flown in brilliant moonlight on 28/29 December from Lindholme, ten senior crews attacking Wilhelmshaven and eight "freshers" attacking Emden. Many bursts were seen from the aiming points and large fires were reported burning in both target areas. No losses were sustained. The Air Officer Commanding No.1 Group forwarded the following message to G/Cpt A.F. Davidson, RAF Station, Lindholme, "Will you please convey my congratulations and thanks to the officers and all ranks of Nos.304 and 305 Squadrons on the very successful operations of their units on the night of 28th/29th December 1941. It is most gratifying that the operations constituted a record since the formation of both units, and contributed largely to the record which was equally set up on that night for the Group as a whole."

1941, the first year of the RAF bomber offensive, was a very trying one for Bomber Command. Grievous losses, as exemplified by those suffered on the night of 7/8 November, and the critical Butt's Report,[30] which assessed that only 66 percent of all bombers dispatched were recorded as attacking the assigned targets and that only one in three of these got within five miles of them, called into question the effectiveness of the Command's operations. Production of new bombers, slower than expected, was sufficient to replace wastage, but did not permit significant expansion of first line strength for months to come.[31]

The development of scientific aids to navigation to increase bombing accuracy at night (and through this the effective

Air and ground crew personnel of No.300 Squadron by BH•B, depicted at Hemswell on 16 December 1941. (T. Kopański)

strength of the attacking force) became crucially important. But despite these difficulties, the bomber offensive remained a fundamental principle of British strategy, fully backed by Churchill himself at every stage. It represented the only means open to the British to carry the war into Germany until such time as the Allied land forces were ready for the invasion of Europe – the point which completely escapes the fashionable present-day critics of that strategy.

In the terms of the proportional Polish contribution to the RAF overall bombing effort, 1941 was the greatest year, with the four Polish Wellington squadrons representing on average 10% of the available bomber force. No.300 flew 441 sorties in 2,484 hours; No.301, 418 sorties in 2,285.5 hours; No.304, 214 in 1,202 hours; No. 305, 284 in 1,479.5 hours – a total of 1,357 sorties in 7,451 hours.

During these 1,793 tons of bombs were dropped on ports, representing 59% of the attacked targets – 802 sorties, with Bremen (204 sorties), Hamburg (153 sorties) and Brest (114 sorties) topping the list; railway and population centers, 31.5% of the targets – 425 sorties, mainly against Cologne (172 sorties), Mannheim (76 sorties) and Düsseldorf (56 sorties); industrial centers, 6.5% of the targets – 91 sorties, with Duisburg (48 sorties) and Essen (23 sorties) the major targets; the remaining 3% (39 sorties) consisted of seven sorties comprising attacks on enemy airfields and AA defences and air-sea rescue searches, and of 32 unspecified sorties. The operational losses amounted to 49 Wellingtons missing, destroyed or damaged beyond repair, and 137 aircrews killed, five missing and 67 made PoW.[32]

The above figures indicate the higher than average operational effectiveness of the Polish bomber units, which was in

[30] Prepared for Lord Cherwell, Churchill's Scientific Advisor, by D.M.B. Butt of the War Cabinet Secretariat in August 1941.

[31] Under the "Target Force E" plan, inaugurated in June 1941, Bomber Command was to possess an operational force of 4000 heavy bombers (250 standard-sized squadrons in eight operational Groups, with a total of 168 airfields) by the spring of 1943, but during the first 15 months of the plan the average daily availability of the bombers only rose from some 400 to just over 510, although the proportion of four-engined aircraft increased from less than 10% to over 40%.

[32] "Summary of the operational effort of the Bomber Force 1940-1945" and "Lists of targets," PAF HQ, Study Department, Ref.LB/1300/Op.Stat/Tjn, 15.03.1946 and 26.03.1946, as amended in April 1946 (Copies: J.B. Cynk, Polish Aircraft Archives), with mathematical errors corrected by the author.

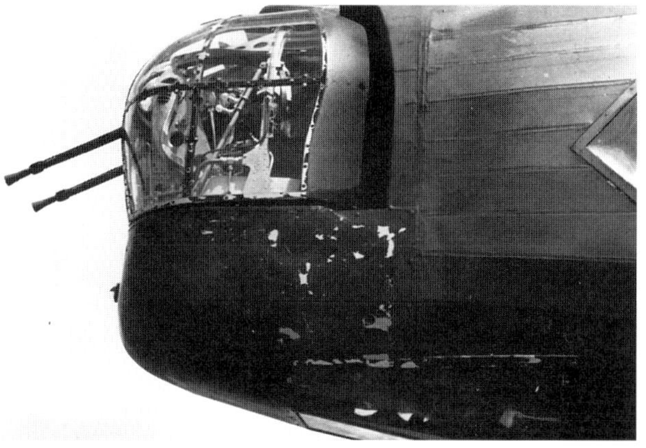

Left: The Frazer-Nash FN 5 power-operated twin-gun nose turret on a Polish Wellington. (IPMS) Right: No.305's Wellington II SM•E before a raid; Lindholme, early 1942. The aircraft carries smaller PAF chessboards on the forward fuselage, introduced towards the end of 1941 to improve its camouflage against detection by enemy night fighters. (K. Chołoniewski)

no small measure contributed to by excellent work and cooperation from the ground crews, whose ingenuity and devotion to duty resulted in five to ten percent better aircraft serviceability as compared with other bomber squadrons. An unusual feature of the Polish bomber scene was the sizable number of older officers, including those of higher ranks and members of squadron HQs, who pleaded to be given a chance to take part in operations whenever the opportunity arose and flew with crews – frequently captained by Sgts – as volunteer gunners.

Bomber Command entered 1942 in a subdued mood and on 23 February received a new AOC-in-Chief, ACM Sir Arthur Harris, who succeeded ACM Peirse. He inherited the problems of his predecessor, which were largely beyond the control of both of them: inaccurate bombing, high wastage of aircraft and disappointing expansion, hindered by deficiencies in production. On the day he took command, he had 51 operational bomber squadrons (out of the total of 58, instead of the 118 envisaged by the "Target Force E" plan), but he states that: "there were 378 aircraft serviceable with crews, and only 69 of these were heavy bombers. About 50 aircraft in the force were not even medium bombers but light bombers ... In effect, this meant that we had an average force of 250 medium and 50 heavy bombers until such time as the Command really began to expand."[33] Some 40 serviceable Wellingtons, or 16 percent of the entire medium bomber force, were those of the Polish squadrons.

The year's first tasks were flown on the night of 5/6 January by 22 senior crews from Hemswell and nine from Lindholme, attacking German warships in Brest, and five "freshers" from Lindholme attacking the docks at Cherbourg,

[33] Sir A. Harris, "Bomber Offensive," Collins, 1947, p. 73.

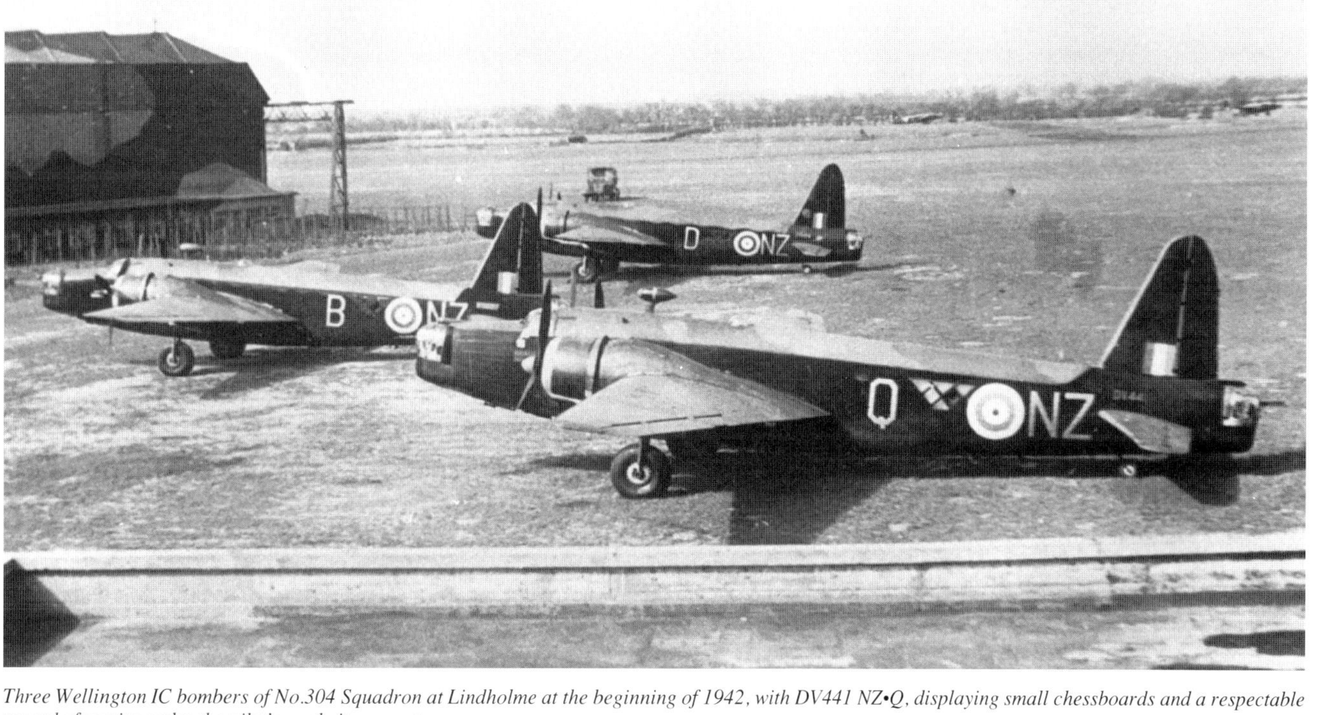

Three Wellington IC bombers of No.304 Squadron at Lindholme at the beginning of 1942, with DV441 NZ•Q, displaying small chessboards and a respectable record of sorties under the pilot's cockpit, nearest.

but cloud cover over both targets affected bombing accuracy. Two nights later 22 Wellington IVs from No.300 and 301 Sqdns raided Brest again, and for the following two nights 13 bombers from Lindholme and 20 from Hemswell took it in turns to strike at the same target, while ten "freshers" were busy over Cherbourg. On the 10th 15 aircraft from Nos.304 and 305 attacked Wilhelmshaven, but two Wellington ICs of No.304 were lost with their crews. These were: R1082, crewed by Sgt (Cpl) Stanisław Obiorek-Sgt (LAC) Bolesław Patek-P/O (Lt) Janusz Kurek-Sgt (Cpl) Andrzej Kwiecień-Sgt (Cpl) Alojzy Sankowski-Sgt (AC1) Adam Rogowski, with which all contact was lost after take-off; and DV432, crewed by P/O (S-Lt) Jan Zając-Sgt (Cpl C-Off) Stanisław Garstka-P/O (S-Lt) Józef Mączyński-Sgt (AC1 C-Off) Jacek Strzyżewski-P/O (Lt) Tadeusz Klewicz-Sgt (LAC) Rajmund Pokrzywa, which was damaged by AA fire and crashed into the sea. On account of thick fog the returning bombers were diverted to various airfields.

On 14/15 January 13 aircraft from Nos.304 and 305 were detailed to attack an aircraft factory in Hamburg. Severe icing compelled three crews to abandon the primary target and F/Lt Edward Zakrzewski experienced an engine failure over Hamburg, but returned safely on one engine. The following night 12 Wellington IVs from Nos.300 and 301 were sent over Hamburg, but three crashed on landings in bad weather. BH•K (Z1265) was completely wracked and all its crew, F/O (Cpt) Jan Bąk-Sgt (LAC) Franciszek Chylewski-F/O (Cpt) Wacław Mosiewicz-Sgt (Cpl) Paweł Krenżel-Sgt (Sgt) Marian Łagodziński-P/O (S-Lt) Stefan Berdys, lost their lives; F/O Śliwiński's GR•O and P/O Liszka's GR•A were damaged and three members of the two crews were slightly wounded. On the night of the 17th 12 senior crews from the Lindholme squadrons raided Bremen and all successfully bombed the target.

Münster was attacked by 15 bombers from Nos.300 and 301 on the night of the 22nd and enemy fighters took their toll. GR•S (Z1285) was shot down into the sea and its crew, Sgt (Cpl) Konstanty Soroko-Sgt (LAC) Piotr Nocoń-P/O (Lt) Wacław Reda-Sgt (AC1) Albert Rokus-Sgt (AC1) Czesław Kosmalski-Sgt (AC1) Bolesław Borecki, were drowned, and P/O Dreja's BH•Z (Z1276) was damaged and P/O Adrian Florial of his crew was wounded. The night of 26 January saw a major effort, with 26 bombers from all squadrons – ten of them from No.305 – taking part in a raid on Hannover. Twenty, which bombed the primary target in perfect visibility, observed bursts and fires. However, the Wellingtons ran into heavy AA fire and two aircraft of No.305 were hit and further damaged in crash landings. A few "freshers" sorties were also made with Emden as the target. The last operations in January were flown on the night of the 28th by ten senior crews from Lindholme, against Münster, and by 13 "freshers" from all four squadrons, against Boulogne. A day earlier, on the 27th, W/Cdr Romuald Suliński took over command of No.300 Squadron from W/Cdr Cwynar.

Alarmingly increasing losses of British shipping resulted in the diversion of the bomber force to attacks on enemy naval targets, French docks, German shipyards and harbours, and mine-laying operations, which in February 1942 absorbed 67% of the Bomber Command effort. The preoccupation of the Command with the German battleships in Brest continued with a heavy attack on the French port on the night of 6 February 1942 in which 16 bombers from Hemswell took part. BH•F (Z1282) was damaged by enemy fighters and crashed on landing; three members of its crew, P/O (S-Lt) Wojciech Veit (2nd pilot)-F/Sgt (Sgt Mjr) Czesław Biały (radio op.)-Sgt (LAC) Stefan Niczewski (gunner), were killed, F/O Henryk Kraciński (1st pilot) and F/O Jan Chomka (nav.) were seriously wounded and only Sgt Zygmunt Blachowski escaped injury. A photoflash flare exploded on board the returning BH•O (R1610) and its pilot, Sgt Dzierzbicki, issued the order to abandon aircraft; two members of the crew and he himself jumped, but the front gunner, Sgt (AC1) Zygmunt Sadurski, collided with an airscrew and was killed. The 2nd pilot, F/O Franciszek Malec, decided to stay with the bomber and helped by Sgt Mieczysław Chanecki (radio operator), landed safely.

Snow and widespread mist made Lindholme unserviceable on several days in the early part of February. Hemswell was less affected and on the night of the 10th 16 bombers raided Bremen. Enemy fighters were encountered and BH•X (Z1220) was slightly damaged and one of its gunners, Sgt Stefan Miniakowski, was wounded. The following night eight "freshers" ventured over Le Havre. On the 12th four Wellingtons from No.300 and eight from No.301 were ordered to carry out their first daylight operation since the formation of the squadrons: an attack on the Scharnhorst and Gneisenau, which had slipped out of Brest into the Channel. Only one section sighted the wake of ships through a break in the clouds and dropped their bombs (the rest did not). The daring escape of the German ships despite a massive effort to intercept and destroy them was a considerable blow to British prestige.

On the night of the 13th Cologne was bombed by 11 senior crews from Hemswell, and Le Havre by 11 "freshers" from all four squadrons. An attack on the Mannheim docks by 16 bombers from Lindholme followed the next night. Seven of the aircraft carried cameras and obtained very good photographs of ground detail. On 16/17 February 19 Wellingtons from Nos.300 and 301 made harassing raids on a number of towns in the area between Bremen and Münster.

On 22/23 February 19 bombers from Nos.300 and 301 opened a series of attacks on German floating docks, beginning with Wilhelmshaven. On the 25th and the two following nights the Kiel floating dock, where the Scharnhorst had taken refuge, was the subject of heavy raids by the Hemswell Wellingtons, interspersed by aircraft from Nos.304 and 305 (one strike on the 26th). F/O Wiktor Jankowiak (navigator) and Sgt Franciszek Porada (1st pilot) (No.301) took excellent night photos of Kiel, which earned them congratulations from AVM R.O. Oxland, ACO No.1 Group. Opposition from ground defences was intense and a number of the bombers were struck by shrapnel. Enemy fighters were also active and on the 26th the rear gunner of Sgt Siciński's crew, Sgt Więckowski (No.304), reported hitting an attacking Bf 110. On the 25th GR•F (Z1377) with the entire crew, F/O (Lt) Jan Jabłoński-F/O (S-Lt) Andrzej Kozieł-F/O (Cpt) Franciszek Fugiel-Sgt (AC1) Karol Droździk-Sgt (AC1) Wincenty Bujak-Sgt (Cpl) Edmund Dolata perished. The following night SM•R (W5423) was damaged over the target and exploded attempting an emergency landing in Denmark, killing F/O (Lt) Jan Orzechowski-P/O (S-Lt) Stanisław Żeromski-F/O (Cpt) Aleksander Golczewski-P/O (S-Lt) Walerian Gidaszewski-F/O (Cpt) Mieczysław Gorzeński-Sgt (LAC) Edward Cegłowski.

(3.) Great Achievements and Decline in Strength (March-December 1942)

In March 1942 Bomber Command, under the leadership of Harris, switched its emphasis back to the original task of building up the offensive on Germany – her industry, transport and population centers. Bad weather interfered with operations at the beginning of the month. On the 8th 13 Wellingtons from No.300 and 301 Sqdns were among the 211 aircraft dispatched to attack the Krupp works at Essen, which became one of the Command's most important targets and the subject of repeat raids on the following two nights. On the 9th 32 bombers (of 35 ordered) from all four Polish units took off to carry out the task – Nos.304 and 305 from alternative airfields (Swanton Morley and Bodney) because of the wet state of Lindholme. Several bursts in the Krupp works were reported and No.301's aircraft caused a violent red explosion and a fire which was seen from 60 miles away. Twenty Polish Wellingtons were sent to Essen on the 10th. Several of the aircraft were damaged by AA fire, but without injury to the crew. The first trials of the new "Shaker" bombing technique[34] and the first large-scale use of the "GEE" navigational aids[35] were made during these raids, and although neither of them was yet working fully satisfactorily, they paved the way for great improvements in bombing accuracy.

On 12/13 March 13 bombers from Nos.300 and 301 took part in a raid on Kiel and seven "freshers" from No.304 and 305 bombed the docks at Emden. GR•J (Z1257) did not return from Kiel and its crew, F/O (Lt) Stefan Zakrzewski-Sgt (AC1) Zbigniew Korzoń-F/O (Lt) Prot Jacek Perkowski-Sgt (AC1) Edmund Kosior-Sgt (LAC) Mieczysław Ziajski-Sgt (AC1) Eugeniusz Siuda, perished. An air search by seven Wellingtons the next day failed to find any trace of the missing bomber. Nos.304 and 305 dispatched 19 senior crews to attack Cologne on the night of the 13th. Sgt Mołata's aircraft, SM•Z (Z8399), carried one 4,000 lb "blockbuster" – colloquially known as "Cookie" – which was dropped from 15,000 ft and caused a large fire after bursting.[36] His bomber was then hit by AA fire, but returned to England on one engine. SM•B (Z8438), also damaged over the target, crashed into a building while attempting to land at base. F/O (Lt) Czesław Rymkiewicz-P/O (S-Lt) Mieczysław Dranicki-Sgt (LAC) Eugeniusz Bala-Sgt (LAC) Wacław Sasin died instantly, F/O (Lt) Stanisław Ostaszewski (nav.) died in hospital and only Sgt Mieczysław Galas (gunner) escaped with minor injuries.

Two large operations against the Krupp works at Essen were mounted on the nights of the 25th and 26th, the former involving the largest Polish effort yet, with 46 bombers from all four squadrons attacking the target. Very heavy AA fire

[34] This consisted of three groups of bombers: the illuminators (the precursors of the later Pathfinder Force) pinpointing the target with lanes of flares; the target markers, loaded with incendiary bombs to create a concentrated area of fire; and the followers, dropping high explosives on the fire zone.

[35] "GEE" was a system giving the exact position of the aircraft by the time taken to receive pulse signals from three different ground stations.

[36] RAF Form 540 (IPMS, Lot. A V,37/11).

Z8399 SM•Z, which dropped the squadron's first 4,000 lb "Cookie" on Cologne on the night of 13 March 1942. (IPMS)

was encountered and many of the aircraft – including seven from No.300 – were damaged, but none of the crews were harmed. However, of the 18 Wellington IVs from Nos.300 and 301 which attacked Essen the next night, three failed to return. These were: BH•L (Z1269), shot down over Holland, whose crew, F/O (Lt) Bogumił Żelaziński-P/O (S-Lt) Bolesław Uszpolewicz-F/O (Cpt) Franciszek Jakubowski-Sgt (Cpl) Jerzy Kwieciński-Sgt (Cpl) Teofil Niemeczek-Sgt (AC1) Władysław Stachurski, were killed; GR•R (Z1262), destroyed by AA fire near Düsseldorf with its crew, P/O (S-Lt) Wiktor Jaroszyk-P/O (S-Lt) Zdzisław Retur-F/O (Lt) Cezary Kowal-P/O (S-Lt) Czesław Grycz-Sgt (LAC) Mieczysław Gołaszewski-Sgt (AC1) Antoni Bochaczek; and GR•H (R1590), shot down by enemy fighters over Holland with the loss of F/Sgt (Sgt) Franciszek Porada-Sgt (Cpl) Emil Horbajczuk-F/O (Lt) Wiktor Jankowiak-Sgt (AC1) Jan Toboła-Sgt (AC1) Jerzy Pietrow; Sgt Ryszard Pukas was made PoW. Reports indicate that three members of that crew jumped, but they were shot at from the ground and only Pukas survived. Two bombers heavily damaged by AA fire, F/O Zygmunt Radecki's BH•F (Z1291) and W/Cdr Piotrowski's GR•C (R1535), returned home. The 2nd pilot of the latter aircraft, F/O Stanisław Mioduchowski, who flew it over the target, was severely wounded by shrapnel, but Piotrowski succeeded in regaining control of the Wellington and brought it back to a good landing. "Freshers" from the two Lindholme units were sent to attack St. Nazaire on the 27th.

On the night of 28/29 March 24 bombers from all four Polish squadrons took part in a big raid – 234 aircraft – on Lübeck, which became the first German town to go up in flames. Almost half of some 300 tons of bombs dropped were incendiaries and almost 50% of the whole city area, including nine factories, was destroyed by fire and some 16,000 people were made homeless. This came as a bad shock to the Germans. Goebbels wrote: "The damage is really enormous. It is horrible ... we can't get away from the fact that the English raids have increased in scope and importance." The Lübeck attack, regarded in Britain as an outstanding success and far exceeding anything previously attained, may be considered as a turning point in the fortunes of Bomber Command. SM•M (W5567) was brought down by AA fire and the whole crew, F/O Józef Jungowski-Sgt Franciszek Wąsiński -F/O Kazimierz Pawluk[37]-Sgt Mieczysław Skubiszewski-Sgt Czesław Krupa-Sgt Jan Krawiec, were taken prisoner.

On 1 April 1942 S/Ldr (Mjr) – on the 15th promoted to W/Cdr – Stanisław Krzystyniak succeeded W/Cdr Piotrowski as C.O. of No. 301 Squadron. Nos.300, 301 and 304 put up 31 bombers for a major attack on Cologne on the night of 5 April, while No.305 flew only a "fresher" sortie against Le Havre. No.304's X9764 was shot down by an enemy fighter over Belgium and its crew, P/O (S-Lt) Alfred Osadziński-P/O (S-Lt) Kazimierz Ziemiański-F/Lt (Cpt) Zygmunt Natkański-Sgt

[37] He was among the 53 RAF officers executed by the Gestapo after a mass escape from Stalag Luft III at Sagan on 24 March 1944.

Lindholme airfield, photographed by a No.304 Wellington in the early part of 1942. (IPMS)

(LAC C-Off) Dominik Grajnert-P/O (S-Lt) Ludwik Assman-Sgt (LAC) Zdzisław Barbaj, lost their lives. Essen was the target of 16 Wellingtons from Hemswell the following night, but very difficult weather conditions greatly reduced their effectiveness. A Hamburg raid by 39 senior Polish crews from four squadrons two nights later was also affected by cloud and icing. A number of aircraft were damaged by AA splinters in these operations.

The night of 10/11 April was another one of maximum effort, with 37 Polish bombers attacking Essen in good visibility and causing large fires at the cost of three Wellingtons. GR•L (Z1333) failed to return and its crew, P/O Jerzy Wasilewski-F/O Edmund Burszewski-S/Ldr Kazimierz Przykorski-Sgt Stefan Mucha-Sgt Marian Zawodny-Sgt Leon Błach, were made PoW. R1230, one of No.304's original issue with 31 sorties to its credit, was shot down by AA fire. Its pilot, Sgt (LAC) Józef Janik, was killed and other members of his crew, Sgt Jerzy Solecki-F/O Bolesław Piwowarek-Sgt Zenon Klus-Sgt Tadeusz Moszyński-Sgt Wacław Walukiewicz, were taken prisoner. SM•U (W5519), badly damaged by a fighter which injured its captain, F/O Tadeusz Czołowski, and the rear gunner, Sgt Kazimierz Wziątek, ditched in the sea off Cromer. No lives were lost, thanks to the exceptional skill of the wounded pilot and the resolute behavior of the crew.

Two nights later the Krupp works at Essen were bombed again, this time with 34 aircraft participating from the four squadrons. Several bombers were damaged by fighters or AA fire, and No.304 suffered most. Its Wellington IC X9687 was shot down and one member of the crew, F/Lt (Mjr) Edward Młynarski (nav.), was killed and the rest, P/O Walery Miśniakiewicz-Sgt Marian Bratkowski-F/Sgt Władysław Graczyk-Sgt Marian Niewolski-Sgt Jan Broda, were made PoW. Sgt Edward Joński's NZ•V (Z1172) was riddled by cannon and machine-gun fire and his rear gunner, F/O Wacław Kossakowski, sustained a bullet wound, but the aircraft made a good landing. Sgt Alojzy Łozowicki had to force-land his damaged NZ•L (X9664) near March, and two of his crew, Sgt Marian Strączek (2nd pilot) and Sgt Ludwik Gibas (radio op.)

were injured and taken to hospital. No.300 lost BH•H (Z1213), which after a long battle with an enemy fighter was shot down over Holland; its crew, Sgt (Cpl) Władysław Żalejko-F/O (Cpt) Piotr Pajer-Sgt (LAC) Konrad Lubojański-Sgt (LAC) Józef Bałucki-Sgt (LAC) Władysław Strzelczyk, were killed, and only Sgt Jan Kuczak (2nd pilot) survived and was made PoW.

On the night of 14/15 April 30 and 19 bombers respectively, on both occasions from all four squadrons, carried out two attacks on Dortmund in poor visibility. GR•J (Z1379), attacked by a fighter during the first raid, crashed into the sea with the loss of P/O (S-Lt) Mieczysław Rzemyk-F/O (Lt) Zdzisław Zamoyski-Sgt (LAC) Janusz Truszkowski-Sgt (AC1) Stanisław Niewiadomski; F/O Stanisław Reyman (2nd pilot) and Sgt Edward Wojan were picked up by the Germans and made PoW. SM•W (Z8586) crash landed out of fuel and its pilot, F/O (Cpt) Stefan Sznidel, and radio op., Sgt (Cpl) Kazimierz Pasich, were killed, and F/O Kazimierz Joszt (2nd pilot) was wounded. Another large operation against Hamburg on the night of 17/18 April involved 26 Polish senior crews from the four squadrons. F/Lt Robert Hirszbandt's aircraft (No.305) carried a 4,000 lb "blockbuster," which caused a tremendous explosion in the old town. Wellington IV BH•P (Z1267) was destroyed by a fighter and its crew, F/O (Lt) Józef Kuśmierz-F/O (Lt) Jan Urbaniak-F/O (Lt) Jan Orsza-Matysek-Sgt (LAC) Jan Orlewski-Sgt (Sgt) Franciszek Bay-Sgt (Cpl) Aleksander Hupało, perished.

Beginning with 23/24 April and for four consecutive nights, Rostock, a town further along the Baltic coast, with the Arado works and the Heinkel factory in the Marienehe suburb, was the subject of precision bombing by the most experienced RAF crews. On the first night four bombers each from Nos.300 and 301, fitted with additional fuel tanks, six from No.304 and five from No.305 were included in the 161 aircraft raiding Rostock. The 19 Polish crews raised conflagrations in the town itself. Wellington IC X9829 (No.304) was shot down by a fighter over the Ems Estuary and its crew, F/O (Cpt) Jan Kwak-P/O (S-Lt) Lech Dzierzbicki-F/Lt (Cpt) Stanisław Wójcik-Sgt (Cpl) Lucjan Woźniak-Sgt (Cpl) Władysław Jankowski-F/O (Lt) Stanisław Zieliński, were drowned. On the night of the 24th seven crews from Hemswell, 14 from No.304 and eight from No.305 – 29 in all – provided almost a quarter of the attacking force of 125 bombers. For the two Lindholme units this mission represented a record operational effort and, flown on the eve of the first anniversary of their first raids – with Gen Sikorski already present at the Station for the next day's celebrations – had a special meaning. NZ•S (Sonia) R1697, having among its crew four members who had taken part in the squadron's first task – S/Ldr Kazimierz Czetowicz, F/Lt Jerzy Iszkowski, F/Lt Bronisław Lewkowicz and F/Sgt Józef Apanasewicz, all except Iszkowski being in the same crew for a year – was attacked by an enemy fighter. Czetowicz, piloting the aircraft, and his gunner, P/O Stanisław Apanasik, were injured but the crew returned to base where they crash-landed with no further injuries. Gen Sikorski pinned the Cross of the order of Virtuti Militari to Czetowicz's jacket, while the pilot was carried on a stretcher to the waiting ambulance. Sgt Stefan Więckowski, a gunner of Sgt Walery Siciński's NZ•J (X9680), destroyed a Bf 110 near Schleswig.[38]

25 April was a day of celebrations at Lindholme. After an open air High Mass, No.304 and 305 Sqdns were inspected by Gen Sikorski in the presence of Gen Kopański, C.-in-C. Polish

Gen Sikorski inspects R1697 NZ•S "Sonia," which, heavily damaged by enemy night fighters during a Rostock raid, crash-landed on its return to Lindholme on the morning of 25 April 1942. (IPMS)

The price to be paid by men and machines – a crippled airman examines the crippled R1697 "Sonia." (IPMS)

[38] RAF Form 540 (IPMS, Lot. A V,36/12).

The first anniversary celebration of bomber operations by No.304 and 305 Squadrons at Lindholme on 25 April 1942, with a Wellington II W5590 SM•A in the foreground.

The open air High Mass celebrated at Lindholme on 25 April 1942, with a Wellington IC NZ•B in the background. (IPMS)

Forces in Libya, Gen Ujejski, Inspector General of the PAF, AVM R.D. Oxland, AOC No.1 Group and many other Polish and British senior officers. Later Gen Sikorski and Gen Kopański presented medals and decorations to members of the squadrons for valour and devotion to duty, these including 27 Crosses of the Virtuti Militari (11 of them to the personnel who were killed or missing) and 68 Crosses of Valour.

On the third night of the attack on Rostock, on the 25th, only seven Polish bombers from Hemswell flew in the operation and P/O Kazimierz Szrajer's crew brought back excellent photographs of the target. The last night (26th/27th) was one of outstanding success. Twenty Polish Wellingtons, drawn from the four squadrons, were among the 107 aircraft dispatched and the strike, made in many instances from between 6,000 and 2,000 ft, was extremely effective. Goebbels described the results as "pretty disastrous" and "more devastating than those before ... in some sections catastrophic ... Seven tenths of the city have been wiped out. More than 100,000 people had to be evacuated ... There was, in fact, panic ... " He recorded: "The Führer is in extremely bad humour about the poor anti-aircraft defence ... the Luftwaffe wasn't adequately prepared, and this alone made the damage to the Heinkel works possible."[39] In 521 sorties flown in the whole four-night operation, Bomber Command lost only 12 aircraft. The Polish bomber force, which contributed 75 sorties (15% of the total), lost two Wellingtons. The second of these, GR•U (Z1317), crippled over the target during the last of the raids and flying on one engine and losing petrol, set course for Sweden. Its pilot made a successful crash landing near Ystad. The crew, F/O Czech Nowicki-P/O Witold Bohuszewicz-F/O Bernard Budnik-Sgt Zdzisław Taczalski-Sgt Krzysztof Grabowski-Sgt Bolesław Bestecki, burned the aircraft before being interned. They all returned later to England.

On the night of 27/28 April the Hemswell and Lindholme squadrons put up 25 senior crews for a heavy raid on the Knocker-Humbolt-Deutz factory in Cologne. BH•W (Z1276) was shot down by a night fighter over Belgium and its crew, P/O Jan Fusiński-P/O Wacław Wąsik-F/O Tadeusz Wawerski-Sgt Andrzej Małecki-Sgt Mieczysław Sierpina-Sgt Stefan Miniakowski, were taken prisoner, but all evaded capture and returned to England helped by the Comet and Dragon Escape Lines.[40] BH•D (Z1264) withstood seven enemy attacks during which F/O Mieczysław Daab (nav.) and Sgt Stanisław

[39] All quotations from Richard and Saunders "The Royal Air Force 1939-1945," p.139, HMSO, 1974.

[40] A resistance escape network, the Comet Line covered the Low Countries and the smaller French Dragon Line, mainly the Ardennes area. The chief of the latter line, L'Abbe Grand Jean died in Dora Concentration Camp.

Gen Sikorski salutes the Polish Air Force Colour during a march past on 25 April 1942, with W5590 SM•A, which crashed into the sea eight days later, providing the setting. (IPMS)

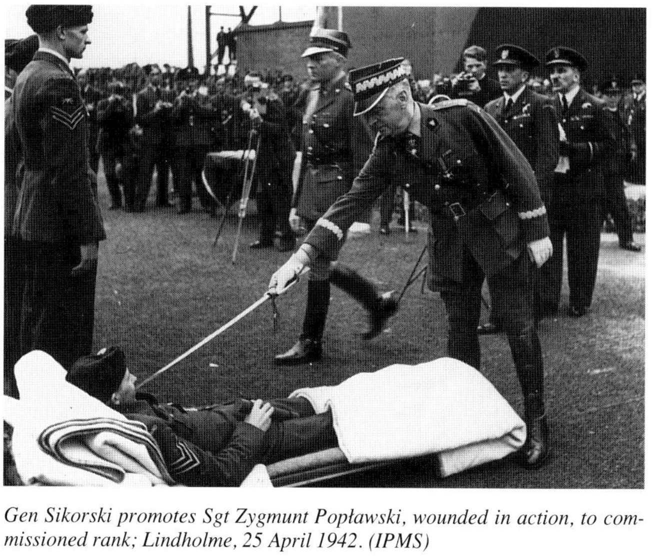

Gen Sikorski promotes Sgt Zygmunt Popławski, wounded in action, to commissioned rank; Lindholme, 25 April 1942. (IPMS)

Kamiński (gunner) were wounded, but the skill and calmness of its pilot, F/O Dominik Mendela, won the day and the crew returned safely. GR•A (Z1280) fought an equally desperate battle with a Ju 88 fighter, which damaged one of its engines, the right tailplane and the hydraulic installation. The rear gunner, Sgt Władysław Nalepa, fired 800 rounds at the attacker, which was hit and dived away (counted as damaged.). The pilot, P/O Kazimierz Szrajer, nursed the aircraft back to base and made a successful crash landing (without flaps), his third in as many weeks, and his crew was uninjured. No.304 was the least fortunate and lost two aircraft in the operation. NZ•D (Z1088) was shot down by AA batteries and crashed in Belgium killing its crew, F/O (Lt) Ryszard Szczurowski-F/Lt (Lt) Edward Kowalski-F/Sgt (Sgt Mjr) Zdzisław Pieczyński-Sgt (Cpl) Stefan Ferenc-Sgt (Cpl) Wincenty Garbacz. NZ•B (W5627) was damaged and lost its bearing, with the crew, F/O Julian Morawski-F/O Jan Waciński-Sgt Bolesław Woźniak-Sgt Bohdan Lipski-Sgt Edward Polesiński, ending up in Spain. They were interned in the Miranda camp, but managed to return to England in August 1942. On 29/30 April six Wellington IVs from No.300 Sqdn took part in a very successful precision raid on the Gnome-Rhone works and a synthetic rubber factory in Paris.

Apart from all these major operations by the senior crews, several tasks by "freshers" from all four squadrons, directed mostly against French ports, were also carried out, bringing the April 1942 total to 352 sorties – with No.300 in the lead with 107 – as compared with 197 in March. However, the Poles suffered heavy losses amounting to 13 aircrews – almost 80 men. No.304 was particularly unfortunate in losing six aircrews. Personnel replacement became the critical factor for the future of the Polish bomber force. Even for the RAF, whose overall personnel strength had almost doubled in 1941 – from about 500,000 to nearly a million[41] – the Bomber Command's April 1942 losses, totaling over 900 aircrew killed and missing on operations and in crash landings, presented a deeply worrying wastage of experienced men, who could not be replaced easily. For the PAF, losses on this scale were catastrophic.

The Polish Armed Forces in Britain, cut off from their natural recruitment base by force of circumstance, could not be expanded and the PAF could not count on any significant increase in human resources. The reduction in the size of Wellington crews from six to five members (by dispensing with the 2nd pilot) which began to be introduced in No.304 Sqdn towards the end of April 1942 – extended by June to all other RAF Wellington squadrons as the standard practice – was not sufficient to make up for the shortage of personnel and a decision was taken to transfer the squadron to Coastal Command – initially on a temporary basis – where casualties were lighter. The operation against Cologne on the night of 27 April became therefore its last one with Bomber Command. The squadron was formally transferred on 7 May and actually joined Coastal Command on the 10th, with the main party leaving Lindholme on the 9th and traveling to Tiree, in the Hebrides, by land and sea and the air party following four days later. The remaining three squadrons had usually only eight to ten complete, fully trained, crews each available for operations in the later part of April. Several new arrivals were received by the squadrons from No.18 OTU, Bramcote, for operational proficiency training. Various personnel interchanges were made. F/

[41] According to the Air Historical Branch Monograph 116/14 "Manning Plans and Policy," RAF Personnel strength on 1 October 1940 was 437,473, including 24,806 officers, and on 1 October 1941, 836,916, including 40,892 officers.

One of the Wellington IIs of No.305 Squadron at Lindholme in the spring of 1942.

O (Lt) Franciszek Fijałkowski succeeded F/O Pijanowski as Engineering Officer of the squadron and held this post for almost a year, passing command of the Technical Flight to F/O (Lt) Tadeusz Papis in April 1943.

May 1942 saw a further escalation of the night bomber offensive. Although the average daily availability of aircraft in Bomber Command during the month was less than 420, Harris cherished the idea of 1,000 aircraft in a single raid against the same target. He thought this was possible to achieve by committing his entire front-line strength and the whole of his reserve (OTUs) and by seeking help from other Commands. He obtained Churchill's warm approval for this great and risky operation, which, code-named "Millennium," was put up at the full moon on 30 May.

Polish bomber squadrons began their May operations on the night of the 3rd with an unfortunate attack on Hamburg in poor weather conditions by nine senior crews from No.300 and nine from No.305. This cost the loss of two aircraft. Wellington IV BH•J (Z1183) was shot down by a fighter over Holland; F/Lt (Mjr) Wacław Wolski-F/O (Cpt) Norbert Lewald-Jezierski-Sgt (LAC) Zygmunt Samuliński-Sgt (LAC) Józef Erdt-F/Sgt (Sgt) Bogusław Nowicki were drowned and Sgt Stanisław Lis (2nd pilot) was taken prisoner. Wellington II SM•G (Z8406) perished with its entire crew, F/O (Cpt) Antoni Jankowski-Sgt (Cpl) Władysław Wieraszka-F/O (Lt) Stanisław Barzdo-Sgt (Cpl) Tadeusz Zagórski-Sgt (Cpl) Mieczysław Nowotarski-Sgt (Cpl) Stanisław Gross.

Stuttgart was the object of attack by six bombers from No.301 on the night of the 4th, and by 13 – seven from No.300 and six from No.305 – on the night of the 5th. No.305 again lost one of its Wellingtons, SM•R (Z8599), whose engines, damaged by the ground fire, failed. Its captain, F/O Stanisław Krawczyk, himself wounded, ordered his crew to jump, but one of them reported that he had no parachute. Krawczyk handed over his own parachute to him and when the crew baled out he made a "dead-stick" emergency landing. Except for Sgt Ludwik Czarnecki, who was made PoW, Krawczyk and his crew evaded capture and, helped by the Comet and Dragon Escape Lines, returned to England. Krawczyk was awarded the Cross of the Virtuti Militari for his actions.

On 8/9 May the two Hemswell squadrons put up 12 Wellington IVs – their number limited by the availability of long-range fuel tanks – for a precision raid on the Heinkel works at Warnemünde, which was highly successful. On 18 May No.300 moved to Ingham, a satellite of Hemswell. The next heavy raid was flown against Mannheim on the night of 19/20 May, with 23 aircraft – seven senior crews from each of the two Hemswell units and nine from No.305 – taking part. Some of the bombers were carrying 4,000 lb "blockbusters" and the attack was pressed home with great vigour. Apart from participation in major operations, a number of sorties were made by "freshers" from all three squadrons during May against easier targets, including the docks and shipping at St. Nazaire and Nantes, and on mine-laying missions.

A Wellington II Z8406 SM•G which was lost with its crew during an attack on Hamburg on the night of 3 May 1942.

Various individual markings embellished Polish bombers. This unidentified badge was probably carried by a Wellington of No.304 Squadron in 1942. (IPMS)

Preparation of as many bombers as possible for the "Millennium" brought about some relaxation in major operations in the period leading to it. Bright moonlight and good weather decided the date (30 May), and Cologne was selected as the most suitable target. To achieve a maximum saturation, the attack was to take no more than 90 minutes. The force used numbered 1,046 bombers[42] – 292 of them four-engined "heavies" – of which 367 were drawn from the Operational Training Groups and four from Flying Training Command. The Polish contribution consisted of 66 Wellingtons (out of 598 employed): 15 from No.300, 14 from No.301, 13 from No.305 and 24 from No.18 OTU.[43] Some icing problems were experienced on the way to Germany, but 898 aircraft claimed to have reached the target and dropped 1,455 tons of bombs. The damage inflicted, as confirmed by the daylight reconnaissance, was greater than in all the previous 70 raids on the city; 600 acres of Cologne laid waste, 36 factories ceased production altogether and some 70 were reduced to half of their normal output. The RAF losses amounted to 41 aircraft missing and nine lost in collision and crash landings on land and in the sea – none of them from the

[42] The Air Historical Branch Narrative 117/1 "The Bombing Offensive Against Germany" and the "Official History"; Harris in "Bomber Offensive" (Collins, 1947) puts its strength at 1,047 bombers.

[43] Contrary to the assertions in various Polish publications, No.304 Sqdn did not participate in this operation. This is quite clear from RAF 540/541 Forms and from the squadron's own combat records.

A Wellington IC bomber of No.18 OTU, depicted at Bramcote in the spring of 1942.

Polish units. About 45 more bombers were seriously damaged, including one from No.300 Sqdn captained by F/O Dreja.

On the night of 1/2 June 1942 28 Polish Wellingtons from Hemswell and Ingham, with 13 from No.305 and two from No.18 OTU were among 957 aircraft launched against Essen, one of the most difficult and heavily defended German targets. An unforeseen change in the weather greatly reduced the effectiveness of the attack. Not all of the bombers were able to reach the target area and night fighters took their toll. GR•G (R1615) was destroyed over Belgium and its crew, P/O (S-Lt) Wincenty Głowacki-F/O (Lt) Stefan Praschill-F/O (Lt) Rajmund Kiwacz-Sgt (Cpl) Jarosław Zalewski-P/O (S- Lt) Jan Halastra-Sgt (LAC) Wacław Żeligowski, were killed. SM•Z (Z8583) crashed and burnt out at Swanton Morley on the return flight, killing F/Lt (Mjr) Robert Hirszbandt-F/O (Lt) Kazimierz Wieliczko-F/O (Lt) Zbigniew Szela-Sgt (LAC) Józef Zawistowski-Sgt (LAC) Bronisław Sędzimir-Sgt (AC1) Józef Ulicki.

After these two missions, Sir Archibald Sinclair, Secretary of State for Air, sent the following message to Gen Sikorski on 4 June 1942: "Polish crews to the number 101 took part in the large-scale operations on Cologne and the Ruhr. The Royal Air Force has learnt to admire the valour, tenacity and efficiency of their Polish Allies. In these operations again they have shown how admirable is their contribution in support of our common cause to the destruction of the war power of the enemy. We are grateful to you and to Poland for these redoubtable squadrons."[44]

Two nights later, on the 3rd, another major operation was launched with Bremen as the target. Ten bombers each from Nos. 300 and 301 Squadrons, eight from No.305 and four from No.18 OTU were dispatched and the raid was regarded as satisfactory by the crews, although strong fighter opposition was encountered. BH•F (Z1291) was lost and its crew, F/O Dominik Mendela-F/Sgt Bronisław Hułas-F/O Mieczysław Malinowski-Sgt Władysław Hauptman-Sgt Stefan Bedełek-Sgt Jan Pakiet, were made PoW. P/O Leopold De Costres' SM•W (Z8581) was attacked seven times by a fighter, which wounded the radio operator, Sgt Józef Rulewski.

Further attacks followed on Essen, on Emden and on Essen again on the nights of 5, 6, and 8 June, with a participation of 23, 20 and 21 Wellingtons respectively from all three Polish squadrons. Heavy losses were sustained. On the 5th two aircraft of No.301 were shot down by enemy fighters and both crews were killed. These were GR•C (Z1331), manned by F/O (Lt) Daniel Pławski-F/O (Lt) Jan Kozak-F/O (Lt) Paweł Wybraniec-Sgt (Cpl) Józef Czerwonka-Sgt (Cpl) Kazimierz Podgórski-Sgt (LAC) Roman Radecki; and GR•L (Z1467), manned by F/O (Mjr) Franciszek Witakowski-Sgt (LAC) Sykstus Lewkowicz-F/O (Cpt) Fidelis Łukasik-Sgt (Cpl) Zygmunt Nogaj-P/O (S-Lt) Andrzej Kułakowski-Sgt (Cpl) Jan Danik. Sgt Józef Przybyła's SM•M (Z8601) was struck by AA fire and was forced to ditch in the sea, but all the crew were

A Wellington IV Z1337 GR•O of No.301 Squadron, carrying the Griffin badge, and its crew, left to right: Cpl Anderman, chief of the •O's maintenance crew; P/O Mieczysław Stachiewicz, pilot; P/O Roman Lipczyński, front gunner; Sgt Piotr Kosin, radio operator; F/O Roman Chmiel, navigator. Sgt Piotr Straub, rear gunner, took the photograph; Hemswell, June 1942. (M. Stachiewicz)

[44] This message is usually wrongly interpreted in Polish publications as referring only to the Cologne raid. In fact, as can be seen from the above text – which is based on RAF 540/541 Forms and unit records in IPMS and Public Records Office – 109 Polish Wellingtons took off on both operations, but 101 actually bombed the targets.

saved. On the 6th GR•E (R1390) was shot down and crashed into the sea; Sgt (Cpl) Mieczysław Bober-F/O (Lt) Stanisław Powierza-Sgt (LAC) Florian Sapieha-W/O (W/O) Józef Podgórzak-Sgt (Cpl) Tadeusz Szykowny, were killed and only P/O Mieczysław Kowalówka survived and was made PoW. On the 8th, BH•Q (R1725) failed to return and its crew, F/O (Lt) Franciszek Malec-F/Lt (Lt) Marian Michałowski-F/O (Lt) Aleksander Paszkowski-Sgt (Cpl) Julian Keryluk-Sgt (AC1) Leon Zaleski[45]-Sgt (LAC) Edmund Polak, perished. Another bomber, BH•V (Z1382), was severely damaged by a Bf 110; its gunner, Sgt Eugeniusz Kwiatkowski, was seriously wounded and radio operator, Sgt Leon Trzabiatowski, slightly wounded. Its pilot, Sgt Edmund Fojud, achieved a remarkable feat by flying back the aircraft with the hydraulic system out of action and a quarter of the port wing covering destroyed by fire, and gliding it to a perfect wheels-up landing from 8,000 ft, after both its engines went dead. On the night of 9/10 June six aircraft from each of the three squadrons took part in a major mine-laying operation in enemy waters.

A heavy attack on the port and town of Emden, involving 23 Polish bombers from all three squadrons, was carried out on the night of 19/20 June. Three of the aircraft fell victim to enemy fighters. BH•A (Z1256) crashed in Germany, killing its crew, F/O (Lt) Józef Tomaszewski-Sgt (LAC) Jan Kropacz-F/O (Lt) Konrad Jeżewski-Sgt (AC1) Stanisław Mielcarek-Sgt (AC1) Stanisław Kaczmarczyk-Sgt (LAC) Kazimierz Błaszczak. BH•E (Z1215) fell into the sea and its crew, Sgt (Sgt) Czesław Dziekoński-Sgt (Cpl) Ludwik Michalski-F/O (Cpt) Eugeniusz Lech-Sgt (Cpl) Stanisław Wardyński-F/O (Cpt) Kazimierz Szyszkowski-P/O (S-Lt) Stefan Neulinger, were drowned. SM•N (Z8339) was shot down over Holland; Sgt (Sgt) Alojzy Gusowski-F/O (Lt) Henryk Jankiewicz-Sgt (Sgt) Kazimierz Ardelli-Sgt (Sgt) Marian Świderski were killed and of the five-man crew only F/O Stefan Madejczyk survived

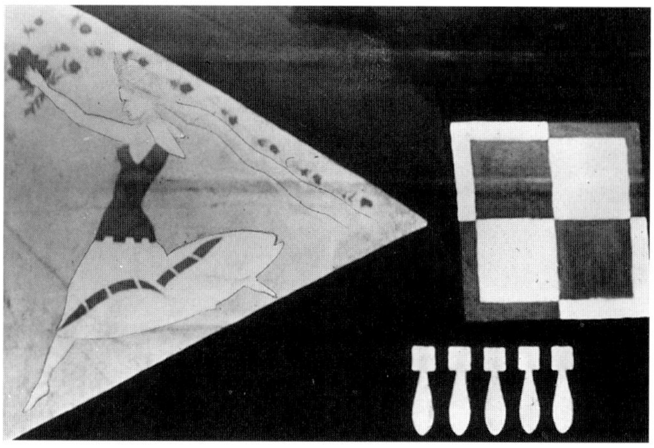

A close-up of individual markings on an unidentified Wellington IV of No.300's 'A' Flight. The badge is exactly similar to that of the pre-war 11th Karaś Eskadra. The photograph was taken at Hemswell in May 1942. (IPMS)

to be taken prisoner. The following night and on the 22nd/23rd, attacks on Emden were repeated with 20 and 21 Polish Wellingtons taking part. On the former operation F/O Gerard Kunowski's SM•Y (Z8537) was compelled to alight in the sea owing to engine failure, but all the crew were saved. On the latter raid, Wellington IV GR•D (Z1345) was shot down into the sea by an enemy fighter and its crew, F/O (Lt) Zbigniew Dubas-F/O (Lt) Wacław Michalik-Sgt (LAC) Jerzy Kapa-Sgt (Cpl) Zdzisław Twarkowski-Sgt (LAC) Józef Meller, were drowned. Photographs of Emden taken on the 23rd showed that severe damage had been inflicted on the town and dock areas and that fires were still burning. Mine-laying sorties were flown by aircraft from Nos. 300 and 305 on the night of the 23rd. St. Nazaire was bombed by three crews from No.305 on the same night and by six from No.301 the following night. S/Ldr (Mjr) Kazimierz Śniegula became the new Commander of No.305 Sqdn on 21 June and was promoted to W/Cdr on 6 July.

On the night of 25/26 June 1942 another "1,000 raid" was launched, this time on Bremen, where, apart from important docks, U-boat construction yards and the Focke-Wulf works

[45] As recorded in available squadron's personnel records (IPMS, Lot. A V.34/9 and 34/62); "Księga Lotników Polskich" spells the name as Leon Zalewski (p. 549).

Left: Apparently the same aircraft (judging by the discolored patches and overpainted earlier chessboard aft of the badge) with the repositioned PAF chessboard and 33 bomber sorties recorded on its forward fuselage, as seen at Ingham some weeks later. (A. Glass) Right: A group of No.300 Squadron airmen, including F/O Narcyz Krupowicz, Chief Technical Officer; F/O Bolesław Solak, radio & radar officer; W/Cdr Romuald Suliński (third, fourth and fifth from the left); and S/Ldr Mieczysław Wolański, 'A' Flight C.O (eighth from the left), by a Wellington IV Z1398 BH•L, with "The Great Bear," star constellation emblem of the pre-war 212th Eskadra and 41 ops log marked on its forward fuselage; Ingham, circa June 1942. (Mrs. Krupowicz/J. Płoszajski)

were located. In addition to Bomber Command's Operational and Training Groups, Coastal Command contributed 102 and Army Cooperation Command, five, to the total of 1,003 aircraft taking part. The Polish share consisted of 71 Wellingtons: 13 from No.300, 14 from No.301, 15 from No.305, seven from Coastal Command's No.304 and 22 from No.18 OTU. The raid was regarded as successful and the Focke-Wulf factory was extensively damaged, but at the cost of 53 aircraft destroyed and 22 seriously damaged. No.301 lost the C.O.'s GR•A (Z1479) and its crew, W/Cdr (Mjr) Stanisław Krzystyniak-F/O (Lt) Jerzy Praśniewski-F/O (Lt) Paweł Tobolski[46]-Sgt (AC1) Marian Przybylski-Sgt (LAC) Jarosław Szymankiewicz-Sgt (LAC) Julian Kwiatkowski, were made PoW. P/O Józef Szybka's SM•R (Z8528) experienced engine failure and ditched in the sea. The navigator launched a dinghy, but G/Cpt (S-Col) Stanisław Skarżyński,[47] flying as a volunteer member of the crew, stepped out of the cockpit from the wrong side, and was hit by a wave and swept away. He was heard calling for help for some time, but the rough sea prevented the dinghy from reaching him and he was drowned. The rest of the crew was picked up eight hours later. No.304's Wellington IC "Q" (DV441) was shot down by a fighter and its crew, Sgt (LAC) Henryk Kuć-P/O (S-Lt) Jan Kramin-F/O (Cpt) Marian Dydziul-Sgt (Sgt) Tadeusz Kamyszek-Sgt (Cpl) Jan Wojas-Sgt (LAC) Nikita Tałach, drowned. The inexperienced crews of No 18 OTU lost three aircraft. These included: Wellington IC T2612, crewed by P/O (S-Lt) Maksymilian Niemczyk-F/O (Lt) Zygmunt Wieczorek-Sgt (LAC) Aleksander Rożdżyński-Sgt (LAC) Stanisław Wolski-Sgt (AC1) Włodzimierz Mikoś; and T2717, crewed by P/O (S-Lt) Roman Balicki-F/O (Cpt) Franciszek Kusek-Sgt (LAC) Kazimierz Cacko-Sgt (Cpl) Józef Jaśkiewicz-Sgt (AC1) Janusz Lech-Pierożyński, all of whom were killed. On 26 June S/Ldr (Cpt) Maksymilian Brzozowski was appointed to command No.301 Sqdn. On the night of the 27th seven aircraft from No.305 made mine-laying sorties, and on the 28th crews from the Hemswell squadrons were involved in an air-sea rescue search in the North Sea. In June No.18 OTU put up seven mine-laying sorties in addition to bombing raids on Essen and Bremen, bringing the number of operational flights during the month to 35.

Bomber Command's operational losses in the second quarter of 1942 totaled 526 aircraft missing or destroyed in crash landings and over 2,600 aircrew members killed or missing, and the July 1942 operations were reduced in scale.

On the night of 2/3 July 1942 27 Wellington IVs from No.300 and 301 Sqdns and ten Wellington IIs from No.305 Sqdn took part in a highly successful attack on Bremen. Excellent visibility over the target assisted the bombers, but also helped enemy fighters, which claimed four of the Polish bombers. P/O Maksymilian Kubacki's BH•B (Z1326), damaged by the fighters, fell into the sea, but its crew was rescued. No.301 was less fortunate. Its GR•M (Z1314) and GR•V (Z1204), with crews of five and six respectively, were shot down. Of the former crew, S/Ldr Maksymilian Brzozowski-F/O Stanisław Ateński (who jumped when wounded)-P/O Michał Benoit-Sgt Stanisław Kowalczyk, were captured – Brzozowski after four days of hiding and walking West in the attempt to escape to England[48] – and made PoW. The fifth member, Sgt (Sgt Mjr) Stanisław Kędzierski, was killed. The latter crew, F/Sgt (Sgt) Teofil Kubiak-Sgt (LAC) Zdzisław Czapski-F/O (Lt) Henryk Krasnodębski-Sgt (AC1) Longin Pokrant-Sgt (LAC) Marian Łoziński-Sgt (LAC) Stanisław Ługowski, were all killed. GR•G (R1650) was badly crippled by a fighter and five members of its crew, Sgt Wacław Niezręcki-Sgt Marian Król- F/O Marian Aduckiewicz-Sgt Stefan Wojteczko-Sgt Zygmunt Wesołowski, were wounded, the last two mortally, with Wesołowski dying the next day and Wojteczko on 7 July. Only Sgt Michał Iwancz (gunner) escaped injury. Niezręcki had great difficulty in keeping the bomber level and made a good belly landing at Swanton Morley – an outstanding accomplishment. S/Ldr (Cpt) Henryk Kołodziejek took over command of No.301 on 3 July and was promoted to W/Cdr on the 12th.

During the next few nights two mine-laying tasks were performed and on 8/9 July. U-boat construction yards at Wilhelmshaven were the prime target of attack: 15 bombers from Hemswell and eight from Lindholme took part. BH•J (Z1489), captained by Sgt Leon Szychowiak, was damaged by AA fire and came down into the sea; its crew were picked up, except for Sgt (LAC) Jan Rawski (gunner) who drowned.

On 9 July W/Cdr Suliński was posted to No.138 Special Duties Squadron and S/Ldr (Mjr) Władysław Dukszto became Acting C.O. of No.300. He was confirmed as its Commander on 1 August. On the night of the 13th, 16 aircraft from Nos.300 and 301 and seven from No.305 bombed Duisburg in very poor weather conditions. Over a period of the following seven nights less demanding mine-laying operations were flown. Some daytime air-sea rescue searches were also flown in the early and middle parts of the month. During a repeat raid on Duisburg on the night of the 21st No.301 lost two aircraft. GR•G (Z1406) received a direct hit by an AA shell and five members of the crew, F/O (Lt) Cezary Lewicki-F/O (Lt) Bolesław Peszkowski-F/O (Lt) Kazimierz Kucza-Sgt (AC1) Robert Maliszewski-P/O (S-Lt) Piotr Kuderski, were killed; Sgt Józef Fuśniak (gunner) jumped and was made PoW. GR•T (Z1405) and its five-man crew, P/O (S-Lt) Harry Bock-F/O (Lt) Wiktor Lipowczan-Sgt (LAC) Jan Pawlak-P/O (S-Lt) Adam Głowacki-Sgt (AC1) Michał Barteczko, vanished without trace. On the 22nd two aircraft from No.301 went to search for the missing crews, and GR•H (Z1472) was engaged by two Fw 190s. After an 11-minute running battle, one of the Wellington's engines was set on fire, the bomber lost height and skimmed the sea with its

[46] Executed by the Gestapo after the mass escape of RAF officers from Stalag Luft III at Sagan on 24 March 1944.

[47] Skarżyński gained international fame in two long-distance flights: the Tour of Africa in the PZL Ł.2 in 1931 and the crossing of the South Atlantic in the single-seat RWD 5bis – the lightest aircraft ever to do so and a World Distance Record (2,224 miles) – on 7-8 May 1933.

[48] He was one of the organizers of the mass escape of RAF officers from the Stalag Luft III at Sagan.

A Wellington II W5455 SM•L (No.305 Sqdn) depicted at Lindholme on 15 July 1942. (IPMS)

fuselage now alight as well. The pilot, P/O Antoni Majcherczyk, succeeded for a time in keeping the aircraft flying, although still under attack, and with his own gunners continuing the fight, but the aircraft eventually flew into the sea. In the circumstances, no preparations could be made for ditching and only the pilot and radio op., Sgt Wiktor Jabłoński, survived; F/O (Lt) Władysław Szymanowski-F/Sgt (Sgt) Andrzej Rudel-Sgt (LAC) Jan Nowak were drowned.[49]

On 23 July No.305 left Lindholme for Hemswell and joined No.301 Sqdn at the station. The move was preceded by a reorganization of technical personnel at Hemswell. In June F/O (Lt) Narcyz Krupowicz was appointed Chief Technical Officer of the station and F/O (Lt) Leszek Chełmiński took over from him as Engineering Officer of No.301 Sqdn. F/O (Lt) Tadeusz Papis became Engineering Officer of No.300 Sqdn at Ingham, replacing F/O Piotrowski. Chełmiński was succeeded at the end of 1942 by F/O (Lt) Stanisław Lipa who held the post until the disbandment of No.301 Sqdn in April 1943 and Papis was replaced by F/O (Lt) Marian Mażyński at about the same time. In May 1942 F/O (Lt) Jerzy Kaliniecki was appointed Engineering Officer of No.305 Sqdn, succeeding F/Lt Wesołowski.

A further attack on Duisburg was made on the night of 25/26 July, with 13 bombers from No.300, six from No.301 and eight from No.305 taking part. Several encounters with enemy fighters were reported. F/O Edward Metler's damaged BH•E (Z1244) crash-landed at Ingham, killing Sgt (AC2) Stanisław Janek and slightly wounding Sgt Henryk Kowalski. GR•Z (Z1329) was attacked by a Ju 88; its pilot, F/O Józef Krasiński and radio op., Sgt Zygmunt Nowicki, were seriously wounded, and the rear gunner, Sgt Eryk Neumann, jumped and was made PoW. Krasiński regained control of the aircraft and brought it back to base with the rest of the crew. He was immediately awarded the DFC.

The following night Hamburg was the target of a big raid. No.300 sent 15 crews – three of them on Wellingtons borrowed from No.301 Sqdn – No.301, seven (all the available crews) and No.305, nine. BH•T (Z1270) was shot down and Sgt Piotr Bakalarski-F/O Kazimierz Faliński-Sgt Zdzisław Kobiela-Sgt Jerzy Olszewski-P/O Bolesław Boguszewski, all wounded, were taken prisoner; Boguszewski died on the way to hospital (Bakalarski later escaped from a camp in Poland, crossed the Russian front line and returned to England via Moscow). Several aircraft returned damaged. On the credit side, F/O

[49] Station Narrative No.8, RAF Station Hemswell, 24 July 1942 (IPMS, Lot. A V, 35/8).

"Ciapek," the mascot of No.305 Squadron, approving the rear guns of a Wellington; Lindholme, 15 July, 1942. (IPMS)

A Wellington II of No.305 Squadron having its port Merlin X engine repaired; Lindholme, 15 July 1942. (IPMS)

Kazimierz Wójcicki's crew (No.305) claimed to have shot down an enemy fighter. On 29/30 July Saarbrücken was successfully bombed by a large force, which included 24 aircraft from the three Hemswell squadrons. On the night of 31 July/1 August 630 bombers were dispatched to attack Düsseldorf, No.300 providing 14 Wellingtons, including three borrowed from No.301, and No.305, six. For the latter squadron this was the last operation to be flown in Wellington IIs.

On the nights of 4 and 5 August large-scale mining operations of the Lorient waters were carried out by No.300 and 301 Sqdns. These were exceptionally long and tiring sorties, lasting some eight hours each and on the second night No.301 lost GR•P (Z1386); its 1st pilot, Sgt Paweł Hur was made PoW and the rest of the crew, Sgt (AC1) Walenty Gohres-F/O (Lt) Jakub Ciołek-Sgt (LAC) Antoni Frankowski-P/O (S-Lt) Ludwik Domański-Sgt (AC1) Ludwik Kurowski, lost their lives. The crippling losses suffered by No.301 within a short space of time left it with only six complete crews available for operations for most of August 1942.

No.305 began re-equipment with the Wellington IV in the early part of August, receiving its first Mk.IV on the 5th and five more on the 8th, and intensive conversion training interrupted its operations for more than two weeks. In the attack on Duisburg on the night of 6 August only No.300 took part, providing ten aircraft for the task. Three nights later Nos.300 and 301 were able to dispatch only six and three Wellingtons respectively for a raid on Osnabrück. Ths same numbers were put up by the two squadrons for operations against Mainz on the nights of the 11th and the 12th, but in addition two "fresher" crews from No.301 flew mine-laying sorties. On the night of 15 August seven senior crews from No.300 and two from No.301 went to bomb Düsseldorf, and the following night four aircraft from the latter squadron laid mines in enemy waters.

Osnabrück was the target of a major operation on the night of 17/18 August, and six Wellingtons (of the seven detailed) from No. 300 and five from No.301 carried out the task. BH•A (Z1409) was shot down by AA fire and crashed in Holland; its pilot, Sgt (Cpl) Izydor Konderak was killed and the rest of the crew, F/O Kazimierz Zakrzewski-Sgt Henryk Kowalski-Sgt Władysław Vandalli-Sgt Jan Hmenia, were made PoW. The following night No.301 laid mines, and No.305 flew its first four sorties on the new Wellington IVs, three of the aircraft being detailed to lay mines and one – F/O Leszek Wielochowski's crew – to bomb the town of Flensburg. On the night of the 20th 12 aircraft from No.300 and 301 Sqdns attacked Brest and mined the French and Dutch coastal waters, and two from No.305 flew mine-laying sorties. The latter squadron suffered the loss of its first Wellington IV, SM•O (Z1471), whose crew, F/O Kazimierz Borowski-Sgt Adam Barszcz-F/O Stefan Lintzel-Sgt Edward Pelc-Sgt Feliks Ratajski-Sgt Witold Raginis, were taken prisoner.

On 24 August five Wellingtons from No.300 and eight from No.301 took part in a night raid on Frankfurt a/M. BH•W (Z1488) failed to return and its crew, Sgt Bolesław Koścień-F/O Mieczysław Daab-Sgt Józef Głodkowski-Sgt Mieczysław Gądek-Sgt Włodzimierz Krzyżanowski, were made PoW. For a big operation against Kassel on the night of 27/28 August the Polish squadrons put up 24 Wellington IVs – ten of them by No.305. The raid was highly successful and crews reported seeing big fires. One of No.305's bombers, SM•D (Z1245), was shot down over Holland and three members of its crew, Sgt (LAC) Jan Pytlak-F/Sgt (Sgt) Feliks Gawlak-Sgt (LAC) Józef Janik, were killed. F/Lt Antoni Kiewnarski was made PoW[50] and Sgt Tadeusz Frankowski evaded capture and succeeded in getting back to England two months later.

Saarbrücken was the object of the next attack the following night in which 21 Polish Wellingtons were involved, losing two of their numbers. GR•Z's (Z1491) crew parachuted to safety and five, F/O Władysław Gandor-F/O Stanisław Świda-Sgt Antoni Wasiak-F/O Edward Mucha-Sgt Jan Uramowski, were taken prisoner, but P/O Józef Tyszko, 1st pilot, dodged his pursuers and found his way back to England in a record time of one week, bringing with him important intelligence information to boot. SM•K (Z1281) was destroyed by a fighter over Belgium and its crew, P/O (Lt) Stanisław Turski-Sgt (Cpl) Jan Korepta-F/O (Lt) Jan Klocek-Sgt (Cpl) Kazimierz Górny-Sgt (Sgt) Czesław Poniatowski, were killed. F/O Tadeusz Czołowski's SM•N (Z1460) was attacked ten times by an enemy fighter which was eventually shot down. Another night fighter took up the fight and killed the rear gunner, Sgt (LAC) Stefan Rueger. The bomber caught fire and F/O Henryk Aleksandrowicz (nav.)-Sgt Mieczysław Ćwikliński (radio op.)-Sgt Franciszek Kula were ordered to bale out; the first two were made PoW, but Kula managed to escape and returned to England five weeks later. Czołowski and the 2nd pilot, Sgt Mieczysław Seredyn, believing that Rueger was only injured, brought the aircraft back and made a crash landing at Manston.[51]

Although the numbers of bombers detailed for night attacks in August 1942 were generally smaller than those employed in the May-June period, the bomber offensive became increasingly threatening to the Germans. On 17 August B-17 bombers, flying with a heavy fighter escort, initiated the daylight precision bombing of enemy targets by the USAAF, at first in occupied France and the Low Countries. At the same time the newly created Pathfinder Force entered the scene, significantly improving the target location and accuracy of night bombing, and this, combined with the increasing availability of four-engined Lancasters and Halifaxes with far greater lift capacities and new radar aids, enhanced the effectiveness of Bomber Command operations.

On the first night of September 1942 seven Wellington IVs from No.300, three from No.301 and four from No.305 took part in a raid on Saarbrücken. The following night senior crews from all three squadrons attacked Karlsruhe and two nights later, 17 were dispatched to bomb Bremen. In addition,

[50] Executed by the Gestapo after the mass escape of RAF officers from Stalag Luft III at Sagan on 24 March 1944.

[51] RAF Form 540 (IPMS, Lot. A V,37/11).

Above: The stripped "Zośka" – Z1407 BH•Z – as she returned home after a close encounter with the German flak over Bremen on the night of 4 September 1942. (IPMS) Below: "Zośka's" lucky crew, left to right: F/Sgt Józef Tomiec, gunner; P/O Zbigniew Brzeziński, gunner; P/O Stanisław Machej, pilot; W/O Klemens Muszyński, radio operator; and F/Lt Bolesław Biliński, navigator, who survived the ordeal unscratched. (IMPS)

one "fresher" crew from each squadron carried out a mine-laying sortie. BH•K (Z1320) did not return from the Bremen operation and its crew, Sgt (Cpl) Leon Szychowiak-F/O (Cpt) Adam Szpak-Sgt (Cpl) Tadeusz Kowalak-Sgt (LAC) Józef Jakubiec-Sgt (LAC) Zygfryd Blachowski, drowned. BH•Z "Zośka" (Z1407), captained by P/O Stanisław Machej, sustained heavy structural damage, its dinghy was blown out and the fabric cover of the rear half of the fuselage completely torn off. Machej, with the cooperation of the whole crew, achieved a remarkable feat in bringing the stricken bomber back to base and in making a perfect landing. No.300 was then left with only six operational crews and three not fully experienced. On the night of the 6th, three senior crews from No.300 and five from No.301 raided Duisburg, and the next night three crews from No.305 performed a similar task, while the former two squadrons sent four aircraft each on mine-laying operations. On the night of the 8th No.305's five crews flew sorties against Frankfurt a/M.

For a major attack on Düsseldorf on the night of 10 September, involving 476 aircraft, the PAF provided 22 crews – six from No.300, five from No.301, four from No.305 and seven from No.18 OTU. Heavy damage was inflicted on the town and its industrial plants, but No.300 Sqdn sustained another loss. Its BH•N (Z1258) fell victim to a night fighter over Holland and the crew, F/O (Cpt) Stanisław Waszkiewicz-F/O (Lt) Jerzy Wierzbicki-Sgt (LAC) Edward Zakielarz-Sgt (Cpl) Borys Sawczuk-Sgt (LAC) Jan Horoch, were killed. Three nights later, three Wellingtons from No.300, five from No.301, four from No.305 and two from No.18 OTU, participated in an equally successful raid on Bremen and the following night, six crews from No.300 and nine from the two other squadrons ventured over Wilhelmshaven. On the night of 15 September nine aircraft from Nos.300 and 301 laid mines and the next night Essen was the target of a heavy raid in which 12 Polish bombers from all three units took part. On 16 September No.300 had only three complete senior crews and eight serviceable bombers available for operations, with six further crews being on rest. The personnel situation in No.301 improved, with the number of available crews rising to nine.

A large-scale mine-laying operation of the Elbe estuary was carried out by 17 Wellingtons from the three Polish squadrons on the night of the 18th. The following night Saarbrücken was the object of a major raid, for which the squadrons put up 15 aircraft, seven of them by No.301. Starting on 23 September, for four successive nights, extensive mine-laying tasks in Frisian waters were performed by all three squadrons. Command of No.301 passed to S/Ldr (Mjr) Adam Dąbrowa on 25 September.

According to Mieczysław Stachiewicz, a former pilot of No.301 Sqdn, around August-September 1942 Polish bomber crews at Hemswell were briefed for a special punitive operation against small German towns and hamlets, which were to be attacked and destroyed by Polish bombers. The action was to be taken in retaliation for the annihilation of many Polish villages by German troops in the Lublin area of occupied Poland, where Polish partisan units were most active. The Emden area was selected and the strikes were to be carried out from a low altitude. Only Polish squadrons were to be involved. The whole operation was eventually canceled.

During the first week of October 1942 three large attacks were mounted on the nights of the 2nd, 5th and 6th, on Krefeld, Aachen and Osnabrück respectively. The first two raids, in which 19 Polish bombers took part on each occasion, were affected by bad visibility, severe icing problems and electrical storms, but the last one, involving 17 bombers, was highly successful. Over the period of 8 to 12 October, on three nights (except for the 9th), a total of 49 mine-laying sorties were logged – 24 by No.300, 15 by No.301 and ten by No.305 – by the bombers, but at the price of two aircraft lost by No.301. On the night of the 8 GR•Z (Z1255) was shot down by a fighter over Holland and except for the rear gunner, Sgt Stanisław Chudy, who was made PoW, the rest of its crew, Sgt (Cpl) Stanisław Bielec-F/O (Cpt) Jan Stąpor-Sgt (LAC) Stanisław Sapeta-Sgt (AC1) Stanisław Szczepański, lost their lives. On the 11th/12th GR•F (R1585) ditched off the Danish coast and its crew, Sgt (LAC) Marian Błachowski, severely wounded-F/O Jerzy Stachurski-Sgt Tadeusz Kowalski-Sgt Zygmunt Turuta-Sgt Karol Molewicz were taken prisoner, Błachowski dying in hospital on the 12th. The following night Sgt Bronisław Wojno's BH•B (Z1244), heavily damaged by AA fire, was written-off in a crash landing, but its crew survived.

On the night of 13/14 October, 13 aircraft – four each from No.300 and 301 Sqdns, and five from No.305 – took off to attack Kiel. Ten bombed the prime target, but BH•K (Z1250) and its entire crew, F/Lt (Lt) Józef Wójcik-Sgt (LAC) Stanisław Brych-P/O (S-Lt) Ryszard Janas-Sgt (LAC) Zenon Wilczyński-Sgt (LAC) Wojciech Pęczkowski-Sgt (AC1) Jerzy Łagoński, perished. A raid on Cologne on the night of the 15th, in which five Wellingtons each from Nos.300 and 301, and four from No.305 took part, cost No.300 another aircraft. BH•X (Z1475) presumably crashed into the sea and its crew, Sgt (LAC) Czesław Szymen-F/O (Lt) Piotr Sobolewski-Sgt (LAC) Stanisław Piątek-Sgt (AC1) Michał Mielnik-Sgt (AC1) Lech Biały, drowned. GR•H (Z1493) was engaged by a Bf 109 and severely damaged – its starboard engine, fuel tank, wing, elevator and undercarriage installation were hit by enemy cannon and gun rounds – but the pilot, Sgt Kazimierz Brochocki, though wounded, brought the aircraft back to a good emergency landing.

Nine crews of No.300 Sqdn were detailed to attack Milan on the night of 15 October, but because aircraft which were to be borrowed from other units were unavailable, only five took off on what was the first operation by Polish bombers against an Italian target and carried out the task. On the same night ten Wellingtons from the other two squadrons were sent to lay mines. No further operational sorties were flown by the squadrons in October, except for mine-laying tasks by nine aircraft from Nos.301 and 305 on the last night of the month, which cost the latter unit one aircraft. SM•O (Z1279) was shot down into the sea and its crew, F/O (Lt) Andrzej Rogowski-F/O (Lt) Jerzy Górecki-F/O (Cpt) Kazimierz Joszt-Sgt (LAC) Bronisław Ossowski-Sgt (Cpl) Wacław Pilarski, drowned. On 31 October S/Ldr (Mjr) Adam Kropiński took over command of No.300 Squadron from W/Cdr Dukszto.

In the first half of November 1942 all three squadrons were engaged in extensive mine-laying operations, No.300 flying 24 sorties, No.301, 25, and No.305, 15. No.300 lost two of its Wellington IVs on these missions. On the night of the 3rd BH•V (Z1421) went missing, presumably shot down by a fighter, and the crew, P/O (S-Lt) Ireneusz Sawicki-F/O (Lt) Walenty Zubrzycki-Sgt (LAC) Mieczysław Ościak-Sgt (LAC) Zenobiusz Kozłowski-Sgt (LAC) Janusz Treutler, drowned. On 8/9 November BH•L (Z1343) crashed into the sea off the Danish coast, killing three members of its crew. F/O (Lt) Ludomir Borowicz (nav.)-Sgt (LAC) Zygmunt Rausiński-Sgt (AC1) Jan Cieszyński-Nałęcz; Sgt Stefan Chmielewski (pilot) and Sgt Mieczysław Dadej (gunner) were made PoW. On the night of the 9th, 14 Wellingtons – four each from Nos.300 and 301, and six from No.305 – took part in a major raid on Hamburg, which, however, was hindered by extremely bad visibility. Sgt Jan Drobny's SM•F (Z1278) was badly holed, but safely returned to base.

More mining tasks followed in the second half of the month with the loss of SM•S (Z1496) with its entire crew, Sgt (LAC) Ludwik Werschner-P/O (S-Lt) Sergiusz Lubojemski-Sgt (LAC) Eugeniusz Szlenk-Sgt (Sgt Mjr) Alfred Psuja-Sgt (Cpl) Józef Gross, on the night of the 19th. The Polish squadrons also made their contribution to three major raids: on Turin, on the night of the 20th (14 aircraft); on Stuttgart, on the night of the 22nd (14 aircraft); and on Essen, in daylight on the 25th (6 aircraft – three each from Nos.300 and 305). On the Stuttgart operation Sgt Kazimierz Brochocki's GR•R (Z1375) was badly shot up by AA fire and its radio op., Sgt Władysław Stachnik, was injured. The Essen raid was a complete failure; owing to unsuitable weather conditions, none of the aircraft succeeded in reaching the target and most brought their bombs back. BH•J (Z1495) was lost over Holland in this unfortunate operation and Sgt (LAC) Czesław Gębaczka (gunner) was killed and the rest of the crew, Sgt (LAC) Władysław Kaźmierczak-F/O (Lt) Jan Gerstel-Sgt (LAC) Józef Skonieczny-Sgt (LAC) Stanisław Abłamowicz, were taken prisoner; Skonieczny was wounded and was at first sent to hospital.[52] During November No.18 OTU flew one operational sortie, dropping leaflets over Paris.

December 1942 operations were impaired by the weather and only three major attacks were launched during the month. Mannheim was the target of a raid on the night of 6/7 December, with 15 aircraft from all three squadrons taking part. Three of the Polish bombers failed to come back. BH•O (Z1401) fell victim to AA batteries near Boulogne and its crew, Sgt (LAC) Józef Furmanik-F/O (Lt) Stanisław Cichowicz-Sgt (LAC) Zdzisław Derulski-Sgt (LAC) Bolesław Sapko-Sgt (AC1) Adam Kozakiewicz, perished. F/O Kazimierz Suszczyński's GR•O (Z1337), damaged by enemy ground fire, ran out of fuel and the crew had to abandon it over the British coast; Suszczyński, F/O Roman Chmiel and Sgt Piotr Kosin landed safely, but the parachutes of F/O (Lt) Alfred Grodzki (2nd pilot), Sgt (LAC) Władysław Ponulak and Sgt (LAC) Augustyn Fenger were blown over the sea and all three drowned. Sgt Wacław Moszczyński's damaged SM•F (Z1278) had to be ditched and two members of his crew, Sgt (AC1) Roman Najda and Sgt (LAC) Władysław Siudak, were drowned. On the 7th No.300 was left with only seven serviceable Wellington IVs and five senior crews available for operations, with two more on rest and six "fresher," or inexperienced, ones.

For an attack on Turin on the night of the 9th, which was reported to be very successful, No.300 provided six aircraft, No.301, four, and No.305, two. On the 17th, the first daylight operation of the series, code-named "Moteing" – harassing raids, specifically aimed to subvert the morale of the German civil population – was attempted against Leer, but because of a lack of the required cloud cover, none of the bombers could reach the primary target. During the early and middle parts of the month several mine-laying tasks were carried out by all three squadrons. For GR•P (Z1263), one of these, flown on 11 December, was its 60th operational sortie – the highest number to be logged by any of the Polish Wellingtons at that time. No further operations were recorded by the Hemswell squadrons between 20 and the end of the month, except for a mine-laying mission by three aircraft of No.305 on the night of the 31st. During that period No.300 began to convert to the new Wellington Mk.III and by 1 January 1943 had taken nine of them on charge.

1942 was still a very difficult year for Bomber Command, and the Allies in general, but it heralded the turning of the tide. The Command, although still below its requisite strength to meet the demands of a sustained all-out air offensive, acquired considerable prestige with its massive night raids, and the USAAF was poised to carry its daylight offensive into the heart of Germany. However, enemy night fighters, assisted by improved airborne radar and the increasingly effective Kammhuber Line defence system,[53] took their toll and lessened the chances of survival of RAF bomber crews. During the year some 1,400 RAF bombers were lost – the equivalent of more than 70 squadrons – and over 2,700 damaged in action. More than half of these were claimed by Luftwaffe fighters.

For the Polish bomber force 1942 was a trying year, combining great achievements with irreplaceable losses which led to a decline of its strength. In the first half of the year the PAF succeeded in putting up a force of the order of more than 60 Wellingtons for the 1000-bomber raids, but after the departure of No.304 Squadron to Coastal Command and the mounting summer losses, the availability of fully experienced Polish crews dropped to an average of about 20 in all three squadrons together in the last quarter of the year. In 1942 No.300 logged 872 sorties in 4,691.5 flying hours, No.301, 706 sorties in 4,066 hours, No.304 (until 10 May), 216 sorties in 1,279 hours and No.305, 587 sorties in 3,273.5 hours – a total of 2,381 sorties in 13,310 hours, the record for the Polish squadrons in Bomber Command, which, after adding 69 combat sorties flown by No.18 OTU, rose to 2,450 sorties (See Appendix 23).[54]

During these operations a total of 3,292 tons of bombs and mines were dropped by the three squadrons (excluding No.18 OTU). Ports represented 34% of the targets attacked – 809 sorties flown, with Bremen (169 sorties), Hamburg (138 sorties) and Emden (106 sorties) heading the list; industrial centers, 33% – 784 sorties, with Essen (296 sorties), Duisburg (95 sorties) and Saarbrücken (75 sorties) most frequently attacked; railway and population centers, 9% – 212 sorties, with Cologne (116 sorties) and Mannheim (42 sorties) on top of the list; and German warships, 3.5% – 86 sorties. Mine-laying operations accounted for 17.5% – 423 sorties; various other ancillary tasks, including air-sea rescue searches and leaflet dropping for 1.5%-34 sorties, and the remaining 33 unspecified sorties for 1.5%. The prime targets were attacked on about two/thirds of the sorties flown and bombs were dropped on

[52] He died suddenly in a PoW camp on 6 February 1943.

[53] Named after Gen Josef Kammhuber, architect of the German air night defence system and the main force behind its expansion. The line by the end of 1942 stretched from Denmark to east of Paris and embraced some 500 night fighters (mostly Bf 110F and G, and Ju 88C and R), 75% of Germany's heavy AA guns and five regiments of radar-guided searchlights – all controlled by central plotting rooms and supported by early-warning radar systems.

[54] The totals stated in the amended PAF HQ Operational Bomber Summary (Ref.LB/1300/Op.Stat/Tjn) – operational flying hours for No.18 OTU were not given separately in available records, but were included in the official tabulations.

secondary targets on most of the other occasions; only some 10% of the sorties were abortive.

The Polish operational losses in 1942 included 89 Wellingtons lost and 290 aircrew members killed, 16 missing and 91 made PoW[55] – the highest yearly total of the war. Among these were several experienced officers and skilled engineers with university degrees, who, instead of being held back for retraining for more demanding and useful functions, were allowed to fly on combat sorties as volunteer gunners. This pointed to a failure on the part of the PAF Command to operate a rational personnel policy, and to exercise effective control of its limited human resources, and deepen the shortage of suitable candidates for navigational training in the later stages of war.

The Polish bomber force never fully recovered from its 1942 losses, partly because of the lack of replacements and partly because of expanding commitments of the PAF clandestine operations to Poland and the rest of occupied Europe. In order to bring its depleted squadrons to the requisite strength of at least 16 operational crews, one of ths units had to be disbanded. This difficult decision was implemented in the early spring of 1943 and affected No.301 Squadron. The two other squadrons still retained the aging Wellingtons. The PAF Inspectorate had already begun to press for rearmament of the force with Lancasters in the second half of 1942, despite some misgivings because of the size of the crews required to man these aircraft. However, Lancasters were in short supply and in January 1943 the British suggested re-equipment of two Polish squadrons with Mosquito bombers. This was rejected and at the beginning of March the RAF offered to provide No.300 Squadron with Halifaxes during April, but the Poles expressed a preference to retain Wellingtons until Lancasters, or perhaps Liberators, were available.

(4.) Aiding the Destruction of Germany with a Slimmer Force (January-August 1943)

After the stalemate of 1942, 1943 was the first year of really bad fortune for the Germans. On top of disasters on the Eastern front, the rout in North Africa, the loss of the Italian ally and the defeat of U-boats in the Battle of the Atlantic, the whole might of strategic air offensive, unleashed by the USAAF and the RAF, brought a havoc of destruction to the heart of Germany. With four-engined "heavies" taking over from the earlier twin-engined bombers, around-the-clock operations and the new bombing aids, Oboe (an audible radar direction beam, first used on 20 December 1942) and H2S (an airborne radar scanner, permitting a "sight" of the ground below through fog, haze, or cloud, first used on 30 January 1943), the operational capacity of the attacking forces was transformed and resulted in intense devastation of targets on a scale that was impossible before.

[55] "Summary of the operational effort of the bomber force 1940-1945," PAF HQ, Study Department, Ref.LB/1300/Op.Stat/Tjn, 15.03.1946 (Copy: J.B. Cynk, Polish Aircraft Archives).

At the Casablanca conference, which culminated in the issue of the Allied "Unconditional Surrender" ultimatum to Germany on 23 January 1943, the Combined Chiefs of Staff set out objectives to achieve this end. The task defined for strategic air power was "the progressive destruction of the German military, industrial and economic system, and the undermining of the morale of the German people to a point where their armed resistance would is fatally weakened."

At the beginning of 1943 Bomber Command was extensively engaged in anti U-boat operations, concentrating its attacks on the Biscay bases, but from early March it resumed its prime task, as restated in the Casablanca directive, switching its main effort to the bombing of Germany's biggest industrial area in what is generally labeled by historians as the Battle of the Ruhr. Although the Ruhr remained a principal target of the Command throughout the year, these operations were interspersed by a concentrated ten-day attack on Hamburg at the end of July/beginning of August, frequently referred to as the Battle of Hamburg, and since the middle of November, by a four-month offensive on the German capital known as the Battle of Berlin.

No.300 Squadron's operations in January 1943 were restricted by the conversion to Wellington IIIs and by the 14th only six mining sorties were logged by its crews, all still on the Wellington IV. Nos.301 and 305 flew respectively 16 and 20 similar sorties during the same period. On the night of the 9th, GR•G (R1535) of the former unit was badly damaged by an enemy fighter and crashed on Dunmoor Hill with the loss of four members of its crew, F/O (Lt) Tadeusz Tyrała-F/O (Lt) Tadeusz Sokoliński-Sgt (AC1) Józef Pasierski-Sgt (LAC) Ernest Tabaczyński; only Sgt Marian Korczyński survived, wounded. Four nights later No.305 sustained the loss of SM•N (Z1272), which crashed on the way back, killing the crew, P/O (S-Lt) Wacław Krawczyk-P/O (S-Lt) Andrzej Płusa-Sgt (Sgt C-Off) Karol Dutkiewicz-Sgt (AC1) Jan Laskowski-Sgt (Sgt) Stanisław Dorosz. On the morning of 14 January two aircraft from each of the three Polish squadrons took off on a "Moteing" operation against Leer, but lack of cloud cover led to its abandonment. The same night four aircraft from No.300 and three each from the other two squadrons laid mines. A similar task was performed by Nos.300 and 301 the following night; for the former squadron this was the last operation to be flown on the Wellington IV. On 17 January W/Cdr (Mjr) Tadeusz Czołowski took over command of No.305 Sqdn from W/Cdr Śniegula.

After a ten-day break, the next attack, a raid on Lorient, was made on the night of 26 January. No.300 put up four of its first six combat-ready Wellington IIIs, while No.301 sent seven Wellington IVs and No.305, nine. GR•L (Z1318) crashed on landing; its pilot, Sgt Marian Król, was slightly wounded and one of the gunners, Sgt Marian Korczyński, more seriously. Nos.300 and 305 were detailed to carry out two more mining tasks each before the month was out. The latter unit's SM•N (Z1415) failed to return from the second of these missions on the 29th/30th and its crew, F/O (S-Lt) Adam Pankiewicz-P/O (S-Lt) Tolimir Kozłowski-Sgt (AC1) Edmund Malinowski-Sgt (Cpl) Tadeusz Główczyński-Sgt (LAC) Wiktor Sroka, drowned. On 31 January No.300 Squadron returned to Hemswell.

The February 1943 operations opened with a major raid on Hamburg on the night of the 3rd/4th, in which five aircraft from each of the three Hemswell squadrons took part. Severe icing conditions forced many aircraft to abandon the task. Two Polish bombers failed to return. No.300 lost its first Wellington III, BH•X (BK511), and all its crew, Sgt (LAC) Walerian Zieliński-Sgt (LAC) Paweł Marciniszyn-F/O (Lt) Stefan Kalina-Sgt (LAC) Józef Matelowski-Sgt (Sgt) Feliks Łazowski-Sgt (LAC) Zenon Sienkiewicz, were killed. No.305's SM•O (Z1392) was shot down over Holland and only its pilot, Sgt Mieczysław Seredyn, survived and was taken prisoner; those killed were F/O (Lt) Edward Lesisz-Sgt (LAC) Henryk Gzell-F/Sgt (Sgt Mjr) Alfons Kowalski-Sgt (LAC) Bernard Kozłowski. On the same night another Wellington III from No.300, BH•Q (BK303), flying a mine-laying sortie over the Lorient waters, was shot down into the sea, and its crew, Sgt (Sgt) Władysław Sowa-P/O (S-Lt) Tadeusz Malicki-Sgt (Cpl) Jakub Mierzwa-Sgt (LAC) Wacław Wydrzerzecki-Sgt (LAC) Teodor Ogrodnik, perished. On the night of the 4th, for a raid on Lorient, No.300 was able to put up only one crew, while No.301 provided four and No.305, five.

During the following fortnight several mining tasks were performed by crews from all three squadrons, two more attacks on Lorient were made and, on the night of the 14th, seven bombers from Nos.300 and 301 raided Cologne and three from No.305, Hamburg. Five nights later six aircraft from No.300, five from No.301 and three from No.305 were detailed to take part in a large operation against Wilhelmshaven. Sgt Kajetan Ignatowski's SM•L (Z1288) had to ditch in the sea owing to the failure of both its engines, but the crew was rescued by a RAF launch. On the night of the 26th Nos.300 and 301 attacked Cologne again and a few mining sorties were flown. More mine-laying flights followed the next night, and on the last night of the month four Wellingtons each from Nos.300 and 301 and three from No.305 Sqdns bombed St. Nazaire. In the later part of February some dinghy searches were also carried out. At the end of the month Nos.300 and 301 had only five fully experienced crews each available for operations, and No.305, eight; No.300 had nine serviceable Wellington IIIs, and Nos.301 and 305, eight and nine Wellington IVs respectively.

On the first night of March 1943 all three squadrons set out to lay mines and the following day dinghy searches were carried out over the North Sea. On the night of the 3rd four bombers each from Nos.300 and 301 and two from No.305 raided Hamburg, but strong enemy opposition and icing conditions hindered the operation.

The opening attack in the Battle of Ruhr was delivered on Essen on the night of 5/6 March by a force of 442 bombers, which included five Wellingtons each from Nos.300 and 301 and four from No.305 Sqdns. The raiders, helped in finding the target – permanently obscured by the industrial fog and haze – by the new Oboe guiding technique, released their loads with a precision never achieved before. Over 600 acres were virtually laid waste, including 13 main buildings of the Krupps works that were gutted by fire and a further 50 seriously damaged, at the cost of 14 bombers missing and 38 damaged. No.300 lost BH•Q (BK150), whose navigator, S/Ldr (S-Col) Jan Jankowski, was killed, and the rest of the crew, F/O Kazimierz Romaniszyn-Sgt Michał Woźniak-Sgt Zygmunt Abramik-Sgt Chester Zełęski, were made PoW. Sgt Kazimierz Artymiuk's SM•S (R1525) from No.305 was attacked by enemy fighters on the return journey and its rear gunner, Sgt Brunon Godlewski – an American Pole born in Chicago, aged 19, who volunteered for the PAF – was severely injured and had to have both his hands amputated. The pilot received an immediate award of the DFM for skillfully evading repeated enemy attacks and for bringing the severely damaged bomber back to base to a successful belly landing. SM•A (R1530), captained by Sgt Wacław Moszczyński, was also engaged by fighters and another DFM award was made to its gunner, Sgt Zdzisław Pisarek, who showed fearless courage and coolness in repelling the Bf 110 attack. He fired at point blank range at the enemy, which was at once engulfed in thick smoke and dived vertically out of control; two parachutes of its crew were seen to open. Essen was the target of another successful attack by 14 Polish bombers from Hemswell on the night of 12 March.

On the night of the 13th/14th the docks at St. Nazaire were bombed and mines were laid by the Polish squadrons. BH•K (BK516) did not return from this sortie and no trace of the aircraft or its crew, Sgt (Cpl) Tadeusz Kuźmiński-Sgt (Cpl) Karol Kramarczyk-Sgt (Sgt) Stanisław Napora-Sgt (LAC) Anastazy Papkow-Sgt (LAC) Franciszek Połom, was ever found. Then widespread fog and generally adverse weather conditions brought operations to a halt for almost a fortnight. They were resumed on the night of the 26th with a heavy raid on Duisburg, in which 14 bombers from all three Hemswell units took part. Two nights later St. Nazaire was attacked and mines were laid in the Lorient waters.

The following night, 29 March, 15 Polish aircraft – five from each squadron – were detailed to bomb Bochum as part of a large force, and although a number of the raiders had to abandon the task owing to icing and technical failures, the mission was generally regarded as very successful. For No.301 this was the last combat mission to be made with Bomber Command, and for No.305, this was its 197th operation and the last one to be flown on Wellington IVs.

At the end of March-beginning of April 1943 the Polish bomber force underwent an extensive reorganization. On 31 March the decision was announced to disband No.301 Squadron, its withdrawal from Bomber Command taking effect on 1 April. At the same time seven of its senior aircrews, under command of S/Ldr (Mjr) Stanisław Król, were formed into a flight which was to be posted to No.138 Special Duties Squadron at Tempsford as 'C' (Polish) Flight. While in Bomber Command, No.301 logged 249 operations – seven on Battles and 242 (190 bomber raids and the rest mine-laying and other tasks) on Wellingtons – totaling 1,260 sorties in 7,367 flying hours, during which 1,428 tons of bombs and 222 tons of mines were dropped; its casualties were 203 killed, missing, and taken prisoner.

No.301 was formally disbanded on 7 April – or rather, reduced to "number only" basis, since the PAF never accepted its dissolution as a permanent measure and the 'C' Flight was unofficially regarded as its continuation and even referred to in some Polish records as "No.301 Flight." Most of the squadron's remaining flying and ground crews were transferred to No.300 Squadron and the rest to No.305. Further reinforcements for both these squadrons were provided by No.18 (Polish) OTU at Bramcote, which – in view of the lack of new Polish trainees – was moved to Finningley and reformed as No.18 (RAF) OTU on 31 March, retaining only one Polish flight.

At the same time the two remaining Polish bomber units began to rearm with the Wellington Mk.Xs – the most potent version of this veteran bomber, with the more powerful, 1,675 hp variants of the Bristol Hercules radial engines – No.300 receiving its first two machines of this type at the beginning of March. While for No.300, already flying the earlier Hercules-powered Wellington III, the conversion to Mk.X did not present any problems, for No.305 the change over from Wellington Mk.IVs fitted with American Pratt & Whitney engines was a long process, which was completed only in the middle of May.

During April 1943 only No.300 Squadron took part in operations, having 13-16 crews and 15-16 serviceable Wellington Mk.IIIs and Xs available for combat on most days of the month. On the night of the 2nd three crews attacked Lorient. Two nights later five aircraft took off to bomb Kiel and one of them fought a long battle with fighters, claiming one Bf 110 destroyed. Eight Wellingtons (of nine detailed) took part in a major raid on Duisburg on the night of 8/9 April, during which Mk.X BH•T (HE148) was shot down and sunk off the Dutch coast with the loss of its entire crew, P/O (S-Lt) Stefan Tomicki-P/O (S-Lt) Jan Rudek-Sgt (LAC) Stanisław Ślusarski-Sgt (LAC) Tadeusz Kniażycki-Sgt (LAC) Władysław Marczuk-Sgt (AC1) Stanisław Stępień. On the night of the 10th No.300 contributed six Wellingtons to an attack on Frankfurt a/M.

S/Ldr Adam Dąbrowa's crew (No.301 Sqdn) saying farewell to their C.O. by his Wellington IV at Hemswell on 29 March 1943; left to right: Sgt Władysław Sibilski, radio operator; F/Sgt Marceli Ostrowski, gunner; S/Ldr Adam Dąbrowa; F/Sgt Stanisław Kondras, gunner, and F/Sgt Stanisław Szymański. (A. Dąbrowa/T. Kwiatkowski)

On the night of 14 April the squadron put up 11 bombers for a very successful major operation against Stuttgart, and, after a mine-laying task by three crews on the following night, 13 of its aircraft took part in a big raid on Mannheim on the night of the 16th. Heavy fighter opposition was encountered on this mission and BH•C sustained substantial damage, but none of the Wellingtons were lost. St. Nazaire and Lorient were the targets of operations on the nights of the 17th and the 22nd. BH•V (HE291) was shot down into the sea by enemy AA fire during the latter sortie and its crew, F/Lt (Cpt) Eugeniusz

A new Wellington X BH•G of No.300 Squadron flying from Hemswell in the spring of 1943. (M.J. Ingham)

No.305's Wellington IV SM•R, still operated by the squadron at Hemswell in the spring of 1943. (M.J. Ingham)

Przysiecki-Sgt (LAC) Janusz Belina-Prażmowski-Sgt (LAC) Czesław Krełowski-Sgt (LAC) Ireneusz Dębowski-Sgt (AC1) Tadeusz Szeremeta, perished.

Eleven Wellingtons from No.300 took off for another large operation against Duisburg on the night of 26/27 April and all reported highly successful results. Mining of St. Nazaire and Lorient, carried out by nine bombers the following night, completed the April missions which included 77 effective sorties, as compared with 35 in March.

On the first night of May 1943 Brest was the target for six Wellingtons of No.300 Sqdn. On 4 May W/Cdr (Mjr) Mieczysław Kucharski took over command of the squadron from W/Cdr Kropiński. On the same night 11 Polish bombers participated in a major attack on Dortmund. After a week's break, No.300 put up 13 aircraft for a big raid on Duisburg on the night of 12/13 May and suffered the loss of BH•P (HE295) with the crew Sgt (LAC) Stanisław Werner-F/O (Lt) Rudolf Tabaczyński-Sgt (LAC) Henryk Szymanowicz-Sgt (LAC)

Wellington X bombers of No.300 Squadron at Hemswell in the late spring of 1943. Note different positions of the PAF chessboards on the aircraft in front, embellished with a dwarf image, and BH•E behind, and the low-vision code letters, improving the night camouflage of the bombers. (IPMS)

Henryk Lercel-F/Sgt (LAC) Mieczysław Gałas, of which no trace was ever found. The following night 12 Wellingtons carried out a very successful attack on Bochum.

On the night of 21/22 May No.305 rejoined operations, with six of its Wellington Xs flying their first sorties to lay mines in Frisian waters, and eight crews of No.300 carried out a similar task.

The night of the 23rd was one of the biggest efforts by No.300. The squadron put up 16 senior crews for a devastating raid on Dortmund – the heaviest of the Battle of the Ruhr in which 826 aircraft took part. No.305 provided eight. All the bombers of No.300 and seven of No.305 identified the target with the help of pathfinders and bombed it successfully, but BH•K (HE374) was shot down by AA batteries and its crew, F/O (Lt) Henryk Piątkowski-F/O (Lt) Witold Jankowski-F/O (Lt) Henryk Lewicki-Sgt (LAC) Czesław Ratajczyk-Sgt (LAC) Eugeniusz Kulikowski, were killed. Three other Wellington Xs of No.300, W/Cdr Kucharski's BH•F (HF464), Sgt Skowron's •G (HE376) and Sgt Rech's •N (NA768), and one of No.305, SM•K (NB691), captained by Sgt Kajetan Ignatowski, returned damaged by the intense ground fire.

Large raids were mounted on Düsseldorf on the night of the 25th and on Essen on the night of the 27th, with 14 bombers from No.300 and six from No.305 taking part in the former operation, and 12 from No.300 and six from No.305 in the latter. On the second of these two nights three other crews from No.300 also laid mines in the Helgoland waters. Wuppertal, a highly industrialized town in the Ruhr, was chosen as a new target for a big attack on the night of 29 May, to which No.300 contributed ten bombers and No.305, five. All crews reported good results and terrific fires, covering the whole of the target area. Sgt Witold Wasiluk's BH•T (HE420) was caught in searchlights and its fuselage was badly holed by ground fire. Mining tasks were carried out by both squadrons on the night of 30/31 May. No. 300 logged 104 successful sorties in May (out of 111 attempted). Three Polish crews from No.18 OTU dropped leaflets over France during the month.

Mine-laying operations by five aircraft from No.300 and three from No.305 on the night of 3/4 June 1943 were hindered by low cloud and bad visibility. On the night of 11/12 June, 16 bombers from No.300 and seven from No.305 set out to take part in a major attack on Düsseldorf, but three of them had to abandon the mission because of icing. The following night another mining task was performed, involving aircraft from both units.

On 21 June No.300's BH•B (HF591) crashed during a high altitude training flight, killing its crew, F/Lt (Lt) Fabian Pokorniewski-F/O (Lt) Zbigniew Krętkowski-Sgt (LAC) Henryk Czarkowski-Sgt (LAC) Tadeusz Szczuryński-Sgt (LAC) Bernard Kot. The same night No.300 put up 15 Wellington Xs and No.305 eight for a heavy raid on Krefeld, which, although generally successful, cost the squadrons three aircraft shot down and one severely damaged. The entire crews of BH•S (HE327) and BH•W (HE985), Sgt (LAC) Mieczysław Bronicki-Sgt (LAC) Mieczysław Glass-Sgt (LAC) Stanisław Nogacki-Sgt (LAC) Stefan Jama-Sgt (AC1) Jan Leżuch, and F/Sgt (Sgt Mjr) Mikołaj Białobrówka-P/O (Lt) Józef Blajda-Sgt (LAC) Jan Ciuchciński-Sgt (AC1) Jerzy Trzebuchowski-Sgt (AC1) Roman Jacenik, respectively, were killed. Two members of the crew of SM•F (HE347), Sgt (Sgt) Stanisław Szpaliński (pilot)-Sgt (LAC) Leopold Makarski (gunner) met the same fate, two, Sgt Rudolf Raczkowski-Sgt Alfred Fried, were made PoW, and the fifth, P/O (S-Lt) Józef Gąsecki (nav.), evaded capture and returned to England ten weeks later. AA batteries crippled F/L Grzegorz Roter's BH•V (..439), which crash landed on the return to base, and its navigator, P/O Bolesław Regeli, was seriously injured. On 22 June both squadrons moved from Hemswell to Ingham. No.300 had at that time 20 crews and 17 aircraft operational and seven further crews and two aircraft not available for action; No.305 had 11 crews operational.

On the night of the 23rd nine bombers from No.300 and five from No.305 laid mines in enemy waters off Lorient and St.Nazaire; Sgt Jan Cholewa's BH•E (HZ373) returned home badly damaged by ground defences. The following night Wuppertal was the target of a major raid, for which No.300 provided 11 crews and No.305, eight. Heavy losses were sustained. BH•C (HF606) and BH•G (HE376) were destroyed over Germany; the crew of the former, Sgt (LAC) Stanisław Jawoszek-F/Lt (Lt) Karol Obrycki-Sgt (LAC) Eugeniusz Bartosiak-Sgt (LAC) Franciszek Skośkiewicz-Sgt (LAC) Sergiusz Jacewicz, were killed. So were three crew members, Sgt (LAC) Antoni Bielski-Sgt (LAC) Jan Bijowski-Sgt (LAC) Kazimierz Żarniewski, of the latter, the remaining two, F/O Władysław Turecki-Sgt (AC1) Roman Kosiarski, being taken prisoner. Three further bombers – W/Cdr Kucharski's BH•F (HF464), Sgt Stefan Zieliński's BH•E (HE373) and Sgt Henryk Anglik's SM•A (HF463) – were extensively damaged. F/Lt Olgierd Łuczkowski, the rear gunner of SM•O (HF490), showed commendable courage and coolness fighting a desperate battle with a Ju 88 fighter, which eventually dived out of control and exploded on striking the ground.

More mine-laying sorties followed in the last week of June and on the night of the 28th a large attack was launched on Cologne. The bombers encountered strong enemy opposition and a number fought heavy engagements with night fighters. BH•J (HZ438) crashed in Belgium and its crew, F/Sgt (Sgt Mjr) Marian Kleinschmidt-F/O (Lt) Stefan Bogusławski-Sgt (LAC) Piotr Mazgaj-Sgt (LAC) Maksymilian Cieślik-Sgt (AC1) Tadeusz Knebloch, perished.

July operations opened with a very successful major attack on Cologne on the night of the 3rd, in which 13 bombers from No.300 and eight from No.305 took part. Five nights later eight aircraft from both Ingham squadrons laid mines off the West coast of France. On the night of 13 July a heavy raid was made on Aachen, with No.300 putting up 15 Wellingtons and No.305, six for the operation. Further mining missions were carried out on the night of the 18th and during these F/O Jan

Mioduchowski's crew, flying SM•J (LP452), observed and took good photographs of the V1 flying bomb site at La Calabasse Rocks.[56] Five leaflet dropping sorties were made by Polish Wellingtons from No.18 OTU in July.

The Battle of Hamburg began on the night of 24/25 July 1943 with a crippling raid by 791 aircraft – 718 four-engined "heavies" and 73 Wellingtons, including nine from No.300[57] and seven from No.305. "Window"[58] was used by the bombers for self-protection for the first time, and the whole Hamburg defence system – 54 radar-directed heavy AA gun batteries, searchlights and night fighters on six airfields within combat range – was thrown into utter confusion. As a result, Bomber Command losses were minimal – only 12 aircraft failed to return and 34 others were classified as damaged. The same night three crews from each of the two squadrons laid mines in the Elbe Estuary. During the following night, 25/26 July, Essen was the target of a big operation, regarded by some historians as the last one in the Battle of the Ruhr. No.300 contributed 17 bombers and No. 305, six, to the attacking force, which dropped 2,000 tons of bombs. The former unit lost BH•Z (HZ486); three members of its crew, F/Sgt (Cpl) Edmund Garczyński-Sgt (LAC) Stanisław Galiczyn-Sgt (LAC) Stanisław Skwarek, were killed, and two, F/O Edward Bemben and F/Sgt Zygmunt Cipirski, were made PoW.

The initial assault of the Hamburg battle, on the night of the 24th, was followed by daylight bombing of the city by the B-17 Flying Fortresses of the USAAF Eighth Air Force on 25 and 26 July and by RAF Mosquitoes, which kept the pot boiling on these two nights. The next major night attack – by 787 aircraft, which included 16 Wellingtons from No.300 and seven from No.305 – came on 27/28 July. A blanket of incendiary and high explosive bombs, which covered the city, caused the phenomenon known as a fire-storm and resulted in horrifying devastation on a scale never before experienced in the air war. The next morning almost a million people fled from the stricken port, leaving behind over 40,000 dead among the burning rubble. Two nights later the RAF bombers returned. No.300 put up 14 aircraft and No.305 six for this attack, during which new marker bombs and flares, nicknamed by the Germans "Christmas Trees," were used for the first time. The final blow in the Battle of Hamburg came on the night of 2/3 August 1943, when a force of 740 aircraft, which included 66 Wellingtons – 15 of them from No.300 and seven from No.305 – completed the destruction of the hapless city. The raid was affected to some extent by poor weather conditions – severe electrical storms and much icing – and German fighters, operating mainly by visual contact (to overcome the Window menace) achieved

[56] RAF Form 540/541 (IPMS, Lot. A V,37/11).

[57] 13 were detailed, but four were prevented from taking off by various technical faults.

[58] Aluminum foil strips, which, when dropped, rendered radar location screens useless. Both sides developed this simple jamming device – known to the RAF as Window and to the Luftwaffe as Düppel – in parallel early in 1942. Neither wanted to use the method for fear of revealing it to the enemy, the Germans being well aware that they had far more to lose. As the possibility of German retaliation bombing rapidly diminished and as Bomber Command losses escalated (1,110 bombers were lost on night raids alone and over 2,350 more were damaged in the first six months of 1943, and it was argued that some 35% of these would have been saved by the used of Window) keeping Window under wraps began to make little sense, and in the middle of July Winston Churchill personally authorized its use over Germany.

Right and below: An armament crew loading mines aboard Wellington X HZ598 BH•E at Ingham in the middle of 1943.

some successes; 30 bombers failed to return, two of them Polish. BH•P (HF605) was shot down near Hannover, F/O (Lt) Witold Smyk and P/O (Lt) Tadeusz Skalisz were killed and the remaining three members of the crew, Sgt Witold Czernichowicz-Sgt Antoni Piłat-Sgt Adam Jaremko, were made PoW. SM•C (HZ467) shared a similar fate and its entire crew, F/Sgt (Sgt) Marian Grześkowiak-F/O (Lt) Marian Runiewicz-Sgt (Cpl) Władysław Paluszkiewicz-Sgt (LAC) Paweł Beeger-Sgt (Sgt) Stanisław Rutkowski, perished.

In these four major actions against Hamburg the RAF put up a total of 3,095 sorties and dropped almost 9,000 tons of bombs – about half of them incendiary – at the cost of just 86 (2.8%) aircraft lost. The PAF's share amounted to 81 aircraft sorties and two bombers lost. The results of the assault were terrible. Some 6,000 acres of the town were gutted, 183 large industrial establishments, including all four major ship-building yards, and over 4,000 small factories were destroyed or put out of action, 61% of the living accommodations were destroyed, making some 800,000 people homeless, and some 50,000 civilians were killed.[59] Hamburg's Gauleiter described the effects as "a catastrophe, the extent of which simply falters imagination." Yet, in spite of all this, Hamburg made a remarkable recovery and various calculations estimated the resulting loss of production at only between seven and nine weeks. Hitler, enraged by the attack and bent on revenge, immediately signed directives for mass production of the A4 (V2) rocket, Germany's latest retaliation weapon.

The participation of the PAF Wellingtons in the Battle of Hamburg provided the ending to another chapter in the operations of the Polish bomber force. With the prospective regrouping of the Allied air forces for the invasion of Europe, PAF

[59] This was not far short of the total British civilian casualties from bombing raids throughout the entire war. The favorite British comment on the Hamburg massacre was that the Germans "have sown the wind and were now reaping the whirlwind" (Hosea, Chapter 8, verse 7).

A groundcrew of No.300 Squadron resting by JA451 BH•J at Ingham in the summer of 1943. (M.J. Ingham)

No.305 crew delivering a 4,000 lb "Cookie" for "Basia" – MS495 SM•B – whose rear fuselage is visible in the background; Ingham, summer 1943. (M.J. Ingham)

No.300's mascot, "Bari," looks out of the port flight-deck window of a Wellington; Ingham, the summer of 1943. (M.J. Ingham)

Obviously a posed photograph of No.305 crew boarding MS495 SM•B at Ingham in the summer of 1943. What is noteworthy is the inscription "Poland" above the PAF chessboard. (M.J. Ingham)

HQ became anxious to include as many Polish squadrons as possible in the emergent 2nd Tactical Air Force (2nd TAF).[60] An agreement was reached with the RAF that only one of the Polish bomber squadrons, No.300, would convert to Lancasters and remain in Bomber Command, and another, No. 305, would be transferred to 2nd TAF. No.305 was expected to begin rearmament with light bombers in June, but owing to the lack of suitable aircraft, this was postponed by a couple of months. Its command passed to W/Cdr (Cpt) Kazimierz Konopasek with the posting of W/Cdr Czołowski for duty with RAF Staff College on 28 July. The squadron flew no more operational missions in Bomber Command after the night of 2/3 August and was withdrawn from the line from 9 August. Between 17 and 31 August the squadron surrendered all its Wellingtons prior to rearmament with North American B-25 Mitchell II bombers and no flying by this unit took place until its transfer from Ingham to 2nd TAF on 5 September 1943.

With the ongoing replacement of Wellingtons in Bomber Command with four-engined "heavies," the few squadrons still flying the bomber began to be assigned mostly to mine-laying tasks. No.300 flew 67 such sorties in August and during one of these, on the night of the 15th, one of its aircraft, BH•N (HE768), was shot down; three members of the crew, F/O (Lt) Józef Roubo-Sgt (Cpl) Józef Martyniec-Sgt (AC1) Henryk Błażejewski, were killed and two remaining ones, Sgt Maksymilian Rech (pilot) and Sgt Czesław Poddany, were taken prisoner. The squadron's next major operation – after the last attack on Hamburg at the beginning of the month – was the participation in a heavy raid on München-Gladbach on the night of 30/31 August, for which No.300 put up 15 crews. BH•Q (JA116), shot down by a night fighter, crashed in Holland, and its crew, F/Lt (Lt) Leon Ośmiałowski, F/Lt (Cpt) Aleksander Steininger-P/O (S-Lt) Stanisław Święch-Sgt (LAC) Mieczysław Ząbal-Sgt (AC1) Franciszek Buda, were killed. During the month of August the squadron had on average 18 crews and 15 aircraft available for operations. For the conversion to Lancasters No.300 had to wait until the end of the year.

(5.) With Coastal Command in the Battle of the Atlantic (May 1942-June 1943)

With Germany's possession of the whole coastline of Western Europe and her Navy and the Luftwaffe waging a grimly effective assault on Britain's supply lines, Coastal Command, pitifully small and ill-equipped, faced varied and enormously difficult tasks – to be performed amidst fogs and storms over distant waters – which stretched its capabilities almost to the breaking point. The German attempts at a "total blockade," aimed to bring about a collapse of England, were as threatening in 1941-42, the early phase of the Battle of the Atlantic, as the Luftwaffe's offensive had been in 1940. At the end of 1942/beginning of 1943 Coastal Command became almost as important to defeat the German U-boat offensive at a crucial stage of the Atlantic battle as Fighter Command to defeat the Luftwaffe during the Battle of Britain. The U-boat "wolf-packs" and the long-range four-engined Fw 200 Condor reconnaissance bombers known as the "Scourge of the Atlantic" inflicted immense damage upon the convoys and sank almost 5 million tons – 1,300 ships – of Allied and neutral shipping in the Atlantic alone between July 1940 and December 1941. After a temporary drop, shipping losses rose again in the early part of 1942, averaging during the February-April period some 730,000 tons a month, of which almost 500,000 tons – a monthly average of 85 ships – were claimed by U-boats in the North Atlantic. The total number of operational U-boats rose from about 90 in January 1942, to some 150 in August and to over 200 by the end of the year.

However, towards the end of 1941 the War Cabinet afforded the greatest priority to expansion of the strategic air offensive on Germany which required the immediate building up of the bomber strength, and some pressure was put on Coastal Command to relinquish a few Wellington and Whitney squadrons to Bomber Command. The proposal was sharply turned down by the Admiralty. When, in the spring of 1942, the PAF personnel crisis enforced the reluctant transfer of No.304 Bomber Squadron to Coastal Command, this was at first regarded as a temporary "loan" of the unit to the Command, and became permanent only over a year later.

The main party of No.304 Squadron, arriving from Lindholme, disembarked at Scarinish, Isle of Tiree, from S.S. Hebrides on 10 May 1942 and was welcomed on the jetty by G/Cpt G.W. Tuttle, Station Commander. The air party landed at Tiree on the 13th. The unit shared the Station with No.224 Sqdn, equipped with Lockheed Hudsons, and was incorporated in No.15 Group, Coastal Command. Living conditions on the desolate, tiny island, washed by the Atlantic and battered by gales and heavy rains, were difficult, and the task entrusted to the squadron, patrolling the Western Approaches, was completely new to the Poles, who had no previous maritime experience and had never seen a U-boat from the air. Yet, despite their misgivings, they settled surprisingly well into the tedious sea-patrol routine, though not without a few initial "misunderstandings," such as attacks on one or two hapless sharks, and even the solitary rock of Rockall, sticking out of the sea some 300 miles west of Scotland, did not escape the combat zeal of one enthusiastic crew. The code letters NZ were soon removed from the aircraft and only their individual letters were retained on the fuselages.

The squadron flew its first operational mission from Tiree – an antisubmarine sweep by seven aircraft – as early as 18 May and this was followed by similar patrols made almost daily, sometimes twice a day. On the 26th S/Ldr Jan Buczma's crew – pilot F/O Tadeusz Skarżyński – sighted the wake of a U-boat in the vicinity of a convoy and carried out the squadron's first attack with depth charges, but without visible results.[61] On the

[60] The first, North African Tactical Air Force, was established in February 1943 within Middle East Command and took over control of the Desert Air Force.

A Wellington IC R1245 NZ•Q at Tiree in May 1942, after the transfer of No.304 Squadron to Coastal Command.

31st F/O Wacław Waltera's crew – pilot Sgt Leopold Matjaszek – had to ditch on the Atlantic owing to engine failure, but they were picked up some seven hours later. By the end of May No.304 put up 13 anti-submarine missions, comprising 60 sorties in over 360 flying hours.[62] In the first nine days of June 1942 the squadron made a further six similar missions, involving 31 sorties in 190 hours. On its last operational flight from Tiree, on 9 June, another U-boat was sighted and attacked with four depth charges by F/Lt Zdzisław Hirsz's crew.

After its first encounter with the Atlantic at Tiree, No.304 was ordered to move to RAF Station Dale, in Wales, in No.19 Group, which kept watch on the Bay of Biscay – a vital sector of operations and the gateway to the Atlantic for German submarines, and protected by long-range fighters. The departure from Tiree was an emotional occasion because the Poles got on very well with No.224's crews and won the heart of G/Cpt Tuttle, who specially arranged for the Coastal Command band to come all the way to the Isle to play the unit on its way when its main party left on 12 June, and he took leave of every airman in person. The AOC, No.15 Group, sent the signal: "On leaving my command I wish you very good luck. Your speedy and successful adaptation to new conditions of operation and your enthusiastic acceptance of new tasks has given me great satisfaction. You leave many friends behind you in No.15 Group." The Station Commander's letter, sent to No.304's C.O. a few days later, read: "I must write and express my appreciation to you and the whole of your squadron for their magnificent work here. I was amazed the whole time at the enthusiasm with which they tackled anything, and at their cheerful acceptance of the living conditions here ... I cannot imagine any station commander having an easier job than having 304 Squadron at his station, and only hope that I shall meet you all again some time."

The main party arrived at Dale on 13 June and on the 15th No. 304's 16 Wellingtons took off from Tiree and, after a farewell flypast over the Isle, flew to their new airfield. On 25 June seven No.304 bombers went to Bircham Newton airfield to take part in a "1,000 raid" on Bremen that night, and – as already related earlier[63] – lost one of its aircraft in this opera-

[61] RAF Forms 540/541 (IPMS, Lot. A V,36/12) – some Polish narratives repeat ill-founded suggestions that this U-boat was "probably damaged."

[62] "Coastal Command's historical data: Facts and figures of the PAF contribution," PAF HQ, Study Department (Copy: J.B. Cynk, Polish Aircraft Archives).

[63] See section "*(3.) Greatest Achievements and Decline in Strength (March-December 1942)*" on page 297.

Tiree airfield, photographed by No.304 Squadron. (IPMS)

tion. The squadron's first operational task from Dale, an anti-submarine patrol by six aircraft, was made on 30 June. The Poles began to fly from Dale as the Battle of the Atlantic approached its climax – in June Dönitz's submarines sunk 700,235 tons (144 ships), the highest monthly total claimed by the U-boats in the North Atlantic theater in the entire war.

On 6 July 1942 the crew of 'L' (Z1112) – piloted by P/O Skarżyński – on the morning patrol, fought an encounter with a Ar 196 seaplane, which was damaged by its rear gunner, P/O Konrad Tettamandi. On the same day in the afternoon Sgt Henryk Gołębiowski's crew noticed a conning tower and attacked the swirl of the submerging vessel with six depth charges in one salvo, but to no visible effect. On 10 July, F/O Ryszard Figura's crew believed they had seen a submerged submarine and released their depth charges, but subsequent analysis of the attack did not confirm the presence of a U-boat.

The following day tragedy struck when Wellington IC Z1072 crashed on take-off on a training flight – P/O (S-Lt) Michał Mazurkiewicz and F/Sgt (Sgt) Antoni Kosturkiewicz (1st and 2nd pilots respectively) were killed instantly, and a passenger, S-Lt Pietraszak of the Polish Army, was fatally wounded and died the next day; four other occupants received lesser injuries and three escaped unhurt. On 16 June, during a routine patrol by nine aircraft, Sgt Gołębiowski's crew exchanged fire with a Fw 200C Condor, which turned towards France and disappeared into cloud. On the 18th a party from the Polish destroyer Ślązak, in port at Pembroke Dock, led by Commander Tymiński, visited the RAF Station Dale and were the guests of the squadron.

Some U-boat sightings were reported by No.304 crews in the later part of July and two attacks on enemy submarines were carried out at the very end of the month. On the 29th F/O Marian Kucharski's crew, flying in 'V' (Z1172), dropped six depth charges on a very large submarine, believed to be Italian, and observed two large dark brown oil patches on the water. The following day F/Lt Hirsz's 'I' (R3215) released depth

Dale airfield in the summer of 1942. A Wellington of No.304 Squadron which is coming in to land, and its shadow on the ground are visible left of the tip of the coast. (IPMS)

charges on a submerged U-boat and the crew saw an oil patch. During July No.304 Sqdn had on average 13 crews and ten Wellingtons available for operations, and recorded ten missions comprising 60 sorties in 521 flying hours; 2.7 tons of depth charges were dropped.

Intensive patrol operations in August 1942 resulted in several sightings and four attacks on submarines. On the 3rd, F/O Edward Zarudzki's crew, flying in 'O' (DV807), made what was described by Air Staff/HQ Coastal Command as "a very good attack"[64] on a submerged U-boat, which was believed to be damaged. On the 9th, Wellington 'T' (DV597), piloted by F/O Figura, detected a submarine, but failure of the release mechanism prevented it from dropping its depth charges. Another attack on a U-boat on the 11th was inconclusive, but two days later, on the 13th, 'R' (Z1181), navigated by F/O Nowicki and piloted by Sgt Gołębiowski, dropped depth charges and anti-submarine bombs on a surfacing U-boat, causing some damage to it.

[64] RAF Form 540 (IPMS, Lot. A V,36/12).

A Wellington IC DV671 NZ•D, setting out on an anti-submarine patrol over the Bay of Biscay, circa June 1942.

Earlier, on 12 August at about 03:20 hours, the squadron suffered the deeply felt loss of Wellington 'L for Love' (HX384) with one of its most experienced crews, F/O (Lt) Tadeusz Siuzdak-Sgt (Cpl) Michał Omieljaszko-F/Lt (Cpt) Ludwik Maślanka-Sgt (LAC) Paweł Droździok-Sgt (LAC) Marian Modrzewski-Sgt (LAC) Stanisław Wojtowicz. At the time of their take-off on an anti-submarine patrol a strong gale was raging and its full force hit the runway, situated on the small Dale peninsula and ending almost at the edge of a cliff, some 200 ft above the sea. Buffeted by the cross wind and unable to gain enough speed, the aircraft plunged down into the sea and its depth charges exploded on impact – none of the crew survived.[65]

On 16 August W/Cdr Poziomek formally handed over command of No.304 to S/Ldr (Cpt) Kazimierz Czetowicz. In the second half of the month the squadron flew a number of uneventful anti-submarine and convoy protection missions and on the 25th carried out its first night operation from Dale, striking at enemy shipping at La Pallice, but severe icing conditions forced four of the seven participating aircraft to abandon the attack. On 20 August another Wellington IC – 'V' (Z1172) – crashed into the sea on a training flight, killing its entire crew, P/O (S-Lt) Antoni Zieliński-Sgt (LAC) Ryszard Jarno-F/O (Lt) Leszek Wasilewski-Sgt (LAC) Ludwik Ruszel-Sgt (AC1) Grzegorz Gramiak-Sgt (AC1) Antoni Tycholis. In August No.304 made 15 missions, involving 80 sorties in 757 flying hours (the highest monthly total in 1942), and dropped 6 tons of depth charges.

At the same time, the aircraft, which hitherto were singularly unsuccessful in their attacks on U-boats, became more effective thanks to a progressive introduction of the improved ASV (Air-to-Surface Vessel) form of radar and development of the Leigh Light airborne searchlight,[66] and the new Torpex explosive. In the first six months of 1942 only 21 U-boats were destroyed, of which only six, and one shared with the Royal Navy, were attributable to air action – three of them to the Americans and none to Coastal Command. In the next three months, July to September 1942, 31 U-boats fell victim to Allied attacks, and 13 of these were sunk by aircraft and four more were shared with naval vessels – Coastal Command destroyed six and shared one. For the whole of the second half of 1942 enemy losses rose to 65 submarines, aircraft accounting for 34 1/2, including 15 (23% of the total) by Coastal Command. Admiral Dönitz, who as late as August exclaimed that "the aeroplane can no more eliminate the U-boat than a crow can fight a mole," suddenly changed his tune, lamenting some six weeks later that U-boat traffic around Scotland and in the Bay of Biscay was "gravely endangered by daily, even hourly, hunts by aeroplanes ... which force U-boat disposition far out into the center of the Atlantic with consequent higher fuel consumption, shorter operational periods, and greater difficulties in finding the enemy convoys ..."; he concluded that "these problems will lead to irreparable losses, to a decline in ship sinking and ... to reduced chances of success in the U-boat warfare as a whole."

To counter this rapidly growing threat from Allied aircraft, the Germans had to commit increasing numbers of long-range fighters to the Atlantic, using the fast and powerfully armed Ju 88Cs for the purpose. By September 1942 encounters of single Coastal Command bombers with formations of the Junkers heavy fighters became commonplace and 44 such combats were reported during the month, adding a new dimension to the Battle of the Atlantic.

On 2 September the crew F/O Mieczysław Kucharski-F/O Stanisław Żurek-F/O Zbigniew Jaroszyński-Sgt Tadeusz Jasiński-Sgt Bolesław Kocyk-Sgt Stanisław Żeśko, in Wellington IC 'A' (HF894) on patrol over the Bay of Biscay, sighted a fully surfaced U-boat and dived to attack, dropping six depth charges. The explosions completely obliterated the U-boat, which was still surfaced. The aircraft then released two anti-submarine bombs and followed up with machine-gun fire. Several of the U-boat crew who were on deck were seen to fall, while others dived into the sea. The submarine, later identified as the Italian "Reginaldo Guiliano," which the previous day had been attacked and slightly damaged by Sunderlands of No.10 RAAF Sqdn, was severely damaged – it was down by the bows with part of her screw showing, had a pronounced list to port and left a very large oil patch.[67] Kucharski's crew was congratulated on "an excellent attack" – the first undoubted success against a submarine by No.304 Sqdn. The crippled "Reginaldo Guiliano" eventually got underway and sought shelter in the Spanish port of Santander.

Two days later, on the 4th, Wellington 'J' (R1413) on an anti-submarine patrol, skippered by Sgt Jan Bakanacz and navigated by S/Ldr Bolesław Nowicki, fought a successful duel with two Ju 88C heavy fighters, which attacked it several times. One of the Germans limped off home after hits had been scored on it with the port engine smoking furiously, and hits were registered on the second Junkers, which also withdrew towards France trailing smoke. The Wellington returned home with only seven harmless bullet holes.

More encounters with enemy aircraft were recorded during patrols over the Bay of Biscay and on 16 September 'E for Ela' (HF836) fought an epic battle with six Ju 88s, which lasted some 15 minutes.

At 16:12 hours, after almost seven hours in the air, the crew, F/O Stanisław Targowski-Sgt Zygmunt Kowalewicz-F/

[65] In September 1991 divers of the Llantrisant Sub-aqua Club discovered the wreck of 'L' HX384 on the seabed and retrieved one of the airscrews and Browning machine guns. The restored relics were presented to the Poles and taken on board the Polish Navy's Training Sail Ship "Iskra" to be delivered to the National Museum in Warsaw and to the Museum of the Polish Air Force Officers' College at Dęblin.

[66] Evolved by S/Ldr H. de V. Leigh, this was a searchlight which operated in conjunction with ASV and illuminated submarines resting on the surface during the hours of darkness, thus making them susceptible to night attack for the first time. Tests with the Leigh Light began in early 1941, but the operational trials started only in the middle of 1942 (on Wellingtons), with the first "kill" reported on 5 July 1942.

[67] RAF Forms 540/541 (IPMS, Lot. A V,36/12).

O Władysław Minakowski-Sgt Zygmunt Piechowiak-Sgt Władysław Młynarski-Sgt Franciszek Kubacik, sighted six aircraft in the cloudless sky – one to starboard, three to port and two far astern. The Wellington jettisoned its load and dived to 30 ft. As the first Junkers began to shadow the Polish bomber without attacking, the first attack by the second enemy was met head-on. The front gunner gave it two short bursts at point-blank range and the rear gunner a longer one. Similar tactics were followed when the third fighter made its pass and again hits were seen from 'Ela's' gun positions. The attacker's port airscrew stopped and moments later the aircraft turned steeply and fell into the sea. The fourth German was fended off in the same manner and his fire missed. Numbers five and six closed in and opened up with cannon and machine-guns, the former hitting the Wellington's starboard wing, damaging one of its tanks and filling the cabin with smoke from the explosions, but there were no flames and nobody was wounded. The guns of the bomber never stopped blazing and found their mark once again, shooting off large pieces of the attacker's tailplane and causing thick black smoke to engulf one of its engines. The sixth fighter was also hit. The second and fourth Ju 88s returned to make a number of consecutive attacks, which were all successfully repelled. Some clouds appeared and the bomber made for them, eventually evading the Germans. Targowski's superb flying skills and the crew's determination saved the aircraft, which landed at Portreath short of fuel after being airborne eight hours and 20 minutes. Congratulations flooded in and the daily press gave a wide coverage to the "bomber's win."[68]

On 24 September Wellington 'H' (R1657), captained by Sgt Bakanacz, fought another battle with two Ju 88C fighters. The front and rear gunners, Sgt Zdzisław Gustowski and Sgt Stanisław Malczyk respectively, fired at close quarters at the second fighter and hit its port engine, wing root and cockpit area, and the enemy, losing height, turned over and dived into the sea. The first Junkers was also hit and abandoned the fight.

During September 1942 No.304 carried out 13 missions, which apart from patrols included also some bombing attacks on enemy shipping. These comprised 81 sorties in 733 flying hours; over 7 tons of bombs and depth charges were dropped. The squadron had on average 15 crews and eight Wellington ICs available for operations.

The October operations followed a similar pattern. On 8 October Sgt Gołębiowski's crew, flying in 'T' (DV597), sighted a U-boat and released six depth charges which exploded a few seconds after the U-boat had submerged. The crew followed up with two attacks with anti-submarine bombs which detonated within yards of the bubbles visible on the surface after the depth-charge strike. On the same day Wellington 'A' (HF894), skippered by P/O Jan Wróblewski, exchanged fire with a Fw 200 Condor which was escorting two vessels. Some encounters with Ju 88s were also reported in the first half of the month, but no combat took place. On 15 October Wellington R1704 came down in the sea while on a low-level training flight and its two occupants, F/O (Lt) Tadeusz Skierkowski and F/O (Lt) Adam Schultz drowned.

The following day, the 16th, F/O Targowski, a veteran bomber pilot who exactly one month earlier fought the memorable battle with six Ju 88s, set out on another patrol in high spirits – he had that morning received a letter from his wife, the first news of her since he had left Poland in 1939. Some two hours after his take-off a signal was received at the station announcing the award of a DFC to Targowski and a party was immediately prepared to celebrate the two pieces of good news upon his return. But his fellow airmen awaited in vain. His Wellington 'J' (R1413), sent a SOS message after being attacked by enemy fighters and no trace was ever found of the aircraft, nor its crew, which apart from F/O (Lt) Stanisław Targowski included Sgt (Cpl) Gerard Twardoch-F/O (Sgt)

[68] RAF Forms 540/541, Squadron's Diary and Combat Reports (IPMS, Lot. A V,36/12 and 36/15).

A Wellington IC DV759 'R', No.304 Squadron, in the white Coastal Command camouflage; Dale, circa October 1942. (IPMS)

DV594 'D', earlier serving as 'F', seen at Dale at the same time. Note the differently camouflaged top of the aircraft, as compared with 'R'. Some Wellingtons still retaining their original Bomber Command camouflage operated alongside the white-camouflaged machines from Dale until the beginning of 1943.

Tadeusz Oleś-Sgt (Cpl) Zygmunt Piechowiak-Sgt (LAC) Władysław Młynarski-Sgt (LAC) Franciszek Kubacik. Several more anti-submarine sweeps were carried out in the later part of the month, but with little to report.

About October 1942 the squadron began to receive modified Wellington ICs, better suited to Coastal Command work and finished in the Command's white camouflage, which made them more difficult to see from the lookouts on the U-boat conning towers. These carried at first only individual aircraft letters (with no unit code markings). In October No.304 Sqdn recorded 17 missions, comprising 81 sorties in 713 flying hours.

On 1 November 1942 Wellington 'L' (R1716), one of six employed on an anti-submarine sweep, failed to return from its flight and its crew, F/O (Lt) Stanisław Krawczyk-F/Lt (Mjr) Ignacy Skorobohaty-F/O (Cpt) Alojzy Szkuta-Sgt (LAC) Jan Rogala-Sobieszczański-P/O (S-Lt) Mieczysław Wodziński-Sgt (LAC) Zygmunt Sasal, perished. Another Wellington on the same mission fought a ten-minute encounter with a Ju 88C. On 3 November No.304 was transferred to Talbenny in Wales.

The squadron established friendly relations with No.311 (Czechoslovak) Coastal Command Squadron stationed on the same airfield. Although the number of patrols from Talbenny greatly increased, their duration was almost halved, and in addition to anti-submarine patrols – the first of which was performed from the new base on the 6th – a number of shipping strikes were also made. Some brief, inconclusive encounters with Ju 88s and Fw 200s were reported, but otherwise these operations were uneventful. On 29 November preparations began for the return to Dale. The move was completed on 1 December, though the Maintenance Flight, which during the earlier stay of the squadron at Dale was stationed at Talbenny, remained on the same airfield. During November No.304 put up 26 missions, comprising 80 sorties in 677 flying hours. In the middle of November the Polish Training Flight was formed

A Wellington IC R3212 'F' with no beam-guns fitted; Dale, December 1942.

at No.6 OTU, at Silloth in Cumbria within No.17 Group, Coastal Command.

On 1 December 1942 two No.304 crews operated on patrol from Dale and during the next three days three-aircraft anti-submarine sweeps were flown. On the 6th six Wellingtons were engaged on patrol but bad weather forced them to land at other airfields and interrupted operational flying for a few days. Six crews which took off on a sweep on 12 December all had to return early for the same reason – one of them experienced engine trouble but reached the base safely. On the 14th a 5,000-ton cargo ship was attacked with bombs with inconclusive results. A few more sweeps followed and one crew reported seeing a Ju 88 and a Fw 190 which avoided combat. On 24 December five aircraft were engaged on patrol, but only one landed at base and the others were directed to Chivenor. Then F/O Pronaszko's 'A' (HF894) suffered W/T communication failure and the crew, after spending almost 12 hours in the air, baled out without injury.

In December the squadron made 14 missions, involving 60 sorties in 482 flying hours. No.304's total operational record in Coastal Command, for the period May to December 1942, amounted to 115 missions, comprising 549 sorties in 4,478 hours, during which some 35.5 tons of bombs, depth charges and mines were dropped, and nine U-boats attacked apart from those forced to dive.

In October-November 1942 the Allied shipping losses climbed again to very high levels; the tonnage sunk by U-boats in November – over 729,000 tons – being the highest of the war,[69] although the North Atlantic sinkings – 509,000 tons (83 ships) – stayed below the record figures. But, in the period December 1942 to May 1943 – called in some official publications "the most prolonged and complex battle in the history of naval warfare,"[70] when the number of operational U-boats reached its peak (nearly 240 in April-May of which 60 were engaged in the North Atlantic theater), the shipping losses began to drop significantly (except for the freak increase in the tonnage sunk in March). At the same time the U-boats' own losses escalated disastrously. In the first four months of 1943 55 U-boats were destroyed, 29 of them by aircraft, but there was soon worse to come. In the later part of May about two-thirds of the U-boats sunk fell victim to Allied aircraft which began to use homing torpedoes, and the total for the month amounted to 41 U-boats destroyed, 23 of them by aircraft, with Coastal Command claiming 16 of the "kills." In two Atlantic U-boat operations between 17 and 23 May, the last of this kind of the whole war, the "wolf-packs" lost eight boats – six of them to aircraft – and not a single ship was sunk from the convoys. On 24 May Dönitz concluded: " ... even heavy losses must be accepted if they are accompanied by proportional successes in tonnage sunk. ... our losses so far in May have reached an intolerable level. The enemy air force played a decisive role in inflicting these high losses." On the same day he called off the North Atlantic convoy operations. The Battle of the Atlantic was won.

For No.304 Sqdn the year 1943 began with an uneventful patrol by four aircraft on 3 January. Two days later two U-boats were spotted by the crew of Wellington IC 'E' (HF836), piloted by Sgt Jan Bakanacz. Both submerged and one was attacked with three depth charges which burst about ten yards ahead of the U-boat's diving swirl. After a lengthy "baiting" procedure[71] the aircraft made further strikes with three more depth charges which exploded close to the conning tower and with machine-guns. A large oil patch came to the surface and the U-boat was deemed to be damaged. Further anti-submarine sweeps followed later in January, but the bad weather generally hampered operations. On 21 January one of the

[69] The total tonnage sunk – by U-boats, surface vessels and aircraft – was 807,750 tons, or somewhat lower than the record June 1942 total.

[70] F.H. Hinsley: "British Intelligence in the Second World War," HMSO, 1981, p.549.

[71] Submarines usually resurfaced after their attackers flew away. "Baiting" procedure was a pretended departure of the aircraft, which returned to the area sometime later to repeat the attack.

Mk.IC X9831 'J' with the bomb-bay doors opened; Dale, December 1942.

Another Wellington IC, Z1112 'W', serving with No.304 Squadron at Dale in December 1942.(IPMS)

squadron's Wellingtons crash-landed on its return to base, owing to failure of its landing gear. On 26 January seven aircraft from No.304 were detailed to a night attack on the port and harbour installations of Bordeaux; the crews observed hits in the target area. During January the squadron put up eight missions, comprising 40 sorties in 302 flying hours.

In February 1943 anti-submarine operations intensified with improvement in the weather conditions and several encounters with enemy fighters took place. On the 4th, the month's second patrol, Wellington 'V' (HE101) evaded two Ju 88s going into cloud, and 'Q' (R1504) escaped a Ju 88 attack in a similar manner the following day. The same 'V' was lost on the 8th, when it ran out of fuel returning from a sweep and F/O Jan Wróblewski's crew had to bale out; F/O Kazimierz Dobrowolski sustained a leg injury and one of the gunners, Sgt Stanisław Malczyk, spent the night suspended in a tree and suffered from shock and exposure. Petrol failure resulted also in the crash of F/Sgt Bakanacz's 'D' (DV671) on its return on the same night, but his crew were uninjured.

The longest and counted among the most outstanding and determined battles ever between a Wellington and a formation of Ju 88C fighters was fought by Wellington IC 'K' (W5718) of No.304 Squadron on 9 February 1943 over the Bay of Biscay. Its crew, S/Ldr (Cpt) Edmund Ladro-S/Ldr (S-Col) Jan Biały-F/O (Lt) Stanisław Płachciński-Sgt (LAC) Kazimierz Chłopicki-Sgt (AC1) Władysław Piskorski-Sgt (Cpl) Antoni Ulicki, were airborne for three hours, when, at 13:40 hours, four Ju 88s suddenly closed in from behind and opened up with cannon rounds. The Wellington returned the fire and dived steeply. The Germans then struck singly in quick succession, making eight separate attacks. One cannon shell exploded inside the Wellington, seriously wounding Sgt Piskorski and putting its intercom out of action. S/Ldr Biały, 2nd pilot, was slightly injured in the leg. More attacks followed and a furious running fight persisted for 58 minutes, until the Ju 88s ran out of ammunition and made off. S/Ldr Ladro, in a subsequent broadcast by the BBC, recalled: "Soon our front gunner was severely wounded, and his turret out of action. A cannon shell exploded in my cabin ... . Another shell hit my port wing, tearing out a hole some two yards in diameter. We were hit many times more. Our tactics were simple: whenever a German fighter dived on us, we flew straight at him, and at the moment of coming within his range we avoided his fire by a violent turn ... ." The damaged bomber landed safely at Predannack and its crew was shortly told that three Beaufighters from No.248 Squadron intercepted the four Ju 88s and shot down three of them.[72]

Subsequent patrols in February were largely uneventful. During the month No.304 carried out 15 missions involving 67 sorties in 590 hours, 77 of which were flown on night operations. A number of new crews arrived from No.6 OTU for flying duties. The first two Wellington Xs, in white Coastal Command camouflage, were delivered directly to the squadron from the Vickers' Squires Gate works in the middle of February, and began operating alongside the old Mk IC bombers during March.

No.304 continued its anti-submarine activities – mainly in the Bay of Biscay – throughout March 1943, exchanging fire with enemy aircraft on some of the patrols.

On the 18th two Wellington ICs bombed two large enemy vessels and reported near misses. On the 22nd a single Wellington was engaged by three Fw 200 Condors and returned fire, hitting one of the Germans, which broke off the combat with smoke pouring from its starboard wing; the other two terminated the battle after two further attacks. On 26 March Sgt Franciszek Targowski's crew in Wellington IC 'F' (R3212) sighted a fully surfaced U-boat about two miles away and attacked it with six depth charges which straddled its track. A large amount of pieces subsequently came to surface, indicating that the U-boat was damaged, and the attack was thought

[72] RAF Forms 540/541 (IPMS, Lot. A V,36/12).

HZ258 'S', one of the newly received Wellington Xs, which attacked a U-boat on 28 March 1943 with inconclusive results.

to be successful. Two days later, one of the squadron's new Wellington Xs, 'S' (HZ258), piloted by Sgt Stanisław Murlowski and navigated by F/O Tadeusz Kołodziejski, spotted a surfaced U-boat and released six depth charges while its periscope was still visible, but nothing else was seen. No.304 Sqdn recorded in March 14 missions, comprising 82 sorties in 703 hours, 100 of which were flown at night. At the end of the month the squadron began preparations to move to the RAF Station at Docking, in No.16 Group.

On 2 April 1943, No.304's main party, consisting of 16 aircraft, their crews and ground personnel arrived at Docking, but a detachment of six crews remained at Dale for the next ten days. The detached crews put up from Dale four anti-submarine sweeps, and on 9 April Sgt Franciszek Rybarczyk's Wellington IC 'H' (R1657) spotted a surfaced U-boat, which dived instantly. Shortly afterwards the periscope was sighted and the aircraft attacked, straddling the submarine's track with five depth charges, and minutes later a large patch of oil and a trail of air bubbles was observed. The detached crews rejoined the squadron at Docking on 12 April.

With the move to Docking the squadron entered a period of reorganization and intensive training to prepare for rearmament with new aircraft equipped with the ASV II or ASV III radar and a Leigh Light. Some experienced aircrew members were posted away from the unit and several crews were sent to various courses to master new operational techniques involving anti-submarine and anti-shipping tactics by day and night and the use of homing torpedoes. At the same time new crews joined the squadron on posting from No.6 OTU. No combat sorties were flown from Docking, but on 17 April six crews went on a four-day detachment to Davidstow Moor in Cornwall, to carry out operational assignments under HQ No.19 Group orders. The crews made three anti-submarine patrols involving three aircraft each. The last of these was flown on 20 April. During April No.304 performed seven operational tasks involving 21 sorties in 175 flying hours; its non-operational flying (training, ferrying and tests) totaled 382 hours.

No operational flights were made in May, except for those carried out by three crews on Wellington Xs, who were briefly detached to Dale and flew anti-submarine patrols on 15, 19 and 20 May, adding seven sorties in 41 flying hours to the squadron's operational record.

On 1 June 1943 the squadron received four more Wellington Xs on transfer from No.311 (Czechoslovak) Squadron. Two days later No.304 surrendered eight of its seasoned Wellington ICs to other units, re-equipping with Wellington Mk X and XI. On 7 June the main party left Docking for Davidstow Moor by rail, returning to No.19 Group. A week later the squadron began operations from Davidstow Moor, opening a new chapter in its combat career. These are described in Part III of this work.

(6.) Assessments and Conclusions

During 1941, and in particular during the early part of 1942, the four Polish bomber squadrons, operating at full strength and for several special operations supplemented by the Polish OTU bombers, constituted a formidable part of the available RAF bomber strength and contributed very significantly to a

A Wellington X, HE488 'V', which began to serve with No.304 Squadron early in 1943. The new Mk Xs were armed with four-gun rear turrets. (IPMS)

revival of the strategic bomber offensive against Germany. Daunting Polish losses and a lack of sufficient replacements, combined with escalating supply needs of the Polish Underground Army and other resistance movements in Europe, made a continued Polish bomber effort on this scale impossible and compelled the PAF Command to reduce the Polish bomber force to only one squadron.

The conclusions from this period of the PAF bomber operations are as follows:

- After very modest beginnings in 1940, the Polish bomber effort grew steadily in 1941 and reached a peak of intensity in 1942. The significance of this effort to Bomber Command – at a time when the Command was striving to mount a meaningful strategic offensive and launch the first "Millennium" raids – was comparable to that of the Polish pilots in Fighter Command in critical stages of the Battle of Britain.
- The price in human lives that had to be paid for this exertion was greater than the PAF could afford. In common with the rest of Bomber Command – which suffered over two-thirds of all the RAF personnel losses during the war – only about 40 percent of the Polish bomber aircrew survived the first tour of operations (30 sorties) and less than 20 percent, the second tour (a further 20 sorties). In 1942, the casualties peaked at between 60 and 90 percent of the normal establishment of the Polish bomber squadrons. Owing to a shortage of reserves, this caused an irreversible decline in the strength of the Polish bomber force.
- The damaging slaughter of the Polish aircrew elite, which included many distinguished officers and technical experts volunteering to fly as substitute crew members, ran contrary to the agreed personnel policy and pointed to a mismanagement of the PAF's limited human resources. The appropriate policy called for the preservation of maximum PAF strength for the final stages of the war, participation of the squadrons in the battle for liberation of Poland and the PAF's return. Implementation of this policy was a responsibility of the Inspectorate General of the PAF.
- The remarkably smooth transition of No.304 Squadron from Bomber Command to considerably different and very tedious Coastal Command operations did credit to the Polish aircrews, highlighting their flexibility and keenness to fight the enemy in every way demanded of them.

NOTES

NOTES

NOTES

NOTES

NOTES

NOTES

NOTES

Also from the Publisher

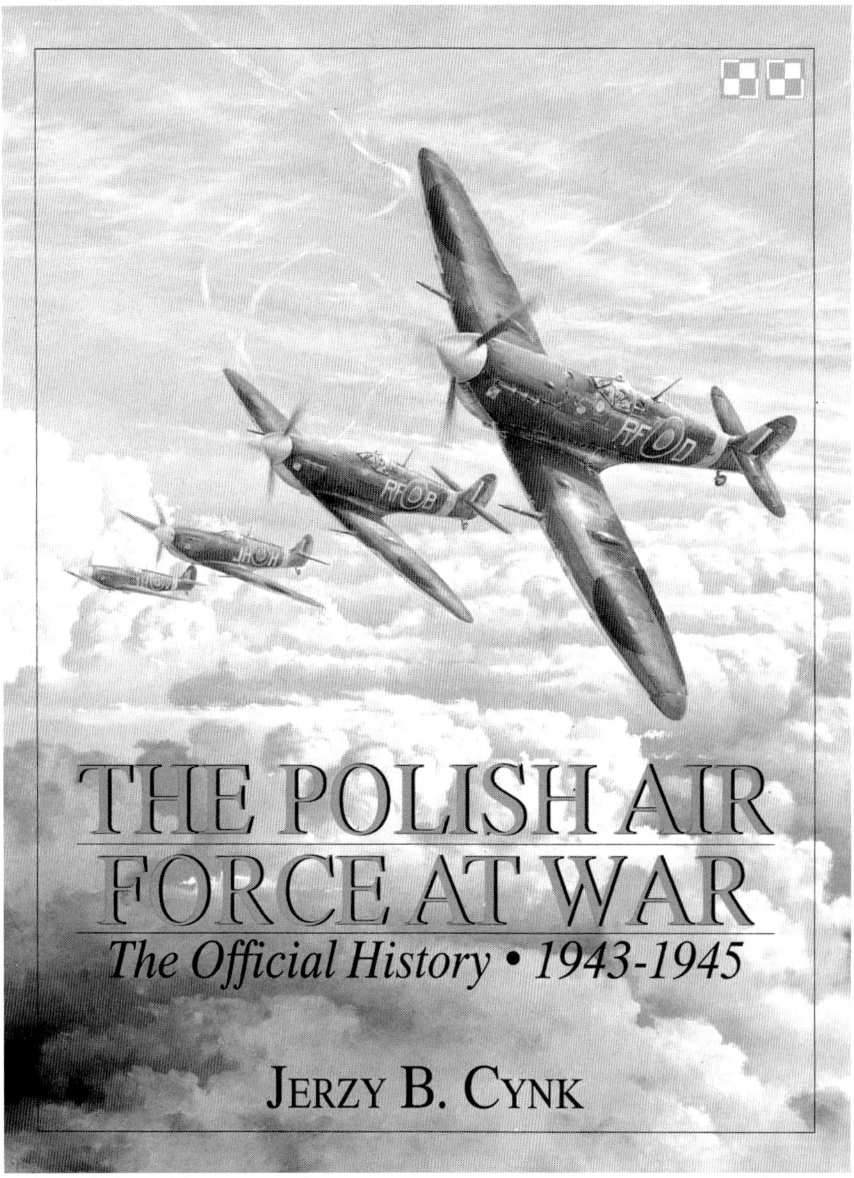

THE POLISH AIR FORCE AT WAR

The Official History

Vol.2 1943-1945

JERZY B. CYNK

Size: 9" x 12" over 270 b/w and color photographs, profiles, maps, appendices, index
352 pages, hardcover
ISBN: 0-7643-0560-3 $59.95